T0303864

WATER AND LIFE

WATER AND LIFE
THE UNIQUE PROPERTIES OF H$_2$O

edited by
RUTH M. LYNDEN-BELL, SIMON CONWAY MORRIS,
JOHN D. BARROW, JOHN L. FINNEY,
and CHARLES L. HARPER, JR.

CRC Press
Taylor & Francis Group
Boca Raton London New York

CRC Press is an imprint of the
Taylor & Francis Group, an **informa** business

CRC Press
Taylor & Francis Group
6000 Broken Sound Parkway NW, Suite 300
Boca Raton, FL 33487-2742

© 2010 by Taylor and Francis Group, LLC
CRC Press is an imprint of Taylor & Francis Group, an Informa business

No claim to original U.S. Government works

Printed in the United States of America on acid-free paper
10 9 8 7 6 5 4 3 2 1

International Standard Book Number: 978-1-4398-0356-1 (Hardback)

Library of Congress Cataloging-in-Publication Data

Water and life : the unique properties of H2O / edited by Ruth M. Lynden-Bell … [et al.] ; foreword by Owen Gingerich.
    p. cm.
Includes bibliographical references and index.
ISBN 978-1-4398-0356-1
    1. Water--Composition. 2. Biomolecules. 3. Life (Biology) 4. Water chemistry. I. Lynden-Bell, Ruth Marion. II. Title.

QD169.W3W33 2010
546'.22--dc22
                               2009044701

Visit the Taylor & Francis Web site at
http://www.taylorandfrancis.com

and the CRC Press Web site at
http://www.crcpress.com

Contents

PART I   This Strange Substance Called "Water"

PART II   The Specific Properties of Water—How and Why Water Is Eccentric

PART V   Water—The Human Dimension

Foreword

In the final decades of the twentieth century, cosmologists became increasingly aware of, and puzzled by, the fact that the physical constants of nature seem singularly tuned to allow the existence of intelligent life on earth. Change the nuclear binding energy by only a few percent, and carbon (a virtually indispensable element for complex life) would become rare instead of being number four in the list of most abundant atoms. Even small alterations in the ratio of the kinetic energy in the Big Bang explosion to the gravitational potential energy of the rest mass then created, a result sometimes referred to as the flatness of the universe, would disrupt an array of physical processes so that stars and galaxies would not form. Martin Rees has called this "the most remarkable feature of our universe."

Consider the huge ratio between electrostatic and gravitational forces, 10^{36}. If gravity were a million times stronger and the ratio 10^{30}, a typical star would last only 10,000 years; if evolution could proceed that fast, the strong gravity would limit the largest creature to something like the size of an insect.

For the physicists, imagining counterfactual universes with alternative physical parameters became an intellectual game. In their final oral exams, candidates for doctoral degrees could expect to face such questions as, "How would the universe be different if the fine structure constant were 20% less?"

Naturally, the physical constants have to be the way they are; otherwise, we wouldn't be here to observe them. But "that's just the way things are" doesn't seem, philosophically, to be a profoundly satisfying answer, even for those who feel the universe is purposeless, signifying nothing. While we are perhaps no further in solving the mystery of precisely what the purpose of the universe is, it does provide a challenge to think about, and science is nothing if not a way of posing questions to the universe itself. Indeed, science has been phenomenally successful by asking questions with answers, whereas the query of why the universe is so congenial for life might well be a question without an answer in the scientific arena. Nevertheless, we can probe further and try to see whether fine-tuning might also exist, for example, in the biochemistry of life.

Consequently, the John Templeton Foundation, with its interest in the big questions of the universe and in queries that might be too daring for the science foundations that specialize in targeting problems with answers, decided to examine that very question—namely, is there evidence of fine-tuning in biochemistry? Thus, in October of 2003, we convened a two-day interdisciplinary symposium of biochemists, cosmologists, and theologians at the Harvard–Smithsonian Center for Astrophysics to consider whether anything comparable to cosmological congeniality existed in the world of biochemistry.*

The "Fitness of the Cosmos for Life" symposium celebrated the ninetieth anniversary of the publication of Lawrence J. Henderson's book *The Fitness of the Environment*[†] and served as a stimulus for developing the subsequent book on the same theme.[‡] The discussions were full of intriguing information, from the folding of proteins and molecules in space to the mysteries of the origin of life itself, all framed with historical and theological insights. For me, the most eye-opening result of that meeting was the recognition of a deep cultural difference between the cosmologists and the biochemists. While the cosmologists made a minor industry of proposing universes with other laws

---

* "Fitness of the Cosmos for Life: Biochemistry and Fine-Tuning," held at the Harvard–Smithsonian Center for Astrophysics October 11–12, 2003. See: http://www.templeton.org/archive/biochem-finetuning/.

† Henderson, L. J. (1913). *The Fitness of the Environment: An Inquiry into the Biological Significance of the Properties of Matter.* New York: Macmillan. Repr. (1958), Boston: Beacon Press, and (1970), Gloucester: Peter Smith.

‡ See: http://www.cambridge.org/us/catalogue/catalogue.asp?isbn=9780521871020; http://www.cambridge.org/uk/catalogue/catalogue.asp?isbn=9780521871020.

of physics, the biochemists had a vast unexplored chemical world that didn't encourage any imaginary tinkering with the physical constants. They felt no incentive to consider what would happen, for example, in a counterfactual world where hydrogen bonding was weaker. Christian de Duve, one of the participants, put it very vividly by pointing out that if we were to make one each of every possible 40-unit protein by taking in turn each of the 20 standard amino acids in each position in the string, the total mass would equal that of 1600 earths. And for a 50-unit protein, the total mass would far exceed the mass of our entire Milky Way galaxy.* This exercise doesn't even begin to consider amino acids that are not in the standard set, which is something chemists now have begun to explore. The atomic building blocks offer so many possibilities that the chemists don't bother to worry about changing the basic set. It's a totally different game the physicists play—they want to fiddle with the blocks themselves.

Given the fascinating but indecisive outcome of the "Fitness of the Cosmos for Life" symposium, the Templeton Foundation decided to focus the question more sharply by convening a further meeting specifically on one of the key components of life as we understand it—that is, on the extraordinary properties of water. The sessions were held on April 29–30, 2005 along the picturesque shores of Lake Como, in the village of Varenna, Italy, to consider the structure, properties, and interactions of water in the context of its ability to support life. Some presenters also looked at other substances that could possibly support life—although perhaps not "life as we know it." It is from that symposium that this volume has emerged.

* * *

Recognition of the essential nature of water is as old as philosophy itself. Thales, the first of the "seven sages" of antiquity and sometimes called the father of philosophy, declared that water was the basic constituent of everything—or at least that is what might be deduced from two brief references in Aristotle's corpus. Eventually, water became, along with earth, air, and fire, one of the four terrestrial elements in Greek cosmology.

"Water, water, everywhere, nor any drop to drink." Thus declared Coleridge's ancient mariner, thereby encapsulating two of water's most essential features. First, water is ubiquitous. Roughly 70% of the earth's surface is covered with it. Human bodies are more than half water. With hydrogen the most abundant element in the cosmos and oxygen number three (after helium), H_2O is one of the universe's most common molecules.

Second, water is the nearest thing we have to a universal solvent, so that the earth's oceans are impotably salty. (This saltiness is, incidentally, a clue to the gradual solubility of rocks and the great age of the earth's oceans.) Of particular interest to earthlings is the way that ocean water can dissolve carbon dioxide. The ability of oceans to dissolve carbon dioxide and to deposit it in the form of limestone is a great boon. If the oceans had not absorbed the gas, it would have remained in the atmosphere, as it has done on the desiccated planet Venus. Imagine a column of limestone twice as high as the Washington Monument crushing down on your shoulders. This matches the atmospheric pressure at the surface of Venus, where the carbon dioxide has not been converted to carbonate rocks.

Not only is the solubility of carbon dioxide critical for human life, but so is its easy release back into gaseous form. Shake vigorously a can of carbonated soda and unzip the top. Watch out! The unpleasant and perhaps unexpected shower is driven by the effervescent CO_2. In our bodies, carbon dioxide is a principal waste product from the "burning" of carbohydrates, the energy source that keeps us functioning. But how to eliminate the waste? The blood cells carry it to the lungs, where a pressure difference allows the dissolved CO_2 to be released. L. J. Henderson, in his classic 1913 book, *The Fitness of the Environment,* devoted an entire chapter to this topic, noting in part,

> In the course of a day a man of average size produces, as a result of his active metabolism, nearly two
> pounds of carbon dioxide. All this must be rapidly removed from the body. It is difficult to imagine by

---

* de Duve, C. *Singularities: Landmarks on the Pathways of Life.* New York : Cambridge University Press, 2005; p. 108.

what elaborate chemical and physical devices the body could rid itself of such enormous quantities of material were it not for the fact that, in the blood, the acid can circulate partly free . . . and in the lungs [carbon dioxide] can escape into the air which is charged with but little of the gas. Were carbon dioxide not gaseous, its excretion would be the greatest of physiological tasks; were it not freely soluble, a host of the most universal existing physiological processes would be impossible. (pp. 139–40)

It is not simply the solubility of the CO_2 in water that matters, but also the interaction with the water to form an acidic ion H^+ and a bicarbonate base, HCO^-. As one of our chapters in this volume concludes, "For we humans, water is clearly a uniquely important solvent because it is our solvent. With each passing year, we learn more about how incredibly complex we humans are, and it seems likely that much of this complexity will be linked in one way or another to water, our solvent."*

It is beyond the scope of the present Foreword to go into further details or to address some of the other remarkable properties of water so wondrously useful for the development and sustainability of life. This I have done in a chapter in *Fitness of the Cosmos for Life*, the volume resulting from the 2003 symposium. In a comparatively elementary fashion, I there described how our knowledge of atomic structure and of hydrogen bonding, unknown to Henderson, gives a modern insight into the underlying physical reasons for many of the unusual properties of water.[†]

In the present volume, you will find a much deeper probing, often with the powerful tools of quantum mechanics, of the subtle and sometimes unexpected features of the water molecule in its various states. The introductory chapter by Simon Conway Morris and Ard Louis provides a guide to these papers. You will find here a group of counterfactual studies, where the chemists have picked up the challenge of the cosmologists to imagine other universes where certain physical constants are different. (Of particular interest is the strength of the hydrogen bond, with its implications not only for the physical behavior of water, but for the zipping or unzipping of the nucleic acid links in the strands of genetic DNA.) Toward the end of the book, a more philosophical approach to these pursuits is taken, searching for possible implications to the "big questions" and asking whether anything from the biochemical laboratories hints at an answer about the purposefulness of the universe. Perhaps not unexpectedly, the answers are ambiguous, and the search goes on.

Owen Gingerich
Cambridge, Massachusetts

---

* See "Fine-tuning protein stability" by Carlos Warnick Pace in this volume.
† Revisiting *The Fitness of the Environment*. In: *Fitness of the Cosmos for Life: Biochemistry and Fine-Tuning*, Edited by J. D. Barrow, S. Conway Morris, S. J. Freeland and C. L. Harper, Jr. Cambridge: Cambridge University Press, 2008, pp. 20–30.

Preface

Life, as we know it, has evolved on a planet where water is ubiquitous. Water plays an essential role for many aspects of life, ranging from processes within cells to behavior of organisms in their environment. This does not necessarily mean, however, that water is essential for all possible forms of life.* In this book, we bring together contributions concerning the properties of water and its interaction with life. The chapters in this volume reflect the rich technical and interdisciplinary exchange of ideas that occurred during a symposium held in Varenna, Italy in April 2005: "Water of Life: Counterfactual Chemistry and Fine-Tuning in Biochemistry—An Inquiry into the Peculiar Properties of Water Related to L. J. Henderson's *The Fitness of the Environment*."†

The fields represented are diverse. From the sciences, the fields of chemistry, biology, biochemistry, planetary and earth sciences, physics, astronomy, and their subspecialties are represented. In addition, a number of essays drawing on humanistic disciplines—history of science and theology— were included to provide additional perspectives. In this collection of essays, the editors sought to develop a variety of approaches that might conceivably illuminate ways in which to address deeper questions with respect to the nature of the universe and our place within it.

The chapters offer a range of themes and questions to reflect the symposium discussions and to cover ongoing key areas of debate and uncertainty. In addition to Owen Gingerich's thoughtful Foreword, twenty-seven authors contributed twenty-two chapters, grouped in five broad thematic areas:

Part I: This Strange Substance Called "Water"
Part II: The Specific Properties of Water—How and Why Water Is Eccentric
Part III: Water in Biochemistry
Part IV: Water, the Solar System, and the Origin of Life
Part V: Water—The Human Dimension

We hope that a variety of readers will find much information and insight in this volume to assist them in their own explorations of the origin and meaning of life and of the possible role of water in its maintenance. In addition, we hope that we have produced a book that will serve to stimulate thinking and new investigations among many scientists and scholars concerned with the fundamental question, *Why can and does life exist in our universe?*

RUTH M. LYNDEN-BELL
Queen's University Belfast;
University of Cambridge

SIMON CONWAY MORRIS
University of Cambridge

JOHN D. BARROW
University of Cambridge;
Gresham College

JOHN L. FINNEY
University College London;
London Centre for Nanotechnology

&

CHARLES L. HARPER, JR.
American University System;
Vision-Five.com Consulting‡

---

* R. M. Daniel, J. L. Finney, and M. Stoneham (editors). The molecular basis of life: is life possible without water? *Phil. Trans. Roy. Soc.* B, 359, 1141–1328, 2004.

† See: http://www.templeton.org/archive/wateroflife/.

‡ Formerly of the John Templeton Foundation.

Acknowledgments

The editors wish to acknowledge the John Templeton Foundation and the late Sir John Templeton* for making this project possible.

- We also wish to thank Pamela Bond Contractor and her staff, working in conjunction with the John Templeton Foundation, the symposium co-hosts/co-organizers, and the volume editors, for co-organizing the symposium at the Villa Monastero in Varenna, Italy in 2005 and for serving as developmental editor of this book with her assistant editor, Matthew P. Bond;
- Elio Sindoni of the University of Milan, working with Donatella Pifferetti and their staff and assistants at the International School of Plasma Physics and "Piero Caldirola" Center for the Promotion of Science, Milan, as well as Tommaso G. Bellini and Marco Bersanelli of the University of Milan and the organization Euresis for co-hosting and co-organizing the symposium at the Villa Monastero in Varenna, Italy in 2005;
- Hyung S. Choi, Director of Mathematical and Physical Sciences at the John Templeton Foundation, for assuming an important role in developing the academic program for the symposium in conjunction with former JTF Senior Vice President and Chief Strategist Charles L. Harper, Jr., currently Chancellor for International Distance Learning and Senior Vice President of the American University System, as well as President of Vision-Five.com Consulting;
- Owen Gingerich of the Harvard–Smithsonian Center for Astrophysics, who contributed the Foreword and hosted the "Fitness of the Cosmos for Life" symposium at Harvard in 2003; and
- Taylor & Francis for supporting this book project and, in particular, Hilary Rowe for her editorial management.

---

* Sir John passed away on July 8, 2008. See http://www.templeton.org/newsroom/sir_john_templeton/.

Contributors

Wesley D. Allen
Department of Chemistry and Center for
    Computational Chemistry
University of Georgia
Athens, Georgia

Philip Ball
(Formerly of) *Nature*
London, United Kingdom

Steven A. Benner
Westheimer Institute for Science and
    Technology
Foundation for Applied Molecular Evolution
Gainesville, Florida

Humberto Campins
Planetary and Space Science Group
Physics Department
University of Central Florida
Orlando, Florida

Martin F. Chaplin
Water and Aqueous Systems Research, and
Food Research Centre
London South Bank University
London, United Kingdom

Simon Conway Morris
Department of Earth Sciences
University of Cambridge
Cambridge, United Kingdom

Pablo G. Debenedetti
School of Engineering and Applied Science
Department of Chemical Engineering
Princeton University
Princeton, New Jersey

Michael J. Drake
Cosmochemistry and Geochemistry
Lunar and Planetary Laboratory
University of Arizona
Tucson, Arizona

John L. Finney
Department of Physics and Astronomy
University College London
and
London Centre for Nanotechnology
London, United Kingdom

Felix Franks
BioUpdate Foundation
London, United Kingdom

Giancarlo Franzese
Department of Fundamental Physics
University of Barcelona
Barcelona, Spain

Branka M. Ladanyi
Department of Chemistry
Colorado State University
Fort Collins, Colorado

Louis Lerman
Westheimer Institute for Science and
    Technology
Foundation for Applied Molecular Evolution
Gainesville, Florida

Ard A. Louis
Rudolf Peierls Centre for Theoretical Physics
Department of Physics
University of Oxford
Oxford, United Kingdom

Ruth M. Lynden-Bell
Queen's University Belfast
Belfast, United Kingdom
and
Murray Edwards College
University of Cambridge

Alister E. McGrath
Centre for Religion and Culture
King's College
London, United Kingdom

Thomas C. B. McLeish
Department of Physics and Vice-Chancellor's
  Office
Durham University
Durham, United Kingdom

Carlos Warnick Pace
Department of Biochemistry and Biophysics
Texas A&M University
College Station, Texas

Wilson C. K. Poon
Scottish Universities Physics Alliance
and
School of Physics and Astronomy
University of Edinburgh
Edinburgh, United Kingdom

Abbas Razvi
Laboratory of Professor William E. Balch
Scripps Research Institute
La Jolla, California

Colin A. Russell
Department of the History of Science,
  Technology and Medicine
The Open University, Milton Keynes
and
Department of History and Philosophy of
  Science
University of Cambridge
Cambridge, United Kingdom

Henry F. Schaefer, III
Center for Computational Chemistry
University of Georgia
Athens, Georgia

J. Martin Scholtz
Texas A&M University
College Station, Texas

H. Eugene Stanley
Center for Polymer Studies
Boston University
Boston, Massachusetts

Adrian F. Tuck
Physics Department
Imperial College London
London, United Kingdom

Veronica Vaida
Department of Chemistry and Biochemistry
University of Colorado at Boulder
Boulder, Colorado

Peter D. Ward
Department of Biology and Department of
  Earth and Space Sciences
University of Washington
Seattle, Washington
and
NASA Astrobiology Institute

Bruce H. Weber
Department of Chemistry and Biochemistry
California State College
Fullerton, California
and
Department of Science and Natural Philosophy
Bennington College
Bennington, Vermont

Peter G. Wolynes
Department of Chemistry and Biochemistry
  and Department of Physics
University of California
San Diego, California

Part I

This Strange Substance Called "Water"

1 Is Water an Amniotic Eden or a Corrosive Hell?
Emerging Perspectives on the Strangest Fluid in the Universe

Simon Conway Morris and Ard A. Louis

CONTENTS

1.1   INTRODUCTION

Colorless, transparent, and tasteless, the substance we call water is ubiquitous and common-place. Arguably, it is also the strangest liquid in the universe with many peculiar counterintuitive properties that, it is widely proposed, are central to the existence of life. In the words of Franks, water is a "strange and eccentric" liquid (p. 11). The anomalies of water, unsurprisingly, have been recruited by those who see an intriguing, if not suspicious, fitness to purpose, so far as life is concerned.

The fact that ice floats because of the hydrogen bonding imposing a perfect tetrahedrally coordinated network, linking them into six-membered rings with much empty space between the molecules (Franzese and Stanley, p. 105), is perhaps the best known of what are widely seen as a long list of curiosities. Water's maximum density at 4°C and its unusually high thermal capacity are also familiar anomalies. Many others, however, are less celebrated but are surely as noteworthy. Both the melting and boiling points of water are unexpectedly high when it is placed in the sequence of group VI hydrides. So Lyndell-Bell and Debenedetti remind us by this extrapolation, although not by this imagery, that ice placed in a gin and tonic would melt at –100°C and a cup of tea should be prepared at –80°C. Not only that, but the effect of supercooling is also remarkable, so that at ambient pressure it can reach –41°C, whereas at 2 kbar it may be as low as –92°C (Franzese and Stanley, p. 102). These authors also remind us that if the supercooling is very rapid the water fails to crystallize and becomes a glass. This is of more than passing interest because in its high density form it is "the most abundant ice in the universe, where it is found as a frost on interstellar grains" (Franzese and Stanley, p. 103). This is not the only regime in which water becomes amorphous. In the hydration layer associated with a peptide, the water again has glasslike properties "with a very rough potential-energy landscape and slow hopping between local potential minima" (Ball, p. 56).

Water also has unexpected properties in more mundane settings. When cooled, water becomes exceptionally compressible, to extent that, if this property did not exist, the oceans would be about 40 m higher. The diffusivity of water is also anomalous, with the counterintuitive observation that initially as the pressure increases so does diffusivity. This list of what are widely seen as anomalies could be greatly extended, and are detailed in the various contributions, but at this juncture we will only give one further example, that of the so-called lyotropic series. This is the perplexing observation that one series of ions is decreasingly soluble and serves to stabilize proteins, whereas the other series is increasingly soluble and will assist in the inactivation of proteins (Franks, p. 17).

To the untutored eye, many anomalies of water make for a striking list but otherwise appear to be without particular rhyme or reason. This sentiment should not be dismissed because, as discussed below, in the counterfactual extrapolations posited by Chaplin, some properties (e.g., viscosity) are much more sensitive to change than others. Nevertheless, it must be emphasized that, at the deeper level of quantum mechanics, all the properties of water are predictable (although, in practice, we must also sound an antitriumphalist note as accurate theoretical prediction can still be surprisingly difficult, as evidenced by the many different competing water models and the contentious literature that surrounds them). This, of course, allows counterfactual speculations such as what might happen if hydrogen bond strengths are altered. It is also important in the general context of asking what the wider origins of an ordered universe rest upon and the extent to which apparently peculiar properties, and here one might think of not only water but other phenomena such as consciousness, are inherent in the basic fabric of the cosmos. This regression (in an entirely positive sense) of inquiry is clearly articulated by Allen and Schaefer when they write "if we marvel at the fitness of water for life, then we should properly marvel not at independent properties of the universal solvent, but at the remarkable richness of the mathematical solutions arising out of the elegant and encompassing form of the Schrödinger equation itself" (p. 124).

Although the intrinsic peculiarities of water are best understood by physicists, our focus of attention mainly revolves around the perceived "biofriendliness" of water. Do its many special properties either directly or indirectly render water uniquely suitable for either biological function or general habitability? Alternately, are there viable alternatives that, when properly explored, will lead to the conclusion that water is very far from ideal? The difficulty in adjudication between these perspectives is compounded by our ignorance of the conditions under which life on earth emerged, as well as the subtleties of delineating the limits of life's robustness and adaptability. Biological function can appear extremely sensitive to water properties. For example, even small doses of heavy water are known in at least some cases to be highly toxic. This is because of subtle changes in reaction rates that destabilize metabolic pathways and signal transduction. But this apparent fragility (or naive fine-tuning) simply reflects the finely poised solutions life searches out to enhance its complexity and masks the deeper robustness to such perturbations evidenced here by the fact that simple organisms have been shown to adapt by evolutionary change to increasing concentrations of D_2O. Similarly, complex life would almost certainly have developed happily in this medium if hydrogen normally appeared with a neutron. So the question of "fine tuning" for biofriendliness does not revolve around these simpler questions of sensitivity, but rather asks whether the whole suite of water properties and related chemistries, taken as a combined whole, are critical to the emergence of life in all its rich diversity and fecundity. As will become apparent, so far as the contributors to this volume are concerned, there is a strong but not unanimous presumption that the collective properties of water are central to both cellular mechanisms and macroscopic properties that range from its viscosity to planetary habitability. Nevertheless, there are important counter-voices, perhaps most notably Benner, and there is general agreement that any sense that water is uniquely fitted to life is tentative. Our suspicion is that as water's interactions with life are further explored, especially in terms of proteins (see below), the more striking will be the match and specificity of properties. The discovery of silicon-based life disporting itself in an ethane ocean should not, even with present-day information, come as a total surprise, but we will rashly predict that although the galaxy houses many bizarre chemistries, only the carbaquist combination will qualify as living. Either way, whether water is uniquely suitable or is just one of at least several solvent molecules employed by life, we have much to learn.

1.2 HOW BIOPHILIC IS WATER?

Given that organisms are largely composed of water, it may seem oxymoronic to inquire why this substance is so biophilic (or so it appears). Yet despite water's oft-quoted fitness to purpose (or at least function), it exhibits many subtleties that are incompletely understood. Before touching on these, especially with respect to water–protein interactions, it is worth emphasizing, as indeed Lyndell-Bell and Debenedetti do, that even when considering "normal" water our thinking is too often pitched in terms of standard pressure and temperature (STP). Yet, as they remind us, even terrestrial life copes with water across a temperature range from well above boiling to below freezing, and ambient pressure from the rarefied heights near the top of the troposphere to the crushing environment of the oceanic trenches (not to mention within the Earth's crust). At elevated pressures and temperatures, water in some sense becomes less anomalous (adopting the parameters more typical of the Lennard-Jones interactions), and so begs the question of how organisms living in high pressure and/or temperature adapt to a liquid that is arguably less biophilic. In any event, such extreme niches are intrinsically interesting for two other reasons. First, they pose questions concerning biochemical, especially enzymatic, adaptation. Second, these environments might provide some insights into how carbaquist life might function on non-Earth-like planets. Consider, for example, large ocean planets with water depths in excess of 100 km. At depth, the extreme pressure would produce ice polymorphs and, although these have no direct equivalent on Earth, at shallower depths we might investigate extensions of terrestrial biochemistry.

Returning to Earth (perhaps in more than one sense), what of the specific interactions between water and biochemistry? It is here, after all, that intuitively it might be felt that evidence for some type of "fine-tuning" would be most evident. Two very important points are emphasized by Ball. The first is that so intimate are the interrelationships between water and the various organic substrates that water itself must be treated as a biomolecule. As Ball writes, "Water is an extraordinarily responsive and sympathetic solvent, as well as being far more than a solvent"; it simply cannot be considered in isolation. The second and related point that Ball makes is that, although the discussion of water is largely focused on its properties as a solvent, it also serves as a ligand. For example, in both hemoglobin and cytochrome oxidase, the binding and subsequent release of water molecules is critical to their proper function.

It is in the context of protein function that the role of water is seen as not only one of increasing subtlety, but also an area that is by no means completely understood. Although it has long been appreciated that the hydrophilic and hydrophobic interactions between water and a protein are crucial in successful folding, it is clear that the process is exceptionally finely balanced between competing demands. Proteins require stable folds, robust to environmental fluctuations, but they must also find solutions that are flexible enough for allostery and complex interactions within the proteome while simultaneously exhibiting a rich and variable designability within the reach of an evolutionary (biased?) random walk through sequence space. So, to achieve this finely poised flexibility, the free energy of the folded protein is remarkably low, being equivalent to about two to four hydrogen bonds, as pointed out by Finney (p. 43). McLeish remarks that the process of folding entails a "very subtle range of weak, local interactions between molecules in an aqueous medium." In commenting on the same phenomena, Pace emphasizes that "we still do not have a good understanding of water and its interactions with other molecules" (p. 200). Assumptions based on routine chemistry would suggest, as McLeish reminds us, that "the ability of water to exchange entropy with a folding protein is one of its more astonishing properties" (p. 208), that is, the process ought to be strongly exothermic. What, in principle, might be highly deleterious to cell function is, in terms of entropy, efficiently compensated and thereby "is a potential repository of information" (McLeish, p. 208) that plays a central role in the apparently mysterious self-assembly of proteins and indeed other biochemical components.

The question of whether any other liquid would be equally fit to purpose as a medium for protein activity is repeatedly raised in this volume. Finney emphasizes how "water appears to be a responsive, sympathetic, yet versatile solvent" (p. 43) whose crucial advantage is considerable latitude in the employment of its hydrogen bonds so as to enable a variable coordination. Finney goes on to note that this flexibility in bonding may well find parallels in other amphiphilic liquids. Even so, although a definitive conclusion is not yet possible, water still holds the trump card in terms of its versatility of bonding arrangements. In addition, the bonding strength between water molecules is enormously high, so explaining why it remains as a liquid at a much higher than expected temperature. The strength also confers a sort of rigidity, yet molecular diffusion is not compromised, even in highly confined environments because of subtle correlations in hydrogen bonding. There is little doubt that this combination of properties is important, perhaps critical, in protein function. The current presumption, yet to be fully tested, is that although alternatives exist, their total effectiveness falls short of water itself.

The exploration of alternative possibilities depends also on whether life is restricted to effectively terrestrial-like environments, close to triple point of water (i.e., where ice, liquid water, and vapor coexist). Poon makes an important point that the importance of water is not only its existence as a liquid, but the central role of vapor/liquid interfaces. It is in this setting that effective intermolecular encounters are guaranteed, not least in the droplets that Vaida and Tuck (see also Lerman) suggest as the site of prebiotic synthesis. By restricting the interactions to two-dimensional liquids, not only do chemical species retain mobility within the interface, but (as famously pointed out by Delbruck) a random walk is significantly faster when compared with a three-dimensional milieu.

1.3   COUNTERFACTUALS

At many points in this volume, the apparent fitness of purpose that water shows with respect to both life itself and planetary habitability begs the question of alternatives, that is, counterfactuals. Such questions can be divided into three categories that actually address rather different issues. The first is very broad and speaks to the question of cosmological "fine-tuning" and with an immediate resonance for anthropic arguments. The other two, however, are certainly wide-ranging, but in the parochial sense inasmuch as they ask whether our thinking as terrestrial scientists has remained cripplingly narrow. The first area of counterfactuals deals, therefore, with cosmological alternatives, where one basic physical parameter is changed so that the properties of water are altered, perhaps radically. Although such *gedanken* experiments might seem to have an element of whimsy, they are important in indicating the extent to which water's fine-tuning is governed by deeper physical constraints. Such cosmological alternatives, however, cover a wide range of possibilities from the mundane (e.g., a change in a bond angle) to deeply alien (e.g., altering the strength of the hydrogen bond or a fundamental physical constant). The second counterfactual approach is to ask what other liquids, perhaps in environments very different from Earth, might be suitable for life. The third avenue, and one that is only lightly addressed in this volume, is to inquire whether the definition of life is too narrow. Can we begin to envisage either different chemistries, such as silicon (and here an alternative liquid to water might be a *sine qua non*), or a physical environment where liquid of any sort is simply not used.

So far as cosmological counterfactuals are concerned, several approaches are possible. Given the fundamental importance of the hydrogen bond in water, it is an almost obvious question to ask how its properties might alter if the bond strength were to be altered so that it increased or became weaker. The results presented by Chaplin are intriguing, as a number of properties are relatively insensitive (e.g., surface tension). Density is also relatively insensitive, but the critically important maximum density of 4°C in freshwater would not preclude the formation of floating ice in a counterfactual environment where the hydrogen bond was weaker, but it could still lead to quite significant consequences for life in rivers and lakes. Just as the freezing point of water could change, the boiling point would shift. Thus, if the hydrogen bond strength was weakened by 22%, water would turn into steam at body temperature. Other properties are much more sensitive to change in bond strength. For example, with increasing bond strength, viscosity would rapidly increase. Although not directly addressed by Chaplin, the corollaries are interesting. Thus the Reynolds number tells us that much larger organisms would live in a laminar-flow environment, while features such as blood circulation might be severely compromised. However, Chaplin's comments on aqueous solubility show that in this context matters are not simple. This is because the increase in viscosity might be offset to some extent by the increased solubility of O_2 and CO_2. Conversely, diffusivity would markedly decrease. As Chaplin points out, juggling these variables makes predictions about how life would fare difficult. In conclusion, it seems likely that bond strengths could vary quite widely and still allow some sort of life, but the denizens might be very different from those of Earth.

Even as a *gedanken* experiment, these speculations would be a fertile area for evolutionary studies, as they might help to refine some areas of functional biology. As noted above, changing viscosity has implications for any organism in a fluid medium. In this context, what might be the effect on water transport in plants? Given the discussion on water as a biomolecule and a ligand, would an investigation of hypothetically changing bond strengths throw light on the apparent goodness of fit in the context of protein function? Such speculations should be set against the likelihood that if bond strengths were much different from the actual values, there would likely be nobody to investigate the outcome simply because intelligent life itself would be precluded.

This is because of a potentially important distinction between what are referred to as cosmotropic as opposed to chaotropic ions. This is a somewhat old-fashioned nomenclature but, arguably, still has its uses and reflects the subtly different ways these ions interact with water. Sodium (Na^+) is one such chaotrope and has a negative entropy of hydration. Cosmotropes show the reverse relationship and so an ion such as potassium (K^+) exhibits a positive entropy of hydration. An increase in the strength of the hydrogen bond would obviously alter the nature of ionic interactions with water. In particular, because Na^+ and K^+ happen to lie on either side of the cosmotrope/chaotrope divide, a shift in hydrogen bond strength could, in principle, have a major impact on membrane physiology. Therefore, the well-known roles of Na^+ and K^+, including nervous conduction, would, in this counterfactual world, be impossible. As Chaplin points out, although there are alternative ions that could be used, they are either rare (cesium, lithium, rubidium) or toxic (ammonia, which is little different from K^+ as a chaotrope).

Lyndell-Bell and Debenedetti also note that if the strength of the hydrogen bond was to weaken then the resulting decrease in tetrahedral order would result in water becoming more "normal." They also investigate the counterfactual effect if the bond angle of the water molecule were to be reduced to either 90° or 60° from 109°. The results are "found to be quite dramatic" (p. 92), especially at 60°, where the tetrahedral structure was lost and there is a marked increase in the diffusivity factor (by almost eight times). Their findings echo those of Chaplin inasmuch as "different properties exhibit varying degrees of sensitivity to changes in water geometry and hydrogen bond strength" (p. 98), but they reiterate the point already touched upon that, for extremophiles, the ambient water differs markedly from familiar STP conditions.

In what is a rather more radical approach to the counterfactual question, Allen and Schaefer ask what would result if the two parameters central to chemistry, that is, the fine-structure constant (α) and the ratio between the mass of the proton and the electron (β), were to change. It has long been appreciated that these two values, respectively, of 1/137 and 1/1836, are crucial components in fine-tuning arguments and so have a direct bearing on the "sensitivity" of water. Significant shifts in the values would lead to the breakdown of chemistry and presumably an uninhabitable universe. Yet the molecular transitions to these bizarre states are, in principle, open to description. Therefore, although in the case of β a "molecular structure catastrophe" occurs somewhere between values of

0.01 and 1, there appears to be no step-function or singularity. If the change in β is relatively small, the effects on, for example, ionization energy are muted, but when β reaches 1, atomic instability threatens. Correspondingly, if α were to increase then atoms such as oxygen would become relativistic and show peculiar features (e.g., color absorption) that are presently associated with heavier elements such as gold and lead. When α reaches 0.2, all chemistry would cease as covalent energies fall. Allen and Schaefer stress (see above) that notwithstanding the anomalies shown by water, the underlying quantum mechanics play true to form. Hence, as they emphasize, investigation of increasing the values of α and β provides a rich field of possibilities as to how the current properties of water might change as we slip from the parameter space of our familiar universe.

A more mundane approach to counterfactuals is to ask what other fluids might be hospitable to life. It needs to be admitted at the outset that speculation runs far ahead of experimentation, but in many ways such investigations are the key test as to whether the biofriendly aspects of water are attractive merely because the investigators are carbaquist. In a brief survey of candidate liquids, Wolynes notes that at high temperatures there is some reason for skepticism. Life forms based on hydrogen plasmas, molten metals, and even neutron fluids within neutron stars are conceivable, but on the assumption that life must possess "long-lived information-bearing molecules" (p. 214), each of these milieu presents daunting problems. Cold (or very cold) fluids, however, are potentially more promising. Nevertheless, in this context a recurrent problem is one of solubility, and whether the van der Waals forces will exceed the ambient thermal energy. As Wolynes notes, this is not, in itself, a fatal objection. This is because some candidate fluids such as supercritical molecular hydrogen are still hypothetically viable, given their presumed ability to participate in the folding sequence of macromolecular arrays. As we have seen, the interaction of water in the folding of proteins has intriguing specificities, and it is likely that any form of life on a planetary surface will require either proteins or an analogous structure to form the basis of the information-bearing cell (or equivalent). Nevertheless, as Wolynes again stresses, a key aspect of the water–protein interactions is not only the hydrophobic interactions but also the role of water molecules interpolated into protein interfaces or interiors. It is, at present, far from clear that alternative solvents would enable such complex and "unexpected" interactions.

1.4   THE DANGEROUS LIQUID WATER

Wolynes' account rightly leaves the matter open-ended, albeit with the hint that water may yet prove to be special. So far as there is a persistently skeptical view of water being uniquely suitable for life's activities, it is through Benner's articulation that, far from water being a benign carrier, it is manifestly far from optimal. In this provocative strategy, he reminds us of the many disadvantages of water. In one sense, this disability has long been recognized by those working on the origin of life. As Benner also reminds us, the hydrolytic activities of water present a severe barrier to the assembly of many molecules abiotically, not least the nucleotides. So what are the alternatives? Benner emphasizes such liquids as formamide and hydrocarbons such as ethane. The former is sensitive to hydrolysis, but on desert planets might provide the substrate for life. The existence of liquid hydrocarbons in Titan-like settings is also a current focus of attention. These deal with quasi-Earth like settings; should our horizons be wider? What of the gas giants? Here, as Benner explains, a habitable zone can be defined with respect to dihydrogen. This possibility crucially depends on the pressure-temperature field and for life is straddled by the constraints of when dihydrogen becomes a supercritical field as against the rise in temperature to a point where carbon–carbon bonds are compromised. (Bio)chemical space is vast, and still largely unexplored. Although not discussed, one could also think of supercritical fluids or ionic liquids as two further classes of potential solvents that are currently in vogue.

So far, if there is a consensus in this volume, it is that water is peculiar, possibly very peculiar. Despite decades of work, this conclusion is still surprisingly provisional. Nevertheless, as our understanding of the biological role of water advances, so the interactions seem to be revealed as increasingly subtle. Moreover, although many fluids are reasonable candidates in one respect or another, there is also the sense that no other fluid possesses *all* the properties that make water so biologically versatile. Not only does it provide a broad and flexible canvas for life to paint its full multifaceted tableaux, it is itself part of the palette. This particularity may also extend to the origin of life itself, if it transpires that the crucible of synthesis was not the notorious "warm, little pond," but the bubbles associated with it.

These views remain provisional. Alternative views insist that water is far from ideal, and that other fluids are strong candidates for alien life forms. Benner, in particular, is forthright in his suggestion that as water-based life-forms we are hobbled in our imagination and too constrained by the familiar. He has a point, and it is as yet uncertain whether life (as we know it) in its earliest stages was faced with almost intractable problems and its apparent fitness to function merely reflects compromise and adaptation. It is, however, equally plausible from our present perspective that just as DNA is arguably the strangest polymer in the universe, so water is the strangest molecule.

1.5 CONCLUSIONS

Although it is the science of water that is very much in the foreground of this volume, it is no accident that a number of authors touch on more general aspects. As an agency or symbol of purity, be it via ancient rites of lustration or perhaps the stoup in a Catholic church, water plays a central role in many areas of religion (see also Russell). So too can it be a source of conflict, be it the squabbles over a desert well or the threatened "water wars" where regional violence may escalate into something even worse. From biochemistry to purification, life would be literally unimaginable without water.

Given this, it is all the more remarkable that there is much about water that remains to be explored. Underlying this relative ignorance is a fertile tension between those who take the view that water is literally unique and those who see it as just one of a series of liquids in which life, albeit of very different sorts, flourishes. As Weber and Woodworth stress, there is much at stake here. Underlying our desire to understand the strange dynamic entity we call life, there is a strong sense that in the absence of a general theory of organization and emergence, our attempts may be frustrated. What is it that explains the complexity, stability, and robustness (and sometimes extreme sensitivity) of life? Curiously, it may transpire that water provides the solution. After all, if it turns out, after decades of experimentation and perhaps recovery of extraterrestrial life, that water appears to be truly unique, then we can advance the argument for precise specificities, if not fine-tuning, and move to a position where definitions of life (and indeed its origins) cannot be considered without water. Alternatively, if it transpires that life is based on many types of chemistry (and possibly even physics), then we would be both on the threshold of a richer universe of possibilities and also, perhaps, able to identify more general conditions that allow a definition of life.

As Weber and Woodworth (and also Ward) point out, the current interest in artificial or *in silico* life (A-life) is potentially important because it may instruct us as to how self-organization occurs. It may also be our shortest (and cheapest) way of moving beyond the perennial problem that, of all life, we only know the one example. These, and the other possibilities already mentioned, are still entirely open-ended. The general prospect is that, not only is life ubiquitous, but may be based on many physicochemical systems. It remains equally possible that systems other than water may generate complex chemistries and organizations, but lack the spark we call vitality.

ACKNOWLEDGMENTS

We thank the John Templeton Foundation for the opportunity to participate in the meeting at Varenna, and the various individuals (especially Charles Harper and Pam Contractor) for facilitating many aspects of both the conference and subsequent publication. We also warmly thank Mrs. Vivien Brown for superbly efficient typing and manuscript presentation. Cambridge Earth Sciences Publication ESC.1274.

2 Water and Life
Friend or Foe?

Felix Franks

CONTENTS

2.1 INTRODUCTION: THE H$_2$O MOLECULE

Even without consideration of its close involvement with life, H$_2$O, particularly in its liquid state, must be classed as "strange" and "eccentric" (Franks, 1972, 2000). Many of its properties are not what might have been predicted if the substance had only just arrived on Earth. It certainly does not appear in its expected place in the periodic table. If it did, then its boiling point should be in the neighborhood of –93°C. The basic molecular properties of the water molecule that are of prime importance are

- The *sp^3* hybridization of the H$_2$O molecular orbitals, which gives rise to an approximately tetrahedral disposition of four possible hydrogen bonds about each central oxygen atom.
- The quadrupolar nature of the molecule (two positive and two negative charges).
- The low energies of the O–H–O (hydrogen) bonds, in comparison to covalent bonds or ionic interactions, make these bonds susceptible to minor structural distortions; most important, however, these are the only interactions in which water molecules can participate.

An exact description of the molecular structure of H$_2$O in all its condensed states illustrates its diversity and its many subtleties (Finney, 2004), but for purposes of establishing and validating the role of water in life processes, the simplified but well-known Bjerrum model (Bjerrum, 1952) will suffice. It is shown in Figure 2.1.

Although the above features are also found in other molecules, their combination in one substance is probably unique. Thus, liquid ammonia, NH$_3$, resembles water in some respects, but with three proton-donor sites and only one acceptor site, it cannot form the type of three-dimensional

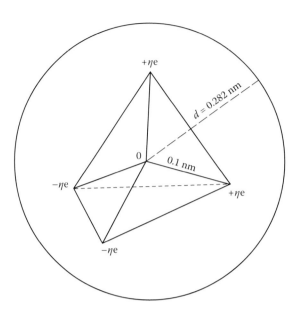

FIGURE 2.1 The Bjerrum model of a water molecule. Orbitals are pointing toward the vertices of a regular tetrahedron. The van der Waals radius is taken as one half of the O–H–O distance, but the protons are usually not located at the center of the hydrogen bond. (From Franks, F. *Water—A Matrix of Life*. Cambridge: Royal Society of Chemistry, 2000. With permission.)

networks that are characteristic of ice and liquid water. On the other hand, germanium dioxide, GeO_2, and silica, SiO_2, possess four-coordinate structures, similar to that of ice, but the Ge–O and Si–O bonds are covalent. Similar limitations exist with other molecules that possess some, but never all, of the above features, for example, HF, H_2O_2, CO_2, *N*-substituted amides, dimethyl sulfoxide (Me_2S=O), etc. The molecules that most closely resemble ^1H_2O are its several isotopic modifications and, more important, carbohydrates (polyhydroxy compounds, PHCs) of general formula $C_m(H_2O)_n$. They feature largely in the water relationships of living organisms (Franks and Grigera, 1990).

Taking the molecular structure of the H_2O molecule, shown in Figure 2.1, as a starting point, it is possible to construct many different types of three-dimensional assemblies based on combinations of squares, pentagons, hexagons, etc., of oxygen atoms, linked by hydrogen bonds. A regular structure, based only on linear hydrogen bonds of identical lengths, is provided by hexagonal ice (ice-1h), the "usual" form of ice, shown in Figure 2.2. Slight distortions in bond lengths and/ or angles, which are likely to take place on melting, give rise to manifold structural possibilities. The physical properties of water indeed suggest that liquid water consists of such ice-resembling

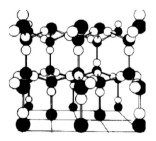

FIGURE 2.2 Local structure of water molecules formed as a system of four tetrahedrally arranged neighbors (black circles) surrounding each central oxygen molecule, characteristic of "ordinary" hexagonal ice. (From Franks, F. *Biophysics and Biochemistry at Low Temperatures*. Cambridge University Press, 1985. With permission.)

structures, but without the crystalline long-range order, and with very short persistence times, on the order of picoseconds.

2.2   LIFE PROCESSES: A MINIMALIST APPROACH

A useful inquiry into how (or if) any of the above mentioned properties of water have played a part in the development of an environment on this planet that is able to support the "creation" and further evolution of life must commence with a careful definition of the word "life." In the context of this chapter, the *Oxford Dictionary* begins its definition of life as follows:

> The active principle peculiar to animals and plants and common to them all, the presence of this in or by the individual, living state, the time for which it lasts or the part of this between its beginning or its end…

Scholarly? Perhaps, but one could take issue with this definition, because no mention is made of reproduction or of microorganisms or other simple forms of life, and some processes are included that are certainly *not* common to *all* animals *and* plants.

To translate the dictionary definition into scientific terms, life processes, as we understand them, must encompass all of the following functions in sequence:

- To control the synthesis of simple, chiral molecules and their reactions to form complex polymers, based mainly, but not exclusively, on carbon, hydrogen, oxygen, nitrogen, and phosphorus
- To program and direct the assembly of such molecules into supramolecular structures, organelles, cells, organs, tissues, and organisms, that is, the achievement of differentiation in the right places and at the right time
- To control cascades of chemical reactions (e.g., metabolism), resulting in growth to maturity, steady-state maintenance, defense against predators and chemical deterioration, energy-conversion processes, reproduction, followed by a more-or-less rapid senescence and expiry

Before discussing the role of water in the above processes, it should be noted that there exists a sizable and growing literature, mainly related to "protein engineering," that is devoted to the in vitro functioning of isolated enzymes, allegedly in the dry state or even in organic solvents. However, bearing in mind the above criteria for life, in vitro processes involving isolated molecules, which have themselves been manufactured from DNA templates in living cells, can hardly in themselves be considered as life processes. In addition, such "dry" enzymes invariably contain some residual water molecules located in strategic positions within the enzyme, where they contribute to the active enzyme structure and function.

The intimate relationship between water and biochemical processes is strikingly demonstrated by substituting the deuteron (^2H) for the proton (^1H), a substitution that most physicists would regard as trivial—merely a change in the zero-point energy. However, even this "minor" isotopic substitution is toxic to most life forms, demonstrating that life processes are so sensitively attuned to the $O–^1H–O$ hydrogen bond energies that the substitution by the heavier deuteron alters the kinetics of biochemical reactions, but to different extents, and will thus interfere with their coupling. Only the lowest forms of life, for example, some protozoa, can tolerate the complete deuteration of their constituent biopolymers, but only if brought about in gradual stages. All higher forms of life will exhibit signs of enhanced senescence and eventual death. Nevertheless, if conditions on this planet were to change, such that some cosmic upset would lead to a major ^2H concentration increase, then evolution theory suggests that some ^2H-tolerant species would survive and eventually become the dominant varieties.

2.3   WATER IN THE UNIVERSE AND IN OUR BIOSPHERE: THE HYDROSPHERE

Water was one of the four Aristotelian elements of earth, air, fire, and water. Early on, the Alexandrian scientists realized that, of the first three, earth and air are mixtures and fire is the manifestation of a chemical reaction. However, water remained an "element" until 1790, when Lavoisier and Priestly demonstrated that water could be decomposed into "air" (oxygen) and "inflammable air" (hydrogen). The question of where water existed in the universe was solved much later: it is now established that most of it is adsorbed on interstellar dust particles that eventually make up the tails of comets. On the other hand, it is not at all certain how the molecule came to be synthesized in outer space, because three-body atomic collisions (two H and one O) in a gaseous medium at very low pressures are extremely rare events. Whatever its origin in the universe, water must have arrived in our region of the solar system at a time when the temperature on Earth was still well above the critical point of water ($374°C$). If all the water that now makes up the oceans had previously existed as a supercritical atmosphere, then the pressure on the Earth's surface would have been 25 MPa/m²! As the earth cooled to subcritical temperatures, there must have taken place a massive and sudden condensation of water that caused major changes to the nature of the Earth's surface. Much of this water will have immediately boiled off again, giving rise to the Earth's present hydrosphere.

Our present water resources total 1.34×10^9 km³, of which 97% make up the oceans.* The major portion, 99.997%, of the remaining freshwater is locked in the Antarctic ice cap. The fresh liquid water immediately accessible for agriculture, domestic, and industrial use therefore amounts to no more than 0.003% of the total freshwater resources. Of that total, 75% is used for irrigation purposes.

The hydrological cycle (transpiration/evaporation, followed by precipitation) ensures that our surface water is recycled 37 times per year, constituting a vast water purification system. Unfortunately, 75% of water precipitation falls into the oceans and thus becomes largely useless, at least to the terrestrial animal and plant kingdoms, unless the salt is removed.

The total volume of "moisture" held in the soil is 25,000 km³. Plants normally grow on what is considered to be dry land, but this is a misnomer, because even desert sand can contain up to 15% water. Apparently, plant growth requires extractable water; thus an ordinary tree withdraws and transpires ca. 190 L/day. Groundwater hydrology has become a subject of extreme importance because less than 3% of the earth's available freshwater occurs in streams and lakes. Now that global warming is believed to cause a major future threat, groundwater hydrology, so that the construction of pipeline networks should become activities of extreme importance. On the other hand, it should be emphasized that global warming is not a new phenomenon. Otherwise, where did Greenland get its name from? Former global warming periods were brought to a halt by a series of ice ages.

The search for water (and therefore also the search for life) has become a popular aspect of space research, and the existence of solid H_2O (not necessarily ice) in many cold stars and meteorites has been firmly established. It is also believed that, next to hydrogen, oxygenated hydrogen, in the form of various free radicals, are the most abundant chemical species in outer space. More surprisingly, however, the characteristic infrared spectrum of nonsolid, oxygenated hydrogen (i.e., liquid water) has been detected in the photosphere of the sun (Wallace et al., 1995).

For a sensible discussion of life on Earth, we must learn when and how molecular oxygen first appeared. Early prokaryotes learned to use water, rather than H_2S, to provide hydride quite early on, but with a devastating side effect: the release of molecular oxygen, which was the enemy of the cell chemistry of primitive life. Eventually, living organisms became able to gain energy from its breakdown:

$$O_2 + C/H/N \text{ compounds} \rightarrow N_2 + CO_2 + \text{energy}$$

---

* One cubic kilometer is the volume of water that would cover a midsize city, such as Florence, to a depth of 1 meter.

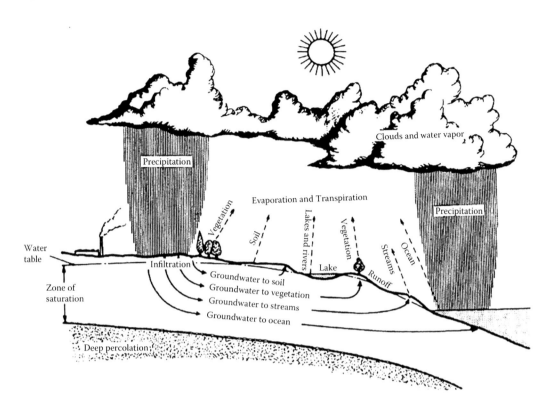

FIGURE 2.3   The hydrological cycle, showing the continuous recycling paths of the Earth's water resources. Until humans learn to control the spatial distribution of rainfall, 75% of all precipitation will presumably continue to fall into the (saline) oceans. (From Franks, F. *Water—A Matrix of Life*. Cambridge: Royal Society of Chemistry, 2000. With permission.)

Aerobic life forms eventually developed and produced several sophisticated technologies, for example, photosynthesis, for the splitting of water. It is amusing to note that we humans have the arrogance of classifying early anaerobic life forms that exist in the hot deep sea trenches as "extremophiles." Because those life forms existed on this planet long before we did, we would be more justified in classifying life forms that not only tolerate but even require molecular oxygen for their existence as extremophiles.

Within the overall hydrological cycle of transpiration and precipitation, shown in Figure 2.3, there exist several subcycles that are equally important for the maintenance of living species. Most of them use water not only as a substrate, but also as a reactant or a product of reactions. Two such coupled subcycles are between food producers (plants) and consumers (animals). The nitrogen and phosphorus cycles are of almost equal importance; they can be easily upset where large quantities of fertilizers and/or detergents find their way into water sources, such as rivers and lakes. The imbalances caused between producers and consumers by excesses of nutrients, such as nitrates and/or phosphates, can have disastrous effects, as witnessed by the eutrophication (overfeeding) and "death" of Lake Erie during the 1950s. Such a scenario might well arise in highly developed countries. Eutrophication from agrichemicals, in combination with warm effluents from a power station, might lead to a gradual death of aquatic species in a lake due to a lack of oxygen and light caused by the overgrowth of algae. The situation will be made worse by an erosion of rocks at the bottom of the lake. The repair of such manmade ecological disasters, if it can be achieved at all, takes several decades. Its result, not repaired, exists today in the Black Sea, where only 50 m down from the surface are now still fit for life.

2.4   LINKS BETWEEN WATER AND LIFE CHEMISTRY

The interactions between water and molecules that govern life processes can be studied at several levels of increasing complexity. Over the past 50 years, much has been written about a so-called "water structure." Many biological and technological phenomena have been ascribed to this ill-defined water structure and to "bound" water—even to "strongly bound" and "weakly bound" water. Such distinctions are never described in detail, but they run counter to the universally accepted laws of physics, which do not even recognize the existence of molecules at all. They certainly do not permit the distinction between "different" molecules, all answering to the formula H_2O. We here want to emphasize that hydration, if accepted as a valid scientific concept, must be rigorously defined in terms of one or more of the following attributes:

• Structure, as expressed by spatial coordinates, distances, and angles
• Energetics, expressed in terms of interaction energies (enthalpies) between water and the hydrated species (ΔH), with hydrogen bonds playing a dominant role
• Dynamics, expressed as lifetimes of water molecules at a given site, exchange rates of water in the hydration shell with water in the bulk, and general diffusive behavior

Despite all the ifs, buts, and caveats about the laws of physics, the notion of "bound" water and all manner of other "special" types of water continues to appear in the scientific literature, the daily newspapers, on radio and television, and, regrettably, in the patent literature. Volumes could be written on this subject. Probably the most long-lived aberration was that of polywater, a late-1950s observation, made in good faith by an unsuspecting physicist in the Soviet Union, of a form of water that did not freeze at 0°C, did not boil at 100°C, and did not exhibit a maximum density phenomenon. Because of its spectroscopic properties, it was later given the name "polywater." A hot debate about its reality quickly developed, which provided a fascinating insight into the sociology of scientists (What makes them do what they do?) and which lasted for 15 years, before it was finally laid to rest as a nondiscovery, but not before millions of dollars had been spent on its study (Franks, 1981). Sadly, the lesson has not been taken on board; new forms of water with well-nigh magical properties continue to make their appearance. The following e-mail arrived on my desk:

SUBJECT: NEW! PATENTED WATER GOES MLM
DATE: 10 JULY 2005, 15:02

Absolutely Amazing! A product unlike anything you've ever seen!
-- Turns Ordinary Water into Extreme Body Fuel --
Millionaires will be created by the
#1 SIZZLE PRODUCT OF THE 21st CENTURY
Instantly transforms H_2O into X_2O, creating a Super Beverage Xtreme X_2O Body Fuel
Outperforms: Gatorade, Orange juice, Red Bull, Apple juice, AMP, Lipton tea, Cranberry juice, Full Throttle, PowerAde, and many more!
Energizes Cleanses Defends
2 US Patents NOT a Filter NASA Lab Tested.
Launching Summer, 2005. This is the BIG money maker you've been waiting for.

2.5   IONIC AND MOLECULAR HYDRATION

Water can interact *directly* only by hydrogen bonding, either with ions or with molecules that, like water itself, possess proton donor and/or acceptor sites. Ionic hydration can, to some extent, be treated by the laws of electrostatics. The development of neutron scattering has enabled "hydration

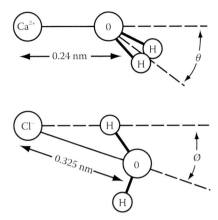

FIGURE 2.4 Disposition of water molecules in the hydration shells of monatomic metal and halide ions. Both hydration shells consist of six water molecules, arranged octahedrally. Angles θ and ϕ define the spatial disposition of water molecules in the primary hydration sphere; both angles increase with increasing solution concentration.

structures" in solution with short lifetimes to be defined in detail. They are shown in pictorial form in Figure 2.4. As expected, the orientation of the water molecule is governed by the ionic charge. The number of water molecules that form the hydration shell depends on the ionic radius. For monatomic ions, for example, Na^+, K^+, Ca^{2+}, Ni^{2+}, Cl^-, Bt^-, etc., this nearest-neighbor hydration shell consists of six water molecules, octahedrally disposed about the central ion. The strengths of ion–water hydrogen bonds depend on the ionic charge and the ionic radius. Many biochemical reactions involve ion–water interactions, but the textbooks pay little attention to basic processes, such as differences in hydration energies of Na^+ and K^+.

There remain several unsolved mysteries related to ionic hydration. Chief among them is the phenomenon known as *lyotropism*.* This term refers to an observation that ions, especially anions, can be arranged in a series (the lyotropic series) based on their influence on the aqueous solubility of molecules. Thus, some ions (sulfate, phosphate, etc.) reduce the solubility, whereas others (nitrate, iodide, etc.) enhance it. The order in which these ions act appears to be the same for all aqueous solutions. Surprisingly, the series was first observed to hold for the solubilities of different proteins, and this led to all manner of complicated explanations in terms of protein–ion interactions (Hofmeister, 1888). However, it was found much later that the same ions, placed in the same order, affect the aqueous solubility of argon or nitrogen in the same manner as they affect proteins. Equally perplexing was the finding that the same ion order also operates in ion effects on the stabilities and biological activities of proteins. Thus, ions that enhance solubility also inactivate proteins, whereas ions that reduce solubility stabilize proteins against inactivation, for example, by temperature or pH.

An abbreviated representation of the lyotropic series is shown below, with the effects on solubility increasing from left to right. It is interesting, perhaps biologically significant, that the chloride ion is found near the middle of the series; it hardly influences solubility or protein stability.

---

* When an experimentally observed phenomenon cannot be satisfactorily explained, scientists are in the habit of giving it a name, for example, *catalysis*, believing that nomenclature can take the place of comprehension. Lyotropism is one example of this custom.

sugars. An example is shown for raffinose in Figure 2.5. Raffinose is a trisaccharide, constructed of a galactose–glucose–fructose chain. In its usual crystalline form it contains five water molecules, which give it the configuration shown in the figure.

Of the 20 oxygen atoms, five belong to the water molecules of hydration, each of which is hydrogen bonded either to a sugar oxygen or to another water oxygen. The water molecules, although they occupy specific locations in the crystal lattice, are labile in the sense that they can fairly easily be removed and are also subject to fairly rapid exchange with other water molecules. When they are removed, the crystal structure collapses (Kajiwara and Franks, 1997).

2.6 HYDROPHOBIC EFFECTS: A UNIQUE PHENOMENON?

Whereas ionic hydration and hydration by direct hydrogen bonding can be treated by classical physical approaches, there remains one type of unique hydration interaction that is still the subject of debate; it was, of course, given a name: "hydrophobic hydration" (Franks, 2000). Its complete description and interpretation are beyond the scope of this chapter, but its basic features need to be mentioned because it forms the basis of many biochemical processes, from controlling the stability of proteins and nucleotides to the spontaneous assembly of supermolecular systems, such as cell membranes and complex enzyme structures.

Essentially, and as the name implies, hydrophobic hydration describes the interaction between water and molecules (or ions) that "hate" water and are incapable of participating in the formation of hydrogen bonds. Such molecules include the noble gases and hydrocarbons, but also atomic groups attached to hydrogen bonding functions. The simplest example is methanol, in which the –OH group favors the interaction with water, but the –CH$_3$ group is hydrophobic and is repelled by water. It becomes a struggle between hydrophobia and hydrogen bonding as to which effect will predominate. In the case of methanol, the –OH group wins, making the alcohol completely miscible with water. On the other hand, molecules forming the cell membranes contain long alkyl chains and only one single polar head group. On balance they are therefore insoluble in water.

When a hydrophobic molecule or residue is "forced" into an aqueous medium, it forces a rearrangement of the water–water hydrogen bonding pattern in its vicinity to a related but not identical water structure, thereby creating a cavity that it then fills. An example of this type of structure is shown in Figure 2.6. Energetically, this type of rearrangement of water molecules may be favorable, but configurationally it perturbs the favored icelike water structure and reduces the number of degrees of freedom that the water molecules could adopt. The opposite effect, the release of some water molecules, which enables them to relax back to their favored structure, can be achieved by the

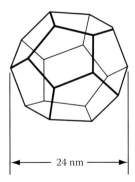

24 nm

FIGURE 2.6 A hydrophobic hydration shell, able to surround a small apolar molecule. The water network has been altered structurally but has maintained O–H–O bond lengths identical to those in ice, but –OH vectors are barred from pointing toward the center of the cage. This reduces their degrees of conformational freedom.

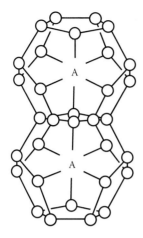 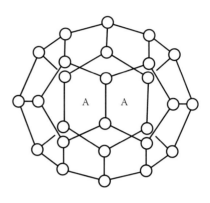

FIGURE 2.7 Two possibilities of accommodating two apolar atoms or molecules A in water hydration cages with a reduction in the total number of constrained water molecules that would be required by two separate cages; mistakenly termed a hydrophobic "interaction."

forcing of two or more "encapsulated" hydrophobic residues to associate; different possibilities—fused and shared water cages—are possible; examples are shown in Figure 2.7. However, what appears to be an attraction between the hydrophobic species is in fact produced by a repulsion of the apolar molecules, or groups of molecules, by water. This is a difficult concept to grasp, one that has frequently been misrepresented in the scientific literature as a net attractive force between the apolar groups. On a larger scale, the alignment of hydrophobic molecules to form a cell membrane involves the large scale repulsion of hydrocarbon chains of polar lipids by water, resulting in the structural organization of the bilayer membrane, in order to minimize the *total* free energy of the assembly.

The hydrophobic effect is also a critically important contributor to protein stability and the assembly of complex peptide-based structures, for example, muscle fibers and viruses. Thus, the linear peptide chain of a protein contains three types of amino acids: ionogenic (e.g., glutamate, lysine), polar (e.g., serine, cysteine), and hydrophobic (e.g., alanine, leucine, phenylalanine) in different amounts and different sequences. Folded, native protein structures are maintained by many stabilizing intrapeptide ionic and hydrogen bond interactions. These are, however, counterbalanced by destabilizing hydrophobic interactions between alkyl residues and water. The net stability margin of a protein in its active state rarely exceeds 50 kJ/mol, an energy equivalent to only three hydrogen bonds in a structure that contains many hydrogen bonds. For a globular protein molecule to form a biologically active structure, it requires a ca. 50% content of (destabilizing) hydrophobic amino acids. It is thus the fine balance, caused by water-promoted interactions that have given us life's workhorses that are responsible for the majority of biochemical functions.

2.7 WATER AS REACTANT AND REACTION PRODUCT

Water biochemistry is not a subject that is found in biology teaching texts. It is, however, one of the basic and fascinating aspects of the close relationship between water and life. It would be no exaggeration to describe biochemistry as the chemistry of water, because water participates in the vast majority of biochemical processes. The H$_2$O molecule acts as proton-transfer medium in four basic types of biochemical reactions: oxidation, reduction, hydrolysis, and condensation. There exist, however, many other reactions involving water, with some mechanisms still shrouded in mystery, for example, the oxidation of water to yield molecular oxygen, a basic component reaction of photosynthesis. Standard biochemistry texts compound the confusion by what has been described as "sloppy proton book-keeping." How could one explain an equation that purports to show that one molecule of glucose (12 H) can supply 12 pairs of protons for the production of 36 molecules of ATP in the citric acid cycle? Only by sloppy proton book-keeping.

Even more misleading, students are taught that the enzymatic oxidation of unsaturated compounds (glycerides, fatty acids, etc.), the famous nutritious omega-3 products, takes place by the single-step addition of H and OH across a double bond between carbon atoms. This is a remarkable statement, because this apparently simple reaction actually proceeds in several steps, with water taking an important part in each one. It illustrates yet again how water is treated by biochemists; a case of "familiarity breeds contempt?"

2.8   WATER AS INTRACELLULAR TRANSPORT FLUID

The average human adult has a daily water turnover of 2.5 kg, of which 300 g is produced endogenously by the oxidation of carbohydrates in the mitochondria. The reaction is accompanied by the generation of ca. 100 mol ATP, which is stored and provides the energy requirements of the many physiological functions of the body. This generation of water (and ATP) proceeds in a cascade of 14 steps, with water participating in each step. If this process were to be carried out in a single step, the body temperature would rise by 26°C, clearly an undesirable outcome of maximum engineering efficiency. The in vivo rates of such coupled reactions, performed isothermally, have become sensitively attuned to the physical properties of water, for example, its ionization equilibrium and its hydrogen bonding pattern. An engineer, taught to maximize yields and reaction rates, finds nature's methods laborious and wasteful. He does not realize that evolution has produced optimized, rather than maximized, reaction sequences. Here, even minor changes in any of the properties of water can cause chaos to the coupling between the biochemical reactions. This is illustrated by changes in temperature, pressure, or a substitution of ^1H by ^2H, the injurious outcomes of which have already been discussed.

Apart from the important role water plays in metabolic reactions, other physiological processes associated with water housekeeping include the kidney. Thus, the "normal" daily 1.4 L of excreted urine is produced by the concentration of 180 L of dilute urine. The remaining water is returned, purified, to the body. The process resembles, qualitatively, the industrial desalination of water. The kidney also regulates the quantity of body water and its salt content. Apart from the kidney, other organs, associated with water regulation include the salivary glands, pancreas, intestines, gall

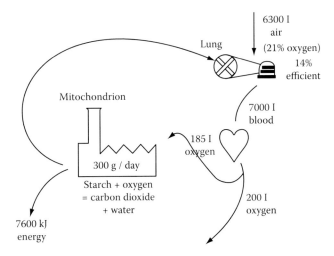

FIGURE 2.8   The daily energy output and consumption of the human ATP factory, arising from the mitochondrial generation of 300 g water. (From Franks, F. *Water—A Matrix of Life*. Cambridge: Royal Society of Chemistry, 2000. With permission.)

bladder, and liver. Even the heart is involved, although only indirectly: each heartbeat pumps 70 mL of blood, so that with 70 beats/min, the heart supplies 7000 L required for the daily generation of energy for carbohydrate oxidation. The "oxygen factory" is shown diagrammatically in Figure 2.8. Among other features shown in the oxygen factory, the above-mentioned principle of optimization rather than maximization is illustrated: evolution achieves optimum rates by sacrificing a degree of engineering efficiency, that is, wasting energy.

Turning briefly to the plant kingdom, all plants on Earth turn ca. 5×10^{10} tons of carbon into carbohydrate annually, requiring 8.5×10^{10} tons of water, with an annual energy requirement of 4×10^{15} kJ. Although solar energy supplies a major part of this energy, the metabolic contribution is not trivial. Of all the raw materials required by the cyclic processes of life, water is the only one that is an abundant and nondepleting resource. The others—nitrogen and carbon dioxide supplies—are much more finely balanced in the ecosphere and can easily be upset by human interference.

2.9   WATER AS THE ORIGINAL AND A NATURAL HABITAT FOR LIFE

The physical properties of water must have played an important role in evolution. Life began in hot water and in the absence of oxygen. Even now, the number of aquatic species far exceeds those that have, more or less successfully, made the journey to "dry" land. But even terrestrial species have been compelled to develop mechanisms, energetically quite inefficient, that would enable them to maintain their correct water balance. Other properties of water, such as density, hydrodynamics, viscosity, diffusion, optical, acoustic, thermal and electrical properties, and surface tension, are all utilized by many species to facilitate motion, awareness of, and defense against predators (Denny, 1993). Here, again, mysteries remain. Outstanding among them is the management of water movement in plants. Another fact, often overlooked by the layman, is that in the oceans, the temperature of maximum density (normally close to 4°C) lies below the freezing point of water. The coldest water is therefore the densest water, which profoundly impacts the buoyancy of aquatic organisms.

A superficial comparison of aquatic and terrestrial species might lead to the conclusion that the former lead a less stressful life. There are, however, other factors, such as the limited oxygen solubility and its slow diffusion in water, that make breathing hard work. A major disadvantage of the saline water habitat is that all organisms exist in a state of osmotic disequilibrium. Energy is therefore required to maintain the body fluids at the correct osmotic concentrations. On the other hand, terrestrial organisms, especially mammals, will still require many millennia to become fully adjusted to the vagaries of liquid water, its supply, its physical limits, and the dangers of freezing and desiccation. Even now, after millions of years on "dry" land, the developing mammalian fetus still begins life in an aqueous environment of a composition similar to that of the ocean, and mammalian red blood also still maintains the high salt osmolarity of seawater.

2.10   WATER: THE FRIEND

It is obvious that water is the friend of all living species on Earth, both as a suitable habitat and as the intracellular fluid that helps in many ways to support the chemistry of life and its correct functioning. Because water is a nondepleting resource in our ecosystem, and because its purification is taken care of by the hydrologic cycle, a global shortage of clean water needs never set the limit to life on this planet, unless such a shortage is produced by human interference.

The warm ocean currents maintain temperate environments, helping to make Earth fit for life without air conditioning. This thermostating effect results from the abnormally high specific heat of liquid water. Thus, the movement of water in the Gulf Stream during its passage from the Gulf of Mexico to the Arctic Circle is accompanied by a 20° temperature drop. The energy released to the atmosphere amounts to 5×10^{13} kJ km^{-3}, which is equivalent to the thermal energy generated by the combustion of 7 million tons of coal. All the coal mined in the world in one year is able to produce this amount of energy for only 12 hours.

Also, as every schoolchild knows, the maximum density of water at 4°C is the phenomenon that causes lakes and rivers to freeze downward from the surface, rather than upward from the bottom, with all its terrifying ecological consequences for aquatic life.

2.11   WATER: THE ENEMY

The support of life has its downside, because water uncritically yields a friendly environment for *all* life forms, including organisms such as the malaria larva and the most toxic pathogens. Coupled with a lack of adequate sewage treatment and other purification technologies, such contaminated water is, at best, the carrier of disease, and at worst, a killer. Unfortunately, water is also capable of assisting in the destruction of life, sometimes on a large scale, but not only by its extreme violence, as witnessed recently by the tsunami in the Indian Ocean and by the action of Hurricane Katrina on the American Gulf Coast. The physical violence of water is awesome, but not nearly as insidious as the longer-lasting consequences of flooding. The destruction by water of places of human habitation provides an ideal breeding ground for pathogens, and the lack of sufficient potable water for months or years to come will give rise to wide-ranging epidemics, even beyond the directly affected regions.

Water shortages and its capricious distribution, leading to droughts and floods in close proximity, also cause severe problems in large parts of the world. Some attempts to regulate precipitation are on record. A well-recorded pilot study to enhance the snow fall over the Rocky Mountains was conducted by the U.S. National Science Foundation during the 1970s. The rationale was to increase the water supplied by the Colorado River to far away places, such as Los Angeles (Weisbecker, 1967). Such attempts, even when technically successful, have invariably failed to be implemented on a practical scale, mainly for economic, nationalistic, or local political reasons. In the case of the Rocky Mountains snow project, it was feared that Kansas might end up in the rain shadow and become a desert. Even more important, the population of the Colorado River basin was not prepared to tolerate the possibility of increased avalanche and mud-slide activity, just so that Los Angeles would have more water for irrigating golf courses in the desert. Threats of litigation caused Congress to ask the National Science Foundation to terminate the snow enhancement project.

It is estimated that 2.7 billion people suffer from severe water shortages, mainly in regions where poverty is already extreme. But where water is plentiful, it forms an ideal breeding ground for all manner of microorganisms, many of them pathogens, carried by insects. Water-borne diseases account for 3.5 million deaths annually, mainly in the developing world, where 66% of the population still lacks access to adequate sanitation.

Other dangers, arising despite an adequate water supply, include inadequate standards of purification. On a planet where 75% of freshwater is used for agricultural purposes, a gradual salination of arable soil is a dangerous scenario for the future of mankind. It has been suggested that India alone annually loses 5% of its high-grade arable soil because of irrigation with insufficiently purified water.

2.12   DEFENSE AGAINST WATER STRESS

Finally, seasonal temperature extremes affect the physical properties of water and hence also disturb the delicate osmotic equilibria of living organisms with their environments, frequently referred to as *water relationships*. The "preferred" aqueous environment for most species lies between relative humidities of 99.9% and 99.999%. Any conditions outside that narrow range will result in pathological symptoms due to drying (Figure 2.9). The extreme conditions are desiccation by freezing or drought, although saline soil conditions are also injurious. Such osmotic disturbances are referred to as "water stresses" and are also depicted in Figure 2.9, where three scales of dryness are included.

To survive, many species have developed the means to either resist the water stress or adapt to it. Examples of freeze resistance are found in some insects and in Antarctic fish species that live

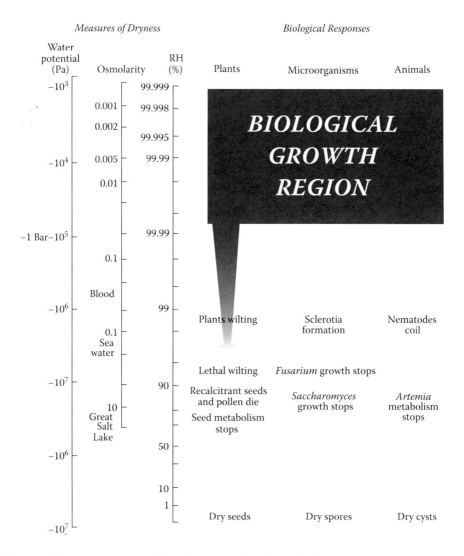

FIGURE 2.9 Effects of drying on the biological state of a variety of desiccation-tolerant organism. (Adapted from Leopold, A. C. (ed.), *Membranes, Metabolism and Dry Organisms*. Ithaca, NY: Cornell University Press, 1986.)

permanently at temperatures below the normal freezing point of blood (−0.8°C). They achieve this via antifreeze peptides, the structures of which interfere with the growth of ice crystals in their blood. This gives them a limited protection against freezing, which is, however, quite adequate because the temperature of the Antarctic Ocean hardly fluctuates and does not reach the freezing point of blood.

Low temperature as a stress generator requires a special mention. In common parlance, "low temperature" and "freezing" are used interchangeably. The two processes are, however, unrelated. Freezing, or the growth of ice, reduces the amount of liquid water in an organism and hence raises the concentrations of all solutes in the cytoplasm. Such freeze-concentration changes are usually irruptive, with ice crystallization rates on the order of 1 m/s. Such rapid changes are injurious, even lethal, unless the organism is equipped with defense mechanisms. Low temperature, on its own, does *not* signify freezing. Its effects are only those that accompany the change of physical properties of water without involving ice crystallization and its concomitant concentration effects.

An example of the difference between low temperature and freezing is graphically illustrated in Figure 2.10, where the two conditions are compared as they affect the specific activity of the enzyme lactate dehydrogenase (LDH) in a dilute solution. At room temperature and under refrigerated conditions (+4°C), the enzyme activity decreases with time. Freezing causes an almost instant loss of activity. Exposure to low temperature (undercooled water, unfrozen at −20°C) preserves 100% of the original enzyme activity for periods of years (Hatley et al., 1987).

The most widespread in vivo mechanism for surviving water stress is tolerance, a process by which the organism can biochemically acclimate to future adverse conditions. This acclimation process usually involves a depolymerization of the organism's intracellular starch reserves to produce a range of low-molecular-weight PHCs or the biosynthesis of free amino acids, most of which do not occur in proteins but are only generated as defense mechanisms against drying stress. They include glycine betaine ($Me_2N^+CH_2COO^-$), strombine ($H_2N^+[CH_3CH_2COO^-]_2$) and α-aminobutyric acid. Their mechanisms of providing defense against desiccation have not as yet been clearly established.

On the other hand, the mechanisms by which PHCs prevent drying injury has been studied in detail and has also been successfully applied to labile molecules in vitro. The involvement of PHCs in stress tolerance is also of particular scientific interest. Of primary importance is their miscibility with water, coupled with their ease of biodegradation. Their –OH groups closely resemble those of water in energy and conformation, so that hydrogen-bond chains, rings, and networks can be formed, either with the incorporation of water or with its separation. Because of their complex crystal structures, PHC molecules are resistant to crystallization in response to water withdrawal (drying); instead, they readily vitrify. It is their ability to form nontoxic in vivo glasses in supersaturated aqueous media that lies at the basis of their protective action during periods of osmotic water stress. The periods required for acclimation range from months for trees, to days for insects, and minutes for microorganisms. Once frozen or dried to the point of vitrification, the organism becomes chemically completely inert (dormant), but it will regain its vegetative (growth) state in contact with water

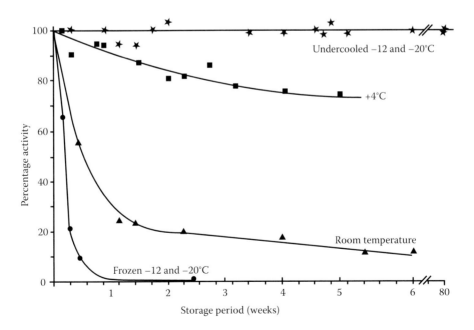

FIGURE 2.10 Maintenance of enzyme activity by LDH solution, stored under differing conditions. LDH concentration: 10 μg mL^{-1} in phosphate buffer (pH 7). (From Hatley, R. H. M., et al. *Process Biochem.* 22(12):169–172, 1987. With permission.)

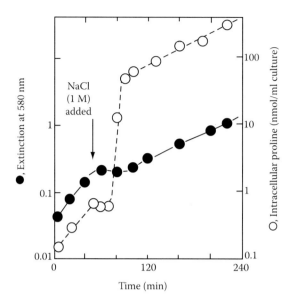

FIGURE 2.11 Acclimation to salt stress by *B. subtilis* in a growing culture, after exposure to a salt shock. The left-hand ordinate is a measure of protein synthesis. (From Gould, G. W. and J. C. Measures, *Philos. Trans. R. Soc. B*, 278, 151, 1977. With permission.)

after recrossing the glass transition. The biochemical acclimation machinery then goes into reverse, with the protecting PHCs being converted back to starch.

Like other glassy materials, they are undercooled liquids on a molecular scale, lacking the long-range order of crystals, but they behave as solids with respect to their mechanical properties. The science of aqueous PHC glasses is a fascinating subject that, however, lies beyond the scope of this chapter. The interested reader is referred to Kajiwara and Franks (1997) and to Levine (2002).

Examples of biological, water-soluble glasses are found in a wide range of seeds, bacterial spores, nematodes, and overwintering insect larvae. Their common characteristic is an ability to respond

TABLE 2.1
Changes in Metabolic Levels in *E. solidaginis* Larvae during Cold Hardening (Concentrations in µmol/g Wet Weight)

	15°C	0°C	−5°C	−30°C
Glycogen	381	258	166	113
Glycerides	167	170	169	159
Protein	81	Unchanged		
Glycerol	154	232	237	238
Sorbitol	1	42	97	147[a]
Glucose	0	16	34	29
Trehalose	55	64	63	70
Proline	32	42	57	56
Glycerol-3-phosphate	0.2	0.9	1.6	1.0
ATP	2.2	2.2	2.2	1.6
Lactic acid	0.2	0.1	0.1	0.5

Source: After Storey, K. B., and J. G. Baust, *J. Comp. Physiol.* 144, 183–190, 1981.
[a] Corresponds to 0.5 g/g dry weight.

to a trigger, be it low temperature, starvation, lowering of the relative humidity, or changes in the photoperiod, when the accumulation of lyoprotectants, usually PHCs, commences. This is shown in Table 2.1 for the gallfly *Eurosta solidaginis*, when exposed to gradually decreasing temperatures. When fully hardened, sorbitol makes up 50% of the total dry weight of the larvae.

Figure 2.11 shows the much more rapid acclimation, under isothermal conditions, of a growing *Bacillus subtilis* culture, after being subjected to a salt shock from exposure to 1 M NaCl. It is seen that protein synthesis is immediately halted and replaced by the synthesis and accumulation of proline to high concentrations. The organism is now hardened against further water stress. Protein synthesis and growth restart, although at a lower rate than initially.

A full understanding, especially of the mechanisms that trigger water stress responses, is still lacking, but the contributing processes involved in acclimation are clearly of a multidisciplinary nature, combining genetics, the biochemistry of starch conversion, physics of crystal nucleation and growth, and the materials science of the vitreous state. All these complex processes are triggered by changes in the properties of water and/or its availability, both globally and to a living organism's ability to detect and act defensively on the signal of a developing water stress.

REFERENCES

Bjerrum, N. 1952. Structure and properties of ice. *Science* 115:385–390.

Collins, K. D., and M. W. Washabaugh. 1985. The Hofmeister effect and the behaviour of water at interfaces. *Q. Rev. Biophys.* 18:323–422.

Denny, M. W. 1993. *Air and Water—The Biology and Physics of Life's Media*. Princeton, NJ: Princeton University Press.

Finney, J. L. 2004. Water? What's so special about it? *Philos. Trans. R. Soc. B* 359:1145–1165.

Franks, F. (ed.) 1972–1982. *Water—A Comprehensive Treatise*, Vols. 1–7. New York, NY: Plenum Press.

Franks, F. 1981. *Polywater*, Cambridge, MA: MIT Press.

Franks, F. 1985. *Biophysics and Biochemistry at Low Temperatures*. Cambridge University Press.

Franks, F. 2000. *Water—A Matrix of Life*. Cambridge: Royal Society of Chemistry.

Franks, F., and J. R. Grigera. 1990. Solution properties of low molecular weight polyhydroxy compounds. *Water Sci. Rev.* 5:187–283.

Hatley, R. H. M., F. Franks, and S. F. Mathias. 1987. The stabilisation of labile biochemicals by undercooling. *Process Biochem.* 22(12):169–172.

Hofmeister, F. 1888. On the understanding of the effect of salts: II. On the regularities in the precipitating effect of salts and their relationship to their physiological behaviour. *Naunyn-Schmiedebergs Arch. Exp. Pathol. Pharmakol. (Leipzig)* 24:247–260.

Kajiwara, K., and F. Franks. 1997. Crystalline and amorphous phases in the binary system water–raffinose. *J. Chem. Soc. Faraday Trans.* 93:1779–1783.

Kajiwara, K., F. Franks, P. Echlin, and A. L. Greer. 1999. Structural and dynamic properties of crystalline and amorphous phases in raffinose–water mixtures. *Pharm. Res.* 16:1441–1448.

Levine, H. (ed.) 2002. *Amorphous Food and Pharmaceutical Systems*. Cambridge: Royal Society of Chemistry.

Storey, K. B., and J. G. Baust. 1981. Intermediate metabolism during low temperature acclimation in the over-wintering gall fly larva, *Eurosta solidaginis*. *J. Comp. Physiol.* 144:183–190.

Wallace, L., P. Bernath, W. Livingston, et al. 1995. Water on the sun. *Science* 268:1155–1158.

Weisbecker, L. W. 1967. *The Impact of Snow Enhancement*. Norman, OK: University of Oklahoma Press.

3 An Introduction to the Properties of Water
Which Might Be Critical to Biological Processes?

John L. Finney

CONTENTS

3.1 INTRODUCTION

The fine-tuning argument with respect to the fitness of the universe for life relates the values of a few fundamental physical constants (Carr and Rees, 1979; Hogan, 2000). In order for the prerequisites for the evolution of the life we know on earth to exist, the precise numerical values of these constants are critical: vary them by even a small amount and the conditions we need will not be able to develop. Examples include the critical values relating to the often-quoted triple alpha resonance that builds ^{12}C from ^4He via ^8Be: the reaction only proceeds because the ^{12}C nucleus has an energy level just above the sum of the energies of ^8Be and ^4He. Furthermore, if the strong coupling constant had a value only a few percent larger, there would be 100% cosmological helium production—not an environment or composition of the universe that would be conducive to the development of even the simplest life form we might be able to imagine.

It is important to note that to be able to make the statements above with any confidence, we need to have a set of theories that can explain the development of the universe from the assumed Big Bang—for example again, the production of carbon via the triple alpha resonance. The existence of such a theoretical structure is essential if we are to be able to explore the hypothetical consequences of varying the values of the fundamental constants. By exploring these consequences, we can identify constants that are reasonable candidates for fine-tuning.

In contrast to the cosmological situation, fine-tuning arguments relating to the fitness of water to host and modulate biologically relevant processes are not straightforward to formulate. The reason is simple: we do not yet know *which* (microscopic and macroscopic) properties of water—let alone their quantitative values—are critical for the wide range of processes that might be labeled as biological. Behind this lack of knowledge of the important properties is the lack of a theoretical framework that would enable us to select candidate properties, or physical values relating to such properties, that might be considered to be candidates for fine-tuning, that is, those that, if their values were perturbed by only a small amount, would no longer be effective in supporting life processes.

Without this basic knowledge of precisely *how* water participates in the many life processes it is clearly involved in, we are somewhat handicapped in trying to decide *which* (if any) of its properties might be fine-tuned for this purpose. We need to make progress in answering this question before we can assess whether water is fine-tuned for life—as found on earth, or in possibly very differently based life forms elsewhere. Put bluntly: if we do not know which properties of water are critical for life—or why—then we do not know which to investigate to see if they are fine-tuned or not.

This paper addresses two issues. First, as background, an introduction is offered to the properties of water itself as an isolated molecule, as a self-interacting entity in both dilute and condensed phases, and as a participant with other molecules in condensed-phase structures and processes. The second part of the paper discusses a few sample biologically relevant processes, and asks questions about which aspects of water are important for those processes. By identifying how water is participating in selected processes, we might be able to identify candidate properties (or values of certain properties) that are critical to these processes, and hence might be possible candidates for fine-tuning. We might also be able to begin a search for another molecule that could substitute for water in some of these processes.

Life on earth developed in an aqueous environment. Hence we would expect evolution to take advantage of the properties of that environment in the development of life. The discussion below is therefore of necessity within a terrestrial context, taking the life forms we find as our working material. The possibility of life of some (perhaps very different) form existing that is based on a nonaqueous environment remains open, and is not addressed here (see Daniel et al., 2004).

3.2   INTRODUCING WATER AND ITS PROPERTIES

3.2.1   THE WATER MOLECULE

3.2.1.1   The Water Molecule as Nuclei

Everyone knows water is H_2O: two hydrogens, each attached to a central oxygen. This tells us by itself very little. More interesting is the basic nuclear geometry of the molecule, with the bonded O–H distance of just less than 1 Å and H–O–H angle of about 104.5° (see Figure 3.1). This angle is close to not only the tetrahedral angle of ca. 109.5°, but also to the internal angle of a pentagon

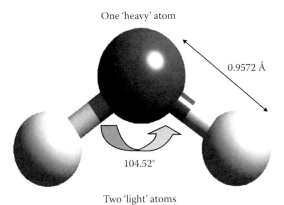

One 'heavy' atom

0.9572 Å

104.52°

Two 'light' atoms

FIGURE 3.1 A schematic representation of the water molecule, emphasizing the constituent nuclei and their geometry.

(108°). This may seem an abstruse observation, but as will be mentioned later, this angle suggests that the puckered hexagonal and pentagonal ring structures found in condensed phases are quite reasonable natural consequences of the basic water molecule geometry.

The water molecule, however, is not the static entity that Figure 3.1 might suggest: the molecule is in continuous internal motion, with the constituent atoms vibrating against each other. The internal motions of an isolated water molecule can be characterized in terms of three normal modes: symmetric and antisymmetric O–H stretch vibrations with respective frequencies of 3657 cm^{-1} (453 meV) and 3756 cm^{-1} (466 meV), and a bending mode with a frequency of 1595 cm^{-1} (198 meV) (Benedict et al., 1956). Interactions between water molecules through the well-known hydrogen bond perturb these frequencies from their isolated-molecule values.

One further observation of potential importance is that the molecule is made up from one heavy atom (oxygen) and two light atoms (the hydrogens). This alerts us to the possibility that quantum effects might be relevant in some processes that involve water; we should not consider ourselves as restricted to the classical world when discussing fine-tuning. This potential quantum aspect also arises when we note the vibrational motion of the molecule mentioned above. Even at absolute zero, there is still zero-point motion; a zero-point vibrational frequency of 4634 cm^{-1} corresponds to a zero-point energy of 55.4 kJ mol^{-1} (13.25 kcal mol^{-1}) (Eisenberg and Kauzmann, 1969). It may seem odd to raise this point in the context of processes that occur at room temperature (where $kT \approx 2.5$ kJ mol^{-1}), but as will be discussed later, there are biological processes in which the zero-point motion of the molecule may be important.

3.2.1.2 The Water Molecule as a Distribution of Charge

Although the picture given in Figure 3.1 is a very familiar one that is often used to describe the water molecule, in terms of the water molecule's chemistry it is quite unrealistic. The molecular chemistry depends on the electrons. So in addition looking at the water molecule as three vibrating nuclei, we need to consider its electronic structure.

Focusing first on the valence electron density, the classical work reported in the PhD thesis of G. A. Jones (Bader and Jones 1963) produced contour maps of electron density that look like a knobby boomerang with high density "peaks" focused close to the hydrogen and oxygen nuclei. Between the nuclei the contours narrow along the oxygen–hydrogen covalent bond, with the whole picture looking a bit like three mountain peaks arranged in a shallow V (looking from the air), with high cols connecting the central peak to the two outliers. Figure 3 of Bader and Jones (1963) shows examples of the valence electron density calculated using slightly different assumptions of the degree of electron delocalization; although these electron density distributions are different in quantitative detail, the overall picture is essentially the same: imagine the ball and stick representation of Figure 3.1 shrouded in clouds of electron density that envelop the spheres representing the nuclei and that link to each other along the bonds joining the hydrogen and oxygen atoms.

Although this kind of picture does illustrate the geometry of the valence-electron distribution, it again tells only part of the truth. In fact, the total electron-density distribution, important with respect to its nonhydrogen-bonded (repulsive as well as attractive) interactions is much closer to being spherical than the boomerang picture of the valence electron density suggests. This was demonstrated by the quantum mechanical calculations of Hermansson on the total electron density done (again in a PhD thesis) about twenty years after Jones's valence electron calculations. This greater sphericity of the total electron density distribution is shown in the section through the hydrogen and oxygen centers shown in Figure 3.2.

The electron density of the molecule can be characterized by a multipole moment expansion. The first term of this expansion, the dipole moment, is, for water, at ca. 1.85 D, relatively large (Clough et al., 1973; Xantheas and Dunning, 1993). A consequence is the high dielectric constant of the liquid, an essentially macroscopic quantity that is relevant to both its specific actions as a molecule and its effectiveness as a charge-screening solvent. But water is far from unusual in this regard: many other molecules have similarly high, or even larger, dipole moments, with their liquids having even higher dielectric constants. The molecule's dipole polarizability—the response of the dipole moment to an applied electric field—is also reasonably, though not excessively or unusually, large. It results in an enhancement of the average molecular dipole moment in condensed phases, and is another reason for the high dielectric constant of the liquid.

The dipole moment is only the first term in the expansion of its charge distribution; it is the higher moments that are potentially important when considering how water molecules interact with each other and with other molecules. The quadrupole moment is probably the most significant in this context, and relates to the simple, almost standard, model of a water molecule as having two regions of positive charge centered close to the hydrogens and two centers of negative charge (lone pairs) tetrahedrally placed with respect to the positively charged regions (see Figure 3.3). However, this picture is yet another simplification, as is shown by detailed quantum mechanical calculations of the electron distribution (Diercksen, 1971; Hermansson, 1984; Baum and Finney, 1985; Buckingham, 1986). It is more realistic to represent the geometry of the charge distribution as *triangular* rather than tetrahedral, with the classical lone-pair regions not separated from each other, but rather forming a single region of negative charge. This trigonal distribution is reflected in some of the more successful potential functions used for computational modeling. It was interestingly also the form

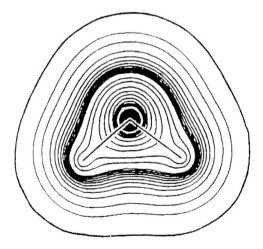

FIGURE 3.2 Contours showing the total electron density in the HOH plane of the isolated water molecule, according to quantum mechanical calculations by Hermansson (1984).

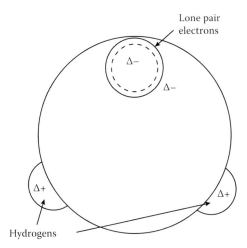

FIGURE 3.3 A schematic of the classical simple picture of the tetrahedrally disposed hydrogens and lone pairs in the water molecule.

of potential favored in the early work of Bernal and Fowler (1933) that was done before computers were dreamed of, let alone complex quantum mechanical calculations possible.

A charge distribution, however, is not only to do with the electrostatic attractions between molecules; it is also relevant to molecular repulsion. Although it is often assumed that the repulsive core of the water molecule is essentially spherical, there is both experimental (Savage and Finney, 1986) and theoretical (Hermansson, 1984) evidence to suggest a significant nonsphericity of the water-water repulsion. The degree of nonsphericity is indicated in Figure 3.4, which also illustrates the near-trigonal arrangement of the charge centers discussed above. One interesting consequence of this recognition has been a much improved ability to rationalize apparently complex water network geometries in a range of situations from the ices to liquid water itself, and in more complex chemically and biologically relevant environments (Savage, 1986a, b; Finney and Savage, 1988).

None of these single-molecule properties seem, of themselves, to be individually particularly remarkable. We could therefore reasonably conclude that water is an unremarkable small molecule.

3.2.2 INTERACTIONS BETWEEN WATER MOLECULES

Water molecules are usually considered to interact with each other through hydrogen bonding. This interaction can be reasonably modeled, both with respect to its strength and directionality, as purely electrostatic, consistent with the charge distribution summarized above, although controversy remains about whether we should consider a significant covalent contribution.

What is not controversial is the strength of the hydrogen bond interaction between two water molecules. A value of ca. 20 kJ mol^{-1} (ca. 5 kcal mol^{-1}) is about 10 times a typical thermal fluctuation at room temperature ($kT_{ambient} \approx 2.5$ kJ mol^{-1}). Being significantly stronger than a typical van der Waals interaction, this explains the apparent anomaly of its liquid phase being stable at ambient temperature—an obviously biologically relevant property for us. As with the other so-called (and often overhyped) anomalies of water, however, it is more useful to consider these anomalies as natural consequences of the basic properties of the molecule rather than focus on the anomalies themselves as some sort of unique (which they are not) set of properties that themselves might be responsible for water's biological potency. In terms of looking for candidate properties for fine-tuning, we should concentrate on the basic molecular properties rather than on the anomalies themselves. The former properties are the more basic; the anomalies are derived properties.

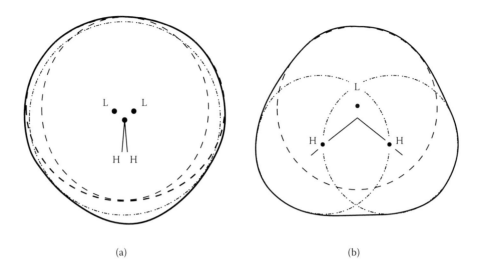

(a)   (b)

FIGURE 3.4 Centers of charge in the water molecule obtained from the quantum mechanical calculations of Hermansson (a) normal to and (b) parallel to the HOH plane. The outer solid lines approximate to the outline—or effective shape—of the molecule. The broken lines are sections of spheres centered on the negative (L, dashed line) and positive (H, dot-dashed line) centers of charge. The negative regions are closer to the center of the molecule than classical lone-pair pictures suggest.

Also of critical importance for the water–water interaction is its directionality. The oversimplified tetrahedral charge-distribution model discussed above appears to rationalize the standard picture of a central water molecule interacting with four neighbors arranged approximately at the corners of a tetrahedron, with the central molecule at the tetrahedron's center (Figure 3.5). Two of the neighboring molecules accept hydrogen bonds from the two hydrogens of the central molecule, while the other two neighbors donate their hydrogens toward the lone pair regions of the central molecule. Even though we have criticized this simple tetrahedral charge distribution model as being oversimplified and overtetrahedral, it is consistent with what we observe.

This approximate tetrahedrality of the water–water interaction is, however, forced by more than the tetrahedral aspect of the charge distribution. Extensive experimental work has shown that polar hydrogens are almost always donated to hydrogen bond acceptor regions of neighboring molecules, even in quite complex systems (e.g., Olovsson and Jönsson, 1976; Finney, 1978, 1979; Savage, 1986a; Bouquiere et al., 1994). It is therefore not surprising that the number of hydrogens in the system, taken together with the available space around the negatively charged region of the molecule, results in the donor/acceptor ratio of 2:2 for an assembly of waters. It is useful to think of the tetrahedral motif being forced by these constraints, rather than the approximate tetrahedrality of the water molecule. This type of structural arrangement would therefore also be expected for the more trigonal geometry that we argue above is more realistic. Indeed, the motif is reproduced in both computer simulations and energy minimization calculations that use clearly trigonal potential functions (e.g., see Finney et al., 1985).

These considerations might lead us to think about what might happen when we modify the donor/acceptor ratio in an aqueous solution or more complex biomolecular situation. We will come back to this point later.

3.2.3 CONDENSED PHASES OF WATER

The four-coordinated motif of Figure 3.5 is central to the structures of water in its condensed phases. It is seen clearly in the familiar form of ice (ice-Ih) that is found on frozen lakes and in your freezer. The two views of the framework of this structure in Figure 3.6 show the fourfold coordination of

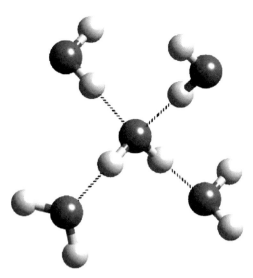

FIGURE 3.5 A four-coordinated arrangement of water molecules. The upper two molecules "donate" hydrogen bonds to the central molecule, while the lower two "accept" hydrogen bonds from the central molecule.

each molecule, but also show that (a) all bonded first-neighbor distances are about the same and (b) all the O...O...O angles are essentially the same. The latter angles are very close to the ideal tetrahedral angle of 109.5°, which is, of course, close to the internal H–O–H angle of the water molecule of 104.5° (Kuhs and Lehmann, 1986; Petrenko and Whitworth, 1999). We can thus conclude that there is little strain in this structure and that the hydrogen bonds between neighboring molecules are not far from being linear. What is also evident is the openness of the structure, which is a consequence of the fourfold coordination. It might therefore seem reasonable that ice is less dense than, and hence can float on, water (another of water's anomalies that can be explained by its molecular properties).

Ice, however, can also be found in different phases with different structures as pressure is raised and temperature is varied (Petrenko and Whitworth, 1999; Finney, 2004). Even in these phases, up to the highest pressures of tens of thousands of atmospheres, the tetrahedral coordination is retained

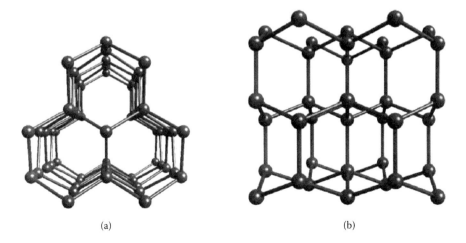

(a)   (b)

FIGURE 3.6 Two mutually perpendicular views of the arrangement of molecular centers (essentially the oxygen atoms) of water molecules in ice-Ih. Each molecule in this structure is four coordinated to other water molecules through hydrogen bonding, and the coordination geometry is very close to tetrahedral.

(Finney, 2001, 2004). The external pressures—and hence the reduced volume—are accommodated in the high pressure structures partly by an increase in variability of the hydrogen bond length. Perhaps more importantly, a significant degree of bond bending is observed, with distortions of up to about 25° apparently acceptable (Finney, 2003). The angular part of the water–water potential function thus seems to be relatively soft, being able to accommodate quite significant variations of O…O…O angle from the tetrahedral, which can lead to strongly bent hydrogen bonds. Similar bond length and angle variation is found in a range of other crystal structures involving water, from simpler ones such as gas hydrates to more complex protein crystals (Savage, 1986b). In the latter systems, however, the donor/acceptor ratio of 2:2 found in the ices and the simpler hydrates is not, in general, maintained. This could perhaps be thought of as reflecting the ability of the water molecule to make use of the trigonality by accepting only a single hydrogen bond from a single neighbor. The potential utility of such donor–acceptor asymmetry is discussed again later.

3.2.3.1 Liquid Water: Structure

The average structure of liquid water is now well known (see, e.g., Soper, 2000). First, we need to recognize that liquid water is not a crystal—there is no lattice to which the average structure can be referred. An instantaneous snapshot of a box of liquid water at room temperature is shown in Figure 3.7: the lack of crystallinity is obvious. Nevertheless, the four-coordinated motif is largely retained locally, as is shown in Figure 3.8, which is a blow up of a small region of Figure 3.7. However, there is a significant population of three-coordinated trigonal local structures, consistent with the trigonal charge distribution discussed earlier, and also a local variability in the donor–acceptor ratio. Bond angles vary, as was found in the high-pressure ices, although the variability is greater in the liquid, as might be expected. Four-, five-, six-, and sevenfold hydrogen-bonded ring structures are found, consistent with the range of bond angles observed in the ice structures.

Two further points might be made about the water structure itself. First, in addition to the variability in donor/acceptor ratio mentioned above, there is at least one other local variation in bonding that might be considered a local defect. In this, a hydrogen appears to be unsure which lone pair region it is pointing to—it has a choice between two neighboring molecules (see inset to Figure 3.8). Such a case is the nearest hydrogen of molecule B in Figure 3.8. This could result in an apparent fivefold local coordination, of which quite a population is found in computer simulations (Sciortino et al., 1991, 1992) and also observed in experimentally determined structures (Finney, 2003). This bifurcation defect is potentially relevant in the context of the dynamics of the liquid (see the following section).

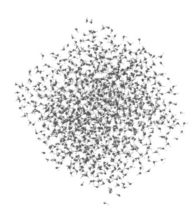

FIGURE 3.7 An instantaneous snapshot of an assembly of water molecules in the ambient temperature and pressure liquid. The model structure shown is fully consistent with state-of-the-art neutron scattering measurements. (With acknowledgment to Dr. Daniel Bowron, ISIS Facility, Rutherford Appleton Laboratory, UK.)

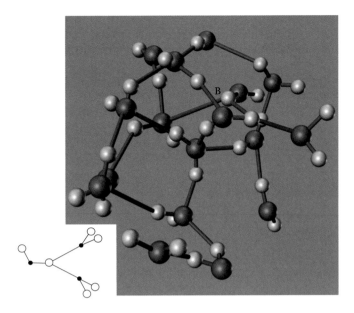

FIGURE 3.8 A detail of Figure 3.8, with lines added to show likely hydrogen bonds between neighboring molecules. Both three and four fold coordination can be seen. Also shown is a bifurcated interaction (the molecule labeled B) in which a hydrogen interacts with the negative regions of two different neighboring waters as illustrated in the inset. (With acknowledgment to Dr. Daniel Bowron, ISIS Facility, Rutherford Appleton Laboratory, UK.)

Second, we noted when previously discussing the charge distribution of the molecule that it has a significant dipole polarizability. As each water molecule will be experiencing the fields of its neighbors, we might expect the average dipole moment of a molecule in the liquid to increase. Although there is no clear experimental information on the magnitude of this enhancement, recent calculations (Silvestrelli and Parrinello, 1999) suggest a range of resultant instantaneous dipole moments of about 2 to 4 D, a significant enhancement with respect to the isolated molecule value of 1.85 D. This is relevant to the liquid's high dielectric constant, and hence to its particularly effective property of being able to dissociate ionic species in solution. But again, we should note carefully that this is not a property that is unique to water—liquids of a number of other polar molecules would be expected similarly to have a higher average dipole moment in the liquid than the isolated molecule.

3.2.3.2 Liquid Water: Dynamics

Experimental measurements tell us that water molecules reorient themselves in the ambient temperature liquid with a characteristic time of ca. 2 ps (Texeira et al., 1985). Furthermore, the time taken for a molecule to diffuse a distance of about one molecular diameter is about 7 ps (Denisov and Halle, 1996). These times are unexpectedly short. With the typical hydrogen bond energy of ca. $10kT_{ambient}$, we would expect water molecules to diffuse or reorient in times that are several orders of magnitude greater than the few picoseconds observed; even the simplest molecular reorientation would be expected to require the breaking and reforming of at least two hydrogen bonds. We are thus led to look for a mechanism that could explain this apparent serious internal inconsistency of our picture.

One plausible mechanism that has been proposed relates to the five-coordinated defects mentioned above: these may offer a transition state between two different configurations without having to cross the much higher barrier of breaking and reforming two or more hydrogen bonds (Sciortino et al., 1991, 1992). Thus, we have a potentially interesting concept of an underlying relatively rigid water network structure, yet one which permits molecular mobility within it so that it can act as the diffusional medium that is essential for its biomolecular functionality.

3.2.3.3   Proton Conduction in Liquid Water

There are always some H$^+$ and OH$^-$ ions—or rather, their hydrated equivalents such as H$_3$O$^+$ or H$_5$O$_2$$^+$—in liquid water. Not only is molecular diffusion unexpectedly high in liquid water, but so also is the conduction of an excess proton. Although much is made of this ability of water, particularly with respect to moving protons as is required for some biomolecular functions, there is continuing controversy concerning the atomic-level mechanism of the conduction process. In the Grotthus mechanism (von Grotthus, 1806) that is usually called upon to explain proton conductivity, a proton is argued to move from an H$_3$O$^+$, rapidly along a hydrogen bond, to recreate an H$_3$O$^+$ on a neighboring water. Another proton from the receiving water then translocates similarly to another neighbor, etc. This is the basic relay mechanism of translocation of protons along a hydrogen bonded chain, in which a proton is the baton that is passed on down the chain—although the proton that is passed on is not the same one as was received. No movement of *molecules* is involved in the process. This proposed mechanism has been refined in a number of ways, some of these invoking proton tunneling along a hydrogen bond. However, experimental evidence is in conflict with all these ideas (e.g., Agmon, 1995).

Recent ab initio computer simulation work has suggested an essentially simpler process that is consistent with the experimental data (Tuckerman et al., 1997; Marx et al., 1999). In this model (shown schematically in Figure 3.9), the proton diffusion is controlled by a thermally induced hydrogen bond breaking process, and can be thought of as being propelled by a hydrogen-bond cleavage in front of a moving proton, with subsequent hydrogen bond reformation in the wake of the disturbance. As in the basic Grotthus mechanism, it is still a proton that moves from one molecule to another, and not the same proton that flows through the system. Thus, the conductivity will be high, although the rate-limiting step in the process does involve thermally induced processes—the breaking and making of hydrogen bonds.

A further point of interest is that these calculations do not support the idea of proton tunneling through a classical barrier as being important. But there does appear to be a significant role for quantum effects in the process: the zero-point motion washes out the barrier to proton motion along a hydrogen bond. That water has two light atoms as part of its makeup thus appears to be important for effective proton transfer through this apparently relatively rigid—although diffusionally soft—molecular network.

3.2.3.4   A Note on the So-Called Anomalies of Water

As suggested earlier, it can be diverting to focus on the so-called anomalies of water in trying to understand what it is about water that fits it for hosting life. These properties are natural consequences of the molecular properties, so it would seem to be more profitable to focus on the molecular properties themselves when thinking about the fine-tuning possibilities. Nevertheless, there is one property of water which, although clearly a consequence of the underlying molecular interaction geometry, is perhaps usefully introduced through a connected macroscopic anomaly. This is the so-called diffusivity anomaly: as pressure is increased, diffusivity initially increases. This is the opposite of what happens in a "normal" liquid, where increase in pressure limits the ability of the molecules to move through a reduction in the specific volume.

In the case of water, however (but in common with other tetrahedral liquids), an increase in pressure tends to break down the local orientational structure (strictly, the local tetrahedral order—see Errington and Debenedetti, 2001) by forcing increased bond bending (as we have already noted happens in the higher pressure ices). Remembering the model for diffusion involving bifurcated defects that was previously discussed, increased pressure will likely increase the concentration of such defects, and thus increase the diffusivity. This could be thought of as another example of

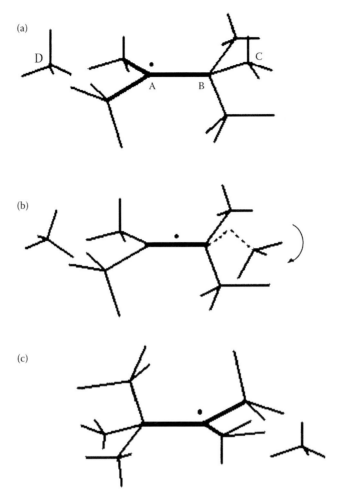#

FIGURE 3.9 A schematic representation of the proton transfer mechanism in water as suggested by recent ab initio computer simulation calculations. The excess proton is initially located on the three coordinated molecule A in snapshot (a). For this proton to move to neighbor B, that molecule must lose one of its coordinated waters, for example, molecule C, as in snapshot (b). Once this transfer has occurred, another water is then able to interact with the original molecule A (as in snapshot (c)). According to this explanation, the proton diffusion rate is controlled by the thermally induced hydrogen bond breaking process in snapshot (b).

water having a relatively thermally rigid and resilient network structure, yet one in which molecular motion is easily possible.

3.2.4 WATER IN DIFFERENT ENVIRONMENTS

Just as we now have an experimentally consistent picture of the structure of water, so we are building up, from direct experimental measurements whose interpretations do not depend on interpretive models, experimentally consistent pictures of the structures and, to a lesser extent, the dynamics of water in interfacial systems. These range from relatively simple aqueous solutions (some of these of major social and economic importance—"water of life" is, after all, an English translation of the Gaelic *uisge beatha*) to relatively complex biomolecular environments.

Without going into the detail (see, e.g., Finney, 2003, and references therein), there are some perhaps unexpectedly simple generalizations that we can draw from these studies that are of potential relevance to our task of identifying biologically critical properties.

3.2.4.1 Structure Making and Breaking—A Problem Concept: Hydrophobic Hydration

There is much in the aqueous solution literature that discusses the "structure making" or "structure breaking" of water by added solutes. These concepts were first used in an environment in which the structure of water itself was unclear, so the reference state against which a made or broken structure would need to be assessed was not available. So what was meant structurally by these ideas was far from clear. Recent developments in experimental and computer simulation techniques have given us a good idea of the detailed structure of water against which these ideas can now be tested. They are found to be wanting.

An example of particular importance with respect to biomolecular processes is that of hydrophobic hydration and the hydrophobic interaction. The standard model of hydrophobicity, which is generally accepted in the chemical and biochemical communities, was first set out by Frank and Evans (1945) more than 60 years ago. To explain the apparent anomalous entropy of mixing of molecules containing nonpolar groups, they argued that when close to a nonpolar group on a molecule, the water was somehow more ordered than in the bulk, and used (perhaps unfortunately with the benefits of hindsight) the term *iceberg* to conceptualize this. Over the past decade, however, neutron scattering measurements have failed to find any evidence to support this model. From the water's point of view, the hydration structure looks like the structure seen in the bulk. There is no measurable ordering (e.g., see Soper and Finney, 1993; Bowron et al., 1998; Finney et al., 2003). This again seems to underline the resilience of the water structure, this time to perturbation by added solutes. In the hydrophobic interaction case, water molecules appear to be able to accommodate nonpolar molecules without a significant perturbation of the local intermolecular structure of the liquid. The structural cause of the thermodynamic behavior needs to be looked for elsewhere than in the first coordination shell of the nonpolar group. "Structure making" is not a useful explanatory concept here.

3.2.4.2 Water Structure at Polar and Mixed Interfaces

We can draw a similar conclusion when considering water at polar, or even mixed polar/nonpolar interfaces. As an illustration of this, Figure 3.10 shows two of the water networks obtained from a high-resolution, low-temperature crystallographic study of coenzyme B$_{12}$ (Bouquiere et al., 1994), which crystallizes, like most relatively complex biological molecules such as proteins and nucleic acids, with a significant amount of water. In this case, there is a single network in the pocket at the left hand side of the figure, but two alternative networks seem to be present (obviously in different unit cells) in the channel on the right.

Both of these networks match the interfaces at which they are attached to, or at which they touch, the coenzyme molecule. Both networks are consistent with the water–water interaction geometry that was discussed earlier. There are examples of both three- and fourfold coordinated water molecules, with the threefold able to mop up local donor–acceptor imbalances at the interface (see, e.g., waters 212, 213, and 216). Had the water been symmetrical in terms of its donor/acceptor interactions, it would not be as easy to accommodate such imbalances. The waters close to nonpolar groups also look normal, forming a cage around the nonpolar groups but being themselves anchored to nearby polar groups (see, e.g., the water chain 212-217-618). If we calculate from the atomic coordinates the radial distribution functions that describe the water structure, we again find a structure that is indistinguishable from bulk water at the same temperature. Similar conclusions can be drawn from water networks in crystals of larger biological molecules such as proteins (Finney, 1977, 1979).

Thus, at polar or mixed interfaces, the observed average water structures apparently fit in comfortably with, or match, the available hydrogen-bonding sites. Through the ease with which it can participate in threefold coordination, the liquid water network can mop up the donor/acceptor imbalances that are inevitable at a protein or other biomolecule interface. The versatility of the water

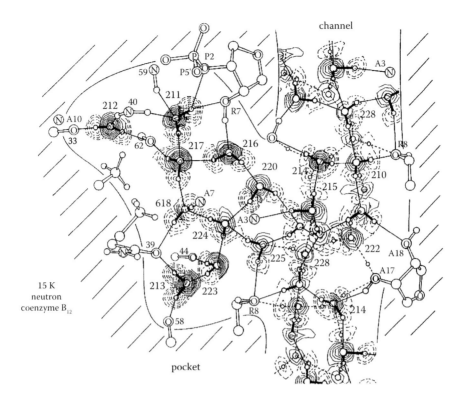

channel

pocket

15 K
neutron
coenzyme B$_{12}$

FIGURE 3.10 The organization of water found in the pocket (left) and channel (right) of crystals of coenzyme B$_{12}$ at 15 K. The contours represent neutron scattering densities of the oxygens (full-line contours) and hydrogens (dashed-line contours) of the water molecules. Solid and dashed straight lines between molecules represent hydrogen bonds in the two different solvent networks identified.

interaction geometry is underlined by the fact that a set of mixed polar/apolar interfaces can be bridged comfortably by more than one different network: in the B$_{12}$ case, four networks have been identified at room temperature (Savage, 1986a), which are reduced to two at low temperature, as shown in Figure 3.10. And from the water's viewpoint, each network looks essentially like bulk water.

3.2.4.3 Water Dynamics at Interfaces

Here our understanding is less advanced. But there are a number of conclusions that can be drawn from experimental work. Experimental evidence is now conclusive in rejecting long-held ideas of water being "bound" to any biological surface. All the water molecules at the surface of a typical protein, for example, appear to have very short lifetimes. The fact that they are apparently localized in particular positions when crystallographic structures (such as Figure 3.10) are examined, the localized waters are, in fact, in rapid exchange with the bulk. Classical NMR work by Halle and others (see, e.g., Halle, 2004) has demonstrated clearly that typical molecular reorientation times of water molecules at protein surfaces are only a factor of a few longer than those in bulk water. Water thus appears to be dynamically, as well as structurally, "comfortable" in a wide range of environments.

3.2.4.4 Confined Water

So far, we have considered only the dynamical behavior of water close to a single interface. What happens when we confine water molecules between two surfaces? Here, an interesting difference occurs between the dynamical behavior of water and "standard" organic liquids that again might be argued relevant to water's fitness for hosting life. Considering how crowded the cell is, if there were significant slowing down of the dynamics of water in a biological environment, the fluid might become too "sticky" to be able to perform its biologically necessary function.

For film thicknesses greater than about 10 molecular diameters, the viscosities for both organic liquids and water are essentially the same as those observed in the respective bulk liquids (Raviv et al., 2001). We see no significant effect of such confinement on the liquid dynamics, for either a normal organic liquid or water. However, as we decrease the separation to around five molecular diameters, the viscosities of the standard organic liquids increase greatly, as we would expect from the known, essentially understood increase in their viscosity with pressure. For water, however, its viscosity remains within a factor of around three of its bulk value.

An explanation of this potentially important behavior takes us back to the water network structure and its response to pressure. As discussed earlier, water diffusivity behaves abnormally as pressure is increased. Below a threshold value, increasing pressure *increases* water diffusivity; for a normal liquid, the increased crowding results in the opposite behavior. These different behaviors thus can be understood. As was previously argued in the short discussion, the explanation of the anomalously high diffusivity of water invoked five-coordinated defects. Applying pressure acts to break down the orientational order in liquid water, and in so doing is likely to increase this defect population—hence the observation that water diffusivity is not strongly reduced in strongly confined environments.

As this behavior relates to the network structure of the liquid—both its geometry and energetics—if the retention of relatively high fluidity in a confined geometry is of biological relevance, then once again the underlying tetrahedral geometry is likely to be a relevant element.

3.3 WHICH PROPERTIES MIGHT BE CRITICAL TO BIOLOGICAL PROCESSES?

The discussion above has already suggested how specific characteristics or properties of the water molecule might be critical in some biologically relevant processes. At this point, it might be worth examining a few of these processes and trying to identify the specific roles in which water may be operating. In each process, a thought experiment might be undertaken to test how far the property of water that is thought to be important might be perturbed before the process would fail to operate. Performing this kind of analysis on a number of different processes could focus attention on a possible window of properties that appear to be essential for the effective operation of the chosen ensemble of biomolecular processes. We might then look for candidate molecules that fit into this window to try to identify another solvent (or mixture of solvents) that mimics that aspect of water that we have tentatively identified as critical. In this context, we consider briefly (1) aspects of the stability of proteins, (2) the diffusion of substrate and product to and from an enzyme active site, and (3) proton transport.

3.3.1 PROTEIN STRUCTURE AND STABILITY

Since the seminal paper of Kauzmann (1959), there has been a general acceptance that hydrophobic interactions drive the folding of proteins to their native, active structures. In this case, water would clearly be a critical element in protein folding, although as mentioned above, the structural basis of the hydrophobic interaction remains poorly understood. With respect to the importance of the hydrophobic interaction in protein folding, however, there is increasing questioning of its dominance. There have been a number of attempts to draw up balance sheets of free-energy contributions to protein stability using what we know from the molecular level structures of proteins (e.g., see Finney et al., 1980, and the chapter by Pace in this volume). As water appears to be involved in almost every one of the several identifiable contributions to the free energy of stability, its properties obviously are critical in both driving the folding process and maintaining the stability of the native structure.

However, controversy remains concerning the balance between these different free-energy contributions. Perhaps the safest assumption is that a number of different kinds of interactions are important (polar, charged, hydrophobic, van der Waals, etc.), and that the involvement of the surrounding fluid in most, if not all, of these interactions is far from negligible.

Let us now focus on one of these contributions: the interaction through hydrogen bonding of water molecules with the polar groups on the protein that are exposed to solvent, such as we have already seen in the case of vitamin B_{12} coenzyme (Figure 3.10). One of the important things to note here is the need to maximize, subject to whatever other constraints are forced on the structure, the enthalpy term. This is particularly important when we remember that the native protein is only marginally stable: its free energy of stability is only of the order of 10–20 kcal mol^{-1}, which amounts to only two to four of the several hundred hydrogen bonds in a typical native protein–solvent system. Lose a small fraction of these hydrogen bonds and, other things being equal, the native structure of the protein falls apart.

Perhaps the prime requirement placed on the solvent in this context is to have an ability to make hydrogen bonds with all those polar groups whose hydrogen bonding capability has not been fully taken up by interactions within the protein. And in doing so, the water itself needs to form structures that themselves satisfy its own hydrogen bonding capability. Looking again at the solvent interactions in Figure 3.10, we see that this is done quite effectively with respect to both the exposed hydrogen bond donor and acceptor groups on the protein surface. But as we have noticed before, not every water molecule is four coordinated: there is a significant population of threefold coordinated molecules, as of course there would also be in the bulk water solvent. To first order, each water molecule is capable of donating two hydrogen bonds through its two protons and accepting *up to* two through its negatively charged (lone pair) region. But it does not necessarily have to make all four bonds: it is happy to drop one of the acceptors if that is made necessary by stereochemical constraints. There is, in fact, some evidence that a single accepted hydrogen bond is likely to be stronger than each of the two that would be made in a fully four-coordinated molecule (Savage, 1986a, b).

In such a complex interfacial system, it is unlikely that all the polar groups on the exposed protein surface would be placed in such a way that their hydrogen bonding requirements would be easily satisfied were the solvent molecule forced to make four hydrogen bonds. The fact that it is happy with only three adds a significant degree of flexibility in satisfying the required polar interactions. In addition to the geometrical constraints of the interface, it is also unlikely that the exposed protein surface would present equal numbers of hydrogen bond donors and acceptors to the surrounding solvent. One of the important properties of the water molecule in this situation therefore seems to be its ability to accommodate variable coordination. Without this ability, it would be unable to either mop up donor/acceptor imbalance, or accommodate itself to a geometrically complex (generally a mixed polar/nonpolar) surface. The water appears to be a responsive, sympathetic, yet versatile solvent that is well matched to this job. It can accommodate itself to varying external hydrogen bonding requirements without its own structure being significantly affected.

We have mentioned earlier the fact that, although a water network appears to be relatively rigid or stiff (in terms of $kT_{ambient}$), it also appears to be a network in which molecular mobility remains "easy"—apparently with the assistance of other bifurcated defects that depart locally from the 2:2 ratio. Such a solvent framework would also appear to be rather useful surroundings for a macromolecule that, although having a similar rigidity (like liquid water, the enzyme is also held together in significant measure by hydrogen bonds), has much less possibility of defect formation that, in its case, could compromise its structural integrity.

We tentatively conclude that the underlying *variable* two donor/two acceptor capability of the water molecule is critical in maintaining the marginal stability of a native protein, yet without seriously compromising the preferred structure of the water region. This aspect of water means that it can minimize the effects of structural frustration by what might appear at first sight to be an "uncomfortable" nonaqueous environment. It may also be relevant that the water network has this interesting characteristic of being relatively rigid, yet in which defect processes allow anomalously rapid molecular diffusion. If these characteristics are important to this aspect of biomolecular function, then what is the window of properties that we can use to try to identify other possible candidates for water's functionality here? Can the water be substituted by some other liquid?

Focusing on the variability aspect of the hydrogen bonding, we should perhaps immediately reject any molecule that insists on having to make a fixed number of interactions. Even though they are not relevant to a life form that has evolved in water, because of their fourfold coordination, systems like liquid silicon, germanium, or silicon or germanium dioxide have often been touted as possible hosts for some form of nonaqueous life. To focus on just this simple parallelism of fourfold coordination as being important would seem to be an oversimplification: as argued above, were water forced to make four, and only four, hydrogen bonds, it would be unlikely to be an effective environment for the biological molecules on which we depend. The variability of the underlying ideal 2:2 donor/acceptor ratio seems to be crucial.

Indeed, there may be nothing particularly critical about the ideal 2:2 ratio itself—it may be the variability that is the key, and such variability is found in other amphiphilic liquids that do not have this particular ideal donor/acceptor ratio. Consistent with this observation, several simple amphiphilic organic liquids and liquid mixtures can indeed substitute for a significant fraction of the water in, for example, satisfying the hydrogen bonding requirements at the interface, with the enzyme remaining functional (sometimes with enhanced activity). However, these molecules do not in general appear to have the kT "rigidity" of water. Nor do they seem to be quite as versatile as water in terms of the donor/acceptor balance and departures from it. Or could some mixture of other small molecules be tuned to work as effectively as water does?

3.3.2   MOLECULAR DIFFUSION

Maintaining an environment in which molecules—including those of the surrounding medium itself—can diffuse sufficiently freely, must be of importance in any system in which substrate is transformed into product. The substrate has to get to the active site, and the product has to be removed to where it is needed. Moreover, this diffusivity should not be significantly compromised in the crowded conditions that are found in the cell.

As we have discussed above, water appears able to remain almost as mobile at an interface as it is in the bulk. Yet it still seems to form a relatively (on the scale of $kT_{ambient}$) robust network. Looking at the molecular level, we have suggested that this very useful property again comes down to both the structure and energetics of its underlying, but usefully defective, tetrahedral geometry. Other organic molecules appear unable to do this—or at least those that have been experimentally tested in this respect.

3.3.3   PROTON TRANSPORT

Many important biological processes require the transport of protons to or from a protein active site. Crystallographic structures in some cases suggest that protons must translocate along a long hydrogen-bonded chain that may involve water molecules. There is also a need to transport protons through the liquid. As observed above, proton mobility in water is anomalously high, with its limiting ionic conductance under ambient conditions about seven times that of Na$^+$. Thus, water seems to be a particularly useful medium for facilitating this biologically important process.

If we accept the proton conduction mechanism as proposed in recent calculations and discussed earlier, the network structure again has a role to play, with thermally induced hydrogen bond breaking being an important part of the mechanism. Moreover, quantum effects seem to play a significant part in the washing out of the barrier to proton motion along the hydrogen bond through zero-point motion. For proton transport, therefore, we do not seem to be able to get away from a thermally resilient hydrogen-bonded liquid that contains a light atom (actually at least two), and the chemistry requires that light atom to be hydrogen—at least in the form of life that we represent.

3.4 CONCLUDING COMMENTS

We started this journey through water and its interactions with the intention of summarizing the main properties of the water molecule and how those controlled its interactions with both other water molecules and other biologically relevant molecules or interfaces. Early on, we noted that the molecule itself appeared to be quite unremarkable as a small molecule, but as we built up networks of molecules, we began to see some quite interesting features that we noted might be relevant to biological functionality. Finally, discussion of the possible role of water in a few biologically important processes focused on particular properties that might be candidates for fine-tuning arguments.

Even before starting this journey, we commented that, unlike the situation in astrophysics, we do not have any theoretical framework that allows us to follow through the consequences for particular processes of changing certain properties of the molecule—either their quantitative values such as the hydrogen-bond strength or their qualitative characteristics such as the hydrogen-bonding geometry. We can, of course, think of modifying each property and then speculate on the way in which a particular process might be affected, as perhaps we have begun to do in the penultimate section above. However, the absence of an underlying theoretical framework that relates such molecular changes to essential biological activity means we can do little more at present than speculate—perhaps semi-quantitatively—with respect to sample processes. Such speculation is, of course, a long way from being able to pin down those properties of water that are essential to life, let alone discuss if they are fine-tuned or not.

The penultimate section did come to one tentative conclusion: that both the structure and energetics of the *defective* tetrahedral network of a simple molecule containing at least two hydrogen atoms do seem to be important for those aspects of biological functionality that have been considered. Whether these characteristics are essential to a biomolecular process that we might envisage as having been built on a different chemistry is arguable. Turning the argument around, we might prefer to note that evolution has operated in an environment in which it had water as perhaps the only candidate for a liquid medium, and that therefore it has tuned life to the properties of the medium that was available. Or perhaps other forms of life developed using different fluid environments that were less successful and could not survive in competition with ours.

Water is indeed a fascinating molecule that has a range of interesting properties. Many of these we can argue are important to specific aspects of biological functionality. But again, most of these properties individually are demonstrated by other molecules. What perhaps is notable in water's case is that it is difficult to think of another molecule in which *all* these properties are found. So perhaps Occam's razor also had a part to play in evolution.

REFERENCES

Agmon, N. 1995. The Grotthuss mechanism. *Chem. Phys. Lett.* 244:456–462.

Bader, R. F. W., and G. A. Jones. 1963. Hydrogen fluoride. *Can. J. Chem.* 46:586.

Baum, J. O., and J. L. Finney. 1985. An SCF-CI study of the water potential surface and the effects of including the correlation energy, the basis set superposition error and the Davidson correction. *Mol. Phys.* 55:1097–1108.

Benedict, W. S., N. Gailar, and E. K. Plyler. 1956. Rotation-vibration spectra of deuterated water vapor. *J. Chem. Phys.* 24:1139–1165.

Bernal, J. D., and R. H. Fowler. 1933. A theory of water and ionic solution, with particular reference to hydrogen and hydroxyl ions. *J. Chem. Phys.* 1:515–548.

Bouquiere, J. P., J. L. Finney, and H. F. Savage. 1994. High-resolution neutron study of vitamin B_{12} coenzyme at 15K: solvent structure. *Acta Crystallogr. B* 50:566–578.

Bowron, D. T., A. Filipponi, M. A. Roberts, and J. L. Finney. 1998. Hydrophobic hydration and the formation of a clathrate hydrate. *Phys. Rev. Lett.* 81:4164–4167.

Buckingham, A. D. 1986. The structure and properties of a water molecule. In *Water and Aqueous Solutions*, eds. G. W. Neilson and J. E. Enderby. Bristol: Adam Hilger.

Carr, B. J., and M. J. Rees. 1979. The anthropic principle and the structure of the physical world. *Nature* 278:605–612.

Clough, S. A., Y. Beers, G. P. Klein, and L. S. Rothman. 1973. Dipole moment of water from stark measurements of H_2O, HDO and D_2O. *J. Chem. Phys.* 59:2254–2259.

Daniel, R. M., J. L. Finney, and A. M. Stoneham, Eds. 2004. The molecular basis of life: is life possible without water? *Philos. Trans. R. Soc. London B* 359:1141–1328.

Denisov, V. P., and B. Halle. 1996. Protein hydration dynamics in aqueous solution. *Faraday Discuss.* 103: 227–244.

Diercksen, G. H. F. 1971. SCF-MO-LCGO studies on hydrogen bonding. *Theor. Chim. Acta* 21:335–367.

Eisenberg, D., and W. Kauzmann. 1969. *The Structure and Properties of Water.* Oxford: Clarendon Press.

Errington, J. R., and P. G. Debenedetti. 2001. Relationship between structural order and the anomalies of liquid water. *Nature* 409:318–321.

Finney, J. L. 1977. The organisation and function of water in protein crystals. *Philos. Trans. R. Soc. B* 276: 3–31.

Finney, J. L. 1978. Volume occupation, environment, and accessibility in proteins. Environment and molecular areas of RNase-S. *J. Mol. Biol.* 119:415–441.

Finney, J. L. 1979. The organisation and function of water in protein crystals. In *Water: A Comprehensive Treatise.* Vol. 6. Edited by F. Franks. New York, NY: Plenum Press.

Finney, J. L. 2001. Ice: structures. In *Encyclopedia of Materials: Science and Technology,* Vol. 5, eds. K. H. J. Buschow, R. W. Cahn, M. C. Flemings, B. Ilschner, E. J. Kramer, and S. Mahajan. Oxford: Elsevier Science.

Finney, J. L. 2003. Water? What's so special about it? *Philos. Trans. R. Soc. London B* 359:1141–1165.

Finney, J. L. 2004. Ice: the laboratory in your freezer. *Interdiscip. Sci. Rev.* 29:339–351.

Finney, J. L., and H. F. J. Savage. 1988. Impenetrability revisited: new light on hydrogen bonding from neutron studies on biomolecule crystal hydrates. *J. Mol. Struct.* 177:23–41.

Finney, J. L., I. C. Golton, B. J. Gellatly, and J. M. Goodfellow. 1980. Solvent effects and polar interactions in the structural stability and dynamics of globular proteins. *Biophys. J.* 32:17–33.

Finney, J. L., J. E. Quinn, and J. O. Baum. 1985. The water dimer potential surface. In *Water Science Reviews,* ed. F. Franks. Cambridge: Cambridge University Press.

Finney, J. L., D. T. Bowron, R. M. Daniel, P. A. Timmins, and M. A. Roberts. 2003. Molecular and mesoscale structures in hydrophobically driven aqueous solutions. *Biophys. Chem.* 105:391–409.

Frank, H. S., and M. W. Evans. 1945. Free volume and entropy in condensed systems. *J. Chem. Phys.* 13: 507–532.

von Grotthuss, C. J. T. 1806. Sur la decomposition de l'eau et des corps qu'elle tient en dissolution à l'aide de l'électricité galvanique. *Ann. Chim.* LVIII:54–74.

Halle, B. 2004. Protein hydration dynamics in solution: a critical survey. *Philos. Trans. R. Soc. London B* 359: 1207–1224.

Hermansson, K. 1984. The electron distribution in the bound water molecule. PhD dissertation, University of Uppsala.

Hogan, C. J. 2000. Why the universe is just so? *Rev. Mod. Phys.* 72:1149–1161.

Kauzmann, W. 1959. Some factors in the interpretation of protein denaturation. *Adv. Protein Chem.* 14:1–63.

Kuhs, W. F., and M. S. Lehmann. 1986. The structure of ice-l_h. In *Water Science Reviews,* Vol. 2, ed. F. Franks. Cambridge: Cambridge University Press.

Marx, D., M. E. Tuckerman, J. Hutter, and M. Parrinello. 1999. The nature of the hydrated excess photon in water. *Nature* 397:601–604.

Olovsson, I., and P.-G. Jönsson. 1976. The hydrogen bond. In *Recent Developments in Theory and Experiments: II. Structure and Spectroscopy,* eds. P. Schuster, G. Zundel, and C. Sandorfy. Amsterdam: North-Holland.

Petrenko, V. F., and R. W. Whitworth. 1999. *Physics of Ice.* Oxford: Oxford University Press.

Raviv, U., P. Laurat, and J. Klein. 2001. Fluidity of water confined to sub-nanometer films. *Nature* 413:51–54.

Savage, H. F. J. 1986a. Water structure in vitamin B_{12} coenzyme crystals: II. Structural characteristics of the solvent networks. *Biophys. J.* 50:967–980.

Savage, H. F. J. 1986b. Water structure in crystalline solids: ices to proteins. In *Water Science Reviews,* Vol. 2, ed. F. Franks. Cambridge: Cambridge University Press.

Savage, H. F. J., and J. L. Finney. 1986. Repulsive regularities of water structure in ices and crystalline hydrates. *Nature* 322:717–720.

Sciortino, F., A. Geiger, and H. E. Stanley. 1991. Effect of defects on molecular mobility in liquid water. *Nature* 354:218–221.

Sciortino, F., A. Geiger, and H. E. Stanley. 1992. Network defects and molecular mobility in liquid water. *J. Chem. Phys.* 96:3857–3865.

Silvestrelli, P. L., and M. Parrinello. 1999. Water molecular dipole in the gas and in the liquid phase. *Phys. Rev. Lett.* 82:3308–3311.

Soper, A. K. 2000. The radial distribution functions of water and ice from 220 to 673 K and at pressures up to 400 MPa. *Chem. Phys.* 258:121–137.

Soper, A. K., and J. L. Finney. 1993. Hydration of methanol in aqueous solution. *Phys. Rev. Lett.* 71:4346–4349.

Texeira, J., M.-C. Bellissent-Funel, S. II. Chen, and A. J. Dianoux. 1985. Experimental determination of the nature of diffusive motions of water at low temperatures. *Phys. Rev. A* 31:1913.

Tuckerman, M. E., D. Marx, M. L. Klein, and M. Parrinello. 1997. On the quantum nature of the shared proton in hydrogen bonds. *Science* 275:817–820.

Xantheas, S. S., and T. H. J. Dunning. 1993. The structure of the water from ab initio calculations. *J. Chem. Phys.* 99:8774–8792.

4 Water as a Biomolecule

Philip Ball

CONTENTS

When Szent-Györgyi[1] called water the "matrix of life," he was echoing an old sentiment. Paracelsus, in the sixteenth century, said that "water was the matrix of the world and of all its creatures."[2] But Paracelsus's notion of a "matrix"—an active substance imbued with fecund, life-giving properties— was quite different from the picture that, until very recently, molecular biologists have tended to hold of water's role in the chemistry of life. While acknowledging that liquid water has some unusual and important physical and chemical properties—its potency as a solvent, its ability to form hydrogen bonds, its amphoteric nature—biologists have regarded it essentially as the canvas on which life's molecular components are arrayed. It used to be common practice, for example, to perform computer simulations of biomolecules in a vacuum. Partly this was because the computational intensity was challenging even without solvent molecules, but it also reflected the prevailing notion that water does little more than temper or moderate the basic physicochemical interactions responsible for molecular biology.

Curiously, this neglect of water as a component of the cell went hand in hand with the assumption that life could not exist without it. That was basically an empirical conclusion derived from our experience of life on Earth: environments without liquid water cannot sustain life, and special strategies are needed to cope with regions that, because of extremes of either heat or cold, are short of the liquid.[3] The recent confirmation that there is at least one world rich in organic molecules on which rivers and perhaps seas are filled with nonaqueous fluid—the liquid hydrocarbons of Titan—might now bring some focus and even urgency to the question of whether water is indeed a unique and universal matrix of life, or just the one that exists on our planet.

Fundamental to that question is the part that water plays in sustaining the biochemistry of the cell. It has become increasingly clear over the past two decades or so that water is not simply "life's solvent," but is indeed a matrix more akin to the one Paracelsus envisaged: a substance that actively engages and interacts with biomolecules in complex, subtle, and essential ways. There is now good reason to regard the active volume of molecules, such as proteins, as extending beyond their formal boundary (the van der Waals surface), as though to activate the shell of water that surrounds them. Moreover, the structure and dynamics of this hydration shell seem to feed back onto those aspects of the proteins themselves, so that biological function depends on a delicate interplay between what we have formally regarded as the molecule and its environment. Many proteins make use of bound water molecules as functional units, like snap-on tools, to mediate interactions with other proteins or with substrate molecules, or to transport protons rapidly to locations buried inside the protein. It seems plain that water must itself be regarded as a biomolecule—and also as a biosupermolecule—in the sense that aggregates of H₂O molecules perform subtle biochemical tasks in a manner comparable to the behavior of lipid membranes or protein assemblies.

Here I review the case for adopting this perspective. I shall try to highlight throughout the distinctions between generic and specific behaviors of biological water. That is, some of its roles and properties may be expected from any small-molecule liquid solvent. Others depend on water's hydrogen-bonding capacity, but not in a way that could not obviously also be fulfilled by other hydrogen-bonded liquids. But some of water's biochemical functions seem to be quite unique to the H₂O molecule. From an astrobiological perspective, the question is then whether these latter roles are optional or essential for any form of life to be tenable.

4.1   WATER AS A SOLVENT

Water is not like other liquids, but neither is it wholly different. The key characteristic that distinguishes it from a typical simple liquid such as liquid argon is that hydrogen bonds link water molecules in a directional manner into a three-dimensional network. The short-ranged structure, as revealed most powerfully by neutron scattering and characterized by the radial distribution function, is thus dominated by the attractive forces between molecules, whereas, in a simple liquid, it is the packing effects due to repulsive interactions that determine the local environment around each molecule. Many of the so-called "anomalous" properties of water, such as its density decrease on freezing and its density maximum at 4°C, result from the consequent tension between the tendency for the molecules to maximize their van der Waals interactions by packing closely together and the tendency of hydrogen bonding to keep them at "arm's reach."

Water's behavior as a solvent hinges, to a large extent, on the question of how this hydrogen-bonded network responds to the presence of an intruding solute particle. The nature of this response is far from obvious and has until rather recently been difficult to probe, and yet it is central to the issue of how intra- and intermolecular interactions in biology are mediated and affected by the veil of water that necessarily surrounds all biomolecules.

4.1.1   SMALL-MOLECULE SOLUTES: HYDROPHILES AND HYDROPHOBES

Water is an extremely good solvent for ions. In part, this is a result of water's high dielectric constant, which enables it to screen the Coulombic potential of ions. By the same token, water is an efficient solvent for biomolecular polyelectrolytes such as DNA and proteins, shielding nearby charges on the backbone from one another.

As a polar species, water can engage in favorable Coulombic interactions with ions and other polar solutes. According to a simplistic picture, water molecules will solvate cations by orienting their oxygen molecules toward the ion, whereas they will adopt the opposite configuration for anions (Figure 4.1a). Neutron scattering studies[4,5] have largely confirmed this picture. Anions such as chloride are coordinated to water in the hydration shell via hydrogen atoms, such that the Cl⁻ · · · H–O

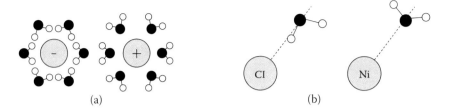

FIGURE 4.1   (a) Hydration of cations and anions: the classical picture. (b) The structures revealed by neutron scattering.

bond is almost linear, whereas for cations like nickel, the water molecules are oriented with the oxygen atoms facing toward the ions (Figure 4.1b).

Hydrophobic solutes in water experience a force that causes them to aggregate. It seems clear that this hydrophobic interaction is in some way responsible for several important biological processes[6]: the aggregation of amphiphilic lipids into bilayers, the burial of hydrophobic residues in protein folding, and the aggregation of protein subunits into multisubunit quaternary structures. The question is whether a single explanation accounts for all these behaviors and, if so, what is it?

We do not yet know how to answer either issue. Part of the reason for that perhaps surprising ignorance might be that there has, for some time, existed an apparent explanation that made such seemingly good sense that it has been hard to dislodge and is still routinely cited in biochemistry textbooks. This is Kauzmann's[7] "entropic" origin of the hydrophobic interaction, which draws on the picture of hydrophobic hydration posited in 1945 by Frank and Evans.[8] They suggested that, to avoid the enthalpic penalty of losing hydrogen bonds in creating the space for a hydrophobe, water molecules arrange themselves around the solute particle in a relatively ordered fashion. This is the "iceberg" model: the hydrophobe is encased in an icelike shell of water.

Kauzmann pointed out that the price of making such a structure is that the rotational and translational freedom of the molecules in the hydration sphere is compromised: there is an entropy decrease. But if two "caged" hydrophobes were to come together, the "structured" water in the region between them is returned to the bulk, leading to an entropy increase. Thus, there is an entropically based force of attraction between these solute particles.

The argument seems sound in principle, but the question is whether hydrophobic hydration really has the semicrystalline character proposed by Frank and Evans. To date, there is no good reason to suppose that it does, and some evidence indicating that it does not. Reviewing the existing data—both structural and thermodynamic, and based on both experiment and computer simulation—on hydrophobic hydration up to that point, Blokzijl and Engberts[9] concluded that there was no reason to suppose that hydrogen bonding is enhanced in the first hydration shell, or that the favorable enthalpy of hydrophobic hydration need be attributed to anything more than normal van der Waals interactions between water and a nonpolar solute. Moreover, they found no evidence that hydration involved any enhancement in the ordering of water molecules around the solute—rather, the hydrogen-bonded network seemed simply to *maintain* its structure, in particular by orienting water molecules such that the O–H bonds are tangential to the solute surface. There is no inconsistency between the occurrence of such orientational effects and the lack of any enhancement in structure because the hydrogen-bonded network introduces directional preferences in local water orientation even in the bulk.

Blokwijl and Engberts suggest that aggregation of hydrophobes results not from Kauzmann's entropic mechanism but from the increasing difficulty in accommodating hydrophobes within the hydrogen-bonded network as their concentration in solution increases. From the perspective of whether there is anything special about water that introduces a hydrophobic interaction, the message is mixed. Yes, the hydrogen-bonded network seems to be important, but not because it becomes more highly structured by hydrophobes; rather, it is because this network is disrupted by too great an accumulation of cavities. On the other hand, Lucas[10] and Lee[11,12] have argued that this disruption is not a function of the network at all, but stems merely from the small size of the water molecules, which creates a high free-energy cost to opening up a cavity to accommodate a hydrophobe. But Southall et al.[13] assert that neither water's structure nor the small-size effect can, by themselves, account for several of the characteristics of the hydrophobic effect, such as its dependence on temperature and on solvent shape.

There is nothing unique to water about solvophobic aggregation. For example, Huang et al.[14] have shown that nonnatural peptide-like molecules with *N*-linked rather than *C*-linked side chains (*N*-substituted oligoglycines or peptoids) will fold into well-defined secondary structures in acetonitrile in which the polar units are buried in the interior: one does not apparently need a strong degree of solvent structure in order for such packing to occur. It will be very interesting to explore the potential complexity of structure and function available to this and other nonaqueous pseudo-protein chemistry.

All the same, one can imagine there being scope for added subtlety in the interactions of a solvophobe and a structured solvent like water. Molecular dynamics (MD) simulations of krypton in aqueous solution[15,16] show no sign, however, of the clathrate-like hydration shells invoked in the iceberg model. This is supported experimentally by Bowron et al.,[17] who used EXAFS to highlight the differences between the hydration shell of krypton in solution and the clathrate cage of krypton in the crystalline clathrate hydrate. The former is unambiguously more disordered, and the tangential orientation of water molecules found in the hydrate is not rigorously maintained in the liquid.

Finney et al.[18] propose that the clustering of small hydophobes might have a more subtle entropic origin. They find that the hydration shells of water molecules are slightly altered by the presence of methanol: the second hydration shell becomes slightly compressed, and the correlations between second-neighbor waters slightly sharpened, leading to a small reduction in entropy. This small decrease in freedom of the water molecules could promote aggregation of the methanols. If so, this is a small effect, and by no means intuitively obvious. In any event, these direct structural probes seem to have diminished any argument for the classic Kauzmann model of the hydrophobic interaction.

4.1.2 LARGE HYDROPHOBIC SOLUTES AND SURFACES

While small hydrophobic species can be accommodated in the hydrogen-bonded network of liquid water without much perturbation of the network, large hydrophobes are another matter. An attraction evidently does exist between such species—for example, the hydrophobic surfaces of proteins. Does this have the same origin as the force that causes small hydrophobes to cluster? That is not obvious.

Close to an extended hydrophobic surface, it is geometrically impossible for the network to maintain its integrity. It has been proposed that this can even lead to "drying" of the interface[19]—the formation of a very thin layer of vapor separating the liquid from the surface. Lum et al.[20] argue that this difference between small and large hydrophobes should lead to qualitatively different behavior, with a crossover length scale somewhere in the region of 1 nm—about the size of small proteins. They suggest that two hydrophobic surfaces in close proximity could expel the water from between them (Figure 4.2), and that the resulting imbalance in pressure would cause the two surfaces to attract: an explanation of the hydrophobic interaction that could be relevant to protein folding and aggregation. In effect, the confined water undergoes "capillary evaporation." This kind of drying transition has been seen in simulations of hydrophobic plates in water when the separation between them falls to just a few molecular layers.[21–23] And simulations by ten Wolde and Chandler[24] suggest that a hydrophobic polymer acquires a compact conformation in water via a process resembling a first-order phase transition in which the rate-limiting step is the nucleation of a sufficiently large vapor bubble—the classical mechanism of heterogeneous nucleation.

But simulations of protein folding show a more complex situation. Zhou and colleagues[25] found that collapse of the two-domain enzyme BphC, which breaks down polychlorinated biphenyls,

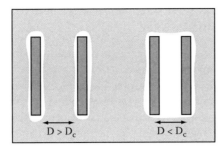

FIGURE 4.2 The mechanism of hydrophobic attraction proposed in Ref. 20, driven by a capillary drying transition between hydrophobic surfaces at nanoscale separations. The hydrophobic surfaces are surrounded by a thin layer of vapor. At some critical separation, D_c, there is a collective transition to the vapor phase in the space between the surfaces.

showed no sign of a sharp dewetting transition as the domains came together. Here the drying seen between hydrophobic plates seems to be suppressed by attractive interactions between the protein and water: dewetting was recovered when the electrostatic protein–water forces were turned off, and was stronger still in the absence of attractive van der Waals forces. Thus, it seems that the inevitable presence of such interactions in proteins complicates the simple picture obtained from hydrophobic surfaces.

On the other hand, Liu and coworkers[26] found a first-order–like dewetting transition in simulations of the association of the melittin tetramer, a small polypeptide found in honeybee venom. Yet single mutations of three hydrophobic isoleucine residues to less hydrophobic ones were sufficient to suppress the dewetting. A survey of the protein data bank[27] suggests that dewetting is rather rare, but is by no means unique to melittin. It seems that a significant number of polar residues in the hydrophobic core (which is common) is generally enough to suppress dewetting.

These results suggest that even if the drying mechanism can operate in the collapse of some proteins, it is nonetheless extremely sensitive both to the precise chemical nature of the protein domains involved and perhaps to the geometry of association—melittin-subunit association forms a tubelike enclosed space, whereas that for BphC is slablike.

Giovambattista et al.[28] have attempted to tease apart the contributions of these factors. They simulated the association of a mutated melittin dimer in which the distribution of hydrophobic and hydrophilic groups is retained but the surface is artificially flattened. The results suggest that flattened melittin behaves as an intermediate case between ideal, flat hydrophobic and hydrophilic surfaces. The drying seen in the case of "real" melittin happens only at very small separations (about one intervening water layer) for the flattened mutant, being localized to a central region where an apolar residue resides. In other words, drying seen for ideal hydrophobic plates is probably stronger than it is for real proteins, where it is likely to be highly sensitive to small variations in surface chemistry and shape.

Experimentally, the nature of water close to a hydrophobic surface has been controversial.[29] Some x-ray and neutron reflectivity measurements[30,31] have seemed to suggest that there is a depletion in water density extending a few nanometers from a hydrophobic monolayer. Even if this is very different from complete drying, the existence of such a depletion layer would be theoretically very perplexing because there is no obvious physical interaction in the system that could introduce a nanometer length scale. But there now seems to be an emerging consensus from other experiments that a depletion layer has a density closer to that of the bulk liquid than of the vapor, and that it extends at most for a few angstroms, and possibly for less than the width of one water molecule.[32–35] It could be important to know whether similar effects are seen for other polar liquids,[34] so that one might elucidate the role (if any) of water's hydrogen-bonded network.

4.2 LONG-RANGE HYDROPHOBIC INTERACTIONS AND THE ROLE OF BUBBLES

As though this picture were not complicated enough, there seems to be more than one type of hydrophobic interaction. In the early 1980s, measurements using the surface-force apparatus revealed that there is an attractive interaction between hydrophobic surfaces that seems to extend over very long distances—up to a few hundred nanometers, greatly exceeding the range of the normal hydrophobic interaction.[36,37] It has been suggested that correlated charge or dipole fluctuations on the two surfaces might lead to a long-ranged electrostatic interaction, in a manner that makes no direct appeal to water structure per se.[38–40] But an alternative explanation for which there is now some experimental evidence invokes the formation of submicroscopic bubbles between the surfaces, whereupon the meniscus pulls them together.[41,42]

There is now persuasive evidence that such bubbles may be formed. Using high-resolution optical microscopy, Carambassis et al.[43] saw bubbles approximately 1 μm in diameter in water in contact with a glass surface coated with fluorinated alkylsilanes. They observed jumps to contact between the surface and a similarly coated glass microsphere as it was brought toward the surface on the tip of an atomic force microscope. Tyrrell and Attard[44,45] have also imaged submicroscopic bubbles, about 100 nm in radius and flattened against the surface, in AFM studies of hydrophobic surfaces in water.

Thus, it has been proposed that there may be two different regimes for the interaction between hydrophobic surfaces: a long-ranged attraction created by bridging bubbles, and a medium-ranged interaction felt at separations of less than 20 nm or so where the attraction is of the same type as that involved in protein aggregation and folding.

We must note with some possible resignation that the interactions between two *hydrophilic* surfaces—such as two well-hydrated proteins—are equally mired in uncertainties and controversy. Measurements have suggested that there is a monotonically repulsive interaction between such surfaces.[46–48] But van der Waals interactions between surfaces would be attractive, and so once again structuring effects unique to water are among the explanations proposed to account for the difference.[49–51] Israelachvili and Wennerström[52] dispute that idea, arguing that in fact the hydration force between two hydrophilic surfaces is indeed either attractive or, because of the layering effects experienced by any liquid close to a sufficiently smooth solid surface, oscillatory. They suggest that the steep repulsion often measured between hydrophilic particles and surfaces at small separations is instead due to the characteristics of the surfaces themselves.

4.3 PROTEIN HYDRATION: NONSPECIFIC EFFECTS

4.3.1 THE HYDRATION SHELL

The pioneering early studies of protein secondary structure by Pauling et al.[53,54] have, perhaps understandably, something of the in vacuo mentality about them, insofar as they stress intra- and intermolecular hydrogen bonding in the polypeptide chains without much consideration of the role of the solvent. Thus, Pauling's iconic α-helices and β-sheets are held together by hydrogen bonds between polar residues, and the apparent implication is that there is an enthalpic penalty to breaking up these structures, without consideration of the hydrogen bonds that can then form between the peptide residues and water. That picture would actually argue for the enhanced stability of these secondary structural motifs in nonpolar solvents, where there would be no such competition for hydrogen bonding from the solvent molecules.

What Pauling's model lacked, of course, is the existence of hydrophobic interactions between the nonpolar protein residues—a feature identified by Langmuir[55] and promoted by Bernal[56] in the late 1930s. Crudely speaking, this notion invokes the burial of hydrophobic residues in the protein interior. It is now generally accepted that protein folding is driven primarily by a balance between these two factors—intramolecular hydrogen bonding and hydrophobic interactions—although there is no general agreement on precisely where the balance lies. Certainly it is an oversimplification to present globular proteins as a kind of "polymer micelle" that is wholly hydrophobic inside and wholly hydrophilic on the surface.

What can be said about the role of water in producing the characteristic folded structures of proteins? And how does water hydrate and decorate those three-dimensional forms? It does not seem sufficient to incorporate water into protein structure prediction merely via some heuristic potential that simply acknowledges the existence of hydrophobic interactions. Atomistic simulations of the folding of the SH3 protein domain have revealed that the process may depend on a rather gradual, molecule-by-molecule expulsion of water from the collapsed interior.[57–59] These studies characterize folding as a two-stage process: first, collapse to a near-native structure that retains a partially hydrated hydrophobic core, followed by slower expulsion of the residual water. This water might play the part of a lubricant to enable the hydrophobic core to find its optimally packed state. Moreover, some water molecules typically remain in the core, hydrogen-bonded to the peptide backbone. This challenges the picture presented by Lum et al.[20] of a collective drying transition as the water is confined between basically hydrophobic surfaces.

Papoian et al.[60] have modified a typical protein structure-prediction calculation to incorporate the possibility of water-mediated interactions between residues. This allows for the formation of relatively long-ranged (6.5–9.5 Å) connections between hydrophilic parts of the folding chain through bridging water molecules—something that is not permitted by the dry model. Such water bridges can be "squeezed dry" in the later stages of the folding process: the water acts as a temporary, loose glue that holds the folded chain together until it is ready for final compaction—in effect, constraining the conformational freedom and smoothing the funnel in the folding energy surface. There is a substantial improvement in the simulation structure predictions for several proteins when these water-mediated contacts are included. Thus, whereas the classical Kauzmann model of interactions between hydrophobic species postulates an attraction resulting from the liberation of "bound" or "ordered" hydration water, the contrasting picture that emerges here in the interactions between *hydrophilic* residues is one in which the attractions are promoted enthalpically by water molecules that are constrained at the macromolecular surface, despite the entropic cost.

4.3.2 DYNAMICS, COOPERATIVITY, AND THE GLASS TRANSITION

The classical picture favored by biochemists posits two distinct classes of water molecule in the hydration shell.[61] Bound waters are hydrogen-bonded to the protein at specific locations, and typically remain in position even in an anhydrous environment (nonaqueous solvent, vacuum, or dry powder). They can thus be identified crystallographically, and are regarded as, in some sense, an intrinsic part of the protein structure. "Free" water, meanwhile, is deemed to remain not only mobile but essentially bulklike even in the immediate hydration layer of the protein. There is no doubt now that this is a highly oversimplified picture: there is a continuum between bound molecules and those that behave dynamically as though they are indeed in the bulk.[62]

Any notion that proteins "only work in water" has now been thoroughly dispelled by evidence that enzymes can retain some functionality both in nonaqueous solvents and in a vacuum.[63,64] But neither of these environments is truly nonaqueous in the sense of being devoid of water: in both cases, some water molecules remain tightly bound to the protein. Although this bound water may not be enough to fully cover the protein surface with a monolayer "aqueous sheath," nonetheless proteins seem to require about 0.4 g of water per gram of protein to achieve their normal functionality.[65]

It is generally considered that a protein needs to maintain a delicate balance between rigidity and flexibility of structure: the specificity of the folded shape is clearly central to an enzyme's substrate selectivity, but it must also remain able to adapt its shape by accessing a range of conformations without getting stuck in local energy minima. It has been suggested that the role of the solvent is to "inject" fluctuations into the protein to give it this conformational flexibility. Olano and Rick[66] find that, for both bovine pancreatic trypsin inhibitor and barnase, which have polar and hydrophobic cavities respectively, the addition of a water molecule into the cavity makes the proteins more flexible by weakening intramolecular hydrogen bonds. Simulations of scytalone dehydratase by Okimoto et al.[67] indicate that water molecules in the protein's binding pocket seem to play a part in the conformational flexibility that is necessary for binding of the substrate: there is evidence for cooperativity in the motions of the bound water molecules and the ligand-free protein.

Both simulations[68] and experiments[69] show that water dynamics in the hydration layer of a peptide are anomalous with respect to the bulk. The hydration water seems to adopt a state akin to that of a glass, with a very rough potential-energy landscape and slow hopping between local potential minima. Thus, the water molecules no longer diffuse independently: their motion is dependent on that of their near neighbors. It is tempting to regard this as a result of the interconnected nature of the hydrogen-bonded network, which is highly constrained close to the protein surface and so develops a degree of cooperativity.

But if the protein impresses glass-like dynamics on its hydration layer, how then might the hydration dynamics feed back on the behavior of the protein? It seems that this kind of anomalous, glassy dynamics is just what a protein needs to attain the kind of conformational flexibility that is intrinsic to its function. If the protein needs to "feed off" the dynamics of its solvation layer, then water looks like the ideal solvent because its hydrogen-bonded network makes it ideally suited to being molded by the protein into a glassy state.

There is some evidence to support the idea that the dynamics of a protein can be "slaved" to those of the solvent. Below about 200–220 K, proteins seem to freeze into a kinetically arrested state that has genuine analogies with a glass[70–72]: the protein atoms undergo harmonic vibrations in local energy minima, but no diffusive motion. Both experiment[73] and simulations[74,75] suggest that this glass-like transition of a protein coincides with dynamical changes in the solvent. Chen et al.[76] have proposed that the transition corresponds to a fragile-to-strong crossover in supercooled water. (A strong liquid shows Arrhenius-like temperature dependence of viscosity, while a fragile liquid deviates from it.) But this interpretation, and indeed the significance of the 220 K transition itself, has been challenged by Khodadadi et al.[77]

The solvent and protein motions appear to be intimately coupled,[78,79] so that as a protein is warmed through its glass-like dynamical transition temperature the dynamics of the hydration shell awaken motions in the protein. Bizzarri and Cannistraro[68] speculate that the dynamics of the protein and solvent are so strongly coupled that they "should be conceived as a single entity with a unique rough energy landscape."

Oleinikova et al.[80] suggest that this dynamical behavior depends on the formation, at a critical "water coverage" on the protein surface, of a fully connected hydrogen-bonded network of water molecules.[81] In other words, the collective dynamics become activated in a two-dimensional percolation transition. This threshold for a single lysozyme molecule appears to require about 50% of the protein surface to be covered with water, which would correspond to about 66% coverage of the purely hydrophilic regions. This is essentially identical to the percolation threshold for clusters formed on two-dimensional square and honeycomb lattices.

4.4 PROTEIN HYDRATION: SPECIFIC ROLES OF WATER IN STRUCTURE AND FUNCTION

4.4.1 SECONDARY STRUCTURE AND PROTEIN–PROTEIN INTERACTIONS

While protein–protein interactions are commonly discussed in terms of the same guiding forces that govern protein folding—hydrogen bonding, polar interactions, and hydrophobic interactions—Fernández and Scott[82] suggest a different way in which hydrogen-bonding groups on a protein backbone might be used to mediate such contacts, via structural units that they call dehydrons.

Dehydrons are intramolecular hydrogen bonds that are imperfectly "wrapped" in hydrophobic residues. It is energetically favorable to remove water from the vicinity of the dehydron, making it an adhesive site for hydrophobic protein chains. Many different proteins possess dehydron units—human myoglobin has 16, for instance, and human ubiquitin has 12. These units appear to be concentrated at sites that engage in complexation with other proteins, and may play an important role in protein–protein interactions such as the association of capsid assemblies in viruses.

Papoian et al.[60] find that, just as water-mediated intramolecular contacts may assist in protein folding, so too can these contacts serve to facilitate selective recognition in protein–protein interactions.[83] In other words, it is not simply the case that water molecules can bridge two proteins: such contacts can be imbued with significant information content that allows the interactions to be discriminating. Thus the protein surfaces extend the range of their influence via their hydration shells.

It is scarcely surprising to find that water molecules may sometimes apparently get "frozen in place" in the folded structure—not as a kind of lubrication that has been imperfectly expelled, but as an element of the secondary structure in their own right. This seems to be the case in internalin B (Inl B), a bacterial surface protein found in *Listeria monocytogenes* that helps to activate the bacterium's phagocytotic defense against the mammalian immune system. The leucine-rich repeat motif of Inl B, which is common to all proteins of the internalin family, contains a series of stacked loops that are held together by water molecules bridging the peptide chains. These waters are organized into three distinct "spines" through the stack, and are an integral part of the secondary structure.[84]

4.4.2 MEDIATION OF TARGET BINDING

Renzoni et al.[85] point out that hydration water can potentially serve divergent purposes in the mediation of target binding by proteins. On the one hand, it can make the binding surface highly adaptable and thus somewhat promiscuous; on the other hand, there is evidence that water molecules occupying crystallographically defined sites in a protein structure through hydrogen bonding to polar residues can act as removable tools or extensions to the peptide chain for assisting in the specificity of substrate binding.

The former role is illustrated in the mechanism by which oligopeptide-binding protein OppA binds very small (2–5 residue) peptides with more or less any amino-acid sequence. This lack of specificity is made possible by the fact that all interactions between the protein and the peptide side chains are mediated by water: hydration of the voluminous binding site creates a highly malleable receptor matrix.[86,87] However, rather than acting as a plastic medium that can be arbitrarily manipulated to accommodate a substrate, this water seems to constitute a well-defined, "bricklike" filler (Figure 4.3).

In contrast, biology also uses water to achieve selectivity. This is evident in the binding mechanism of some antibodies,[88,89] indicating that nature has mastered the rules of incorporating water into the binding site sufficiently to use them for essentially ad hoc challenges of molecular recognition. Water molecules play a similar role in the binding of some protease inhibitors to their target enzymes.[90]

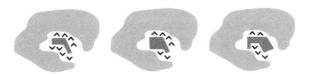

FIGURE 4.3   Water molecules act as removeable bricks that give the binding site of OppA a remarkable promiscuity illustrated schematically here for three substrates of different shape (with V-shaped water molecules in the binding site).

A combination of these seemingly contradictory roles of water—specificity and plasticity—was revealed in a crystallographic study of the binding specificity of the bacterial L-arabinose binding protein for various related sugars.[91] The water molecules here serve as flexible adhesive filling that contributes a degree of selectivity of binding while also allowing the ADP binding pocket to adapt to a different substrate.

Clarke et al.[92] found that displacement of ordered water from carbohydrate-binding sites can have subtle effects on substrate selectivity in such interactions that depend on a delicate balance between entropic and enthalpic effects. Despite the entropic advantage of expelling bound water from a binding cleft, one cannot generalize about the consequent free energy change, and thus about the role of water in protein–substrate interactions and specificity. As a general rule, the question of whether or not it is advantageous to incorporate a water molecule at the binding interface revolves around a delicate balance. Confining a water molecule clearly has an entropic penalty, but this might be repaid by the enthalpic gains of hydrogen-bond formation—an issue that must itself be weighed against the average number of hydrogen bonds that a bulk water molecule engages in. It is not obvious which way the scales will tip in any instance—and so the purposive use of water has been barely explored so far in the tailoring of binding sites for drug molecules.[85,93]

4.4.3 ALLOSTERY

Protein–protein contacts mediated by water molecules can not only serve to assist in recognition and docking but may also play a mechanistic role in function—for example, in the allosteric regulation of oxygen binding to hemoglobin. The hemoglobin of the mollusc *Scapharca inaequivalvis* is dimeric, and the interface of the subunits contains a cluster of 17 well-ordered water molecules. Oxygenation is accompanied by loss of six of the ordered interfacial water molecules. Royer et al.[94] found that these waters have a central role in cooperative oxygen binding, enabling allosteric interactions between the subunits by acting as a kind of transmission unit. The water cluster helps to stabilize the low-affinity form of the protein, whereas a mutant form that lacks two of the hydrogen bonds from this cluster tends to adopt the high-affinity conformation instead. Thus, loss of interfacial water occasioned by oxygen binding to one of the wild-type subunits helps to promote the transition to the high-affinity conformation of the other subunit. Something similar is observed at the interface between the cytochrome c_2 redox protein (cyt c_2) of *Rhodobacter sphaeroides*, which facilitates photosynthetic electron transfer, and the photosynthetic reaction center. Here the interfacial water appears to act as a kind of reversible latch.[95]

4.4.4 INVOLVEMENT OF BOUND WATER IN CATALYTIC ACTION

Water in the active site of a protein can play more than a purely structural role. As a nucleophile and proton donor, it can be a reagent in biochemical processes. Water molecules might engage in nucleophilic attack of a ligand,[96] or provide routes for electron[97] or proton transport.

An example of these processes is found in the action of so-called Clade 3 heme catalases, which convert hydrogen peroxide to water and oxygen.[98] A bound water molecule assists an NADPH molecule in protecting one of the reaction intermediates of the catalytic porphyrin-ferryloxo group against deactivation to an inactive form, by supplying a hydroxyl group that binds covalently but temporarily to side chain on the porphyrin group and then assists the fast two-electron reduction of the intermediate ferryloxo species by NADPH via a series of proton rearrangements.

Proton relays are particularly common. This is a feature of the mechanism by which the cytochrome P450 StaP catalyses the formation of staurosporine, an antitumor agent. A critical step in this process is the abstraction of a proton from an N–H group on the substrate (chromopyrrolic acid) by an iron-oxo species in the enzyme. This seems to happen with the concerted assistance of two water molecules in the binding site.[99]

Bound waters can serve many roles at once. Rodriguez et al.[100] found that a hydrogen-bonded network of waters in the active site of a bacterial heme oxygenase has three functions. First, it conducts protons to the iron-dioxygen complex during catalysis. Second, it propagates changes in electronic structure at the active site during the course of the reaction to remote parts of the polypeptide. Third, it modulates the conformational freedom of the enzyme, allowing it to accommodate and adapt to the changes in conformation required during the catalytic process. If the hydrogen-bonded network is disrupted by mutation, the necessary coordination in dynamics of different parts of the protein is lost and the motions become almost globally chaotic, lowering the efficiency of the enzyme significantly. All this underlines the immense versatility that bound waters offer in enzyme action.

4.4.5 PROTON WIRES AND WATER CHANNELS

One of the most striking consequences of the extended hydrogen-bonded structure of liquid water is the rapid diffusion rate of protons, which is considerably higher than that of other monovalent cations. The traditional explanation, the so-called Grotthus mechanism,[101,102] invokes the fact that

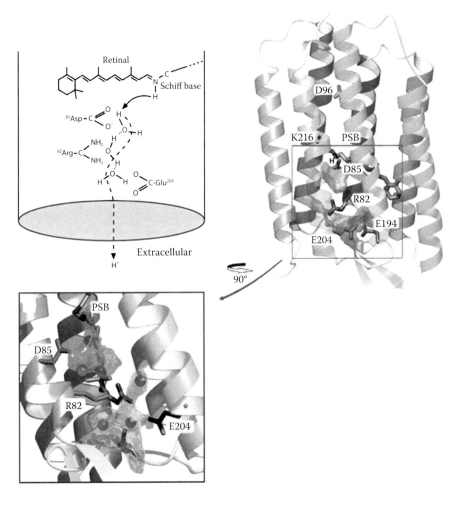

FIGURE 4.4 A chain of water molecules in bacteriorhodopsin appears to provide a proton-conducting pathway from the retinal chromophore to the extracellular medium. The water molecules are shown as grey spheres in the lower left-hand frame. (Courtesy of Klaus Gerwert, with permission.)

protons moving through the network do not, like other cations, have to drag a solvent shell with them. Rather, the water molecules solvating a hydronium (H$_3$O$^+$) ion can actually facilitate proton transport by shuttling it to another molecule. This intermolecular transfer of the proton in fact entails a preemptive rearrangement of the hydrogen-bond network,[102,103] dubbed the "Moses mechanism."

Proton-hopping along a hydrogen-bonded chain of water molecules—"proton wires"—has been postulated to play a functional role in a variety of proteins, such as the photosynthetic reaction center of *R. sphaeroides*,[104] cytochromes,[105–107] gramicidin A,[108] and bacteriorhodopsin[109–113] (Figure 4.4).

The water-conducting protein channel aquaporin provides an interesting contrast to these proton wires. Aquaporin proteins effect water transport[114] via a chain of nine hydrogen-bonded molecules. But if this chain were to permit rapid transmembrane proton motion, that would disturb the delicate charge balance across the membrane. So aquaporin must somehow disrupt the potential proton wire that threads through it. It has been proposed that this is achieved by the introduction of a defect into the hydrogen-bonded chain,[115–117] or by electrostatic blocking of proton motion through the narrowest point of the constriction.[118,119] Spassov et al.[111] and Chen et al.[120] have now shown that electrostatics, rather than water-wire defects, seem to predominate.

Disruption of water channels through transmembrane pores has been proposed as a rather general mechanism for gating behavior. Several gated channels contain constrictions lined with hydrophobic residues, where a water channel could quite easily be pinched off by a conformatonal change. Jiang et al.[122,123] have suggested such a gating process in potassium channels, and Sukharev et al.[124] propose a similar mode of action in the bacterial mechanosensitive channel MscL.

4.5   WATER AND NUCLEIC ACIDS

Compared with the attention given to hydration in determining protein structure and function, the role that water plays in the properties of nucleic acids has been surprisingly neglected. Indeed, it is often overlooked that the famous double-helical structure of DNA is not intrinsic to that molecule but relies on a subtle balance of energy contributions present in aqueous solution. Without water to screen the electrostatic repulsions between phosphate groups, the classic, orderly helix is no longer viable. Thus DNA undergoes conformational transitions, and even loses its double helix, in some apolar solvents[125,126]; and even though both experiments[127] and MD simulations[128] suggest that the double helix is not lost entirely in the gas phase, it has none of the elegance and order of the classic imagery.

As with proteins, DNA in the crystalline state preserves a pronounced degree of ordering in its hydration shells. Dickerson and coworkers[129,130] have reported that in the solid state, A–T segments of DNA have a "spine of hydration" in which one layer of water molecules bridges the nitrogen and oxygen atoms of bases in the minor groove, while a second layer bridges water molecules in the first layer. Shui et al.[131] have identified further two crystallographically defined hydration layers, which produce a series of hexagonal rings of water molecules in the minor groove. As this structure is sequence-specific, they argue that it effectively transmits sequence information to locations remote from the bases themselves.

The folding of RNAs into their functional forms resembles in many ways that of proteins: both macromolecules have hydrophilic and hydrophobic segments in their chainlike structures, and both may engage in intramolecular hydrogen-bonding in the folded state. But the distribution of the two types of component is more regular in RNA—all the bases are, aside from their hydrophilic substituents, hydrophobic, while the sugar–phosphate backbone is uniformly polar. Sorin et al.[132] find that this regular structure leads to correlated collapse of RNA strands into a compact form, which is more likely to trap water molecules between hydrophobic bases than is the less cooperative collapse of proteins, where hydrophobic residues are more sparse. Their simulations suggest that this trapped water is expelled late in the folding process, so that there remains considerable potential for water-mediated interactions as compaction proceeds. In this respect, the results argue that explicit water molecules buried within the folding macromolecule can play an important role in mediating compaction, as proposed for proteins by Cheung et al.[58] and Papoian et al.[60]

4.6 CONCLUSIONS

There can surely be no doubt that water is a biomolecule, insofar as it can be found playing a wide variety of roles in biochemical processes. It maintains macromolecular structure and mediates molecular recognition, it activates and modulates protein dynamics, it provides a switchable communication channel across membranes and between the inside and outside of proteins. Many of these properties do seem to depend, to a greater or lesser degree, on the special attributes of the H_2O molecule, in particular its ability to engage in directional, weak bonding in a way that allows for reorientation and reconfiguration of discrete and identifiable three-dimensional structures. Thus, although it seems entirely likely that *some* of water's functions in biology are those of a generic polar solvent rather than being unique to water itself, it is very hard to imagine any other solvent that could fulfill all of its roles—or even all of those that help to distinguish a generic polypeptide chain from a fully functioning protein. The fact that fully folded proteins moved from an aqueous to a nonaqueous environment may retain some of their functionality does not alter this, and does not detract from the centrality of water for life on earth.

That, however, is not the same as saying that all life must be aqueous. At least with our present (incomplete) state of knowledge about pivotal concepts such as the hydrophobic interaction, it is not obvious that any one of the functions of water in biology has to stand as an irreducible aspect of a living system. It is certainly possible to imagine, and even to make,[133] artificial chemical systems that engage in some form of information transfer—indispensable for inheritance and Darwinian evolution—in nonaqueous media.

Moreover, life in water has some well publicized drawbacks[134]—perhaps most notably the solvent's reactivity, raising the problem of hydrolysis of polymeric structures and of basic building blocks such as sugars. How the first pseudo-biological macromolecules on the early earth avoided this problem is still something of a puzzle. It is also unclear whether a solvent capable of engaging in hydrogen-bonding might initially help or hinder the use of this valuable, reversible supramolecular interaction for defining complex structures in macromolecules and their aggregates. Certainly, there is now reason to believe that such molecules can utilize both hydrogen bonding and solvophobic effects in acquiring well-defined structures without needing water as their solvent.

Attempts to enunciate the irreducible molecular-scale requirements for something we might recognize as life have so far been rather sporadic,[134,135] and are often hampered by the difficulty of looking at the question through anything other than aqua-tinted spectacles. From the point of view of thinking about nonaqueous astrobiological solvents, a review of water's roles in terrestrial biochemistry surely raises one key consideration straight away: it is not sufficient, in this context, to imagine a clear separation between the molecular machinery and the solvent. There is a two-way exchange of behaviors between them, and this literally erases any dividing line between the biological components and their environment. Water is an extraordinarily responsive and sympathetic solvent, as well as being far more than merely a solvent. If living systems depend on that kind of exchange, for example so that molecular information can be transmitted beyond the boundaries of the molecules that embody them, it is tempting to conclude that it would need to make use of water.

It is not just in its molecular-scale structure that water has been characterized as "biophilic"—when Henderson[136] first raised the intriguing notion of water's fitness as life's matrix in 1913, he had in mind the unusual macroscopic properties such as its high heat capacity and density anomalies. Nonetheless, even these have their origins in water's more or less unique set of molecular characteristics. Barring some unforeseen revelation from the exploration of Titan, however, it is likely that we will have to rely on experiment rather than discovery to put Henderson's hypothesis to the test.

Rather excitingly, with the advent of synthetic biology,[137–139] along with chemical and biological systems for exploring alternative biochemistries,[139–144] it is now far from inconceivable that this test can be arranged.*

REFERENCES

1. Szent-Györgyi, A. Welcoming address. In *Cell-Associated Water*. Edited by W. Drost-Hansen and J. S. Clegg. New York, NY: Academic Press, 1979.
2. Jacobi, J. (ed.). *Paracelsus: Selected Writings*. Princeton, NJ: Princeton University Press, 1979.
3. Marchand, P. J. *Life in the Cold*, 3rd ed. Hanover, NH: University Press of New England, 1996.
4. Soper, A. K., Neilson, G. W., Enderby, J. E., and Howe, R. A. A neutron diffraction study of hydration effects in aqueous solutions. *J. Phys. C: Solid State Phys.* 10, 1977:1793–1802.
5. Soper, A. K. The quest for the structure of water and aqueous solutions. *J. Phys.: Condens. Matter* 9, 1997:2717–2730.
6. Tanford, C., *The Hydrophobic Effect*, 2nd ed. New York, NY: Wiley, 1980.
7. Kauzmann, W. Some factors in the interpretation of protein denaturation. *Adv. Protein Chem.* 14, 1959:1–63.
8. Frank, H. S., and Evans, M. W. Free volume and entropy in condensed systems III. Entropy in binary liquid mixtures; partial molal entropy in dilute solutions; structure and thermodynamics in aqueous electrolytes. *J. Chem. Phys.* 13, 1945:507–532.
9. Blokzijl, W., and Engberts, J. B. F. N. Hydrophobic effects. Opinions and facts. *Angew. Chem. Int. Ed.* 32, 1993:1545–1579.
10. Lucas, M. Size effect in transfer of nonpolar solutes from gas or solvent to another solvent with a view on hydrophobic behavior. *J. Phys. Chem.* 80, 1976:359–362.
11. Lee, B. The physical origin of the low solubility of nonpolar solutes in water. *Biopolymers* 24, 1985: 813–823.
12. Lee, B. Solvent reorganization contribution to the transfer thermodynamics of small nonpolar molecules. *Biopolymers* 31, 1991:993–1008.
13. Southall, N. T., Dill, K. A., and Haymet, A. D. J. A view of the hydrophobic effect. *J. Phys. Chem. B* 106, 2001:521–533.
14. Huang, K., Wu, C. W., Sanborn, T. J., Patch, J. A., Kirshenbaum, K., Zuckermann, R. N., Barron, A. E., and Radhakrishnan, I. A threaded loop conformation adopted by a family of peptoid nonamers. *J. Am. Chem. Soc.* 128, 2006:1733–1738.
15. Ashbaugh, H. S., Asthagiri, D., Pratt, L. R., and Rempe, S. B. Hydration of krypton and consideration of clathrate models of hydrophobic effects from the perspective of quasi-chemical theory. *Biophys. Chem.* 105, 2003:323–338.
16. LaViolette, R., Copeland, K. L., and Pratt, L. R. Cages of water coordinating Kr in aqueous solution. *J. Phys. Chem. A.* 107, 2003:11267–11270.
17. Bowron, D. T., Filipponi, A., Roberts, M. A., and Finney, J. L. Hydrophobic hydration and the formation of a clathrate hydrate. *Phys. Rev. Lett.* 81, 1998:4164–4167.
18. Finney, J. L., Bowron, D. T., Daniel, R. M., Timmis, P. A., and Roberts, M. A. Molecular and mesoscale structures in hydrophobically driven aqueous solutions. *Biophys. Chem.* 105, 2003:391–409.
19. Stillinger, F. H. Structure in aqueous solutions of nonpolar solutes from the standpoint of scaled-particle theory. *J. Solution Chem.* 2, 1973:141–158.
20. Lum, K., Chandler, D., and Weeks, J. D. Hydrophobicity at small and large length scales. *J. Phys. Chem. B* 103, 1999:4570–4577.
21. Wallqvist, A., and Berne, B. J. Computer simulation of hydrophobic hydration forces on stacked plates at short range. *J. Phys. Chem.* 99, 1995:2893–2899.
22. Huang, X., Margulis, C. J., and Berne, B. J. Dewetting-induced collapse of hydrophobic particles. *Proc. Natl Acad. Sci. USA* 100, 2003:11953–11958.
23. Koishi, T., Yoo, S., Yasuoka, K., Zeng, X. C., Narumi, T., Susukita, R., Kawai, A., Furusawa, H., Suenaga, A., Okimoto, N., et al. Nanoscale hydrophobic interaction and nanobubble nucleation. *Phys. Rev. Lett.* 93, 2004:185701.
24. ten Wolde, P. R., and Chandler, D. Drying-induced hydrophobic polymer collapse. *Proc. Natl Acad. Sci. USA* 99, 2002:6539–6543.

---

* A longer version of this paper has been published in Ref. 145.

25. Zhou, R., Huang, X., Margulis, C. J., and Berne, B. J. Hydrophobic collapse in multidomain protein folding. *Science* 305, 2004:1605–1609.
26. Liu, P., Huang, X., Zhou, R., and Berne, B. J. Observation of a dewetting transition in the collapse of the melittin tetramer. *Nature* 437, 2005:159–162.
27. Hua, L., Huang, X., Liu, P., Zhou, R., and Berne, B. J. Nanoscale dewetting transition in protein complex folding. *J. Phys. Chem. B* 111, 2007:9069–9077.
28. Giovambattista, N., Lopez, C. F., Rossky, P. J., and Debenedetti, P. G. Hydrophobicity of protein surfaces: Separating geometry from chemistry. *Proc. Natl Acad. Sci. USA* 105, 2008:2274–2279.
29. Ball, P., How to keep dry in water. *Nature* 423, 2003:25–26.
30. Jensen, T. R., Jensen, M. Ø., Reitzel, N., Balashev, K., Peters, G. H., Kjaer, K., and Bjørnholm, T. Water in contact with extended hydrophobic surfaces: Direct evidence of weak dewetting. *Phys. Rev. Lett.* 90, 2003:086101.
31. Schwendel, D., Hayashi, T., Dahint, R., Pertsin, A., Grunze, M., Steitz, R., and Schreiber, F. Interaction of water with self-assembled monolayers: Neutron reflectivity measurements of the water density in the interface region. *Langmuir* 19, 2003:2284–2293.
32. Mezger, M., Reichert, H., Schöder, S., Okasinki, J., Schröder, H., Dosch, H., Palms, D., Ralston, J., and Honkimäki, V. High-resolution in situ x-ray study of the hydrophobic gap at the water–octadecyltrichlorosilane interface. *Proc. Natl Acad. Sci. USA* 103, 2006:18401–18404.
33. Poynor, A., Hong, L., Robinson, I. K., Granick, S., Zhang, Z., and Fenter, P. A. How water meets a hydrophobic surface. *Phys. Rev. Lett.* 97, 2006:266101.
34. Maccarini, M., Steitz, R., Himmelhaus, M., Fick, J., Tatur, S., Wolff, M., Grunze, M., Janecek, J., and Netz, R. R. Density depletion at solid-liquid interfaces: A neutron reflectivity stud. *Langmuir* 23, 2007: 598–608.
35. Granick, S., and Bae, S. C. A curious antipathy for water. *Science* 322, 2008:1477–1478.
36. Pashley, R. M., and Israelachvili, J. N. A comparison of surface forces and interfacial properties of mica in purified surfactant solutions. *Colloids Surf.* 2, 1981:169.
37. Israelachvili, J. N., and Pashley, R. M. The hydrophobic interaction is long range, decaying exponentially with distance. *Nature* 300, 1982:341–342.
38. Attard, P. Long-range attraction between hydrophobic surfaces. *J. Phys. Chem.* 93, 1989:6441.
39. Podgornik, R. Electrostatic correlation forces between surfaces with surface specific ionic interactions. *J. Chem. Phys.* 91, 1989:5840–5849.
40. Tsao, Y. H., Evans, D. F., and Wennerström, H. Long-range attraction between a hydrophobic surface and a polar surface is stronger than that between two hydrophobic surfaces. *Langmuir* 9, 1993:779–785.
41. Parker, J. L., Claesson, P. M., and Attard, P. *J. Phys. Chem.* 98, 1994:8468–8480.
42. Attard, P. Bridging bubbles between hydrophobic surfaces. *Langmuir* 12, 1996:1693–1695.
43. Carambassis, A., Jonker, L. C., Attard, P., and Rutland, M. W. Forces measured between hydrophobic surfaces due to a submicroscopic bridging bubble. *Phys. Rev. Lett.* 80, 1998:5357.
44. Tyrrell, J. W. G., and Attard, P. Images of nanobubbles on hydrophobic surfaces and their interactions. *Phys. Rev. Lett.* 87, 2001:176104.
45. Tyrrell, J. W. G., and Attard, P. Atomic force microscope images of nanobubbles on a hydrophobic surface and corresponding force-separation data. *Langmuir* 18, 2002:160–167.
46. Rand, R. P., and Parsegian, V. A. Hydration forces between phospholipid bilayers. *Biochim. Biophys. Acta* 988, 1989:351–376.
47. Horn, R. G., Smith, D. T., and Haller, W. Surface forces and viscosity of water measured between silica sheets. *Chem. Phys. Lett.* 162, 1989:404–408.
48. Israelachvili, J. N. Measurement of hydration forces between macroscopic surfaces. *Chemica Scripta* 25, 1985:7–14.
49. Marcelja, S., and Radic, N. Repulsion of interfaces due to boundary water. *Chem. Phys. Lett.* 42, 1976: 129–130.
50. Attard, P., and Batchelor, M. T. A mechanism for the hydration force demonstrated in a model system. *Chem. Phys. Lett.* 149, 1988:206–211.
51. Kornyshev, A. A., and Leikin, S. Fluctuation theory of hydration forces: The dramatic effects of inhomogeneous boundary conditions. *Phys. Rev. A* 40, 1989:6431–6437.
52. Israelachvili, J., and Wennerström, H. Role of hydration and water structure in biological and colloidal interactions. *Nature* 379, 1996:219–225.
53. Mirsky, A. E., and Pauling, L. On the structure of native, denatured, and coagulated proteins. *Proc. Natl Acad. Sci. USA* 22, 1936:439–447.

54. Pauling, L., Corey, R. B., and Branson, R. H. The structure of proteins. *Proc. Natl Acad. Sci. USA* 37, 1951:205–210.
55. Langmuir, I. The properties and structure of protein films. *Proc. R. Inst.* 30, 1938:483–496.
56. Bernal, J. D. Structure of proteins. *Nature* 143, 1939:663–667.
57. Shea, J. E., Onuchic, J. N., and Brooks, C. L. Probing the folding free energy landscape of the src-SH3 protein domain. *Proc. Natl Acad. Sci. USA* 99, 2002:16064–16068.
58. Cheung, M. S., Garcia, A. E., and Onuchic, J. N. Protein folding mediated by solvation: Water expulsion and formation of the hydrophobic core occur after the structural collapse. *Proc. Natl Acad. Sci. USA* 99, 2002:685–690.
59. Garcia, A. E., and Onuchic, J. N. Folding a protein in a computer: An atomic description of the folding/unfolding of protein A. *Proc. Natl Acad. Sci. USA* 100, 2003:13898–13903.
60. Papoian, G. A., Ulander, J., Eastwood, M. P., Luthey-Schulten, Z., and Wolynes, P. G. Water in protein structure prediction. *Proc. Natl Acad. Sci. USA* 101, 2004:3352–3357.
61. Otting, G., Liepinsh, E., and Wüthrich, K. Protein hydration in aqueous solution. *Science* 254, 1991: 974–980.
62. Marakov, V., Pettitt, B. M., and Feig, M. Solvation and hydration of proteins and nucleic acids: A theoretical view of simulation and experiment. *Acc. Chem. Res.* 35, 2002:376–384.
63. Klibanov, A. M. Enzymatic catalysis in anhydrous organic solvents. *Trends Biochem. Sci.* 14, 1989: 141–144.
64. Dunn, R. V., and Daniel, R. M. The use of gas–phase substrates to study enzyme catalysis at low hydration. *Philos. Trans. R. Soc. B* 359, 2004:1309–1320.
65. Rupley, J. A., and Careri, G. Protein hydration and function. *Adv. Protein Chem.* 41, 1991:37–172.
66. Olano, L. R., and Rick, S. W. Hydration free energies and entropies for water in protein interiors. *J. Am. Chem. Soc.* 126, 2004:7991–8000.
67. Okimoto, N., Nakamura, T., Suenaga, A., Futatsugi, N., Hirano, Y., Yamaguchi, I., and Ebisuzaki, T. Cooperative motions of protein and hydration water molecules: Molecular dynamics study of scytalone dehydratase. *J. Am. Chem. Soc.* 126, 2004:13132–13139.
68. Bizzarri, A. R., and Cannistraro, S. Molecular dynamics of water at the protein-solvent interface. *J. Phys. Chem. B* 106, 2002:6617–6633.
69. Russo, D., Hura, G., and Head-Gordon, T. Hydration dynamics near a model protein surface. *Biophys. J.* 86, 2004:1852–1862.
70. Knapp, E. W., Fischer, S. F., and Parak, F. Protein dynamics from Moessbauer spectra. The temperature dependence. *J. Phys. Chem.* 86, 1982:5042–5047.
71. Rasmussen, B. F., Stock, A. M., Ringe, D., and Petsko, G. A. Crystalline ribonuclease A loses function below the dynamical transition at 220 K. *Nature* 357, 1992:423–424.
72. Tilton, R. F., Dewan, J. C., and Petsko, G. A. Effects of temperature on protein structure and dynamics: x-ray crystallographic studies of the protein ribonuclease-A at nine different temperatures from 98 to 320K. *Biochemistry* 31, 1992:2469–2481.
73. Reat, V., Dunn, R., Ferrand, M., Finney, J. L., Daniel, R. M., and Smith, J. C. Solvent dependence of dynamic transitions in protein solutions. *Proc. Natl Acad. Sci USA* 97, 2000:9961–9966.
74. Bizzarri, A. R., Paciaroni, A., and Cannistraro, S. Glasslike dynamical behavior of the plastocyanin hydration water. *Phys. Rev. E* 62, 2000:3991–3999.
75. Wong, C. F., Zheng, C., and McCammon, J. A. Glass transition in SPC/E water and in a protein solution: A molecular dynamics simulation study. *Chem. Phys. Lett.* 154, 1989:151–154.
76. Chen, S.-H., Liu, L., Fratini, E., Baglioni, P., Faraone, A., and Mamontov, E. Observation of fragile-to-strong dynamic crossover in protein hydration water. *Proc. Natl Acad. Sci. USA* 103, 2006:9012–9016.
77. Khodadadi, S., Pawlus, S., and Sokolov, A. P. Influence of hydration on protein dynamics: Combining dielectric and neutron scattering spectroscopy data. *J. Phys. Chem. B* 112, 2008:14273–14280.
78. Arcangeli, C., Bizzarri, A. R., and Cannistraro, S. Role of interfacial water in the molecular dynamics-simulated dynamical transition of plastocyanin. *Chem. Phys. Lett.* 291, 1998:7–14.
79. Tournier, A. L., Xu, J., and Smith, J. C. Translational hydration water dynamics drives the protein glass transition. *Biophys. J.* 85, 2003:1871–1875.
80. Oleinikova, A., Smolin, N., Brovchenko, N., Geiger, A., and Winter, R. Formation of spanning water networks on protein surfaces via 2D percolation transition. *J. Phys. Chem. B* 109, 2005:1988–1998.
81. Brovchenko, I., and Oleinikova, A. *Interfacial and Confined Water*. Amsterdam: Elsevier, 2008.
82. Fernández, A., and Scott, R. A structurally encoded signal for protein interaction. *Biophys. J.* 85, 2003:1914–1928.

83. Papoian, G. A., Ulander, J., and Wolynes, P. G. Role of water mediated interactions in protein-protein recognition landscapes. *J. Am. Chem. Soc.* 125, 2003:9170–9178.

84. Marino, M., Braun, L., Cossart, P., and Ghosh, P. Structure of the InlB leucine-rich repeats, a domain that triggers host cell invasion by the bacterial pathogen *L. monocytogenes*. *Mol. Cell* 4, 1999:1063–1072.

85. Renzoni, D. A., Zvelebil, M. J. J. M., Lundbäck, T., and Ladbury, J. E. In *Structure-Based Drug Design: Thermodynamics, Modeling and Strategy*. Edited by J. E. Ladbury and P. R. Connelly. Austin, TX: Landes Bioscience, 1997.

86. Tame, J. R. H., Murshudov, G. N., and Dodson, E. J. The structural basis of sequence-independent peptide binding by OppA protein. *Science* 264, 1994:1578–1581.

87. Tame, J. R. H., Sleigh, S. H., and Wilkinson, A. J. The role of water in sequence-independent ligand binding by an oligopeptide transporter protein. *Nature Struct. Biol.* 3, 1996:998–1001.

88. Bhat, T. N., Bentley, G. A., Fischmann, T. O., Boulot, G., and Poljak, R. J. Small rearrangements in structures of Fv and Fab fragments of antibody D 1.3 on antigen binding. *Nature* 347, 1990:483–485.

89. Bhat, T. N., Bentley, G. A., Boulot, G., Greene, M. I., Tello, D., Dall'Acqua, W., Souchon, H., Schwarz, F. P., Mariuzza, R. A., and Poljak, R. J. Bound water molecules and conformational stabilization help mediate an antigen-antibody association. *Proc. Natl Acad. Sci. USA* 91, 1994:1089–1093.

90. McPhalen, C. A., and James, M. N. G. Structural comparison of two serine proteinase-protein inhibitor complexes: Eglin-C-subtilisin Carlsberg and CI-2-subtilisin Novo. *Biochemistry* 27, 1988:6582–6598.

91. Quiocho, F. A., Wilson, D. K., and Vyas, N. K. Substrate specificity and affinity of a protein modulated by bound water molecules. *Nature* 340, 1989:404–407.

92. Clarke, C., Woods, R. J., Gluska, J., Cooper, A., Nutley, M. A., and Boons, G.-J. Involvement of water in carbohydrate-protein binding. *J. Am. Chem. Soc.* 123, 2001:12238–12247.

93. Ladbury, J. E. Just add water! The effect of water on the specificity of protein-ligand binding sites and its potential application to drug design. *Chem. Biol.* 3, 1996:973–980.

94. Royer, Jr., W. E., Pardanani, A., Gibson, Q. H., Peterson, E. S., and Friedman, J. M. Ordered water molecules as key allosteric mediators in a cooperative dimeric hemoglobin. *Proc. Natl Acad. Sci. USA* 93, 1996:14526–14531.

95. Autenrieth, F., Tajkhorshid, E., Schulten, K., and Luthey-Schulten, L. Role of water in transient cytochrome c_2 docking. *J. Phys. Chem. B* 108, 2004:20376–20387.

96. Erhardt, S., Jaime, E., and Weston, J. A water sluice is generated in the active site of bovine lens leucine aminopeptidase. *J. Am. Chem. Soc.* 127, 2005:3654–3655.

97. Van Amsterdam, I. M. C., Ubbink, M., Einsle, O., Messerschmidt, A., Merli, A., Cavazzini, D., Rossi, G. L. and Canters, G. W. Dramatic modulation of electron transfer in protein complexes by crosslinking. *Nat. Struct. Biol.* 9, 2001:48–52.

98. Sicking, W., Korth, H.-G., de Groot, H., and Sustmann, R. On the functional role of a water molecule in clade 3 catalases: A proposal for the mechanism by which NADPH prevents the formation of compound II. *J. Am. Chem. Soc.* 130, 2008:7345–7356.

99. Wang, Y., Hirao, H., Chen, H., Onaka, H., Nagano, S., and Shaik, S. Electron transfer activation of chromopyrrolic acid by cytochrome p450 en route to the formation of an antitumor indolocarbazole derivative: Theory supports experiment. *J. Am. Chem. Soc.* 130, 2008:7170–7171.

100. Rodriquez, J. C., Zeng, Y., Wilks, A., and Rivera, M. The hydrogen-bonding network in heme oxygenase also functions as a modulator of enzyme dynamics: Chaotic motions upon disrupting the H-bond network in heme oxygenase from *Pseudomonas aeruginosa*. *J. Am. Chem. Soc.* 129, 2007: 11730–11742.

101. De Grotthuss, J. T. Memoir on the decomposition of water and of the bodies that it holds in solution by means of galvanic electricity. *Ann. Chim.* 58, 1806:54–73.

102. Agmon, N. The Grotthuss mechanism. *Chem. Phys. Lett.* 244, 1995:456–462.

103. Day, T. J. F., Schmitt, U. W., and Voth, G. A. The mechanism of hydrated proton transport in water. *J. Am. Chem. Soc.* 122, 2000:12027–12028.

104. Ermler, U., Fritzsche, G., Buchanan, S. K., and Michel, H. *Structure* 2, 1994:925–936.

105. Martinez, S. E., Huang, D., Ponomarev, M., Cramer, W. A., and Smith, J. L. *Protein Sci.* 5, 1996: 1081–1092.

106. Schlichting, I., Berendzen, J., Chu, K., Stock, A. M., Maves, S. A., Benson, D. E., Sweet, R. M., Ringe, D., Petsko, G. A., and Sligar, S. G. The catalytic pathway of cytochrome P450cam at atomic resolution. *Science* 287, 2000:1615–1622.

107. Akeson, M., and Deamer, D. W. Proton conductance by the gramicidin water wire. Model for proton conductance in the F_1F_0 ATPases? *Biophys. J.* 60, 1991:101–109.

108. Pomès, R., and Roux, B. Structure and dynamics of a proton wire: A theoretical study of H$^+$ translocation along the single-file water chain in the gramicidin A channel. *Biophys. J.* 71, 1996:19–39.

109. Rammelsberg, R., Huhn, G., Lübben, M., and Gerwert, K. *Biochemistry* 37, 1998:5001–5009.

110. Garczarek, F., and Gerwert, K. Functional waters in intraprotein proton transfer monitored by FTIR difference spectroscopy. *Nature* 439, 2006:109–112.

111. Spassov, V. Z., Luecke, H., Gerwert, K., and Bashford, D. pK_a calculations suggest storage of an excess proton in a hydrogen-bonded water network in bacteriorhodopsin. *J. Mol. Biol.* 312, 2001:203–219.

112. Rousseau, R., Kleinschmidt, V., Schmitt, U. W., and Marx, D. Modeling protonated water networks in bacteriorhodopsin. *Phys. Chem. Chem. Phys.* 6, 2004:1848–1859.

113. Lee, Y.-S., and Krauss, M. J. Dynamics of proton transfer in bacteriorhodopsin. *Am. Chem. Soc.* 126, 2004:2225–2230.

114. Agre, P. Aquaporin water channels (Nobel lecture). *Angew. Chem. Int. Ed.* 43, 2004:4278–4290.

115. Tajkhorshid, E., Nollert, P., Jensen, M. O., Miercke, L. J. W., O'Connell, J., Stroud, R. M., and Schulten, K. Control of the selectivity of the aquaporin water channel family by global orientational tuning. *Science* 296, 2002:525–530.

116. De Groot, B. L., and Grubmüller, H. Water permeation across biological membranes: Mechanism and dynamics of aquaporin-1 and GlpF. *Science* 294, 2001:2353–2357.

117. Chakrabarti, N., Tajkhorshid, E., Roux, B., and Pomès, R. Molecular basis of proton blockage in aquaporins. *Structure* 12, 2004:65–74.

118. De Groot, B. L., Frigato, T., and Grubmüller, H. The mechanism of proton exclusion in the aquaporin-1 water channel. *J. Mol. Biol.* 333, 2003:279–293.

119. Warshel, A., and Burykin, A. What really prevents proton transport through aquaporin? Charge self-energy versus proton wire proposals. *Biophys. J.* 85, 2003:3696.

120. Chen, H., Wu, Y., and Voth, G. A. Origins of proton transport behavior from selectivity domain mutations of the aquaporin-1 channel. *Biophys. J.* 90, 2006:L73–L75.

121. Chen, H., Ilan, B., Wu, Y., Zhu, F., Schulten, K., and Voth, G. A. Charge delocalization in proton channels, I: The aquaporin channels and proton blockage. *Biophys. J.* 92, 2007:46–60.

122. Jiang, Y., Lee, A., Chen, J., Cadene, M., Chait, B. T., and MacKinnon, R. Crystal structure and mechanism of a calcium-gated potassium channel. *Nature* 417, 2002:515–522.

123. Jiang, Y., Ruta, V., Chen, J., Lee, A., and MacKinnon, R. The principle of gating charge movement in a voltage-dependent K$^+$ channel. *Nature* 423, 2003:42–48.

124. Sukharev, S., Betanzos, M., Chiang, C.-S., and Guy, H. R. The gating mechanism of the large mechanosensitive channel MscL. *Nature* 409, 2001:720–724.

125. Herkovits, T. T., and Hattington, J. P. Solution studies of the nucleic acid bases and related model compounds. Solubility in aqueous alcohol and glycol solutions. *Biochemistry* 11, 1972:4800–4811.

126. Turner, D. H. In *Nucleic Acids: Structure, Properties and Functions.* Edited by V. A. Bloomfield, D. M. Crothers, and I. Tinoco. Sausalito, CA: University Science Books, 2000.

127. Hofstadler, S. A., and Griffey, R. H. Analysis of noncovalent complexes of DNA and RNA by mass spectrometry. *Chem. Rev.* 101, 2001:377–390.

128. Rueda, M., Kalko, S. G., Luque, F. J., and Orozco, M. The structure and dynamics of DNA in the gas phase. *J. Am. Chem. Soc.* 125, 2003:8007–8014.

129. Drew, H. R., Wing, R. M., Takano, T., Broka, C., Itakura, K., and Dickerson, R. E. Structure of a B-DNA dodecamer: Conformation and dynamics. *Proc. Natl Acad. Sci. USA* 78, 1981:2179–2183.

130. Kopka, M. L., Fratini, A. V., Drew, H. R., and Dickerson, R. E. Ordered water structure around a *B*-DNA dodecamer: A quantitative study. *J. Mol. Biol.* 163, 1983:129–146.

131. Shui, X., Sines, C. C., McFail-Isom, L., VanDerveer, D., and Williams, L. D. Structure of the potassium form of CGCGAATTCGCG: DNA deformation by electrostatic collapse around inorganic cations. *Biochemistry* 37, 1998:16877–16887.

132. Sorin, E. J., Rhee, Y. M., and Pande, V. S. Does water play a structural role in the folding of small nucleic acids? *Biophys. J.* 88, 2005:2516–2524.

133. Tjivikua, T., Ballester, P., and Rebek, Jr., J. Self-replicating system. *J. Am. Chem. Soc.* 112, 1990: 1249–1250.

134. Benner, S. A., Ricardo, A., and Carrigan, M. A. Is there a common chemical model for life in the universe? *Curr. Opin. Chem. Biol.* 8, 2004:672–689.

135. *Philos. Trans. R. Soc.* 359(1448), 2004: Special issue, "The molecular basis of life: is life possible without water?"

136. Henderson, L. *The Fitness of the Environment.* New York, NY: Macmillan, 1913.

137. Benner, S. Synthetic biology: Act natural. *Nature* 421, 2003:118.

138. Ferber, D. Microbes made to order. *Science* 303, 2004:158–161.
139. Benner, S. A. (ed.). *Redesigning the Molecules of Life.* Berlin: Springer-Verlag, 1988.
140. Piccirilli, J. A., Krauch, T., Moroney, S. E., and Benner, S. A. Enzymatic incorporation of a new base pair into DNA and RNA extends the genetic alphabet. *Nature* 343, 33–37.
141. Wang, L., Brock, A., Herberich, B., and Schultz, P. G. Expanding the genetic code of *Escherichia coli.* *Science* 292, 2001:498–500.
142. Chin, J. W., Cropp, T. A., Anderson, J. C., Zhang, Z., and Schultz, P. G. An expanded eukaryotic genetic code. *Science* 301, 2003:964–967.
143. Kool, E. T. Replacing the nucleobases in DNA with designer molecules. *Acc. Chem. Res.* 35, 2002: 936–943.
144. Liu, H., Gao, J., Saito, D., Maynard, L., and Kool, E. T. Toward a new genetic system with expanded dimensions: Size-expanded analogues of deoxyadenosine and thymidine. *J. Am. Chem. Soc.* 126, 2004: 1102–1109.
145. Ball, P. Water as an active constituent in cell biology. *Chem. Rev.* 108, 2008:74–108.

5 Water's Hydrogen Bond Strength

Martin F. Chaplin

CONTENTS

5.1 INTRODUCTION TO THE HYDROGEN BOND IN WATER

Latimer and Rodebush first described hydrogen bonding in 1920. It occurs when an atom of hydrogen is attracted by strong forces to two atoms instead of only one, as its single valence electron implies. The hydrogen atom thus acts to form a divalent bond between the two other atoms (Pauling, 1948). Such hydrogen bonds in liquid water are central to water's life-providing properties. This paper sets out to investigate the consequences if the hydrogen bond strength of water was to differ from its natural value. From this, an estimate is made as to how far the hydrogen bond strength of water may be varied from its naturally found value but still be supportive of life, in a similar manner to the apparent tuning of physical cosmological constants to the existence of the universe (Rees, 2003).

Hydrogen bonds arise in water where each partially positively charged hydrogen atom is covalently attached to a partially negatively charged oxygen atom from a water molecule with bond energy of about 492 kJ mol^{-1} and is also attracted, but much more weakly, to a neighboring partially negatively charged oxygen atom from another water molecule. This weaker bond is known as the hydrogen bond and is found to be strongest in hexagonal ice (ordinary ice) where each water molecule takes part in four tetrahedrally arranged hydrogen bonds, two of which involve each of its two hydrogen atoms and two of which involve the hydrogen atoms of neighboring water molecules. There is no standard definition for the hydrogen bond energy. In liquid water, the energy of attraction between water molecules (hydrogen bond enthalpy) is optimally about 23.3 kJ mol⁻¹ (Suresh and Naik, 2000) and almost five times the average thermal collision fluctuation at 25°C. This is the energy required for breaking and completely separating the bond, and equals about half the enthalpy of vaporization (44 kJ mol⁻¹ at 25°C), as an average of just under two hydrogen bonds per molecule are broken when water evaporates. It is this interpretation of water's hydrogen bond strength that is used in this chapter. Just breaking the hydrogen bond in liquid water, leaving the molecules essentially in the same position and still retaining their electrostatic attraction, requires only about 25% of this energy, recently estimated at 6.3 kJ mol⁻¹ (Smith et al., 2004). This may be considered as an indication of extra directional energy caused by polarization and covalency of the hydrogen bond. However, if the excess heat capacity of the liquid over that of steam is assumed attributable to the breaking of the bonds, the attractive energy of the hydrogen bonds is determined to be 9.80 kJ mol⁻¹ (Muller, 1988). This may be considered as an indication of the total extra energy caused by polarization, cooperativity, and covalency of the hydrogen bond. Two percent of collisions between liquid water molecules have energy greater than this.

The Gibbs free energy change (ΔG) presents the balance between the increases in bond strength ($-\Delta H$) and consequent entropy loss ($-\Delta S$) on hydrogen bond formation (i.e., $\Delta G = \Delta H - T\Delta S$) and may be used to describe the balance between formed and broken hydrogen bonds. Several estimates give the equivalent Gibbs free energy change (ΔG) for the formation of water's hydrogen bonds at about −2 kJ mol⁻¹ at 25°C (Silverstein et al., 2000), the difference in value from that of the bond's attractive energy being due to the loss in entropy (i.e., increased order) on forming the bonds. However, from the equilibrium concentration of hydrogen bonds in liquid water (~1.7 per molecule at 25°C), ΔG is calculated to be more favorable at −5.7 kJ mol⁻¹. Different estimations for ice's hydrogen bond energy, from a variety of physical parameters including Raman spectroscopy, self diffusion, and dielectric absorption, vary from 13 to 32 kJ mol⁻¹. It is thought to be about 3 kJ mol⁻¹ stronger than liquid water's hydrogen bonds, as evidenced by an approximately 4 pm longer, and hence weaker, O–H covalent bond (Pimentel and McClellan, 1960).

Although the hydrogen atoms are often shown along lines connecting the oxygen atoms, this is now thought to be indicative of time-averaged direction only and unlikely to be found to a significant extent, even in ice. Various studies give average parameters, as found at any instant, for liquid water at 4°C (Figure 5.1).

Bond lengths and angles will change, due to polarization shifts in different hydrogen-bonded environments and when the water molecules are bound to solutes and ions. The oxygen atoms typically possess about 0.7e negative charge and the hydrogen atoms about 0.35e positive charge giving rise to both an important electrostatic bonding but also the favored *trans* arrangement of the hydrogen atoms as shown in Figure 5.1. The atom charges effectively increase in response to polarization (Table 5.1). Hydrogen bond strength varies with the hydrogen bond angle (O–H····O, shown as 162° in Figure 5.1). If the hydrogen bond is close to straight (i.e., 180°), the hydrogen bond strength

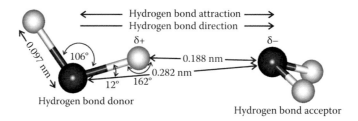

FIGURE 5.1 The average parameters for the hydrogen bonds in liquid water with nonlinearity, distances, and variances all increasing with temperature (Modig et al., 2003). There is considerable variation between different water molecules and between hydrogen bonds associated with the same water molecules. It should be noted that the two water molecules are not restricted to perpendicular planes and only a small proportion of hydrogen bonds are likely to have this averaged structure.

depends almost linearly on its length, with shorter length giving rise to stronger hydrogen bonding. As the hydrogen bond length of water increases with temperature increase but decreases with pressure increase, hydrogen bond strength also depends almost linearly, outside extreme values, on the temperature and pressure (Dougherty, 1998).

There is substantial cooperative strengthening of hydrogen bonds in water, which is dependent on long-range interactions (Heggie et al., 1996). Breaking one bond generally weakens those around, whereas making one bond generally strengthens those around. This encourages cluster formation where all water molecules are linked together by three or four strong hydrogen bonds. For the same average bond density, some regions within the water form larger clusters involving stronger hydrogen bonds while other regions consist mainly of weakly hydrogen-bonded water molecules. This variation is allowed with the water molecules at the same chemical potential (i.e., $\Delta G = 0$) as there is compensation between the bond's attractive energy (ΔH) and the energy required for creating the orderliness apparent in cluster formation (ΔS). Ordered clusters with enthalpically strong hydrogen bonding have low entropy whereas enthalpically weakly linked water molecules possess high entropy. The hydrogen-bonded cluster size in water at 0°C has been estimated to be about 400 (Luck, 1998). Weakly hydrogen-bonding surfaces and solutes restrict the hydrogen-bonding potential of adjacent water so that these make fewer and weaker hydrogen bonds. As hydrogen bonds strengthen each other in a cooperative manner, such weak bonding also persists over several layers and may cause locally changed solvation. Conversely, strong hydrogen bonding will be evident over several molecular diameters, persisting through chains of molecules. The weakening of hydrogen bonds, from about 23 kJ mol^{-1} (in liquid water at 0°C) to about 17 kJ mol^{-1} (in liquid water under pressure at 200°C), is observed when many bonds are broken in superheated liquid water, reducing the cooperativity (Khan, 2000). The breakage of these bonds is due not only to the more energetic conditions at high temperature, but also to a related reduction in the hydrogen bond donating ability by about 10% for each 100°C increase (Lu et al., 2001). The loss of these hydrogen bonds results in a small increase in the hydrogen bond accepting ability of water, due possibly to increased accessibility (Lu et al., 2001).

Liquid water contains by far the densest hydrogen bonding of any solvent, with almost as many hydrogen bonds as there are covalent bonds. These hydrogen bonds can rapidly rearrange in response to changing conditions and environments (e.g., the presence of solutes). Water molecules, in liquid water, are surrounded by about four randomly configured hydrogen bonds. They tend to clump together, forming clusters, for both statistical (Stanley and Teixeira, 1980) and energetic reasons. Hydrogen-bonded chains (i.e., O–H····O–H····O) are cooperative (Dannenberg, 2002), both in formation and rupture; the breakage of the first bond is the hardest and then the next one is weakened, and so on. This is particularly true for cyclic water clusters where ring closure is energetically favored and ring breakage energetically costly. Such cooperativity is a fundamental property of liquid water and it implies that acting as a hydrogen bond acceptor strengthens the hydrogen bond donating ability of water molecules. However, there is an anticooperative aspect, as acting as a donor weakens the capability to act as another donor (e.g., O····H–O–H····O) (Luck, 1998). It is clear, therefore, that a water molecule with two hydrogen bonds, where it acts as both donor and acceptor, is somewhat stabilized relative to one where it is either the donor or acceptor of two hydrogen bonds. This is the reason behind the first two hydrogen bonds (donor and acceptor) giving rise to the strongest hydrogen bonds (Peeters, 1995).

Every hydrogen bond formed increases the hydrogen bond status of two water molecules, and every hydrogen bond broken reduces the hydrogen bond status of two water molecules. The network is essentially complete at ambient temperatures (i.e., almost all molecules are linked by at least one unbroken hydrogen-bonded pathway). Hydrogen bond lifetimes are 1–20 ps, whereas broken bond lifetimes are about 0.1 ps (Keutsch and Saykally, 2001). Broken bonds generally re-form to give the same hydrogen bond; particularly if water's other three hydrogen bonds are in place. If not, breakage usually leads to rotation around one of the remaining hydrogen bonds (Bratos et al., 2004) and not to translation away, as the resultant free hydroxyl group and lone pair are both quite reactive. Also important is the possibility of the hydrogen bond breaking, as evidenced by physical techniques such as IR, Raman, or NMR and caused by loss of hydrogen bond "covalency" due to electron rearrangement, without any angular change in the O–H····O atomic positions but due to changes within the local environment. Thus, clusters may persist for much longer times (Higo et al., 2001) than data from these methods indicate, as evidenced by the high degree of hydrogen bond breakage seen in the IR spectrum of ice (Raichlin et al., 2004), where the clustering is taken as lasting essentially forever.

5.2 SUMMARY OF THE CONTRIBUTIONS TO WATER'S HYDROGEN BOND

The hydrogen bond is part (approximately 90%) electrostatic and part (approximately 10%) covalent (Isaacs et al., 2000), although there is still some controversy surrounding this partial covalency with for (e.g., Guo et al., 2002, favor mixing of bonding orbitals), against (e.g., Ghanty et al., 2000, favor charge transfer to antibonding orbitals), and neutral (Barbiellini and Shukla, 2002) support in the recent literature. If the water hydrogen bond is considered within the context of the complete range of molecular hydrogen bonding then it appears most probable that it is not solely electrostatic (Poater et al., 2003); indeed, the continuous transformation of ice VII to ice X would seem to indicate a continuity of electron sharing between water molecules. Also, although N–H····N and N–H····O hydrogen bonds are known to be weaker than the O–H····O hydrogen bonds in water, there is clear evidence for these bonds' covalent natures (Dingley and Grzesiek, 1998; Cordier and Grzesiek, 1999).

There is a trade-off between the donor O–H and hydrogen bond H····O strengths in a O–H····O hydrogen bond: the stronger the H····O attraction, the weaker the O–H covalent bond, and the shorter the O····O distance. The weakening of the O–H covalent bond gives rise to a good indicator of hydrogen-bonding energy; the fractional increase in the O–H covalent bond length determined by the increasing strength of the hydrogen bonding (Grabowski, 2001). Factors contributing to the hydrogen bonds are given in Table 5.1.

On forming the hydrogen bond, the donor hydrogen atom stretches away from its oxygen atom and the acceptor lone pair stretches away from its oxygen atom and toward the donor hydrogen atom (Kozmutza et al., 2003), both oxygen atoms being pulled toward each other. The hydrogen bond may be approximated by bonds made up of covalent HO–H····OH$_2$, ionic HO$^{\delta-}$–H$^{\delta+}$····O$^{\delta-}$-H$_2$ and

TABLE 5.1
Attractive and Repulsive Components in Water's Hydrogen Bonds

Component	Attraction/Repulsion
Electrostatic attraction; long-range interaction (<3 nm) based on point charges, or dipoles, quadrupoles, etc. They may be considered as varying inversely with distance.	++
Polarization; due to net attractive effects between charges and electron clouds, which may increase cooperatively dependent on the local environment (<0.8 nm). They may be considered as varying inversely with distance[4].	++
Covalency; highly directional and increases on hydrogen bonded cyclic cluster formation. It is very dependent on the spatial arrangement of the molecules within the local environment (<0.6 nm).	+
Dispersive attraction; interaction (<0.8 nm) due to coordinated effects of neighboring electron clouds. They may be considered as varying inversely with distance[8].	+
Repulsion; very short range interaction (<0.4 nm) due to electron cloud overlap. They may be considered as varying inversely with distance[12].	−

long-bonded covalent $HO^-\cdots H-O^+H_2$ parts with $HO-H\cdots OH_2$ being very much more in evidence than $HO^-\cdots H-O^+H_2$, where there is much extra nonbonded repulsion. Contributing to the strength of water's hydrogen bonding are nuclear quantum effects (zero point vibrational energy) that bias the length of the O–H covalent bond longer than its "equilibrium" position length, so also increasing the average dipole moment (Chen et al., 2003). Nuclear quantum effects are particularly important in the different properties of light (H_2O) and heavy water (D_2O) where the more restricted atomic vibrations in D_2O reduce the negative effect of its van der Waals repulsive core and increasing its overall hydrogen bond energy.

Generally, most of the hydrogen bond attraction is due to the electrostatic effects. These are increased by mutual polarization. The van der Waals effects are repulsive within the hydrogen bond as the nearest O$\cdots$O distances are about 0.04 nm shorter than the van der Waals core. The covalency is very important where there are local tetrahedral arrangements and particularly where these allow extensive intermolecular orbitals such as occurs in cyclic pentameric water clusters (Speedy, 1984; Chowdhury et al., 1983).

5.3 CONSEQUENCES OF WATER'S NATURAL HYDROGEN BOND STRENGTH

The hydrogen bonding in water, together with its tendency to form open tetrahedral networks at low temperatures, gives rise to its characteristic properties, which differ from those of other liquids. Such properties are often described as anomalous, although it could be argued that water possesses exactly those properties that one might deduce from its structure.

An important feature of the hydrogen bond is that it possesses direction. When the hydrogen bonding is strong, the water network expands to accommodate these directed bonds and where the hydrogen bonding is weak, water molecules collapse into the spaces around their neighbors. Such changes in water's clustering give rise to the so-called anomalies of water, particularly the different behaviors of hot, which has weaker hydrogen bonding, and cold (e.g., supercooled) water, which has stronger hydrogen bonding. The cohesion of water due to hydrogen bonding is responsible for water being a liquid over the range of temperatures on Earth where life has evolved and continues to thrive. However, it is the clustering of the water, due to the directed characteristics of the hydrogen bonding that is responsible for the very special properties of water that allow it to act in diverse ways under different conditions.

It has often been stated (Luck, 1985) that life depends on these anomalous properties of water. In particular, the large heat capacity, high thermal conductivity, and high water content in organisms contribute to thermal regulation and prevent local temperature fluctuations, thus allowing us to more easily control our body temperature. The high latent heat of evaporation gives resistance to dehydration and considerable evaporative cooling. Water is an excellent solvent due to its polarity, high dielectric constant, and small size, particularly for polar and ionic compounds and salts. It has unique hydration properties that determine the three-dimensional structures of proteins, nucleic acids, and other biomolecules and, thus, control their functions in solution. This hydration allows water to form gels that can reversibly undergo the gel-sol phase transitions that underlie many cellular mechanisms (Pollack, 2001). Water ionizes and allows easy proton exchange between molecules, contributing to the richness of the ionic interactions in biology.

At 4°C, pure liquid water expands on heating or cooling. This density maximum, together with the low ice density, results in (1) the necessity that all of a body of fresh water, not just its surface, is close to 4°C before any freezing can occur; (2) the freezing of rivers, lakes, and oceans is from the top down, so permitting survival of the bottom ecology, insulating the water from further freezing, reflecting back sunlight into space, and allowing rapid thawing; and (3) density-driven thermal convection causing seasonal mixing in deeper temperate waters carrying life-providing oxygen into the depths.

The large heat capacity of the oceans and seas allows them to act as heat reservoirs such that sea temperatures vary only a third as much as land temperatures and so moderate our climate (e.g., the Gulf Stream carries tropical warmth to northwestern Europe, moderating its winters). Water's high surface tension plus its expansion on freezing encourages the erosion of rocks to give soil for agriculture. No other material is commonly found as solid, liquid, and gas.

Notable among the anomalies of water are the differences in the properties of hot and cold water, with the anomalous behavior more accentuated at low temperatures, where the properties of supercooled water often widely diverge from those of frozen ice. As very cold liquid water is heated it shrinks, it becomes less easy to compress, gasses become less soluble, it is easier to heat, and it conducts heat better. In contrast, as hot liquid water expands as it is heated, it becomes easier to compress, gasses become more soluble, it is harder to heat, and it is a poorer conductor of heat. With increasing pressure, cold-water molecules move faster but hot-water molecules move slower.

5.4 CONSEQUENCES OF CHANGES IN WATER'S HYDROGEN BOND STRENGTH

Central to how close the properties of water are to those required for life is the question of the strength of its hydrogen bond. How much variation in water's hydrogen bond is acceptable for life to exist? A superficial examination gives the range of qualitative effects as indicated in Table 5.2.

Intriguingly, liquid water acts in subtly different ways as circumstances change, responding to variations in the physical and molecular environments and occasionally acting as though it were present as more than one liquid phase. Sometimes water acts as a free flowing molecular liquid while at other times, in other places, or under subtly different conditions, it acts more like a weak gel. Shifts in the hydrogen bond strength may fix water's properties at one of these extremes to the detriment of processes requiring the opposite character. Evolution has utilized the present natural responsiveness and variety in the liquid water properties such that it is now required for life as we know it. DNA would not form helices able to both zip and unzip without the present hydrogen bond strength. Enzymes would not possess their 3-D structure without it, nor would they retain their controlled flexibility required for their biological action. Compartmentalization of life's processes by the use of membranes with subtle permeabilities would not be possible without water's intermediate hydrogen bond strength.

In liquid water, the balance between the directional component of hydrogen bonding and the isotropic van der Waals attractions is finely poised. Increased strength of the hydrogen bond directionality gives rise to ordered clustering with consequential effects on physical parameters tending toward a glass-like state, whereas reducing its strength reduces the size and cohesiveness of the clusters with the properties of water, then tending toward those of its isoelectronic neighbors methane and neon, where only van der Waals attractions remain. Quite small percentage changes in the strength of the aqueous hydrogen bond may give rise to large percentage changes in such physical properties as melting point, boiling point, density, and viscosity (see Table 5.3 and Figure 5.2). Some of these potential changes may not significantly impinge on life's processes (e.g., compressibility or the speed of sound), but others are of paramount importance.

TABLE 5.2
Effect of Variation of Hydrogen Bond Strength

Water Hydrogen Bond Strength	Main Consequence
No hydrogen-bonding at all	No life
Hydrogen bonds slightly weaker	Life at lower temperatures
No change	Life as we know it
Hydrogen bonds slightly stronger	Life at higher temperatures
Hydrogen bonds very strong	No life

TABLE 5.3
Potential Changes in the Properties of Liquid Water Relevant to Life Processes

Property	Change on H-bond Strengthening	Change on H-bond Weakening
Melting point	Increase	Decrease
Boiling point	Increase	Decrease
State, at ambient conditions on Earth	→ Solid glass	→ Gas
Adhesion	Decrease	Decrease
Cohesion	Increase	Decrease
Compressibility	Increase	Decrease
Density	Decrease	Increase
Dielectric constant	Increase	Decrease
Diffusion coefficient	Decrease	Increase
Enthalpy of vaporization	Increase	Decrease
Glass transition	Increase	Decrease
Ionization	Decrease	Increase → Decrease
Solubility, hydrophile	Decrease	Decrease
Solubility, small hydrophobe	Increase	Decrease → Increase
Specific heat	Increase	Decrease
Surface tension	Increase	Decrease
Thermal conductivity	Decrease	Increase → Decrease
Viscosity	Increase	Decrease

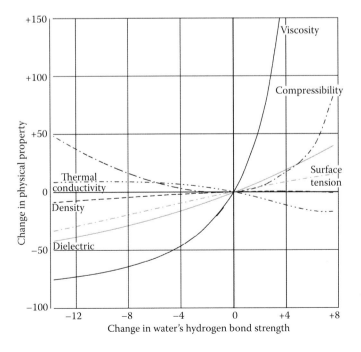

FIGURE 5.2 Variation of water's physical properties with changes in its hydrogen bond strength.

Although in most cases, opposite changes in hydrogen bond strength cause contrary effects on the physical properties, this is not always the case if the hydrogen bond strength tends toward high or low extremes. Adhesion and hydrophilic solubility both decrease on hydrogen bond strengthening due to increased water–water interactions reducing water's ability to bind to the hydrophilic surface or molecule. On hydrogen bond weakening, they both decrease due to the reduced water–surface or water–solute interactions. Strong hydrogen bonding eases the formation of expanded cavities as evidenced in the clathrate ices, and which can accommodate small hydrophobic molecules, increasing their solubility. However, such small hydrophobic molecules will also be more easily dissolved when weak hydrogen bonding allows more facile cavity formation to allow their entry.

5.5 METHODS FOR ESTIMATING THE EFFECT OF CHANGE IN WATER HYDROGEN BOND STRENGTH

Clearly, estimates of the physical consequences due to variations in the hydrogen bond strength may vary from one method to another, but the data in Figure 5.2 indicates that relatively small changes in hydrogen bond strength may have some relatively large effects. Strengthening hydrogen bonding has particularly important effects on viscosity and diffusion as indicated by the large changes occurring in supercooled normal water.

It is possible to investigate the effect that changes in hydrogen bonding strength of water make in its properties by examination of the actual properties of water or other molecules with different hydrogen bond strength. However, different methods, materials, or conditions have weaknesses in their utility. The possibilities for examining the effects of varying hydrogen bond strengths are:

- Changing the physical environment of water such as temperature or pressure and examining the consequential changes in the physical parameters, if changes are assumed to be solely due to the variation in the hydrogen bond strength. However, varying the temperature also changes the heat energy content and some compensation may be required to negate effects other than hydrogen bond strength changes, such as density effects. Also, changing the pressure increases density and reduces hydrogen bond lengths, which increases hydrogen bond strength, but also bends the bonds so reducing their tetrahedrality. A simple way of assessing the average hydrogen bond strength is the enthalpy of evaporation calculated from the difference in the enthalpy of the liquid and gaseous phases (Verma, 2003).
- Modeling water as an equilibrium mixture of low-density and high-density clusters (Vedamuthu et al., 1994) and examining the consequences of hydrogen bond strength variation on the cluster equilibrium with resultant effects on the physical properties. This concept has been shown to explain qualitatively and quantitatively most anomalies of liquid water. The free energy change for the equilibrium between dense and less dense clusters is very small due to compensation between enthalpic and entropic effects. Just a small shift in the enthalpic component, due to changes in hydrogen bond energy, may shift the equilibrium position decisively one way or the other.
- Examine the physical properties of the isoptomers of water, HDO or D_2O. These have apparently stronger hydrogen bonds than H_2O due to their reduced van der Waals core, consequent upon nuclear quantum effects. The hydrogen bond strength differences found using this method are small.
- Examine the physical properties of the hydrides of neighboring elements, NH_3, HF, or H_2S, which possess differing hydrogen bond strengths. The hydrogen bond strength differences encountered by this method are rather large.

In the following discussion, the effects of varying hydrogen bond strength on individual physical properties, and the consequences for life, are initially independently discussed without regard to other changes that might also be occurring at the same time, such as changes in the physical state of water.

5.6 EFFECT OF WATER HYDROGEN BOND STRENGTH ON MELTING AND BOILING POINT

In ice, all water molecules participate in four hydrogen bonds (two as donor and two as acceptor) and are held relatively static. In liquid water, some of the weaker hydrogen bonds must be broken to allow the molecules to move around. The large amount of energy required for breaking these bonds must be supplied during the melting process and only a relatively minor amount of energy is reclaimed from the change in volume. The free energy change ($\Delta G = \Delta H - T\Delta S$) must be zero at phase changes such as the melting or boiling points. As the temperature of liquid water decreases, the amount of hydrogen bonding increases and its entropy decreases. Melting will only occur when there is sufficient entropy change to compensate for the energy required for the bond breaking. The low entropy (high organization) of liquid water causes this melting point to be high. If the hydrogen bond strength (i.e., enthalpy change) in water is raised, then the melting point must rise for the free energy change to stay zero.

At the temperature of the phase change, this free energy is zero, so on melting (solid → liquid) $\Delta H_m = T_m\Delta S_m$, and on vaporization (liquid → gas) $\Delta H_v = T_v\Delta S_v$. To calculate the hydrogen bond strength, it is assumed that the entropy changes, during the phase changes, remain constant with respect to the temperature range. The enthalpy change required to equal the temperature times this entropy change is regarded as the hydrogen bond strength required at the melting point. Thus, the percentage increase in the hydrogen bond strength is given by $100 \times (T\Delta H_m/T_m - \Delta H_T)/\Delta H_T$, where ΔH_T is the bond enthalpy at temperature T and ΔH_m is the bond enthalpy at its normal melting point T_m. Figure 5.3 shows how the bond strength increases affect the melting point.

There is considerable hydrogen bonding in liquid water resulting in high cohesion that prevents water molecules from being easily released from the water's surface. Consequentially, the vapor pressure is reduced and water has a high boiling point. Using a similar argument to that used above for melting point, the percentage reductions in the hydrogen bond strength that result in lower boiling points are given by $100 \times (T\Delta H_v/T_v - \Delta H_T)/\Delta H_T$, where ΔH_v is the bond enthalpy at its normal boiling point T_v under a pressure of 1 atm. Figure 5.3 shows how bond strength decreases affect the boiling point.

The resulting relationship (Figure 5.3) shows that water would freeze at the average surface temperature of Earth (15°C) with a 7% strengthening in water's hydrogen bond or it would boil on a 29% weakening. At our body temperature (37°C) the strengthening required for freezing is 18%

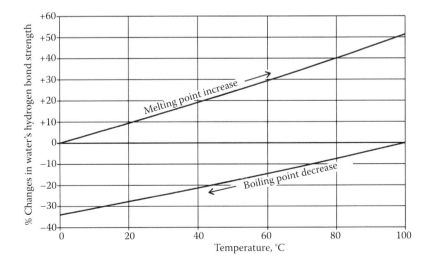

FIGURE 5.3 The effect that changes in water's hydrogen bond strength may have on water's boiling and melting points.

and the weakening required to turn water into steam is 22%. The melting and boiling points of other liquids shows that these values are reasonable. D_2O has a melting point almost 4°C higher than H_2O with a bond strength 2% higher, the values of which fit on the melting point line in Figure 5.3. Hydrogen sulfide, which does form hydrogen bonds with strong bases but is a poor proton donor, has a boiling point of −60°C with intermolecular interactions only 20% of that of water (Govender et al., 2003). Hydrogen fluoride and hydrogen cyanide both possess hydrogen bond interactions slightly greater than 50% of that of water and boil at 20°C and 26°C, respectively.

5.7 EFFECT OF WATER HYDROGEN BOND STRENGTH ON THE TEMPERATURE OF MAXIMUM DENSITY

The high density of liquid water is due mainly to the cohesive nature of the hydrogen-bonded network. This reduces the free volume and ensures a relatively high-density, compensating for the partial open nature of the hydrogen-bonded network. It is usual for liquids to expand when heated, at all temperatures. However, at 4°C water expands on heating or cooling. The density maximum is brought about by the opposing effects on increasing temperature, causing (1) structural collapse of the tetrahedral clustering evident at lower temperatures, increasing density, and (2) thermal expansion, creating extra space between unbound molecules, reducing density.

As expanses of water are cooled, stratification of water occurs that depends on density. In freshwater lakes, the densest water is that at about 4°C. This water sinks to the bottom, circulating its contained oxygen and nutrients. Further cooling causes the surface temperature to drop toward 0°C but has no immediate effect on deep water temperatures, which remain at 4°C. When the surface water reaches 0°C, it may rapidly freeze as only molecules at the surface have to be cooled further. The ice forms an insulating layer over the liquid water underneath and so slows down any further surface cooling. The water at the bottom of ice-covered lakes remains at 4°C throughout the winter, thus preserving animal and plant life there. In spring, the warming rays of the sun melt the surface ice layer first. Seawater behaves differently as the salt content lowers the temperature of maximum density below its freezing point and the maximum density is no longer observed. As seawater density increases with pressure, due to depth, convection only involves about the top hundred meters. A major part of this must be cooled to the freezing point (−2°C) before saltwater surface ice may form.

There would be clear consequences for aquatic life if the temperature of maximum density was not observed in freshwater lakes and rivers. Cooling would result in most of the water being at 0°C before ice formation is initiated. Such changes in hydrogen bond strength would not significantly affect the low density of ice, which would still float on water. However, subsequent ice formation may give rise to slushy ice formation without a well-formed insulating upper surface layer of ice. More ice would form, however, due to the lack of the insulation and this ice would take far longer to thaw as, additionally, more water would have to warm first. Much larger volumes of the fresh water would thus be affected and the greater ice formation may more easily reach the bottom of shallow lakes. The resultant situation would have both positive and negative consequences for the aquatic life as any remaining liquid surface would allow favorable surface gas exchange but there would be less liquid water. The end result for life would therefore be important but not overwhelmingly life-threatening, except for shallow lakes, due to the loss of this density maximum.

The weakening of the hydrogen bond strength required to remove the maximum density property may be estimated in a number of ways. A 2% decrease in the hydrogen bond energy reduces the maximum density by the 4°C required (Figure 5.3). The decrease calculated from the cluster equilibrium of Wilse Robinson (Vedamuthu et al., 1994; Urquidi et al., 1999; Cho et al., 2002), where the free energy change between their proposed water clusters is zero close to 0°C, also agrees with this value. D_2O has a raised temperature of maximum density (11.185°C) due to its stronger hydrogen bonds, but if this bond strengthening is used as an estimate of that required to lower the temperature of maximum density of H_2O below 0°C, this also requires a hydrogen bond energy weakening close to 2%.

5.8 EFFECT OF WATER HYDROGEN BOND STRENGTH ON KOSMOTROPES AND CHAOTROPES

Ions cause considerable changes to the structuring of water. The difference in their effects depends on the relative strength of ion–water and water–water interactions. Ionic chaotropes are large singly charged ions, with low charge density [e.g., SCN^-, $H_2PO_4^-$, HSO_4^-, HCO_3^-, I^-, Cl^-, NO_3^-, NH_4^+, Cs^+, K^+, $(NH_2)_3C^+$ (guanidinium), and $(CH_3)_4N^+$ (tetramethylammonium) ions] that exhibit weaker interactions with water than water with itself and thus interfere little in the hydrogen bonding of the surrounding water. Small or multiply charged ions, with high charge density, are ionic kosmotropes (e.g., SO_4^{2-}, HPO_4^{2-}, Mg^{2+}, Ca^{2+}, Li^+, Na^+, H^+, OH^-, and HPO_4^{2-}). Ionic kosmotropes exhibit stronger interactions with water molecules than water with itself and therefore are capable of breaking water–water hydrogen bonds. If the water–water hydrogen bond energy were to increase, the kosmotropic ions would become chaotropic and if the water–water hydrogen bond energy were to decrease, chaotropic ions would become kosmotropic. At present, the biologically important ions Na^+ and K^+ lie on opposite sides of the chaotropic/kosmotropic divide, facilitating many cellular functions by virtue of their differences. If they both had similar aqueous characteristics, cellular membrane function would have had to evolve differently and it is difficult to suppose how this might occur with the present natural availability of the ions.

The different characteristics of the intracellular and extracellular environments manifest themselves particularly in terms of restricted diffusion and a high concentration of chaotropic inorganic ions and kosmotropic other solutes within the cells, both of which encourage intracellular low density water structuring. The difference in concentration of the ions is particularly apparent between Na^+ (intracellular 10 mM, extracellular 150 mM) and K^+ (intracellular 159 mM, extracellular 4 mM); Na^+ ions create more broken hydrogen bonding beyond their inner hydration shell and prefer a high aqueous density, whereas K^+ ions prefer a lower density aqueous environment. The interactions between water and Na^+ are stronger than those between water molecules, which, in turn, are stronger than those between water and K^+ ions.

The hydration enthalpies for Na^+ and K^+ are known to be −413 and −331 kJ mol^{-1} (Hribar et al., 2004), straddling the kosmotrope/chaotrope divide. Using the mildly chaotropic Cl^- ion, with hydration enthalpy −363 kJ mol^{-1}, as a marker, the division point between these ion types may be estimated as close to halfway between the K^+ and Na^+ hydration enthalpies (−372 kJ mol^{-1}). The changes in the water hydrogen bond energy required to convert the chaotrope K^+ to a kosmotrope is thus estimated as $331/372 = 11\%$ weakening and for converting the kosmotrope Na^+ to a chaotrope is $413/372 = 11\%$ strengthening.

The consequences of changes to the properties of Na^+ and K^+ ions in aqueous environments are difficult to quantify but are clearly far reaching. There is no cation that could easily replace K^+ inside cells as the more chaotropic alkali metal cations Rb^+ and Cs^+ are rare and NH_4^+ is toxic and little different from K^+ as a chaotrope. Although other ions could replace Na^+ as a cationic kosmotrope, Li^+ is rare and divalent ions (e.g., Mg^{2+}) may cause other effects, such as chelation. Life as we know it could not exist without the present balance between Na^+ and K^+ ions. The weakening of hydrogen bond strength shifting K^+ to become chaotropic would either cause K^+ ions to remain outside cells with consequences on the cell membrane potential or would cause intracellular water to be too disorganized to support present intracellular processes.

5.9 EFFECT OF WATER HYDROGEN BOND STRENGTH ON ITS IONIZATION

No amount of liquid water contains only H_2O molecules due to self-ionization producing hydroxide and hydrogen ions.

$$2H_2O \rightleftharpoons H_3O^+ + OH^- \quad K_W = [H_3O^+] \times [OH^-]$$

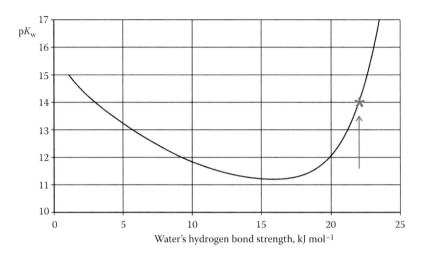

FIGURE 5.4   Variation of the pK_w (=$-\log_{10}(K_w)$) with the hydrogen bond strength of water. The data are cal-culated from the variation with temperature of the enthalpy of vaporization and pK_w (International Association for the Properties of Steam, 1980). Water at 25°C has a pK_w of about 14, as indicated on the right-hand side of the graph.

This ionization of water is followed by the utilization of further water molecules to ease the move-ment of the ions throughout the liquid. Such functions are key to biological processes and do not arise, to a significant extent, in any nonaqueous liquid except hydrogen fluoride. Aqueous ionization depends on both H$_3$O$^+$ and OH$^-$ formation and their physical separation to prevent the rapid reverse reaction reforming H$_2$O. H$_3$O$^+$ and OH$^-$ formation is greater when the hydrogen bonds are strongest, whereas ionic separation requires the hydrogen bond networks connecting the ions to be weak in order to prevent the ions reforming water. Thus, both strong and weak hydrogen bonding lead to lesser ionization (Figure 5.4).

Changes in hydrogen bond strength between water molecules alter its degree of ionization (Figure 5.4). Strengthened hydrogen bonding increases the (Grotthuss) rate of transfer of these ions in elec-trical fields but slows down their diffusion otherwise. Acid strength of biomolecular groups is deter-mined by the competition between the biomolecules and water molecules for the hydrogen ions. The strength by which the water molecules hold on to the hydrogen ion depends on their hydrogen bonding strength as a distributor of the charge. Biomolecular ionization, therefore, also depends on hydrogen bond strength. Since all biological processes have a dependence on charge, a completely new evolutionary perspective is required if water ionization is suppressed by water hydrogen bond strengthening. At intermediate hydrogen bond strength, ionization increases, reducing the pH of neutral solutions. The acidity (pK_a) of biomolecular groups, such as phosphate, also shows complex behavior with decreasing water hydrogen bond strength and often produces a pK_a minimum. Here, there are opposite effects of (1) reduced dielectric, at lower hydrogen bond strength, reducing ionic separation so tending to increase the pK_a and (2) increased water reactivity, also at lower hydrogen bond strength, increasing hydration effects and enabling the ionization, tending to reduce the pK_a.

5.10   EFFECT OF WATER HYDROGEN BOND STRENGTH ON BIOMOLECULE HYDRATION

Water is critical, not only for the correct folding of proteins but also for the maintenance of this structure. The free energy change on folding or unfolding is due to the combined effects of both protein folding/unfolding and hydration changes. Contributing enthalpy and entropy terms may, however, individually be greater than the equivalent of twenty hydrogen bonds but such changes compensate each other, leaving a free energy of stability for a typical protein as just equivalent to one or two hydrogen bonds. There are both enthalpic and entropic contributions to this free energy that change with temperature and so give rise to the range of stability for proteins between their hot and cold denaturation temperatures.

The free energy, on going from the native (N) state to the denatured (D) state, is given by $\Delta G_N^D = \Delta H_N^D - T\Delta S_N^D$. The overall free energy change (ΔG_N^D) depends on the combined effects of the exposure of the interior polar and nonpolar groups and their interaction with water, together with the consequential changes in the water–water interactions on ΔG_N^D, ΔH_N^D, and ΔS_N^D (Figure 5.5). Denaturation is only allowed when ΔG_N^D is negative; the rate of denaturation is then dependent on the circumstances and may be fast or immeasurably slow.

The enthalpy of transfer of polar groups from the protein interior into water is positive at low temperatures and negative at higher temperatures (Makhtadze and Privalov, 1993). This is due to the polar groups creating their own ordered water, which generates a negative enthalpy change due to the increased molecular interactions. Balanced against this is the positive enthalpy change as the preexisting water structure, and the polar interactions within the protein both have to be broken. As water naturally has more structure at lower temperatures, the breakdown of the water structure makes a greater positive contribution to the overall enthalpy at lower temperatures. Weakening of water's hydrogen bonds reduces the enthalpy of transfer of polar groups at all temperatures, as less energy is required to break down water's structure.

In contrast, the enthalpy of transfer of nonpolar groups from the protein interior into water is negative below about 25°C and positive above (Makhtadze and Privalov, 1993). At lower temperatures, nonpolar groups enhance preexisting order such as the clathrate-related structures (Schrade et al., 2001), generating stabilization energy but this effect is lost with increasing temperature, as any preexisting order is also lost. At higher temperatures, the creation of these clathrate structures requires an enthalpic input. Thus, there is an overall positive enthalpy of unfolding at higher temperatures. An equivalent but alternative way of describing this process is that at lower temperatures the clathrate-type structure optimizes multiple van der Waals molecular interactions whereas at higher temperatures such favorable structuring is no longer available. The extent of these enthalpy changes with temperature is reduced if water's hydrogen bonds are weakened, as the enthalpy change is raised at low temperatures and decreased at higher temperatures.

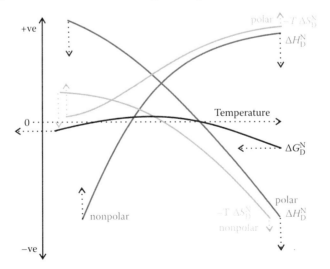

FIGURE 5.5 Variation in ΔG_N^D (———), ΔH_N^D (———), and $T\Delta S_N^D$ (———) with temperature due to protein's polar and nonpolar groups moving from a native compact structure to a denatured extended structure. The lines are meant to be indicative only. The length and direction of the arrows indicate the changes consequent upon weakening of water's hydrogen bond strength.

At ambient temperature, the entropies of hydration of both nonpolar and polar groups are negative (Privalov and Makhtadze, 1993), indicating that both create order in the aqueous environment. However, these entropies differ with respect to how they change with increasing temperature. The entropy of hydration of nonpolar groups increases through zero with increasing temperature, indicating that they are less able to order the water at higher temperatures and may, indeed, contribute to its disorder by interfering with the extent of the hydrogen-bonded network. Also, there is an entropy gain from the greater freedom of the nonpolar groups when the protein is unfolded. In contrast, the entropy of hydration of polar groups decreases, becoming more negative with increasing temperature, as they are able to create ordered hydration shells even from the more disordered water that exists at higher temperatures. Weakening of water's hydrogen bonds raises the entropic cost due to polar group hydration, as there is less natural order in the water to be lost.

Overall, protein stability depends on the balance between these enthalpic and entropic changes. For globular proteins, the free energy of unfolding is commonly found to be positive between about 0°C and 45°C. It decreases through zero when the temperature becomes either hotter or colder, with the thermodynamic consequences of both cold and heat denaturation. The hydration of the internal nonpolar groups is mainly responsible for cold denaturation as their energy of hydration (i.e., $-\Delta H_N^D$) is greatest when cold. Thus, it is the increased natural structuring of water at lower temperatures that causes cold destabilization of proteins in solution. Heat denaturation is primarily due to the increased entropic effects of the nonpolar residues in the unfolded state. The temperature range for the correct folding of proteins (ΔG_N^D in Figure 5.5) shifts toward lower temperatures if water's hydrogen bonds are weaker and higher temperatures if they are stronger. Typically, if the strength of the hydrogen bond increased equivalent to the difference in strength between 0°C and 100°C (i.e., raised cold denaturation) or decreased equivalent to the difference in strength between 45°C and 0°C (i.e., lowered heat denaturation) then present proteins would not be stable in aqueous solution. The shifts required may be calculated from the enthalpy and entropy of water to be a 51% increase or an 18% weakening in water's hydrogen bond strength.

As the degree of interaction between water molecules and biological molecules and structures depends on a competition for the water's hydrogen bonding between the molecules and water itself, such processes would change on varying the water–water hydrogen bond strength. Increasing strength causes water to primarily bond with itself and not be available for the hydrating structuring of proteins or DNA, or for dissolving ions. On the other hand, if the water–water hydrogen bond strength reduces, then the information exchange mechanisms operating within the cell, such as hydrogen-bonded water chains within and between proteins and DNA, will become nonoperational. Evolutionary pressures might be expected to compensate for only some of these effects.

5.11 EFFECT OF WATER HYDROGEN BOND STRENGTH ON ITS OTHER PHYSICAL PROPERTIES

Changes in water's hydrogen bond strength are expected to affect many of water's physical properties (Figure 5.2, Table 5.2). Some of these alterations only make insignificant changes to whether water can act as the medium for life. Pressure dependent properties such as compressibility have unimportant consequences as we live under relatively constant pressure. Some physical properties such as the speed of sound or refractive index impinge little on life's processes. Other physical properties change relatively little, such as surface tension, but even such small changes may affect some processes. Without strong hydrogen bonding, there would not be the cohesion necessary for trees to manage to transport water to their tops.

Viscosity is particularly affected on strengthening of water's hydrogen bonds, increasing tenfold from the value at 37°C for an increase in hydrogen bond strength of only 8%. An alternative calculation using the Wilse Robinson equilibrium model (Cho et al., 1999) gives the higher value of 30% hydrogen bond strengthening required to shift the equilibrium temperature sufficiently to achieve

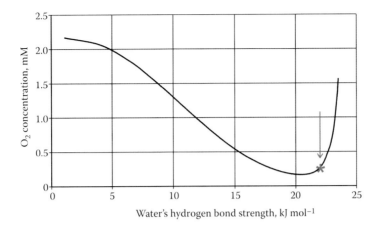

FIGURE 5.6 The dependence of the solubility of oxygen (at atmospheric pressure and composition) on water's hydrogen bond strength. The data for oxygen solubility are taken from Tromans (1998), with the hydrogen bond strength at 25°C indicated.

this viscosity alteration. However, comparing the data from D_2O shows a 23% increase in viscosity at 25°C, or 34% at 0°C, for only a 2% increase in hydrogen bond strength, showing the major effect of hydrogen bonding on the viscosity. As diffusivity varies inversely with viscosity, molecular movements slow down as viscosity increases. This would be expected to have consequences for the speed with which life processes could proceed.

Although D_2O only has 2–3% stronger hydrogen bonds than H_2O, as calculated from their enthalpy of evaporation, it has crucial effects on mitosis and membrane function. In most organisms it is toxic, causing death at high concentrations. It may be assumed however that life generally could adapt to its use, as found for some microorganisms.

TABLE 5.4
Estimates of Effects Consequent on Varying Water's Hydrogen Bond Strength

% Change in Hydrogen Bond Strength	Effect at 37°C
Decrease 29%	Water boils
Decrease 18%	Most proteins heat denature
Decrease 11%	K^+ becomes kosmotropic
Decrease 7%	pK_w up 3
Decrease 5%	CO_2 70% less soluble
Decrease 5%	O_2 27% less soluble
Decrease 2%	No density maximum
No change	No effect
Increase 2%	Significant metabolic effects
Increase 3%	Viscosity increase 23%
Increase 3%	Diffusivity reduced by 19%
Increase 5%	O_2 270% more soluble
Increase 5%	CO_2 440% more soluble
Increase 7%	pK_w down 1.7
Increase 11%	Na^+ becomes chaotropic
Increase 18%	Water freezes
Increase 51%	Most proteins cold denature

Note: The effects are considered individually without consideration of the effects on other physical parameters.

The solubility of gaseous oxygen and carbon dioxide are important features of life's processes. In particular, the solubilities increase steeply as the hydrogen bond strength increases from its natural value (Figure 5.6). Carbon dioxide solubility shows greater sensitivity due to the complex equilibria involved. However, its main behavior is an even steeper rise in solubility at high hydrogen bond strengths than that for oxygen; showing a fourfold increase (Duan and Sun, 2003) for a 5% hydrogen bond strengthening at 37°C. The full consequences of these changes are complex and difficult to assess. Oxygen concentrations cannot be lowered below the threshold necessary for complex circulatory life, ~0.1 mM (Catling et al., 2005). With higher oxygen solubility, circulatory animals would be capable of being larger, but more efficient anti-oxidant detoxification pathways would be necessary. Nevertheless, it is likely that life could adapt to these changes.

5.12   CONCLUSIONS

The major effects of changes to water's hydrogen bond strength are summarized in Table 5.4. It is apparent that small changes of a few percent would not be threatening to life in general, but changes in excess of 10% (equivalent to just 2 kJ mol⁻¹) may cause a significant threat. The overall conclusion to be drawn from this investigation is that water's hydrogen bond strength is poised centrally within a narrow window of its suitability for life.

REFERENCES

Barbiellini, B., and A. Shukla. 2002. *Ab initio* calculations of the hydrogen bond. *Phys. Rev. B* 66:235101.

Bratos, S., J.-Cl. Leicknam, S. Pommeret, and G. Gallot. 2004. Laser spectroscopic visualization of hydrogen bond motions in liquid water. *J. Mol. Struct.* 708:197–203.

Catling, D. C., C. R. Glein, K. J. Zahnle, and C. P. McKay. 2005. Why O_2 is required by complex life on habitable planets and the concept of planetary "oxygenation time." *Astrobiology* 5:415–438.

Chen, B., I. Ivanov, M. L. Klein, and M. Parrinello. 2003. Hydrogen bonding in water. *Phys. Rev. Lett.* 91:215503.

Cho, C. H., J. Urquidi, and G. W. Robinson. 1999. Molecular-level description of temperature and pressure effects on the viscosity of water. *J. Chem. Phys.* 111:10171–10176.

Cho, C. H., J. Urquidi, S. Singh, S. C. Park, and G. W. Robinson. 2002. Pressure effect on the density of water. *J. Phys. Chem. A* 106:7557–7561.

Chowdhury, M. R., J. C. Dore, and D. G. Montague. 1983. Neutron diffraction studies and CRN model of amorphous ice. *J. Phys. Chem.* 87:4037–4039.

Cordier, F., and S. Grzesiek. 1999. Direct observation of hydrogen bonds in proteins by interresidue $^{3h}J_{NC}$ scalar couplings. *J. Am. Chem. Soc.* 121:1601–1602.

Dannenberg, J. J. 2002. Cooperativity in hydrogen bonded aggregates. Models for crystals and peptides. *J. Mol. Struct.* 615:219–226.

Dingley, A. J., and S. Grzesiek. 1998. Direct observation of hydrogen bonds in nucleic acid base pairs by internucleotide $^2J_{NN}$ couplings. *J. Am. Chem. Soc.* 120:8293–8297.

Dougherty, R. C. 1998. Temperature and pressure dependence of hydrogen bond strength: a perturbation molecular orbital approach. *J. Chem. Phys.* 109:7372–7378.

Duan, Z., and R. Sun. 2003. An improved model calculating CO_2 solubility in pure water and aqueous NaCl solutions from 273 to 533 K and from 0 to 2000 bar. *Chem. Geol.* 193:257–271.

Ghanty, T. K., V. N. Staroverov, P. R. Koren, and E. R. Davidson. 2000. Is the hydrogen bond in water dimer and ice covalent? *J. Am. Chem. Soc.* 122:1210–1214.

Govender, M. G., S. M. Rootman, and T. A. Ford. 2003. An *ab initio* study of the properties of some hydride dimers. *Cryst. Eng.* 6:263–286.

Grabowski, S. J. 2001. A new measure of hydrogen bonding strength—*ab initio* and atoms in molecules studies. *Chem. Phys. Lett.* 338:361–366.

Guo, J.-H., Y. Luo, A. Augustsson, J.-E. Rubensson, C. Såthe, C. Ågren, H. Siegbahn, and J. Nordgren. 2002. X-ray emission spectroscopy of hydrogen bonding and electronic structure of liquid water. *Phys. Rev. Lett.* 89:137402.

Heggie, M. I., C. D. Latham, S. C. P. Maynard, and R. Jones, 1996. Cooperative polarisation in ice Ih and the unusual strength of the hydrogen bond. *Chem. Phys. Lett.* 249:485–490.

Higo, J., M. Sasai, H. Shirai, H. Nakamura, and T. Kugimiya. 2001. Large vortex-like structures of dipole field in computer models of liquid water and dipole-bridge between biomolecules. *Proc. Natl. Acad. Sci. USA.* 98:5961–5964.

Hribar, B., N. T. Southall, V. Vlachy, and K. A. Dill. 2004. How ions affect the structure of water. *J. Am. Chem. Soc.* 124:12302–12311.

International Association for the Properties of Steam. 1980. *Release on the ion product of water substance.* Washington, D.C.: National Bureau of Standards.

Isaacs, E. D., A. Shukla, P. M. Platzman, D. R. Hamann, B. Barbiellini, and C. A. Tulk. 2000. Compton scattering evidence for covalency of the hydrogen bond in ice. *J. Phys. Chem. Solids* 61:403–406.

Keutsch, F. N., and R. J. Saykally. 2001. Water clusters: untangling the mysteries of the liquid, one molecule at a time. *Proc. Natl. Acad. Sci. USA.* 98:10533–10540.

Khan, A. 2000. A liquid water model: density variation from supercooled to superheated states, prediction pf H-bonds, and temperature limits. *J. Phys. Chem.* 104:11268–11274.

Kozmutza, C., I. Varga, and L. Udvardi. 2003. Comparison of the extent of hydrogen bonding in $H_2O–H_2O$ and $H_2O–CH_4$ systems. *J. Mol. Struct. (Theochem.)* 666–667:95–97.

Latimer, W. M., and W. H. Rodebush. 1920. Polarity and ionization from the standpoint of the Lewis theory of valence. *J. Am. Chem. Soc.* 42:1419–1433.

Lu, J., J. S. Brown, C. L. Liotta, and C. A. Eckert. 2001. Polarity and hydrogen-bonding of ambient to near-critical water: Kamlet-Taft solvent parameters. *Chem. Commun.* 7:665–666.

Luck, W. A. P. 1985. The influence of ions on water structure and on aqueous systems. In *Water and Ions in Biological Systems*, ed. A. Pullman, V. Vasileui, and L. Packer. New York, NY: Plenum, p. 95.

Luck, W. A. P. 1998. The importance of cooperativity for the properties of liquid water. *J. Mol. Struct.* 448:131–142.

Makhtadze, G. I., and P. I. Privalov. 1993. Contribution of hydration to protein folding thermodynamics: I. The enthalpy of hydration. *J. Mol. Biol.* 232:639–657.

Modig, K., B. G. Pfrommer, and B. Halle. 2003. Temperature-dependent hydrogen-bond geometry in liquid water. *Phys. Rev. Lett.* 90:075502.

Muller, N. 1988. Is there a region of highly structured water around a nonpolar solute molecule? *J. Solution Chem.* 17:661–672.

Pauling, L. 1948. *The Nature of the Chemical Bond*, 2nd ed. New York, NY: Cornell University Press.

Peeters, D. 1995. Hydrogen bonds in small water clusters: a theoretical point of view. *J. Mol. Liquids* 67:49–61.

Pimentel, G. C., and A. L. McClellan. 1960. *The Hydrogen Bond.* San Francisco, CA: W. H. Freeman and Company.

Poater, J., X. Fradera, M. Solà, M. Duran, and S. Simon. 2003. On the electron-pair nature of the hydrogen bond in the framework of the atoms in molecules theory. *Chem. Phys. Lett.* 369:248–255.

Pollack, G. H. 2001. Is the cell a gel—and why does it matter? *Jpn. J. Physiol.* 51:649–660.

Privalov, P. L., and G. I. Makhtadze. 1993. Contribution of hydration to protein folding thermodynamics: II. The entropy and Gibbs energy of hydration. *J. Mol. Biol.* 232:660–679.

Raichlin, Y., A. Millo, and A. Katzir. 2004. Investigation of the structure of water using mid-IR fiberoptic evanescent wave spectroscopy. *Phys. Rev. Lett.* 93:185703.

Rees, M. J. 2003. Numerical coincidences and 'tuning' in cosmology. *Astrophys. Space Sci.* 285:375–388.

Schrade, P., H. Klein, I. Egry, Z. Ademovic, and D. Klee. 2001. Hydrophobic volume effects in albumin solutions. *J. Colloid Interface Sci.* 234:445–447.

Silverstein, K. A. T., A. D. J. Haymet, and K. A. Dill. 2000. The strength of hydrogen bonds in liquid water and around nonpolar solutes. *J. Am. Chem. Soc.* 122:8037–8041.

Smith, J. D., C. D. Cappa, K. R. Wilson, B. M. Messer, R. C. Cohen, and R. J. Saykally. 2004. Energetics of hydrogen bond network rearrangements in liquid water. *Science* 306:851–853.

Speedy, R. J. 1984. Self-replicating structures in water. *J. Phys. Chem.* 88:3364–3373.

Stanley, H. E., and J. Teixeira. 1980. Interpretation of the unusual behavior of H_2O and D_2O at low temperature: tests of a percolation model. *J. Chem. Phys.* 73:3404–3422.

Suresh, S. J., and V. M. Naik. 2000. Hydrogen bond thermodynamic properties of water from dielectric constant data. *J. Chem. Phys.* 113:9727–9732.

Tromans, D. 1998. Temperature and pressure dependent solubility of oxygen in water: a thermodynamic analysis. *Hydrometallurgy* 48:327–342.

Urquidi, J., C. H. Cho, S. Singh, and G. W. Robinson. 1999. Temperature and pressure effects on the structure of liquid water. *J. Mol. Struct.* 485–486:363–371.

Vedamuthu, M., S. Singh, and G. W. Robinson. 1994. Properties of liquid water: origin of the density anoma-
    lies. *J. Phys. Chem.* 98:2222–2230.
Verma, M. P. 2003. Steam tables for pure water as an ActiveX component in Visual Basic 6.0. *Comput. Geosci.*
    29:1155–1163.

BIBLIOGRAPHY

Chaplin, M. F. 1999. A proposal for the structuring of water. *Biophys. Chem.* 83:211–221.
Chaplin, M. F. 2001. Water: its importance to life. *Biochem. Mol. Biol. Educ.* 29:54–59.
Chaplin, M. F. 2005. Water structure and behaviour. http://www.lsbu.ac.uk/water/.

Part II

---

*The Specific Properties of Water—
How and Why Water Is Eccentric*

6 Properties of Liquids Made from Modified Water Models

Ruth M. Lynden-Bell and Pablo G. Debenedetti

CONTENTS

6.1 INTRODUCTION

Life on earth developed in liquid water, and its properties are therefore critically important to the physical and chemical processes necessary for the maintenance and perpetuation of living organisms (Franks, 2000). Water participates in four major classes of biological reactions: oxidation, photosynthesis, hydrolysis, and condensation. It is key to the stabilization of biologically significant structures of proteins, nucleotides, carbohydrates, and lipids (Franks, 1983, 1985). In most animals, blood—an aqueous medium—transports oxygen and nutrients to cells and disposes of carbon dioxide. In plants, a watery fluid, sap, transports mineral nutrients and water upward from the root system. Cells being roughly 80% water by weight, processes such as protein synthesis occur in aqueous solution.

Clearly then, life on earth is "fine-tuned" to water. But is water the sole possible medium for life? How essential are water's distinctive properties for the maintenance of life, including extraterrestrial life (Ball, 2005)? One way of addressing this question is to modify water's structure and observe the resulting changes in properties. This allows one to understand which among water's properties are particularly sensitive to its structure and, hence, how "fine-tuned" is water for life. The modification of water's structure is possible in computer simulations and this is the approach we take in this work.

In addition to the microscopic perspective implied by the previous line of inquiry (i.e., that of the relationship between water's properties and the molecular level chemical and physical processes responsible for life), it is also instructive to note the profound influence that those same distinguishing properties have on events occurring over vastly greater time and length scales. From climate to travel, and from agriculture to public health, the properties of water affect virtually every aspect of our daily lives. Excellent discussions of this topic exist in the literature (Franks, 1983, 2000; Ball, 1999; Denny, 1993), and only a few selected examples need to be introduced to illustrate this point.

The thermohaline ocean circulation system consists of a surface flow of warm, low-salinity water and a deeper return flow of cold, high-salinity water (Ball, 1999). Evaporation from the ocean's surface increases salinity, so that when the warmer, upper-level current releases heat in colder regions, the resulting colder and saltier water sinks. For example, the warm northward surface flow of the North Atlantic component of this circulation system carries thermal energy that is released at higher latitudes as the water cools, becomes denser, and sinks near the Labrador Sea, south of Greenland. The heat released to the North Atlantic Ocean is estimated to be 25% of the amount delivered to the earth by direct sunshine (Ball, 1999). The thermohaline circulation is thus a vast heat exchange system that is responsible for maintaining a temperate climate over much of the earth's surface (Franks, 2000). Underlying the oceans' ability to store and exchange vast amounts of energy is water's large (for a molecule of its size) heat capacity, one of its distinctive anomalies.

Below 4°C liquid water expands when cooled. This, combined with the low density of ice relative to the liquid, causes water masses to freeze from the surface downward. Aquatic life can thus survive in winter under the protective cover of a frozen surface. Water is the only naturally occurring inorganic liquid on earth, and the only substance that occurs naturally in the solid, liquid, and vapor states of aggregation on Earth (Franks, 2000). Whereas methane and ammonia have molecular weights similar to water's, they are gases at ambient conditions. In contrast, water is not only liquid at ambient conditions, but it has unusually high boiling and melting points, relative to comparable nonmetallic hydrides. Extrapolation of the melting and boiling points of group VI hydrides (H_2Te, H_2Se, H_2S) yields values of −100°C and −80°C for the expected melting and boiling points of water (Pauling, 1970). These large differences, of 100°C (freezing) and 180°C (boiling), between extrapolation and reality are vivid indicators of water's distinctive physical properties. Cold liquid water becomes less viscous upon compression, whereas most liquids become more viscous when compressed. At ambient temperature, water is distinctively less compressible than typical liquids; however, upon cooling, water's compressibility increases sharply and it becomes an exceptionally compressible substance.

The source of water's unusual properties lies in the interactions between molecules. Water molecules are "sticky," but this stickiness is specific and directional. The result is that the local structure around a water molecule is more or less tetrahedral. This tetrahedral symmetry is permanent and long-ranged in ordinary hexagonal ice, and fluctuating and short-ranged in the liquid, and it is weakened as the liquid is heated or compressed. This structure is the result of the ability of each water molecule to form strong, directional hydrogen bonds: noncovalent interactions between the electropositive hydrogen atom of one molecule and the electronegative oxygen atom of another molecule. Because each water molecule has two hydrogen atoms and a lone pair of electrons, it can form four hydrogen bonds (it donates two bonds through its two hydrogens and accepts two bonds from hydrogen atoms in two other molecules). The resulting structure, illustrated in Figure 6.1, is found to be locally tetrahedral. Consequently, whereas in ordinary liquids each molecule has twelve nearest neighbors, in water this number is only four.

Computer simulation has been used for many years to study model liquids in order to understand the relation of their properties to the molecular interactions (Frenkel and Smit, 2002). The emphasis in most studies is either to try to model a particular liquid as accurately as possible or to look at properties of simplified models in order to study generic features. Water has been studied extensively and many models with varying degrees of sophistication have been proposed (Stillinger and Rahman, 1974; Berendsen et al., 1987; Mahoney and Jorgensen, 2000). The simplest models have been rather successful in describing water. In these "simple point charge" models, the intermolecular potential consists of two terms. The first term is a spherically symmetric interaction between sites on the oxygen atoms, which is repulsive at short range and attractive at longer range. In the water models this prevents the molecules from getting too close to each other, but if there are no other terms in the intermolecular interactions the resulting liquid has similar properties to normal liquids such as carbon tetrachloride or cyclohexane. The second term models the hydrogen bonding and determines the shape of the molecule. It consists of positive electrical

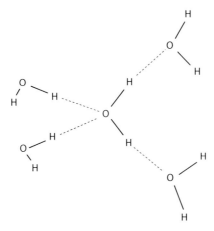

FIGURE 6.1   Local tetrahedral network formed by hydrogen bonds in liquid water.

charges on each hydrogen atom and a balancing negative charge on the oxygen atom. This is shown in Figure 6.2.

The computer liquids formed from such models show many of the characteristics of real water, including negative thermal expansion, increased fluidity upon compression (Errington and Debenedetti, 2001), and expansion upon freezing. They are, of course, not perfect, but seem to contain enough of the essential physics of the intermolecular interactions to give a good understanding of the way in which water works and the trends in its properties with changing temperature and pressure.

Computer representations of water such as the one shown in Figure 6.2 can be used to investigate how changes in molecular structure affect liquid-state behavior. This can be done by modifying selected aspects of the molecular geometry and studying the resulting changes in liquid-state properties. In this way one can understand which among water's properties are particularly sensitive to its structure. Models of the class illustrated in Figure 6.2 can be described by five parameters: intramolecular bond angle, intramolecular bond lengths, charges on the atomic sites, and two so-called Lennard-Jones parameters that describe the characteristic length and energy scales associated with the spherically symmetric, dispersive interaction. Two of these parameters can be used to determine the length and energy scaling, leaving three others as completely independent variables. By varying these parameters one can generate liquids of "quasi-water" and investigate how sensitive various liquid properties are to the precise values of the intermolecular potential parameters. How important is the molecular geometry? How important is the hydrogen-bonding strength? Which properties of water are particularly sensitive to the tetrahedral structure and which are more generic? This can help to answer the question as to how "fine-tuned" for life the liquid properties are.

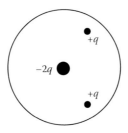

FIGURE 6.2   A simple point charge model of water used in computer simulation.

6.2 MODIFIED WATER MODELS

To tackle the above questions, two types of alterations in water's intermolecular potential were investigated: changes in the HOH angle (Bergman and Lynden-Bell, 2001), and changes in the relative magnitude of the electrostatic and dispersive interactions (Lynden-Bell and Debenedetti, 2005). The SPC/E model (Berendsen et al., 1987) was taken as the water reference. It consists of a Lennard-Jones interaction centered at the oxygen atom, with size and energy parameters $\sigma = 3.169$ Å and $\varepsilon = 0.65$ kJ/mol. Charges of $-0.8476e$ and $+0.4238e$ are placed on the oxygen and hydrogen atoms respectively (see Figure 2). The HOH angle is 109.5°, and the OH distance is 1 Å, resulting in a dipole moment of 2.35 Debye. The SPC/E model gives a good description of liquid water's distinctive properties, including negative thermal expansion and increased fluidity upon compression (Errington and Debenedetti, 2001).

Changes in the HOH angle were investigated by defining two modified water models: the 60° and 90° models (Bergman and Lynden-Bell, 2001). In both models the charges were unchanged and the OH distance was shortened so as to maintain the value of the SPC/E dipole moment. The value of the Lennard-Jones σ was adjusted to give zero pressure at 298 K and 1 g/cm³. The resulting changes in local liquid structure were found to be quite dramatic. In a tetrahedral liquid, the distance between the first and second solvation shells is less than the distance between the central molecule and its first solvation shell, whereas these two distances are about equal in an unstructured liquid with spherically symmetric interactions. By this measure, the 90° liquid is still tetrahedral, albeit its local structure is less pronounced. In contrast, the 60° liquid loses its tetrahedral character completely. Three-dimensional representations of the spatial distribution of molecules showed the 60° liquid to have a linear local structure caused by the ability of the two hydrogen atoms on one molecule to hydrogen-bond to the same oxygen atom on a second molecule. At the single state point investigated, the average number of hydrogen bonds per molecule dropped significantly upon changing the HOH angle, from 3.7 in SPC/E to 2.8 and 2.3 in the 90° and 60° modified water models, respectively (Bergman and Lynden-Bell, 2001). The interaction energy between pairs of hydrogen-bonded molecules decreased going from SPC/E to the 90° model, but was found to be strongest for the 60° model, due to the above-mentioned ability of two protons from the same molecule to form hydrogen bonds with the same oxygen of a second molecule. The progressive loss of local tetrahedral structure has pronounced effects on dynamics: the diffusion coefficient increased by a factor of 4.5 upon changing the HOH angle from 109.5 (SPC/E) to 90°, and by a factor of 3.2 upon changing this angle from the SPC/E reference value to 60°. That the dependence of the diffusion coefficient upon the HOH angle was found to be nonmonotonic is likely due to the strong intermolecular interactions in the 60° model associated with the above-mentioned possibility that one water molecule can donate two hydrogen bonds to the same acceptor molecule.

In order to investigate the effects of electrostatic interactions on the structure, thermodynamics, and transport properties of liquid water, a second type of "not-water" model was introduced, whose intermolecular potential energy is defined by

$$\Phi = \gamma V^{\text{LJ}} + V^{\text{ES}} \qquad (6.1)$$

where V^{LJ} is the Lennard-Jones component of the pair potential, V^{ES} is the electrostatic component, and γ is a scaling factor. By increasing γ from 1 to ca. 3 the liquid changes smoothly and continuously from SPC/E to Lennard-Jones–like. This change in the intermolecular potential corresponds to a reduction in the strength of the hydrogen bonds. In the range of values of γ studied in this work, the relative hydrogen-bond strength is reduced by 15% at $\gamma = 1.5$, 25% at $\gamma = 2$, and 30% $\gamma = 3$.

Figure 6.3 shows the evolution of the structure as measured by the oxygen–oxygen pair correlation function as γ was changed from 1 to 2.78 at 272 K and 1 g/cm³. The lowest curve (SPC/E) is typical of a tetrahedral liquid, with the second shell located at a distance roughly 1.7 times that of the first shell. The top curve on the other hand corresponds to a Lennard-Jones–like liquid, with

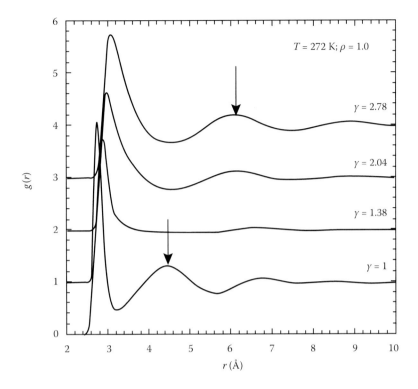

FIGURE 6.3   Radial distribution functions for modified water models. As the hydrogen bond strength decreases (bottom to top) the structure changes. The arrows show how the second peak moves.

the second shell located at twice the distance to the first shell. Interestingly, at intermediate values of γ (e.g., 1.23 and 1.38) there is virtually no structure beyond the first peak. This means that as the liquid changes from tetrahedral to Lennard-Jones it first becomes remarkably destructured; at the risk of anthropomorphizing a physical phenomenon, one can say that at intermediate values of γ the liquid is confused between its tetrahedral and Lennard-Jones limits.

6.3   LIQUID ANOMALIES

We have used two order parameters, t and q (Errington and Debenedetti, 2001), to track the change from a tetrahedral (or water-like) liquid to a more normal liquid such as cyclohexane. q is a measure of the local tetrahedral order and decreases steadily as γ is changed from 1 to 2.78. t is a measure of translational order and quantifies the extent to which pairs of molecules adopt preferential separations, such as exist in a crystal. The changes in t are more interesting, as the value of this parameter depends on the shape of the oxygen–oxygen pair correlation function shown in Figure 6.3. It provides a measure of the amount of radial structure in the liquid and, as expected from Figure 6.3, it first goes through a minimum as a function of γ and then increases again as the structure becomes more Lennard-Jones–like. The behavior of t and q as a function of γ is shown in Figure 6.4 for the same state point as the radial distribution functions of Figure 6.3.

The changes that we find as the liquid is transformed from water-like to Lennard-Jones–like by tuning γ are similar to those found when water is compressed. When this is done the radial distribution functions show the same changes as in Figure 6.3; q decreases and t goes through a minimum value. This shows that under high pressure water becomes more like a normal Lennard-Jones liquid. Similar changes occur if the temperature is increased. We know that life can exist under conditions of high pressure and high temperature. For example, growth phases of the hyperthermophile

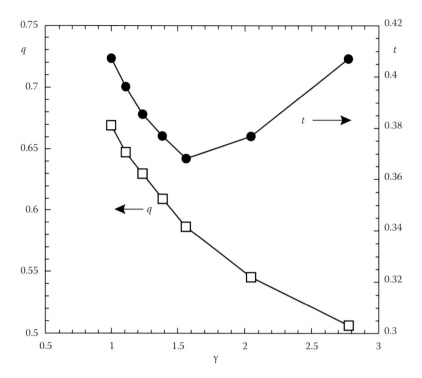

FIGURE 6.4   Variation of the tetrahedral order parameter, q, and the translational order parameter, t, as a function of γ. As the hydrogen bonds weaken (γ increases) the tetrahedral order decreases, but the translational order passes through a minimum and then increases.

archaeon *Pyrococcus abyssi* have been observed at temperatures as high as 105°C (Erauso et al., 1996). *Holothurians* (sea cucumbers, members of the phylum *Echinodermata*) abound in the Mariana Trench Challenger Deep in the West Pacific at a depth of 11 km (pressure of 1.1 kbar) (Kato, 1997). This gives evidence that life can adapt to conditions where the properties of liquid water are less "water-like" than under ambient conditions.

The next question that we can investigate is how far these changes in structure are reflected in the properties of the bulk liquid. How sensitive are water's bulk properties to details of, and changes in, the local structure? Initial investigations of this question are reported in the rest of this section. In the context of the water and life question, the changes in the properties of the liquid as a solvent are also of great interest. Preliminary work addressing this question is reported in a following section.

Among the bulk properties of interest are the well-known anomalies, which are relevant for the role of water in life as we know it. In particular the density maximum is associated with the presence of liquid water below a crust of ice on frozen ponds and lakes. The SPC/E model is rather under-structured and shows this anomaly at lower temperatures than does real water. We found that at 220 K the density range where entropy increases under compression at constant temperature decreases as γ increases (i.e., the liquid becomes less water-like). There is a thermodynamic relationship which shows that the anomalous region where the density increases with temperature at constant pressure is identical to this region of anomalous entropy behavior, so that the disappearance of the entropy anomaly is coincident with the disappearance of the density anomaly. We conclude that the disappearance of these anomalies is closely associated with the decreased structure as measured by the t parameter. This means that one can destroy order by compressing the liquid as well as by changing the intermolecular potential. This is only true for tetrahedral liquids and the property is lost when γ approaches the point of minimum translational order (minimum t).

Another manifestation of the loss of order is an increase in the motional freedom of individual molecules. In this respect, water's anomalies are shown both as an increase in the reorientational rate with density and as an increase of the translational diffusion constant under compression. Both quantities reach a maximum and then behave in a normal manner, decreasing with compression. This has been seen experimentally in real water (Prielmeier et al., 1987). In our numerical experiments, we find that both measures of molecular mobility show an initial increase as the liquid becomes less tetrahedral when the hydrogen bonds are weakened by increasing γ.

Figure 6.5 shows the changes in the translational diffusion constant (D) and the reorientation rate (τ^{-1}) as γ is increased. For comparison, the variation of t is also shown. These results are for liquids at normal density (1 g/cm³) and 272 K. It can be seen that the rate of molecular motion increases as the liquids become less tetrahedral, even beyond the point where the Lennard-Jones order increases. In fact, in the range of values of γ shown, the reorientational rate shows no maximum and the translational diffusion reaches a maximum well past the point of maximum disorder. There is not a close relationship between the degree of translational order and the molecular mobility. In fact, the restrictions on molecular reorientation depend mainly on the strength of the hydrogen bonds, which is reduced steadily as the liquids become less tetrahedral, while the translational diffusion depends more on all aspects of the local structure.

In many ways the effects of compressing the liquid, heating it up or reducing the hydrogen bond strength by increasing the parameter γ are similar. All these changes lead to a reduction of the unique tetrahedral structure of water and make it more like a normal Lennard-Jones liquid with properties similar to those of cyclohexane. This idea is illustrated in Figure 6.6, which shows a space of possible liquids as a function of temperature, density, and hydrogen-bond strength (γ^{-1}). In this diagram the central volume, labeled W corresponds to tetrahedral liquids while the outer

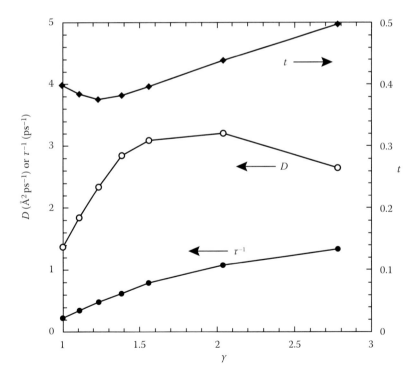

FIGURE 6.5   Plots of the translational order parameter t, the diffusion constant D, and the rate of reorientation $1/\tau$ as a function of the parameter γ. Note that initially the molecules become more mobile as the hydrogen bond strength decreases (increasing γ), but eventually the diffusion constant reaches a maximum value and begins to fall off.

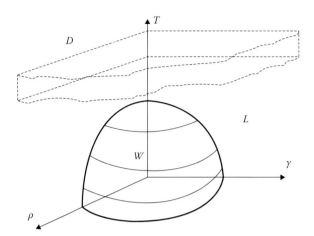

FIGURE 6.6 Three-dimensional representation of liquid states. ρ and T are the density and temperature, while γ is the modification parameter. The origin of the axes corresponds to water under normal conditions. In the region W the liquids are tetrahedral, but increasing density, temperature or γ causes a breakdown of the tetrahedral structure. (Reproduced from Lynden-Bell, R. M., and Debenedetti, P. G., *J. Phys. Chem. B*, 109, 6527, 2005. With permission.)

volume (L) corresponds to liquids that are more Lennard-Jones–like in their structure. The boundary is the locus of minimum values of t. At sufficiently high temperatures, all the liquids reach the supercritical region (D), which does not concern us here. The region near the W/L boundary is where the translational structure breaks down. The anomalous properties of bulk water occur as this boundary is approached and are associated with the breakdown of structure upon densification.

6.4 SOLVATION PROPERTIES

Our investigations of the properties of modified water models as solvents are only just beginning. One property that is thought to be very important for biological processes such as protein folding is the hydrophobic effect (Tanford, 1980). This term covers a range of phenomena from the separation of oil and water to the preference of a macromolecule to adopt conformations in which hydrophilic groups (such as hydroxyl or carbonyl groups) are exposed to an aqueous solvent, while hydrophobic groups (such as alkyl side chains) are folded into the interior of the molecule. The free energy differences involved need not be large, but most of the phenomena can be interpreted in terms of the properties of water rather than of the dissolved molecules themselves (Ashbaugh et al., 2002).

The first investigation that we have performed relating to solvation properties is to examine the potential of mean force between two methane molecules dissolved in some of our liquids as a function of the distance between the molecules. The potential of mean force is an effective interaction potential whose derivative gives the mean force between methane molecules, averaged over all possible water molecule configurations. The graphs of this quantity are well known for models of water and show two clear minima, each of which corresponds to a favorable arrangement of the methane molecules. In the first minimum, the molecules are adjacent to each other in a cavity in the water. In the second minimum, they are separated by a solvent molecule. One also expects this pattern of two minima (or possibly a third one) in a Lennard-Jones liquid, but one would expect the minima in water (and especially the first minimum) to be distinctly deeper (more stable) than in a Lennard-Jones solvent, implying that the concentration of dimers of methane is higher in water than in a Lennard-Jones liquid, reflecting the tendency of oily, nonpolar molecules to aggregate when in water. Figure 6.7 shows the graphs of the potential of mean force measured in our simulations for methane in SPC/E water under ambient conditions ($\gamma = 1$) and in one of our hybrid liquids with $\gamma = 3$, which has more Lennard-Jones than tetrahedral character in the radial distribution function.

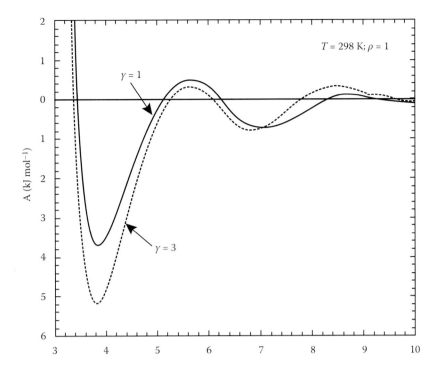

FIGURE 6.7   Potential of mean force for the separation of two methane molecules in normal water ($\gamma = 1$) and modified water ($\gamma = 3$). The first minimum corresponds to adjacent molecules while the second corresponds to a pair of molecules separated by a solvent molecule. Note that there is little change between these two liquids.

Surprisingly, there is very little difference between the two curves, with even a suggestion of a deeper first minimum in the less tetrahedral liquid.

These preliminary results may be compared with those found by Ghosh et al. (2001) for pairs of methane molecules in water at different pressures. They studied pressures up to 8000 atm, and found that the oxygen–oxygen radial distribution function showed similar changes to those in Figure 6.3. At the highest pressure the radial distribution function was clearly of the Lennard-Jones type while at 1 atm it was tetrahedral in character. At intermediate pressures the second shell disappeared. This represents the same range of change that we see in our modified water models. However, the changes in the methane–methane potential of mean force were small; the principal change being an increase in the barriers to desolvation and an increase in the depth of the solvent-separated minimum. The depth of the contact minimum was hardly changed. These results together with our own suggest that the solvation properties of water are less sensitive to its local structure and molecular geometry than the anomalous properties of the pure, bulk liquid. However, more investigations are needed to resolve this issue. The calculation of the methane–methane potential of mean force is probably not the most sensitive way to probe the hydrophobicity of the liquid.

Lynden-Bell and Head-Gordon (2006) studied the solubility of small hard-spheres and small Lennard-Jones particles in various models in order to try to understand the small length-scale hydrophobic effect better. They found that the solubility increased in bent models and decreased in hybrid models. This could be understood in terms of the distribution of cavities in the different liquids at the same density. As the hybrid models become more Lennard-Jones–like they are more close packed with fewer large cavities and so less able to dissolve hard spheres of radii of a few angstroms. The bent models, on the other hand, have a looser network structure and so more large cavities and so dissolve similar hard spheres more easily. One may conclude that the main reason for the lack of solubility of small hydrophobic solutes in water is primarily the small size of the molecule and only secondarily to changes in local structure. This conclusion is supported by scaled particle calculations of Graziano (2008). This work was extended to a temperature study by Chaterjee et al. (2008) for a series of bent models. The characteristic change of sign of the hydrophobic solvation enthalpy from a negative value at low temperatures to a positive value at high temperatures occurs for models with a range of angles. Results are not yet available for hybrid models.

One manifestation of the hydrophobic effect at larger length scales is the drying of a hydrophobic surface. Lynden-Bell and Head-Gordon (2008) investigated the critical separation for wetting to occur between a pair of plates. There is a strong negative correlation between this distance and the liquid–vapor surface tension of various bent and hybrid model liquids. Alejandre and Lynden-Bell (2007) investigated the surface properties of the same set of liquids and found that the surface tension was higher in hybrid models and lower in bent models and is correlated with higher critical temperatures for hybrid models and lower ones for bent models.

An example of further use of this type of modified water to further the understanding of processes in solution is provided by the recent work of Gu et al. (2008), who studied the effect of solvent polarity on the conformation of DNA. They found a change from the B form in SPC/E water to the A form as the electrostatic interaction was reduced relative to the Lennard-Jones interaction.

6.5 CONCLUDING REMARKS

We have provided a progress report on an ongoing computational investigation of the properties of liquids made from modified water models. Water's tetrahedral character persists, albeit distorted, when the HOH angle is changed from 109.5° to 90°, but collapses upon further reducing this quantity to 60°. Decreasing the strength of the hydrogen bond causes the structure to evolve from water-like to that of a normal liquid. This evolution is accompanied by progressive loss of local tetrahedrality, whereas the radial ordering shows nonmonotonic behavior and exhibits a minimum along the path from water to Lennard-Jones ordering. Translational and rotational mobility increase markedly as hydrogen bonds are weakened, and are only weakly coupled with the evolution of radial ordering. The solvation properties that we have investigated so far, namely the potential of mean force between methane molecules and the entropy of ion hydration, are surprisingly insensitive to perturbations in water's geometry.

The picture that emerges from our work is one in which different properties exhibit varying degrees of sensitivity to changes in water geometry and hydrogen bond strength. To this should be added the empirical evidence of the existence of life under extreme conditions (e.g., high pressure and high temperature) where water's local structure and bulk properties are quite different from those existing at ambient conditions. In other words, life occurs in water, but under a very wide range of water behaviors. Furthermore, the distinctive properties that we commonly associate with water are rather resilient to changes in molecular geometry and hydrogen bond strength.

Some of the distinctive features commonly associated with water's solvent properties are derivative quantities, such as the transfer heat capacity (Ashbaugh et al., 2002). Accordingly, it will be interesting to explore the temperature dependence of the entropy and entropy of solvation of solutes in modified water models. It is also important to quantify the effects of γ in terms of equivalent pressure or temperature changes: for example, what is the equivalent increase in pressure that causes the same increase in translational diffusion as a given change in γ? Another interesting avenue of future inquiry is the study of void distributions in modified water models across broad ranges of temperature, density, and changes in molecular geometry. The statistical geometry of voids is intimately related to water's solvation properties (Reiss et al., 1959).

ACKNOWLEDGMENTS

One of the authors (RMLB) gratefully acknowledges financial support from the Leverhulme Trust.

REFERENCES

Alejandre, J., and R. M. Lynden-Bell. 2007. Phase diagrams and surface properties of modified water models. *Mol. Phys.* 105:3029.

Ashbaugh, H. S., T. M. Truskett, and P. G. Debenedetti. 2002. A simple molecular thermodynamic theory of hydrophobic hydration. *J. Chem. Phys.* 116:2907.

Ball, P. 1999. *Life's Matrix. A Biography of Water.* New York, NY: Farrar, Straus and Giroux.

Ball, P. 2005. Water and life: seeking the solution. *Nature* 436:1084.

Berendsen, H. J. C., J. R. Grigera, and T. P. Straatsma. 1987. The missing term in effective pair potentials. *J. Phys. Chem.* 91:6269.

Bergman, D. L., and R. M. Lynden-Bell. 2001. Is the hydrophobic effect unique to water? *Mol. Phys.* 99:1011.

Chaterjee, S., P. G. Debenedetti, F. H. Stillinger, and R. M. Lynden-Bell. 2008. A computational investigation of thermodynamics, structure, dynamics and solvation behavior in modified water molecules. *J. Chem. Phys.* 128:124511.

Denny, M. W. 1993. *Air and Water. The Biology and Physics of Life's Media.* Princeton, NJ: Princeton University Press.

Erauso, G., S. Marsin, N. Benbouzid-Rollet, M.-F. Baucher, T. Barbeyron, Y. Zivanovic, D. Prieur, and P. Forterre. 1996. Sequence of plasmid pGT5 from the archaeon *Pyrococcus abyssi*: evidence for rolling-circle replication in a hyperthermophile. *J. Bacteriol.* 178:3232.

Errington, J. R., and P. G. Debenedetti. 2001. Relationship between structural order and the anomalies of liquid water. *Nature.* 409:318.

Franks, F. 1983. *Water.* Chap. 1. London: Royal Society of Chemistry.

Franks, F. 2000. *Water: A Matrix for Life,* 2nd ed. Cambridge: Royal Society of Chemistry.

Franks, F. 1985. *Biochemistry and Biophysics at Low Temperatures.* Cambridge: Cambridge University Press.

Frenkel, D., and Smit, B. 2002. *Understanding Molecular Simulation. From Algorithms to Applications.* 2nd ed. San Diego, CA: Academic Press.

Ghosh, T., A. E. Garcia, and S. Garde. 2001. Molecular dynamics simulations of pressure effects on hydrophobic interactions. *J. Am. Chem. Soc.* 123:10997.

Graziano, G. 2008. Hydrophobicity in modified water models. *Chem. Phys. Lett.* 452:259.

Gu, B., F. S. Zhang, Z. P. Wang, and H. Y. Zhou. 2008. Solvent-induced DNA conformational transition. *Phys. Rev. Lett.* 100:088104.

Kato, C. 1997. Deep-sea animals living at world's deepest bottom at Mariana Trench Challenger Deep, depth of 11000m. *Umi-ushi Lett.* 14:12.

Lynden-Bell, R. M., and P. G. Debenedetti. 2005. Computational investigation of order, structure and dynamics in modified water models. *J. Phys. Chem. B* 109:6527.

Lynden-Bell, R. M., and T. Head-Gordon. 2006. Solvation in modified water models: toward understanding hydrophobic solvation. *Mol. Phys.* 104:3593.

Lynden-Bell, R. M., and T. Head-Gordon. 2008. Hydrophobic solvation of Gay-Berne particles in modified water models. *J. Chem. Phys.* 128:104505.

Mahoney, M. W., and W. L. Jorgensen. 2000. A five-site model for liquid water and the reproduction of the density anomaly by rigid, nonpolarizable potential functions. *J. Chem. Phys.* 112:8910.

Pauling, L. 1970. *General Chemistry.* New York, NY: Dover.

Prielmeier, F. X., E. W. Lang, R. J. Speedy, and H.-D. Lüdemann. 1987. Diffusion in supercooled water to 300 MPa. *Phys. Rev. Lett.* 59:1128.

Reiss, H., H. L. Frisch, and J. L. Lebowitz. 1959. Statistical mechanics of rigid spheres. *J. Chem. Phys.* 31:369.

Stillinger, F. H., and A. Rahman. 1974. Improved simulation of liquid water by molecular dynamics. *J. Chem. Phys.* 60:1545.

Tanford, C. 1980. *The Hydrophobic Effect: Formation of Micelles and Biological Membranes.* 2nd ed. New York, NY: Wiley.

7 Understanding the Unusual Properties of Water

Giancarlo Franzese and H. Eugene Stanley

CONTENTS

7.1 INTRODUCTION

Water is commonly associated with the existence of life. It is the main component of living organisms: the human body is by weight roughly 75% water in the first days of life and roughly 60% water in the adult age. The majority of this water (roughly 60%) is inside the cells, while the rest (extracellular water) flows in the blood and below the tissues.

Many living beings can survive only a few days without water. This is because water participates in the majority of the biological processes [1], such as the metabolism of nutrients catalyzed by enzymes. To be effective, the enzymes need to be suspended in a fluid to adopt their active three-dimensional structure. The main reason why we need water is that it allows the processes of elimination of cellular metabolic residues. It is through the water that our cells can communicate and that oxygen and nutrients can be brought to our tissues.

However, despite all these considerations, there is no clear reason why water should be the only liquid in which life could form and survive. The debate about the essential role in biological processes played by water and its properties is open.

The first remarkable thing about water is that it has a large number of anomalies with respect to other substances. This fact has fascinated scientists for centuries. It was, probably, L. J. Henderson in 1913 [2] who first asked about the relation between the unusual properties of water and the existence of life.

Since then, scientists have tried to reach the answer to this question, step by step. Here we will follow the first steps of that long path. Those steps can be summarized by the following questions:

- What is unusual about water?
- Why does water have these anomalies?
- What are the full implications of these unusual properties?
- Are these anomalies exclusive properties of water?

Answering these questions is a scientific challenge. We will organize the discussion by identifying some interesting clues, formulating a working hypothesis, and testing that hypothesis. In this process, we will use, among other theoretical tools, computer simulations. Their applicability to the case of water has been demonstrated by many studies and the dramatic increase of computational power makes them more and more feasible.

7.2   WHAT IS UNUSUAL ABOUT WATER?

There are many anomalies associated with water. Their relevance for biological processes is sometimes difficult to determine, but there is one that is clearly evident: liquid water can exist at temperatures far lower than the freezing temperature.

This property allows living systems to survive at very cold temperatures, far below 0°C. At these temperatures, liquid water is said to be in a *supercooled* phase. This phase is metastable, i.e., it cannot last forever; sooner or later, the water will transform into ice.

7.2.1   WATER AT VERY LOW TEMPERATURES

Water in organic cells can avoid freezing at temperatures as low as −20°C in insects, and −47°C in plants, for a time long enough to be compared to their lifetime [3]. In a laboratory, supercooled water has been observed at −41°C at ambient pressure [4] and *in situ* observation has confirmed the presence in clouds of droplets of liquid water at −37°C [5]. The lowest measured supercooled liquid water temperature is −92°C at 2 kbar [6].

Below these temperatures water cannot exist as a liquid and forms crystalline ice, by means of a process known as *crystal homogeneous nucleation*. But water is unusual also in its crystalline phase. Indeed, water is a *polymorph*, i.e., has many crystalline forms, like a few other unusual substances, e.g., carbon, which may form cheap graphite or a very expensive diamond. In the case of water, the known polymorphs are as many as 13.

At roughly 80°C below the homogeneous nucleation temperature, water can exist also in a glassy form, i.e., in a frozen, amorphous, solid form [7], which crystallizes to cubic ice at approximately −123°C at ambient pressure. Again, unusually, instead of just one amorphous phase, at these temperatures there is more than one *polymorph* of glassy water. The first clear indication of this was a discovery by Mishima in 1984. At low pressure there is one form, called low-density amorphous (LDA) ice [7], while at high pressure there is a new form called high-density amorphous (HDA) ice [8], with a volume discontinuity between these two phases of 27%, comparable to that separating low-density and high-density polymorphs of crystalline ice [9–12].

Recently, a new form of amorphous water has been experimentally observed by compressing HDA more than 0.95 GPa, the very-high-density amorphous (VHDA) ice [13, 14], with a volume discontinuity between HDA and VHDA of 11%. At 125 K (−148°C), the three polymorphs can be formed by compression in a stepwise process LDA–HDA–VHDA [15]. However, it is not yet clear from experimental data if the transformations LDA–HDA and HDA–VHDA are real discontinuous transitions or just very sharp increases of densities. Additional investigations are needed to clarify this point.

Remarkably, HDA ice is the most abundant ice in the universe, where it is found as a frost on interstellar grains [16]. HDA ice has been proposed as the cradle where small inorganic molecules combine into the large organic molecules at the origin of life [17].

7.2.2 VOLUME FLUCTUATIONS

The unusual behavior of water is not limited to the supercooled phase. For example, its isothermal compressibility, K_T, that is the departure $\delta \bar{V}$ of the volume per particle $\bar{V}$ from its mean value in response to an infinitesimal pressure change δP, is anomalous at high temperatures.

For a typical liquid, K_T decreases when one lowers the temperature (Figure 7.1). In statistical physics, we learn that K_T is proportional to the average value of the fluctuations of $\bar{V}$, hence we expect that K_T decreases with the temperature T, because the fluctuations decrease with T. For water, however, K_T is anomalous in three respects: (1) it is larger than one would expect; (2) below 46°C,

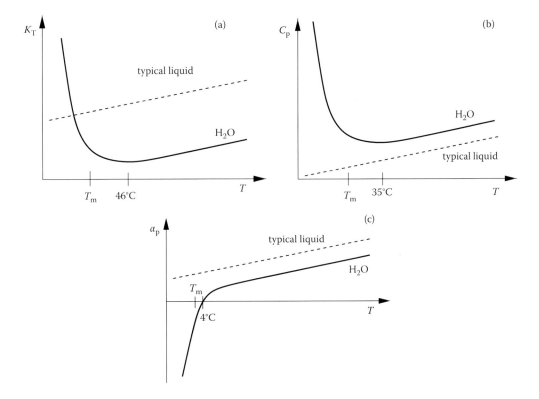

FIGURE 7.1 Schematic dependence on temperature, at ambient pressure, of (a) the isothermal compressibility K_T proportional to volume fluctuations, (b) the constant-pressure specific heat C_P proportional to entropy fluctuations, and (c) the thermal expansivity α_P proportional to volume–entropy cross fluctuations ($\delta V \delta S$). The behavior of a typical liquid is indicated by the dashed line, which, very roughly, is an extrapolation of the high-temperature behavior of liquid water. Note that while the anomalies displayed by liquid water are apparent above the melting temperature T_m, they become more striking as one supercools below T_m.

instead of decreasing as a usual liquid, it increases, doubling its value before reaching the homogeneous crystal nucleation temperature in the supercooled phase; (3) it appears as if it might diverge to infinity, at a temperature of about −45°C, increasing like a power law and hinting at some sort of critical behavior [18].

7.2.3 ENTROPY FLUCTUATIONS

Another anomalous thermodynamic function is the heat capacity at constant pressure C_P, which is the response $\delta \overline{S}$ of the entropy (or disorder) per particle to an infinitesimal temperature change δT. This quantity decreases with T for a typical liquid, because it is proportional to the fluctuations of entropy and these fluctuations decrease with T for normal substances. But water is, again, anomalous in the same three respects: (1) its C_P is larger than expected; (2) below approximately 35°C, the specific heat increases; (3) this increase is approximated by a power law [19]. It is because of the high heat capacity of water that we can easily regulate our body temperature by transpiration.

7.2.4 VOLUME–ENTROPY CROSS FLUCTUATIONS

The last anomalous thermodynamic function we will consider is the thermal expansivity α_P, which is the response $\delta \overline{V}$ of the volume to an infinitesimal temperature change δT. In typical liquids, this quantity decreases with T and is always positive because it is proportional to the cross-fluctuations of entropy and volume $(\delta \overline{V} \delta \overline{S})$. Indeed, in normal liquids, this quantity is positive because when there is a local increase of fluctuation of the volume $\delta \overline{V}$, the particles in that region shuffle around, increasing the disorder and the associated entropy fluctuation $\delta \overline{S}$. Water is anomalous in the same three respects: (1) its α_P is three times smaller than expected; (2) below 4°C, its α_P is negative and grows rapidly in absolute value; (3) this absolute value increases as a power law [20].

The fact that α_P is negative below 4°C shows that water is more disordered when it is more dense. For this reason, ice melts if its density increases, e.g., for an increase of pressure, because only by disordering, or liquefying, can it reach the desired density. This behavior is related to the most famous anomaly of water: its density maximum at 4°C. Below this temperature the volume of water expands, explaining why frozen water pipes break or why ice cubes float on water once liquid water cools down. In contrast, solid forms of typical substances are denser than their liquid form. It is thanks to this anomaly that lakes and seas start to freeze from the top, allowing fishes to survive in the liquid water below the ice.

7.3 WHY WATER HAS ANOMALIES

After this brief introduction on the unusual properties of water, any scientist would ask why water is so special. Perhaps the first clue dates back to Linus Pauling [21], who recognized that the distinguishing feature of water, compared to other chemically similar substances, is the preponderance of hydrogen bonds.

Each water molecule has two hydrogen atoms and two lone electrons. Each hydrogen atom has a partial positive charge, forming an O–H bond with an lone electron on the oxygen side. Each lone electron tends to form a hydrogen bond with the hydrogen of a nearby H$_2$O molecule, whose O–H bond points to the lone pair. Hence, in a simplified view, each molecule can form four hydrogen bonds attracting four nearby molecules.

Because the four electron pairs (two lone electrons and two pairs of the O–H bond) repel each other, the four hydrogen bonds point along the vertexes of an almost perfect tetrahedron. In liquid water, many of the possible hydrogen bonds between nearby molecules are formed, giving rise to a hydrogen bond tetrahedral network. The network is not static because hydrogen bonds have a very short lifetime, on the order of picoseconds, allowing the rotation and the diffusion of H$_2$O molecules [22]. This dynamic hydrogen bond network slows down when the temperature is decreased and freezes in a full hydrogen bonded network, below the homogeneous nucleation temperature, transforming into ice. The reason why water expands when ice forms at ambient pressure is that the distance between nearby molecules in the ice tetrahedral network is larger than the average intermolecular distance in the liquid, giving rise to an *open* structure (hexagonal ice or ice I_h).

Experiments [23] show that this open tetrahedral structure is preserved, even in the liquid at T as low as 268 K (−5°C) up to the second shell of molecules, and that this open structure can coexist with a more compact structure in which the second shell collapses, locally increasing the density of the liquid. These two local structures are called low-density liquids (LDL), for the open structure, and high-density liquids (HDL) for the collapsed structure, in analogy with the LDA and HDA ices. They were first observed in computer simulations [24] and have suggested two ways to develop a coherent picture of the unusual behavior of water: (1) the *liquid–liquid phase transition* hypothesis [24] and (2) the *singularity-free* interpretation [25, 26, 27].

7.3.1 THE LIQUID–LIQUID PHASE TRANSITION HYPOTHESIS

In the liquid–liquid phase transition scenario [24], the LDA–HDA transition is supposedly discontinuous and marked by a line in the pressure–temperature (P–T) phase diagram below the temperature T_H of crystallization to cubic ice (Figure 7.2). It is assumed that this line does not terminate when it reaches the region of spontaneous crystallization at T_X, but extends into it, with LDA transforming without discontinuity into LDL, and HDA into HDL, giving an LDL–HDL phase transition line. According to this hypothesis, along this line the two kinds of liquids coexist as two separate phases, both forming macroscopic droplets of a phase into the other. At pressures above the liquid-liquid coexistence line, the only existing phase is HDL, and below there is only LDL. The collapsed structure of HDL is more disordered than the open structure of LDL, hence the entropy of HDL is larger than that of LDL and this implies, for thermodynamics relations, that the liquid–liquid coexistence line has a negative slope in the P–T phase diagram.

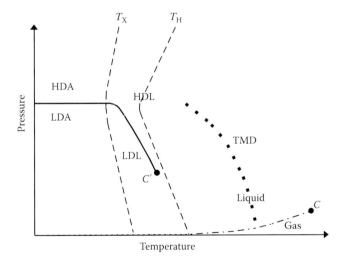

FIGURE 7.2 Schematic representation of the liquid–liquid phase transition hypothesis in the pressure P, temperature T plane. Below the temperature of spontaneous crystallization T_X, HDA ice and LDA ice are separated by a first-order phase transition (continuous line). When this line reaches T_X the slope of the crystallization line changes and HDA and LDA transform continuously in HDL and LDL, respectively. The coexistence line between the two liquids ends in a critical point C' below the homogeneous crystallization temperature T_H. The dotted line denotes the temperature of maximum density (TMD) in the liquid phase. The point-dotted line represents the gas–liquid coexistence line ending in the critical point C.

An experiment able to cross the liquid–liquid coexistence line should measure a sudden disconti-nuity in the local density of the liquid. Theoretical and computational estimates of the liquid–liquid coexistence line locate it in the region below the homogeneous nucleation of the crystal at T_H and above the spontaneous crystallization line at T_X. In this range of temperatures, the experiments on the liquid cannot be performed, but the phase diagram can be explored with the help of comput-ers and sophisticated water models. These simulations [24, 28] show the discontinuity in density which supports the hypothesis. They also show that the density difference between the two phases decreases by increasing the temperature along the line and goes to zero in a liquid–liquid *critical* point C' where the line ends. At higher T, the two phases are indistinguishable and the two kinds of structures (open and collapsed) are found only at the microscopic scale of a few molecules, just as the gas and liquid phases are indistinguishable above the gas–liquid critical point C.

The existence of a critical point induces large (*critical*) fluctuations in a region that extends to temperatures and pressures far away in the phase diagram. For example, experiments show that the effect of the gas–liquid critical point C on the response functions is evident even at temperatures two times higher than the C temperature and that these functions diverge as power laws at C.

In a similar way, the anomalous increase of the response functions is, in this hypothesis, the effect of approaching the liquid–liquid coexistence line, with a genuine divergence at the critical point C'. At T, above the liquid–liquid critical temperature, the thermodynamic response functions appear to diverge to infinity approaching the Widom line, defined as the analytic extension of the liquid–liquid phase transition line in the P–T phase diagram. When the system is *extremely* close to the Widom line, the functions will round off and ultimately remain finite—as experimentally observed in the adiabatic compressibility [29].

This hypothesis is consistent with recent experiments [12]. The melting line of metastable ice IV and stable ice V show an abrupt change in their slope, as predicted if it would intersect the meta-stable liquid–liquid phase transition line. However, the experiment resolution does not allow the user to conclude if the sharp change is a real discontinuity, as required by the liquid–liquid phase transition hypothesis. Therefore, Mishima and Stanley interpolated the experimental data to calcu-late the Gibbs free energy of the liquid at equilibrium with the different ice polymorphs along their melting line and estimated a liquid–liquid critical point at 1 kbar and 220 K.

7.3.2 THE SINGULARITY-FREE INTERPRETATION

In the singularity-free scenario [26, 27] (Figure 7.3), the LDL and HDL are still smoothly connected to LDA and HDA, respectively, but do not represent two distinct phases separated by a discontinu-ous transition. Instead, they represent local fluctuations of densities and, by increasing the pressure, one can pass from LDL to HDL observing a sharp, but continuous, increase of density, occurring in a limited region. Hence, there is no liquid–liquid phase transition and no liquid–liquid critical point in this scenario. The large increase of response functions seen in the experiments represents only an *apparent* singularity that eventually rounds off in a maximum, instead of giving rise to a real divergence, in correspondence to the continuous increase of density.

This can be understood because, when LDL regions form at low T, the volume per particle increases $\delta \overline{V} > 0$, while the entropy (disorder) per particle decreases $\delta \overline{S}$ (the molecules are ordered in a more extended tetrahedral structure), giving negative $\delta \overline{V} \delta \overline{S}$. As we have seen, this cross-fluctuation is proportional to the thermal expansivity α_P that, as a consequence, becomes negative at low temperature. In the same way, the anticorrelated fluctuations of volume and entropy for LDL regions reduce α_P at high T with respect to the expected value for an usual liquid. Therefore, the unusual behavior of α_P can be easily interpreted in this scenario.

The temperature where $\alpha_P = 0$ corresponds to the temperature of maximum density (TMD) and, by changing P, it forms in the P–T phase diagram a continuous line that bends toward low T when P increases, as shown by experiments. All the other anomalies we have mentioned so far can be inter-preted as consequences of the presence of the TMD line. Models and calculations have been shown

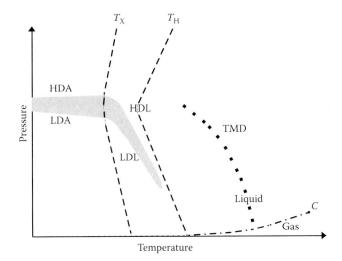

FIGURE 7.3 Schematic representation of the singularity-free interpretation in the pressure P, temperature T plane. The gray area represents the region where there occurs a sharp but continuous variation of density between HDA ice and LDA ice, below T_X, and between HDL and LDL above T_X and below T_H. The dotted line denotes the temperature of maximum density (TMD) in the liquid phase. The point-dotted line represents the gas–liquid coexistence line ending in the critical point C.

to be consistent with this scenario under some hypotheses, such as neglecting the correlations in the hydrogen-bond formation and breaking.

7.3.3 THE DIFFERENCE BETWEEN THE TWO SCENARIOS

The liquid–liquid phase transition hypothesis and the singularity-free interpretation are both thermodynamically consistent and only differ in the experimentally unaccessible region, between T_X and T_H. Away from this region, these two interpretations coincide, giving the same description of the TMD line and the unusual behavior of the response functions.

However, the presence of a new liquid phase at low T would represent a relevant feature in understanding the water dynamics, not only in the supercooled and glassy region, but also at moderately low temperatures, and it could be relevant for biological processes, such as the cold denaturation of protein [30]. For example, the enzymatic activity of most proteins ceases below 220 K [31, 32], the same temperature estimated for the liquid–liquid critical point of water [12], showing a possible relation between the two phenomena. This is in agreement with the idea that the dynamics of interfacial water and protein are highly coupled at low temperatures [33, 34]. Further investigation is needed to understand this relation.

7.4 WHAT ARE THE FULL IMPLICATIONS OF THE UNUSUAL PROPERTIES OF WATER?

Because it is difficult to reach a conclusive picture of the origin of the anomalies of water from experiments, we adopt a different approach, quite common in physics: we develop a schematic model that reproduces the water properties and explores, by theoretical calculations and computer simulations, its phase diagram, focusing on the supercooled region that is difficult to investigate through experimentation. The answers we can get in this way are not definitive, because they depend on the inevitable approximations included in the model; hence, different models can give different answers [35]. However, by studying how the answers change when we modify the features of the model, we can help in clarifying what the implications of the properties of water are.

7.4.1   THE CORRELATION OF THE HYDROGEN BOND NETWORK

We consider a model for water that (1) has the density anomaly, (2) forms a hydrogen bond network, and (3) has a parameter that describes how strong the correlation is among the hydrogen bonds formed by the same molecule [36–38]. The latter feature is introduced because experiments show that the relative orientations of the hydrogen bonds of a water molecule are correlated, with the average H–O–H angle equal to 104.45° in an isolated molecule, 104.474° in the gas, and 106° in the high-T liquid [39–41]. Therefore, there is an (intramolecular) interaction between the hydrogen bonds formed by the same molecule. This interaction depends on T, because the H–O–H angle changes with temperature, consistent with *ab initio* calculations [42] and molecular dynamics simulations [43, 44]. The strength of this intramolecular interaction represents the parameter we use in the model to regulate the correlation among the hydrogen bonds of the same molecule.

We first check, by theoretical calculations [36, 37] (Figure 7.4) and computer simulations [38] (Figure 7.5), that our model reproduces the known phase diagram of fluid water, with the liquid–gas phase transition ending in the critical point C and with the TMD line. As shown in the experiments, we find that the TMD line decreases with increasing P [45, 46].

If we fix the intramolecular interaction between hydrogen bonds to a positive value, we find that, in the deeply supercooled region, the liquid has a liquid–liquid phase transition ending in the critical point C' and following a line with a negative slope in the P–T phase diagram. Therefore, we recover the liquid–liquid critical point scenario.

7.4.2   CONNECTING THE TWO INTERPRETATIONS

At this point, it is natural to ask if the liquid–liquid critical point is a necessary consequence of the properties of water. With the schematic model we use, it is possible to answer this question by changing the values of (a) the parameters determining the density anomaly and (b) the parameter determining the correlation among the hydrogen bonds.

We first show that by decreasing the parameter in (b), associated with the correlation in the hydrogen bond network, the liquid–liquid critical point C' moves to lower T and higher P. When

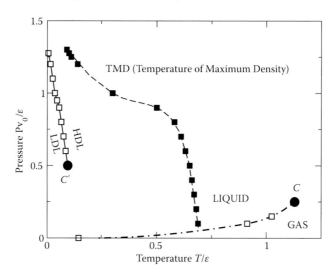

FIGURE 7.4   The mean field P–T phase diagram for the water model with intramolecular interaction. The phase diagram shows the gas–liquid first-order phase transition (dot-dashed) line ending in the critical point C, the (dashed) line of temperatures of maximum density (TMD), and the LDL–HDL first order phase transition (low-T continuous) line ending in the critical point C'. The quantities v_0 and ε are internal parameters of the model, as described in ref [38].

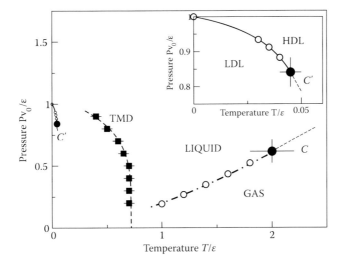

FIGURE 7.5 *P–T* phase diagram calculated by computer simulations for the model with intramolecular interaction with the symbols (see Figure 7.4). Thin dashed lines indicate the position of maxima of K_T (Widom line) emanating from the critical points C and C'. Inset: blowup of the HDL–LDL phase transition region.

C' "retraces" the line of the liquid–liquid phase transition, the divergence of the response functions occurring on this line is replaced by a maximum occurring in a region around a locus (the Widom line in Figure 7.5) in the *P–T* phase diagram.

The case with the parameter in (b) set to zero [27, 47, 48] is reached with continuity, with C' disappearing at $T = 0$ and at the corresponding pressure on the liquid–liquid phase transition line in Figures 7.4 and 7.5, leaving behind the Widom line with the maxima of the response functions, necessary in the singularity-free scenario.

Therefore, we first show that the liquid–liquid critical point is, in this model, a consequence of the hydrogen bond correlation and then show that the two scenarios presented above are related to each other. By decreasing the correlation of the hydrogen bond network, we pass with continuity from the liquid–liquid phase transition scenario to the singularity-free interpretation. The latter, within our schematic model, is only possible in the limiting case of no hydrogen bond correlation, a situation that is not consistent with the experimental data for water.

7.4.3 The Consequence of the Density Anomaly

In all the considered cases, we find that the density anomaly and the TMD line are not affected by the variation of the parameter in (b), consistent with the fact that these properties are also reproduced when this parameter is set to zero, i.e., in the singularity free scenario [27, 47, 48]. Instead, it is possible to see that, in our model, the parameter in (a) determining the density anomaly and the occurrence of the TMD line is associated with the local volume expansion caused by the hydrogen bond formation. This local expansion is also responsible for the open structure of the low-density liquid phase. We, therefore, conclude that the liquid–liquid phase transition is a necessary consequence of the density anomaly (because of the open structure) and of the hydrogen bond correlation characterizing the tetrahedral network.

7.5 ARE THESE ANOMALIES EXCLUSIVE PROPERTIES OF WATER?

The change of the local structure of the liquid water, related to its anomalies, has been proposed as a possible mechanism for biological processes [3, 30, 49–51]. Therefore, the debate about the essentiality of water for life could be relevant to understanding which liquids have the density anomaly.

Because we have shown that the density anomaly implies the liquid–liquid phase transition when the liquid forms a correlated network of bonds, it is natural to ask if we can use the occurrence of a liquid–liquid phase transition as an indicator of density anomalies. For example, recent experiments have shown a liquid–liquid phase transition in phosphorous [52–54] and in triphenyl phosphite [55, 56].

Apart from the clear evidence in phosphorous and triphenyl phosphite, experiments show data consistent with the possibility of a liquid–liquid phase transition in silica [45, 46, 57], carbon [58], aluminate liquids [59–61], selenium [62], and cobalt [63], among others [59–61, 64]. Moreover, computer simulations predict a liquid–liquid critical point for specific models of carbon [65], phosphorous [66], supercooled silica [45, 46, 67, 68], and hydrogen [69], besides the commonly used models of water [24, 28].

The strategy we follow to explore if the occurrence of a liquid–liquid phase transition is sufficient for the density anomaly in a liquid consists in looking at a simple model liquid with the liquid–liquid critical point and test for the density anomaly. To realize this investigation, we use a model where the interaction between molecules depends only on their relative distance and not on their relative orientation.

This kind of model has been used to represent materials such as liquid metals [70–78], colloids, or biological solutions [79]. They have also been proposed to study anomalous liquids, such as water [3, 80–85].

As we have seen, water has an open (at low P and T) and a closed (at high P and T) local structure [23, 86]. The existence of these two structures with different densities suggests to us a pair interaction with two characteristic distances. For example, in water, the shortest distance could be associated with the minimum distance between two non-hydrogen–bonded molecules (closed structure). The largest distance would correspond to the average distance when the hydrogen bond is formed (open structure), as in Figure 7.6. The tendency to have an open structure would correspond to a weak repulsion at lengths between the smallest and the largest distance. The energy gain for the hydrogen bond formation would attract the molecules at the largest distance. We analyze this simple model with two characteristic interaction distances by means of numerical simulations and theoretical calculations.

7.5.1 CRYSTAL OPEN STRUCTURE AND POLYMORPHISM

The first surprising result of our simulation is that this simple model is able to reproduce one of the most relevant features of water, i.e., a crystal open structure (Figure 7.7) [87]. The crystal has a complex unit cell of ten molecules, some at the smallest distance, others at the largest distance, with 8-fold and 12-fold symmetries.

This structure appears to be quite stable at a wide range of pressures and temperatures and is the only one we find by equilibrating the system with molecular dynamics simulations. However, because it does not correspond to the most compact configuration, we know that by increasing the pressure the system will ultimately take a (cubic or hexagonal) close packed form. Therefore, the model is able to reproduce the polymorphism typical of water-like substances. It has, at least, two crystals: a low-density crystal at low and moderate densities, and a high-density crystal at high densities (or pressures).

7.5.2 THE LIQUID–LIQUID CRITICAL POINT

By supercooling the liquid, we can avoid the crystal and study the metastable phase diagram below the melting temperature. Below the liquid–gas phase transition, ending in the liquid–gas critical point C_1, we observe two liquids in the supercooled phase, roughly corresponding to the two crystals mentioned above (Figure 7.8).

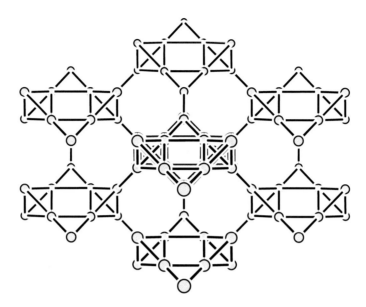

FIGURE 7.7 Crystal open structure of the spherical model with two characteristic distances. Bonds connect particles at the attractive distance. The radius of the particles is *not* in scale with the distances. Larger particles are closer to the observation point, reflecting the eight different levels represented in the figure. The configuration contains 15 cells. The central cell is emphasized by darker bonds. The cell is formed by 10 particles.

7.5.4 THE DENSITY BEHAVIOR

Once we fix the parameters of the model in such a way that satisfies the condition for the liquid-liquid phase transition, we test whether the liquid has an anomalous behavior in density. Surprisingly, we find, by means of thermodynamic integration based on our simulation results, that the liquid has no density anomaly [87, 89].

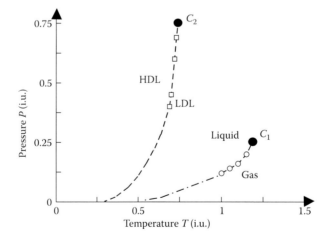

FIGURE 7.8 The *P–T* phase diagram for the spherical model with two characteristic distances, showing the gas–liquid critical point C_1 and the LDL–HDL critical point C_2. The two critical points (full circles) are located at the end of the gas–liquid coexistence (dot-dashed) line and the LDL–HDL coexistence (dashed) line, respectively. Open squares and circles are results of computer simulations. Lines are schematic guides to aid visualization. Pressure and temperatures are shown in internal units (i.u.), given by ten times the attractive energy divided by the hard-core volume for *P*, and the attractive energy divided by the Boltzmann constant for *T*.

This result excludes the possibility that the coexistence of two liquid phases in a substance, one with an open local structure and the other with a closed local structure, is sufficient for determining the anomalous behavior of density and the presence of a TMD line. Therefore, even if the density anomaly is not an exclusive property of water, because other substances such as silica show it, the sole presence of more than a liquid phase does not signal the occurrence of density maxima.

7.6 CONCLUSIONS

Many open questions remain and many experimental results are of potential relevance to the task of answering the question about the importance of water for life. For example, the dynamics of water could play a fundamental role in biological processes, such as determining the protein folding rate. Hydrophobic collapse and sharp turns (or bends) in polypeptide chains (groups of 50–100 amino acids) in the (secondary) structure of proteins involve the mediation of the water molecules in proximity to the amino acids [91]. Protein association could be determined by the dynamics in the water hydration layer around appropriate amino acid residue sites [92]. Interfacial water could rule the rate of recognition of binding sites of proteins by ligands, inhibitors, and other proteins [93] and the water dynamics at the surface of DNA and macromolecules is a promising field of research [33, 34].

In this context, it is unclear what could be the effect of the hypothesized second critical point. It is intriguing to observe that the estimated temperature of 220 K for the liquid–liquid critical point of water [12] coincides with the temperature below which the enzymatic activity of most proteins ceases [31, 32]. This liquid–liquid phase transition is probably hindered by inevitable freezing [38]. Indeed, it appears that the liquid–liquid phase transition is below, or at least close to, the glass transition temperature. A recent simulation analysis of the orientational dynamics of water at fixed density [94] has shown that the temperature of dynamical arrest of the system, defined by the (mode coupling) theory [95, 96], is relatively close in temperature and density to recent estimates of the liquid–liquid critical point [28].

The anomalous properties of water could be relevant for life in many ways. For example, trees survive arctic temperatures because the water in the cell does not freeze, even though the temperature is below the homogeneous nucleation temperature of $-38°C$. This effect, due to the confinement of water, is related to the difficulty for the molecules to form an ordered hydrogen bond tetrahedral network.

The correlation of the hydrogen bonds forming the network is, indeed, the reason of the occurrence of the liquid–liquid phase transition, whose necessity for water is supported by theoretical and numerical calculations [38, 97]. We find that the position of the liquid–liquid critical point depends on the strength of the hydrogen bond correlation. For example, if we assume uncorrelated hydrogen bonds [98, 99], we find that the liquid–liquid critical point disappears at $T = 0$, giving rise to the singularity-free scenario. However, the hypothesis of a vanishing hydrogen bond correlation does not apply to water [43], ruling out the singularity-free scenario for H_2O.

The ability of water to have an open and a closed structure determines the existence of polymorphs, the LDA, the HDA, and the VHDA ice, each in a possible correspondence with, respectively, the LDL, the HDL, and the very HDL (VHDL) shown by computer simulations [28, 100–103]. Their different, and anticorrelated, specific volume and entropy, appear to be the features that give rise to the anomalous properties of water.

However, these properties are not exclusive to water. Other tetrahedrally coordinated liquids have, at low temperature and low pressure, anticorrelated entropy and specific volume fluctuations. Examples of such systems are SiO_2 and GeO_2, known for their geological and technological importance. Recent simulations show, for example, that silica and silicon have a liquid–liquid critical point [67, 68], and experiments show the occurrence of a liquid–liquid critical point in phosphorous [52–54] and in triphenyl phosphite [55, 56].

Interestingly, some properties of water, such as the polymorphism or the existence of a low-density open crystal, could also be present in substances without density anomaly, but with two liquids with different local structures. Because the change of local liquid structure could be relevant in biological processes [3, 30, 34, 49–51], the possibility of a wide class of liquids with this property could help in understanding whether water is essential for life.

ACKNOWLEDGMENTS

We thank our collaborators, S. V. Buldyrev, N. Giovambattista, P. Kumar, G. Malescio, M. I. Marqués, F. Sciortino, A. Skibinsky, and M. Yamada. We thank Chemistry Programs CHE 0908218 and CHE0911389 for support. G. F. thanks the Spanish MICINN-FEDER grant FIS2009-10210.

REFERENCES

[1] F. Franks. 1985. *Biochemistry and Biophysics at Low Temperatures*. Cambridge: Cambridge University Press.

[2] L. J. Henderson. 1913. *Fitness of the Environment: An Inquiry into the Biological Significance of the Properties of Matter*. Basingtoke, Hampshire: Macmillan.

[3] P. G. Debenedetti. 1996. *Metastable Liquids: Concepts and Principles*. Princeton, NJ: Princeton University Press.

[4] B. M. Cwilong. 1947. Sublimation in Wilson chamber. *Proc. R. Soc. A*.190:137–143.

[5] D. Rosenfeld and W. L. Woodley. 2000. Deep convective clouds with sustained supercooled liquid water down to –37.5 degrees C. *Nature* 405:440–442.

[6] H. Kanno, R. J. Speedy, and C. A. Angell. 1975. Supercooling of water at –92°C. *Science* 189:880–881.

[7] P. Brügeller and E. Mayer. 1980. Complete vitrification in pure liquid water and dilute aqueous solutions. *Nature* 288:569–571.

[8] O. Mishima, L. D. Calvert, and E. Whalley. 1985. An apparently first-order transition between two amorphous phases of ice induced by pressure. *Nature* 314:76–78.

[9] O. Mishima. 1994. Reversible first-order transition between two H₂O amorphs at –0.2 GPa and 135 K. *J. Chem. Phys.* 100:5910–5912.

[10] O. Mishima. 1996. Relationship between melting and amorphization of ice. *Nature* 384:546–549.

[11] Y. Suzuki and O. Mishima. 2002. Propagation of the polyamorphic transition of ice and the liquid-liquid critical point. *Nature* 491:599–603.

[12] O. Mishima and H. E. Stanley. 1998. The relationship between liquid, supercooled and glassy water. *Nature* 396:329–335; 1998. Decompression-induced melting of ice IV and the liquid-liquid transition in water. *Nature* 392:164–168.

[13] T. Loerting, C. Salzmann, I. Kohl, E. Mayer, and A. Hallbrucker. 2001. A second distinct structural "state" of high-density amorphous ice at 77 K and 1 bar. *Phys. Chem. Chem. Phys.* 3:5355–5357.

[14] J. L. Finney, D. T. Bowron, A. K. Soper, T. Loerting, E. Mayer, and A. Hallbrucker. 2002. Structure of a new dense amorphous ice. *Phys. Rev. Lett.* 89:205503.

[15] T. Loerting, W. Schustereder, K. Winkel, C. G. Salzmann, I. Kohl, and E. Mayer. 2006. Amorphous ice: Stepwise formation of very-high-density amorphous ice from low-density amorphous ice at 125 K. *Phys. Rev. Lett.* 96:025702.

[16] P. Jenniskens, D. F. Blake, M. A. Wilson, and A. Pohorille. 1995. High-density amorphous ice, the frost on interstellar grains. *Astrophys. J.* 455:389–401.

[17] P. Jenniskens and D. F. Blake. July 7, 1997. High density amorphous ice. http://exobiology.nasa.gov/ice/high.html.

[18] R. J. Speedy and C. A. Angell. 1976. Isothermal compressibility of supercooled water and evidence for a thermodynamic singularity. *J. Chem. Phys.* 65:851–858.

[19] C. A. Angell, M. Oguni, and W. J. Sichina. 1982. Heat capacity of water at extremes of supercooling and superheating. *J. Phys. Chem.* 86:998–1002.

[20] D. E. Hare and C. M. Sorensen. 1987. The density of supercooled water. 2. Bulk samples cooled to the homogeneous nucleation limit. *J. Chem. Phys.* 87:4840–4845.

[21] L. Pauling. *The Nature of the Chemical Bond, and the Structure of Molecules and Crystals*. 3rd ed. Ithaca, NJ: Cornell University Press, 1960.

[22] D. Laage and J. T. Hynes. 2006. A molecular jump mechanism of water reorientation. *Nature* 311:832–835.

[23] A. K. Soper and M. A. Ricci. 2000. Structures of high-density and low-density water. *Phys. Rev. Lett.* 84:2881–2884.

[24] P. H. Poole, F. Sciortino, U. Essmann, and H. E. Stanley. 1992. Phase-behavior of metastable water. *Nature* 360:324–328.

[25] H. E. Stanley. 1979. A polychromatic correlated-site percolation problem with possible relevance to the unusual behavior of supercooled H_2O and D_2O. *J. Phys. A* 12: L329–L337.

[26] H. E. Stanley and J. Teixeira. 1980. Interpretation of the unusual behavior of H_2O and D_2O at low-temperatures — tests of a percolation model. *J. Chem. Phys.* 73:3404–3422.

[27] S. Sastry, P. G. Debenedetti, F. Sciortino, and H. E. Stanley. 1996. Singularity-free interpretation of the thermodynamics of supercooled water. *Phys. Rev. E.* 53:6144–6154.

[28] I. Brovchenko, A. Geiger, and A. Oleinikova. 2005. Liquid-liquid phase transitions in supercooled water studied by computer simulations of various water models. *J. Chem. Phys.* 123:044515.

[29] E. Trinh and R. E. Apfel. 1980. Sound velocity of supercooled water down to –33°C using acoustic levitation. *J. Chem. Phys.* 72:6731–6735.

[30] M. I. Marqués, J. M. Borreguero, H. E. Stanley, and N. V. Dokholyan. 2003. A possible mechanism for cold denaturation of proteins at high pressure. *Phys. Rev. Lett.* 91:138103.

[31] B. F. Rasmussen, A. M. Stock, D. Ringe, and G. A. Petsko. 1992. Crystalline ribonuclease-a loses function below the dynamic transition at 220-K. *Nature* 357:423–424.

[32] M. Ferrand, A. J. Dianoux, W. Petry, and G. Zaccai. 1993. Thermal motions and function of bacteriorhodopsin in purple membranes — effects of temperature and hydration studied by neutron-scattering. *Proc. Natl. Acad. Sci.* 90:9668–9672.

[33] J. Higo, M. Sasai, H. Shirai, H. Nakamura, and T. Kugimiya. 2001. Large vortex-like structure of dipole field in computer models of liquid water and dipole-bridge between biomolecules. *Proc. Natl. Acad. Sci.* 98:5961–5964.

[34] B. Bagchi. 2005. Water dynamics in the hydration layer around proteins and micelles. *Chem. Rev.* 105:3197–3219.

[35] P. G. Debenedetti. 2003. Supercooled and glassy water. *J. Phys. Condens. Mater.* 15:R1669–R1726.

[36] G. Franzese and H. E. Stanley. 2002. Liquid-liquid critical point in a Hamiltonian model for water: analytic solution. *J. Phys. Condens. Mater.* 14:2201–2209.

[37] G. Franzese and H. E. Stanley. 2002. A theory for discriminating the mechanism responsible for the water density anomaly. *Physica A.* 314:508–513.

[38] G. Franzese, M. I. Marqués, and H. E. Stanley. 2003. Intramolecular coupling as a mechanism for a liquid-liquid phase transition. *Phys. Rev. E.* 67:011103.

[39] C. W. Kern and M. Karplus. In *Water: A Comprehensive Treatise.* Vol. 1, pp. 21–91. Edited by F. Franks. New York, NY: Plenum Press, 1972.

[40] J. B. Hasted. In *Water: A Comprehensive Treatise.* Vol. 1, pp. 255–309. Edited by F. Franks. New York, NY: Plenum Press, 1972.

[41] K. Ichikawa, Y. Kameda, T. Yamaguchi, H. Wakita, and M. Misawa. 1991. Neutron-diffraction investigation of the intramolecular structure of a water molecule in the liquid-phase at high-temperatures. *Mol. Phys.* 73:79–86.

[42] P. L. Silvestrelli and M. Parrinello. 1999. Structural, electronic, and bonding properties of liquid water from first principles. *J. Chem. Phys.* 1999. 111:3572–3580.

[43] P. Raiteri, A. Laio, and M. Parrinello. 2004. Correlations among hydrogen bonds in liquid water. *Phys. Rev. Lett.* 93:087801.

[44] P. A. Netz, F. Starr, M. C. Barbosa, and H. E. Stanley. 2002. Translational and rotational diffusion in stretched water. *J. Mol. Liq.* 101:159–168.

[45] C. A. Angell, S. Borick, and M. Grabow. 1996. Glass transitions and first order liquid-metal-to-semiconductor transitions in 4-5-6 covalent systems. *J. Non-Cryst. Solids.* 207:463–471.

[46] P. H. Poole, M. Hemmati, and C. A. Angell. 1997. Comparison of thermodynamic properties of simulated liquid silica and water. *Phys. Rev. Lett.* 79:2281–2284.

[47] L. P. N. Rebelo, P. G. Debenedetti, and S. Sastry. 1998. Singularity-free interpretation of the thermodynamics of supercooled water. II. Thermal and volumetric behavior. *J. Chem. Phys.* 109:626–633.

[48] E. La Nave, S. Sastry, F. Sciortino, and P. Tartaglia. 1999. Solution of lattice gas models in the generalized ensemble on the Bethe lattice. *Phys. Rev. E* 59:6348–6355.

[49] F. Authenrieth, E. Tajkhorshid, K. Schulten, and Z. Luthey-Schulten. 2004. Role of water in transient cytochrome c_2 docking. *J. Phys. Chem. B* 108:20376–20387.

[50] P. M. Wiggins. 1990. Role of water in some biological processes. *Microbiol. Rev.* 54:432–449.

[51] P. M. Wiggins. 2001. High and low density intracellular water. *Cell. Mol. Biol.* 47:735–744.

[52] Y. Katayama, T. Mizutani, W. Utsumi, O. Shimomura, M. Ya-makata, and K. Funakoshi. 2000. A first-order liquid-liquid phase transition in phosphorus. *Nature* 403:170–173.

[53] Y. Katayama, Y. Inamura, T. Mizutani, M. Yamakata, W. Utsumi, and O. Shimomura. 2004. Macroscopic separation of dense fluid phase and liquid phase of phosphorus. *Science* 306:848–851.

[54] G. Monaco, S. Falconi, W. A. Crichton, and M. Mezouar. 2003. Nature of the first-order phase transition in fluid phosphorus at high temperature and pressure. *Phys. Rev. Lett.* 90:255701.

[55] H. Tanaka, R. Kurita, and H. Mataki. 2004. Liquid-liquid transition in the molecular liquid triphenyl phosphite. *Phys. Rev. Lett.* 92:025701.

[56] R. Kurita and H. Tanaka. 2004. Critical-like phenomena associated with liquid-liquid transition in a molecular liquid. *Science* 306:845–848.

[57] D. J. Lacks. 2000. First-order amorphous-amorphous transformation in silica. *Phys. Rev. Lett.* 84:4629–4632.

[58] M. van Thiel and F. H. Ree. 1993. High-pressure liquid-liquid phase change in carbon. *Phys. Rev. B* 48:3591–3599.

[59] S. Aasland and P. F. McMillan. 1994. Density-driven liquid-liquid phase-separation in the system Al_2O_3-Y_2O_3. *Nature* 369:633–636.

[60] M. C. Wilding and P. F. McMillan. 2001. Polyamorphic transitions in yttria-alumina liquids. *J. Non-Cyst. Solids* 293:357–365.

[61] M. C. Wilding, C. J. Benmore, and P. F. McMillan. 2002. A neutron diffraction study of yttrium- and lanthanum-aluminate glasses. *J. Non-Cryst. Solids* 297:143–155.

[62] V. V. Brazhkin, E. L. Gromnitskaya, O. V. Stalgorova, and A. G. Lyapin. 1998. Elastic softening of amorphous H_2O network prior to the *HDA-LDA* transition in amorphous state. *Rev. High Pressure Sci. Technol.* 7:1129–1131.

[63] M. G. Vasin and V. I. Ladýanov. 2003. Structural transitions and nonmonotonic relaxation processes in liquid metals. *Phys. Rev. E* 68:051202.

[64] P. F. McMillan. 2004. Polyamorphic transformations in liquids and glasses. *J. Mater. Chem.* 14:1506–1512.

[65] J. N. Glosli and F. H. Ree. 1999. Liquid-liquid phase transformation in carbon. *Phys. Rev. Lett.* 82:4659–4662.

[66] T. Morishita. 2001. Liquid-liquid phase transitions of phosphorus via constant-pressure first-principles molecular dynamics simulations. *Phys. Rev. Lett.* 87:105701.

[67] I. Saika-Voivod, F. Sciortino, and P. H. Poole. 2001. Computer simulations of liquid silica: Equation of state and liquid-liquid phase transition. *Phys. Rev. E* 63:011202.

[68] S. Sastry and C. A. Angell. 2003. Liquid-liquid phase transition in supercooled silicon *Nat. Mater.* 2:739–743.

[69] S. Scandolo. 2003. Liquid-liquid phase transition in compressed hydrogen from first-principles simulations. *Proc. Natl. Acad. Sci. USA* 100:3051–3053.

[70] G. Stell and P. C. Hemmer. 1972. Phase transition due to softness of the potential core. *J. Chem. Phys.* 56:4274–4286.

[71] F. H. Stillinger and T. Head-Gordon. 1993. Perturbational view of inherent structures in water. *Phys. Rev. E* 47:2484–2490.

[72] K. K. Mon, M. W. Ashcroft, and G. V. Chester. 1979. Core polarization and the structure of simple metals. *Phys. Rev. B* 19:5103–5118.

[73] M. Silbert and W. H. Young. 1976. Liquid metals with structure factor shoulders. *Phys. Lett. A* 58:469–470.

[74] D. Levesque and J. J. Weis. 1977. Structure factor of a system with shouldered hard sphere potential. *Phys. Lett. A* 60:473–474.

[75] J. M. Kincaid and G. Stell. 1978. Structure factor of a one-dimensional shouldered hard-sphere fluid. *Phys. Lett. A* 65:131–134.

[76] P. T. Cummings and G. Stell. 1981. Mean spherical approximation for a model liquid metal potential. *Mol. Phys.* 43:1267–1291.

[77] E. Velasco, L. Mederos, G. Navascués, P. C. Hemmer, and G. Stell. 2000. Complex phase behavior induced by repulsive interactions. *Phys. Rev. Lett.* 85:122–125.

[78] A. Voronel, I. Paperno, S. Rabinovich, and E. Lapina. 1983. New critical point at the vicinity of freezing temperature of K_2Cs. *Phys. Rev. Lett.* 50:247–249.

[79] M. M. Baksh, M. Jaros, and J. T. Groves. 2004. Detection of molecular interactions at membrane surfaces through colloid phase transitions. *Nature* 427:139–141.

[80] T. Head-Gordon and F. H. Stillinger. 1993. An orientational perturbation theory for pure liquid water. *J. Chem. Phys.* 98:3313–3327.

[81] E. A. Jagla. 2001. Liquid-liquid equilibrium for monodisperse spherical particles. *Phys. Rev. E* 63:061509.

[82] M. R. Sadr-Lahijany, A. Scala, S. V. Buldyrev, and H. E. Stanley. 1998. Liquid state anomalies for the Stell-Hemmer core-softened potential. *Phys. Rev. Lett.* 81:4895–4898.

[83] M. R. Sadr-Lahijany, A. Scala, S. V. Buldyrev, and H. E. Stanley. 1999. Water-like anomalies for core-softened models of fluids: One dimension. *Phys. Rev. E* 60:6714–6721.

[84] A. Scala, M. R. Sadr-Lahijany, N. Giovambattista, S. V. Buldyrev, and H. E. Stanley. 2001. Water-like anomalies for core-softened models of fluids: Two dimensional systems. *Phys. Rev. E* 63:041202.

[85] A. Scala, M. R. Sadr-Lahijany, N. Giovambattista, S. V. Buldyrev, and H. E. Stanley. 2000. Applications of the Stell-Hemmer potential to understanding second critical points in real systems [Festschrift for G. S. Stell]. *J. Stat. Phys.* 100:97–106.

[86] O. Mishima. 2000. Liquid–liquid critical point in heavy water. *Phys. Rev. Lett.* 85:334–336.

[87] G. Franzese, G. Malescio, A. Skibinsky, S. V. Buldyrev, and H. E. Stanley. 2002. Metastable liquid-liquid phase transition in a single-component system with only one crystal phase and no density anomaly. *Phys. Rev. E* 66:051206.

[88] G. Franzese, G. Malescio, A. Skibinsky, S. V. Bulderev, and H. E. Stanley. 2001. Generic mechanism for generating a liquid-liquid phase transition. *Nature* 409:692–695.

[89] A. Skibinsky, S. V. Buldyrev, G. Franzese, G. Malescio, and H. E. Stanley. 2004. Liquid-liquid phase transitions for soft-core attractive potentials. *Phys. Rev. E* 69:061206.

[90] G. Malescio, G. Franzese, A. Skibinsky, S. V. Buldyrev, and H. E. Stanley. 2005. Liquid-liquid phase transition for an attractive isotropic potential with wide repulsive range. *Phys. Rev. E* 71:061504.

[91] R. Baron and J. A. McCammon. 2007. Dynamics, hydration, and motional averaging of a loop-gated artificial protein cavity: The W191G mutant of cytochrome c peroxidase in water as revealed by molecular dynamics simulations. *Biochemistry* 46:10629–10642.

[92] C. J. Camacho, S. R. Kimura, C. DeLisi, and S. Vajda. 2000. Kinetics of desolvation-mediated protein-protein binding. *Biophys. J.* 78:1094–1105.

[93] S. Zou, J. S. Baskin, and A. H. Zewail. 2002. Molecular recognition of oxygen by protein mimics: Dynamics on the femtosecond to microsecond time scale. *Proc. Natl. Acad. Sci. USA* 99:9625–9630.

[94] P. Kumar, G. Franzese, S. V. Buldyrev, and H. E. Stanley. 2006. Molecular dynamics study of orientational cooperativity in water. *Phys. Rev. E* 73:041505.

[95] W. Götze. 1999. Recent tests of the mode-coupling theory for glassy dynamics. *J. Phys. Condens. Matter.* 11:A1–A45.

[96] W. Götze and L. Sjogren. 1992. Relaxation processes in supercooled liquids. *Rep. Prog. Phys.* 55:241–376.

[97] F. Sciortino, E. La Nave, and P. Tartaglia. 2003. Physics of the liquid-liquid critical point. *Phys. Rev. Lett.* 91:155701.

[98] A. Luzar and D. Chandler. 1996. Hydrogen-bond kinetics in liquid water. *Nature (London)* 379:55–57.

[99] F. W. Starr, J. K. Nielsen, and H. E. Stanley. 1999. Fast and slow dynamics of hydrogen bonds in liquid water. *Phys. Rev. Lett.* 82:2294–2297.

[100] I. Brovchenko, A. Geiger, and A. Oleinikova. 2003. Multiple liquid-liquid transitions in supercooled water. *J. Chem. Phys.* 118:9473–9476.

[101] S. V. Buldyrev and H. E. Stanley. 2003. A system with multiple liquid-liquid critical points. *Physica A* 330:124–129.

[102] N. Giovambattista, H. E. Stanley, and F. Sciortino. 2005. Relation between the high-density phase and the very high-density phase of amorphous solid water. *Phys. Rev. Lett.* 94:107803.

[103] J. A. White. 2004. Multiple critical points for square-well potential with repulsive shoulder. *Physica A* 346:347–357.

8 Counterfactual Quantum Chemistry of Water

Wesley D. Allen and Henry F. Schaefer, III

CONTENTS

8.1   INTRODUCTION

In physics and cosmology, anthropic principles and fine-tuning investigations regarding the physical laws of our universe have been established as fertile domains of inquiry over the past half-century.[1–3] Spectacular discoveries of apparent fine-tunings of the physical constants and cosmological boundary conditions have stimulated new research in areas such as M-theory and inflationary Big Bang cosmology. Of appeal to wider audiences, these discoveries have modernized and greatly enriched the ancient debate over purpose in human existence. Now, remarkable advances in computational molecular quantum mechanics afford the opportunity to rigorously extend classic fine-tuning inquiries into the realm of chemistry, and hence biology. By mathematical determination of the chemistry of counterfactual universes, governed by precise but alternative physical laws, the structure of our current existence would be profitably illuminated. Of particular significance, such investigations promise to reveal the realm of possibilities for the chemical fabric of biological complexity, i.e., whether life could exist if chemistry were appreciably different in various ways. More traditional scientific fruits are also likely from this broadening of perspective, because deeper theoretical frameworks and new experimental hypotheses for chemistry should be stimulated.

The general question of "counterfactual chemistry" and "fine-tuning" in biochemistry is controversial, not merely because it has philosophical and theological implications, but also because there is no consensus on how to approach the issue. A first impulse of biologists might be to conceive (explicitly or implicitly) of some abstract fitness landscape within the realm of the *fixed* chemistry of our own universe. Let us call this idea the *constrained chemistry* approach. Suppose that improving life fitness corresponds to descending on the fitness landscape, similar to optimizing a potential energy function for a physical system. Life as we know it would exist in some basin near a local minimum of the fitness landscape. The difficulties of precisely defining the life fitness function are severe, but the analogy is nonetheless highly instructive. The question is then whether alternative basins exist that are stable enough to support complex life. In other words, is the [H, C, N, O] biochemistry of our existence the only chemical solution to living complexity, or are there realms of undiscovered "counterfactual chemistry" that could equally support life? In this sense "fine-tuning" could mean that there are few, if any, other stable basins in the life-fitness landscape of chemistry; our basin would either be the only minimum, or simply the global minimum of superior depth to all competitors. In contrast, a fitness landscape with a cornucopia of basins of comparable depth would seem to vitiate fine-tuning interpretations. Of course, one could take the idea of fine-tuning in a highly restrictive and less profitable sense to merely mean "optimal adaptation," which in the fitness landscape analogy would simply be that our familiar biochemistry has settled down into the very bottom of the one basin that we know.

The constrained chemistry approach is evident in the proliferation of recent literature[4–7] addressing questions such as: Is water the only possible matrix of life, or could solvents such as ammonia and formamide support viable systems of complex biochemistry? Why are 20 standard amino acids exclusively used in Terran proteins? Why does Terran genetics use ribose and deoxyribose? Is a three-biopolymer system (DNA–RNA-proteins) essential for life? Even these fascinating and far-reaching questions probe a vanishingly small fraction of the chemical possibilities within our known universe. For example, silicon chemistry in environments drastically altered from our terrestrial conditions is not considered.

The variables comprising the life-fitness landscape in the constrained chemistry approach are the innumerable molecules possible within the known universe, as well as the all-important external environmental conditions. A proper analysis of this type must also take into account the historical development of the universe and the solar systems within, including the nucleosynthesis of the elements in ancient stars. To illustrate this point, consider some hypothetical life system based on more electropositive elements such as boron and carbon that would be destroyed in the presence of electronegative elements such as nitrogen and oxygen. Even if the electropositive life system were to appear as a local minimum on the life-fitness landscape, it is not likely that it would be actualized naturalistically, because the source of the life-bearing elements would presumably also yield the corrosive electronegative elements adjacent in the periodic table.

It is paramount to understand that use of the term "counterfactual chemistry," within the constrained chemistry approach, does not mean that any laws of physics that dictate chemical behavior and reaction possibilities have been changed. Rather, the term refers to the execution of life processes on a chemical scaffolding or within a complex chemical network that has yet to be discovered or actualized, although possible within existing physics. Such an approach to counterfactuals is quite unlike the classic fine-tuning investigations that have borne fruit in physics and cosmology. In essence, the fine-tuning variables have been tacitly changed from the mathematical parameters governing the structure of the underlying physics to the molecules and external conditions that support life. Until the lexicon is universally established, it is important to explicitly define terms and state one's conceptual approach before entering into fine-tuning arguments.

Consistent with the classic fine-tuning ideas of physics and cosmology, we take an *unconstrained chemistry* approach in this essay by considering on a more fundamental level what would happen to the most basic possibilities of chemical structure and reactivity if the laws of physics themselves were different. Of greatest merit for consideration are counterfactual universes that are derived from our known universe by some type of limited change or perturbation, whether in fundamental physics, or in the release of constraints on molecular properties imposed by the quantum mechanics of electrons. This perturbative approach to counterfactual universes is somewhat akin to the stability analyses that are pervasive in mathematical physics, computational chemistry, and myriad other scientific fields. In our approach the life-fitness landscape includes as variables the fundamental constants and physical laws of the universe, and thus genuine counterfactual chemistry is considered, that which cannot be investigated in the laboratory but which provides compelling cosmological elucidation. The starting point of such an analysis is the characterization of the reference system, in this case the quantum chemistry of our known universe.

8.2 *AB INITIO* QUANTUM CHEMISTRY

In a famous quote[8] from 1929, Nobelist P. A. M. Dirac summarized the gains in theoretical physics afforded by the scientific revolutions of quantum mechanics and relativity:

> The underlying physical laws necessary for the mathematical theory of a large part of physics and the whole of chemistry are thus completely known, and the difficulty is only that the exact application of these laws leads to equations much too complicated to be soluble.

Dirac succeeded by essentially pure cerebration in unifying relativity and quantum mechanics in the four-component wave equation that bears his name. Fortunately, the full relativistic mathematical structure of the Dirac equation is only necessary for the heaviest elements in the periodic table. Thus, the chemistry of [H, C, N, O], constituting the basis for life as we know it, is governed to very high accuracy by the one-component Schrödinger equation, the simplified nonrelativistic limit of the Dirac equation, provided that electron spin is imposed *ad hoc*. For a general system of nuclei and electrons, the time-independent Schrödinger equation for the many-particle wavefunction Ψ and the system energy E is

$$\left[-\frac{h^2}{8\pi^2} \sum_I \frac{1}{M_I} \nabla_I^2 - \frac{h^2}{8\pi^2 m_e} \sum_i \nabla_i^2 + e^2 \sum_{I>J} \frac{Z_I Z_J}{r_{IJ}} - e^2 \sum_{I,i} \frac{Z_I}{r_{Ii}} + e^2 \sum_{i>j} \frac{1}{r_{ij}} \right] \Psi = E\Psi, \quad (8.1)$$

where (I, J) and (i, j) are indices for nuclei and electrons, respectively, with attendant masses M_I and m_e, the Z_I are nuclear charges, and r_{IJ}, r_{Ii}, and r_{ij} denote nuclear–nuclear, nuclear–electron, and electron–electron distances, in order. In principle, all knowable properties of the system can be derived within quantum mechanics from Ψ and E. In popular accounts quantum mechanics has been dubbed the "champion" of physical theories. As described by Herbert,[9] "heaping success upon success, quantum theory boldly exposes itself to potential falsification on a thousand different fronts. Its record is impressive: quantum theory passes every test we can devise."

Of the four fundamental forces of nature, Equation (8.1) only involves electromagnetic interactions,* and the only parameters that appear are Planck's constant (h), the elementary charge (e), and the particle masses. These simplifications are possible because chemistry is remarkably uncoupled from the short-range nuclear forces that constitute the elements and the long-range gravitational forces that shape the universe. Nonrelativistic, first principles (*ab initio*) quantum chemistry mathematically determines the structures, properties, thermochemistry, and reactivities of molecules by directly converging on numerical solutions of Equation (8.1) without empirical parameterization. While Dirac's claim that the equations of quantum chemistry are much too complicated to be soluble is still true with respect to analytic mathematics, the dramatic advances of recent decades in electronic structure methods, numerical algorithms, and raw computing power have allowed us to determine solutions very close to the *ab initio* limit for molecular systems of reasonable size. Consequently, we not only have an indisputable epistemology for understanding real chemistry in full quantitative detail, but also have computational means of exploring virtual chemistry in counterfactual universes.

In order to factor out the dependence of the Schrödinger equation on fundamental constants, a scaling of (distances, energies) in Equation (8.1) to atomic units $[a_0 = h^2/(4\pi^2 m_e e^2), E_h = 4\pi^2 m_e e^4 h^{-2}]$ is ubiquitously employed, yielding

---

* In fact, only electrostatic terms are included in Equation (8.1). While magnetic terms give rise to important fine structure for spectroscopic characterization, for [H, C, N, O] chemistry they constitute only a negligible fraction of the total energy of interaction if external fields are not present.

$$\left[-\frac{1}{2}\sum_I \frac{\beta}{\mu_I}\nabla_I^2 - \frac{1}{2}\sum_i \nabla_i^2 + \sum_{I>J}\frac{Z_I Z_J}{\rho_{IJ}} - \sum_{I,i}\frac{Z_I}{\rho_{Ii}} + \sum_{i>j}\frac{1}{\rho_{ij}}\right]\Psi = \varepsilon\Psi, \tag{8.2}$$

in which the energy (ε) and distance (ρ) variables are now unitless, μ_I is the ratio of nuclear mass I to the mass of the proton (m_p), and $\beta = m_e/m_p \approx 1 / 1836$. Because $\beta \ll 1$ (and $\mu_I \geq 1$), nuclear and electronic motions have disparate classical time scales, a wide separation in velocities that has profound consequences for chemistry. In effect, the light, fast electrons instantaneously adjust to the motions of the slow, heavy nuclei, a phenomenon often described via a "flies on an ox" analogy. As a concrete example, consider the H$_2$ molecule. If root-mean-square classical velocities are extracted from the quantum mechanical kinetic energies of the nuclei and electrons in the ground state of the system, one finds the electrons moving at 2370 km s^{-1} while the nuclei are vibrating at 3.6 km s^{-1}, on average.

The electron/nuclear time scale separation is mathematically embodied in the highly accurate Born–Oppenheimer approximation, whereby the electronic part of the Schrödinger equation is first solved with clamped nuclei,

$$\left[-\frac{1}{2}\sum_i \nabla_i^2 + \sum_{I>J}\frac{Z_I Z_J}{\rho_{IJ}} - \sum_{I,i}\frac{Z_I}{\rho_{Ii}} + \sum_{i>j}\frac{1}{\rho_{ij}}\right]\Psi_e = \varepsilon_e\left(\boldsymbol{\rho}_{\text{nuc}}\right)\Psi_e, \tag{8.3}$$

yielding a potential energy surface $\varepsilon_e(\boldsymbol{\rho}_{\text{nuc}})$ for nuclear motion, and the resulting nuclear wave equation is then solved for the final rovibronic energy levels (ε):

$$\left[-\frac{1}{2}\sum_I \frac{\beta}{\mu_I}\nabla_I^2 + \varepsilon_e\left(\boldsymbol{\rho}_{\text{nuc}}\right)\right]\Psi_{\text{nuc}} = \varepsilon\Psi_{\text{nuc}}. \tag{8.4}$$

The topography of the surface $\varepsilon_e(\boldsymbol{\rho}_{\text{nuc}})$ provides the basis for ascribing geometric structures to molecules. The local minima occurring on this multidimensional surface correspond to the pervasive three-dimensional molecular structures of chemistry, on which virtually all chemical intuition is built. The implicit assumption made in drawing static molecular frameworks is that the nuclei are localized in wells centered about such structures and execute only small-amplitude vibrations away from their equilibrium positions.

The Pauli principle imposes a fundamental and far-reaching requirement on the electronic wavefunction (Ψ_e), namely, that all electrons (spin ½ fermions) are indistinguishable and Ψ_e must change sign when any two electrons are interchanged. In order to satisfy this mathematical requirement, many-electron wavefunctions are pervasively written as superpositions of Slater determinants, which comprise antisymmetrized products of single-particle functions called orbitals. For water and most biomolecules, Ψ_e is well approximated as a single Slater determinant constructed from the lowest-energy molecular orbitals of the system. The essential consequence of the Pauli principle is that no more than two electrons can occupy a given spatial orbital. This occupancy limit prevents all the electrons from condensing into the lowest-energy orbital of a chemical species. Instead, the familiar Aufbau Principle must be followed, whereby electrons are successively added in pairs to the molecular orbitals of the system in order of increasing energy. Even partial relaxation of the Pauli principle would generate counterfactual chemistry of strikingly different character.

In summary, the quantum chemistry of life in our universe rests on the following foundations: (1) the nonrelativistic* Schrödinger equation, (2) electromagnetic forces alone, (3) the fundamental

---

* Although of very small abundance in living systems, first-row transition metals do play critical roles in biochemistry, especially in the active sites of proteins. The classical speed of the 1s core electron in the first Bohr orbit ranges from 0.15c to 0.22c in this series of atoms. Accordingly, they are on the periphery of elements for which relativistic effects can be reasonably neglected.

indistinguishability of all electrons, and (4) the Born–Oppenheimer separation of nuclear and electronic motion that gives rise to molecular structure.

8.3 MOLECULAR AND THERMODYNAMIC PROPERTIES OF WATER

Recognizing the firm epistemological foundations of chemistry in our universe is essential to understanding which variables may be fine-tuned in fitness for life arguments, or in particular how properties of water may be logically unconstrained in considering counterfactual universes. Because the extreme quantitative accuracy and broad validity of the Schrödinger equation has been demonstrated repeatedly, one should not think of rigorous *ab initio* quantum mechanical methods as mere chemical models, but rather essentially exact embodiments of chemical reality. Therefore, the unusual molecular and thermodynamic properties of water that intrigued Henderson[10,11] in his biocentricity hypothesis of 1913 are inescapable and regular consequences of Equations (8.3) and (8.4), as well as the statistical mechanical principles that connect macroscopic observables to underlying microscopic phenomena. Because the Schrödinger equation was not discovered until the 1920s, Henderson could not have fully appreciated this fact.

Although there is nothing mysterious about the molecular quantum mechanics of water, it is indeed "one of the strangest substances known to science" from an empirical macroscopic perspective.[12] Water boils at a temperature about 200°C higher than expected from linear extrapolation from the (H_2S, H_2Se, H_2Te) series of congeners. By weight, water has a higher heat of vaporization than any known substance. Without these properties, liquid water would not exist in appreciable quantities on earth, and there would be no plentiful solvent to support life. Water also exhibits an anomalously high specific heat capacity, thermal conductivity, and surface tension. The dielectric constant of water exceeds that of almost all simple pure liquids (hydrogen cyanide and formamide being exceptions), and thus water has an exceptional capacity for stabilizing polar and ionic species in solution. The high polarity of water as a solvent forms the basis for the intricate interplay of hydrophobic and hydrophilic interactions that shape biochemical macromolecules such as enzymes and nucleic acids into their biologically active forms. The water–lipid disparity in polarity drives the self-organization of cell membranes that provide essential compartmentalization for biological processes. The solvent properties of water also engender CO_2 acid–base chemistry that serves effectively both to distribute carbon atoms and to buffer living tissues near pH neutrality. Finally, water expands upon freezing, a highly unusual property that causes ice to float and prevents the lake and oceans on earth from solidifying from the bottom up.[10,12]

The peculiar macroscopic properties of water result directly from the microscopic potential energy surfaces of Equation (8.3) upon application of standard methods of statistical mechanics and molecular dynamics. We are not claiming that exact computations of this sort are yet feasible. For example, with current technology we cannot directly observe liquid water expand upon freezing into its ice-I_h crystal structure by performing molecular dynamics simulations of a large ensemble of molecules moving on explicitly computed, converged *ab initio* potential energy surfaces. However, the *ab initio* computation of highly accurate surfaces for small water clusters is now possible, as well as large-scale dynamical simulations of freezing involving thousands of water molecules moving on physically correct, parameterized model surfaces. In brief, there are no gaps in the link between the quantum mechanical solutions of Equation (8.3) and the macroscopic properties of water. Continuing technological improvements will serve to increase the scale, rigor, and quantitative accuracy of first principles simulations of liquid water.

The microscopic origins of the unusual macroscopic properties of water can be traced* principally to the directionality and strength $[D_0(H_2O-H_2O) = 3.3 \pm 0.7$ kcal $mol^{-1}]^{13,14}$ of the hydrogen bond between two H_2O molecules, an H_2O equilibrium bond angle $[\theta_e(H_2O) = 104.50 \pm 0.02°)]^{15}$ not too far removed from the idealized tetrahedral value (109.47°), and a large molecular dipole moment

---

* See other essays in this volume, particularly the contributions of J. Finney and M. Chaplin.

$[\mu_0(H_2O) = 1.85498 \pm 0.00009 \text{ D}]$.[16] The best purely *ab initio* quantum mechanical computations at the time of this writing give $D_0(H_2O\text{–}H_2O) = 3.3 \pm 0.1$ kcal mol^{-1},[14,17] $\theta_e(H_2O) = 104.485°$,[18,19] and $\mu_0(H_2O) = 1.85$ D, in essentially exact agreement with experiment, and with a much smaller uncertainty for the hydrogen bond strength.

In recent years a virtually complete triumph of *ab initio* theory has been achieved for the molecular properties of H$_2$O. Before 2001, the bond dissociation energy $D_0(H\text{–}OH)$ was firmly believed to be 118.08 ± 0.05 kcal mol^{-1}, an empirical number principally based on a short-range Birge–Sponer spectroscopic extrapolation of the vibrational levels of diatomic OH in its $A^2\Sigma^+$ electronic state.[20] Electronic structure computations brought the experimental analysis into dispute, and eventually three separate experiments utilizing positive-ion cycles gave a consensus value of $D_0(H\text{–}OH) = 117.59 \pm 0.07$ kcal mol^{-1}.[21] In 2004, a purely *ab initio* thermochemical study,[22] employing state-of-the-art methods and incorporating all relevant physical effects, produced $D_0(H\text{–}OH) = 117.62$ kcal mol^{-1}, with an uncertainty comparable to the revised experiments. In reflecting on these developments, it is remarkable that the world's thermochemical database had an error of 0.5 kcal mol^{-1} for a quantity as fundamental as the O–H bond strength of water. It should strike us with even greater impact that pure molecular quantum mechanics was able to identify this disparity and pinpoint the bond dissociation energy of water to within 0.1 kcal mol^{-1}. An additional landmark occurred in 2003 with the publication of a theoretically based semiglobal potential energy surface of water of unprecedented accuracy.[18] This surface reproduced 17,795 spectroscopic rovibrational levels of water isotopologs up to 25,000 cm^{-1} in excitation energy with a standard deviation less than 1 cm^{-1}.

Our point in discussing the theoretical mastery over the properties of water is to show that these features, although in cases unusual in comparison to related compounds, are not independently fine-tuned if the known laws of physics are retained. If we are to marvel at the fitness of water for life, then we should properly marvel not at independent properties of the universal solvent, but at the remarkable richness of the mathematical solutions arising out of the elegant and encompassing form of the Schrödinger equation itself.

8.4   CHEMICAL TUNING OF THE FUNDAMENTAL CONSTANTS

The solutions to the scaled electronic Schrödinger equation [Equation (8.3)] are pure numbers independent of the fundamental constants. Ostensibly then, the chemical properties of universes with the same mathematical laws but altered fundamental constants can be trivially determined once the dimensionless surfaces $\varepsilon_e(\rho_{nuc})$ have been mapped. In the case of water, Equation (8.3) yields an equilibrium O–H bond distance of 1.8100 and an atomization energy of 0.349960,[18,19,22] corresponding to $r_e = 95.782$ pm and AE$_0 = 110,509$ K in our current universe and system of measurement. In a counterfactual universe with (h, e, m_e) scaled by factors of $(10, 10^{-1}, 10^{-1})$ to $(6.62607 \times 10^{-33}$ J s, 1.60218×10^{-20} C, 9.10938×10^{-32} kg$)$, the atomic units (a_0, E_h) change by factors of $(10^5, 10^{-7})$, and suddenly $(r_e, AE_0) = (9.6 \ \mu\text{m}, 0.011$ K$)$! We could conceive of concurrently scaling c by 10^{-3} in order to leave the fine structure constant $\alpha = 2\pi e^2/(hc)$ invariant. In such a universe [H, C, N, O] chemistry would still be nonrelativistic ($\alpha \ll 1$), and the Born–Oppenheimer approximation would still be valid ($\beta \approx 1/18,360$). Because $r_e = 9.6 \ \mu\text{m}$ and AE$_0 = 0.011$ K, we might naively think that quantum effects could be observed visually under a microscope, provided we locked ourselves in a freezer very near absolute zero to prevent the water molecules from dissociating! This conundrum is resolved when we realize that *all* distances, including the dimensions of the observer and any measuring devices, would also be scaled by 10^5 in such a universe. In brief, both quantum and classical entities would increase in size proportionately, and there would be no observational differences in such a counterfactual universe. The same conclusion arises for energies and temperatures. As Barrow[23] has pointed out, only by changing dimensionless ratios of fundamental constants will observational distinguishability from our known universe occur.

The only dimensionless ratios that have consequences for chemistry are the fine structure constant (α) and the electron–proton mass ratio (β). Conventional, nonrelativistic quantum chemistry within the Born–Oppenheimer approximation assumes that both ratios are negligibly small. However, via perturbation theory we can now determine the consequences of these simplifications. For example, in our world the (bond distance, bond angle, barrier to linearity) features of H_2O are (95.782 pm, 104.50°, 16,004 K), and the relativistic and non-Born–Oppenheimer effects on these quantities are (+0.016 pm, –0.074°, +83 K) and (+0.003 pm, +0.015°, –20 K), respectively.[19,24] Clearly, these contributions to water properties are minuscule. In fact, [H, C, N, O] chemistry in our world is already very near the limit of vanishing α and β.

While varying α and β by factors less than 10 might have no appreciable effect on chemistry, allowing these quantities to significantly increase toward unity would have drastic and profound consequences. Barrow has already sketched out the habitable zone in the (α, β) plane for the existence of life-supporting complexity.[25] There are a number of cosmological constraints to be considered. If β exceeds roughly $0.005\alpha^2$, there would be no stars, because their centers would not be hot enough to initiate nuclear reactions. Moreover, modern unified gauge theories suggest that α must lie between 1/180 and 1/85 to prevent protons from decaying long before star formation.[26] Here we are most concerned with *chemical* conditions on α and β. A salient aspect is the molecular structure catastrophe that occurs somewhere between $\beta = 0.01$ and 1. In this region we encounter a complete breakdown of the Born–Oppenheimer separation of motions, and molecular structure is lost. We should emphasize that there are no mathematical singularities in this "catastrophe," but rather a smooth transformation of existence into a form deleterious for life. Contemplating such a universe leaves practicing chemists in bewilderment, because all intuition built on the ability to draw molecular frameworks is invalidated. Nuclei could be as delocalized within the system as any of the electrons. Life as we know it would not be possible. There would be no DNA double helix, no protein folds, no active sites of enzymes, and blurred solvent/solute distinctions; indeed, all the familiar cellular structures would erode. We would only be left with a murky quantum soup of particles. It is fascinating that as β increases significantly above 1, structure reemerges from the quantum soup, but with the bizarre characteristic of slow-moving electrons surrounded by a cloud of fast, delocalized nuclei.

The consequences of the molecular structure catastrophe have not been seriously considered by the quantum chemistry community. There is certainly merit in discerning what functional and information-carrying possibilities would exist in the quantum soup near $\beta = 1$. Nevertheless, preferring the perturbative conceptualization advocated above in our Introduction, we would first advocate research to trace out the β-evolution of chemistry from its existing form up to the threshold of molecular structure breakdown, with the intent of more precisely defining and characterizing the boundaries for living complexity. In the past few years, *ab initio* methods based on first- and second-order corrections to the Born–Oppenheimer approximation[27–30] have attracted interest from high-level computational molecular spectroscopy. Indeed, such techniques have very recently been used to construct the aforementioned groundbreaking semiglobal potential energy surface for the rovibrational spectroscopy of water.[18] In addition, in recent years a number of small-molecule investigations have advanced rigorous non-Born–Oppenheimer variational methods that do not resort to perturbation theory and provide definitive means of determining the characteristics of the quantum soup near $\beta = 1$. These developing non-Born–Oppenheimer methodologies could be applied to β-counterfactual chemistry to better understand how sensitive the existence of chemical structure is to the electron–proton mass ratio.

A few rough calculations serve to illustrate the scale of chemical changes that would result if the electron and proton were of similar mass. As a first example, consider the hydrogen atom, whose exact, nonrelativistic wavefunction for the ground state yields

$$E(\text{H atom}) = -\frac{1}{2}\left(\frac{1}{\beta+1}\right) \quad E_h = -313.75\left(\frac{1}{\beta+1}\right) \text{ kcal mol}^{-1} \quad (8.5)$$

and

$$\langle r \rangle = \frac{3}{2}(\beta+1)\, a_0 = 79.38(\beta+1) \ \text{pm}, \tag{8.6}$$

for the total energy and the expectation value of the electron/nuclear distance, respectively. From these equations, increasing β from its observed value ($\beta_0 \approx 1/1836$) to 1, reduces the ionization energy of the hydrogen atom from 313.58 to 156.88 kcal mol^{-1} and increases the average particle separation from 79.42 to 158.76 pm. Therefore, the atom itself becomes much less stable, an effect that should generally occur for molecules as well. If β is increased only to 0.1, then the changes in the ionization energy and average separation are less dramatic, -9.0% and $+9.9\%$, respectively. Rigorous molecular examples of β dependence would require extensive computations, but as a starting point one could consider the computationally tractable (first-order) Born–Oppenheimer diagonal correction (BODC).[27,30]

$$E_{\text{BODC}} = -\frac{1}{2}\sum_I \frac{\beta}{\mu_I}\left\langle \Psi_e \left| \nabla_I^2 \right| \Psi_e \right\rangle, \tag{8.7}$$

written in the same notation as Equation (8.3) and Equation (8.4). *Ab inito* computations of the BODC have been reported for a number of small molecules with the implicit assumption of $m_e/m_p = \beta_0 \approx 1/1836$. Scaling these results by a factor β/β_0, we can obtain first-order corrections for counterfactual chemistry. For the O–H bond energy in water discussed above, the BODC effect is $+0.11$ kcal mol^{-1} with the standard masses.[22] If β is increased to 0.1, D_0(H–OH) increases by a substantial 20 kcal mol^{-1}, or 17%. Using the BODC for values of β much larger than 0.1 is probably not prudent, because the first-order approximation is not likely to be sufficiently valid.

The generation of counterfactual [H, C, N, O] chemistry by increasing the fine-structure constant α should also add to fine-tuning arguments. Because α enters into the Dirac equation as a product with the nuclear charge Z, artificially increasing α ushers in relativistic chemistry earlier in the periodic table. If [C, N, O] become relativistic atoms in a counterfactual universe, then a plethora of questions arises: Will the bond angle, dipole moment, or hydrogen-bond strength of H$_2$O change in adverse ways for maintaining the unusual macroscopic properties of water? Will carbon maintain its ability to concatenate by forming stable bonds with itself? Will changes in relative [H, C, N, O] bond strengths result in the breakdown of protein backbones? Clearly there is an abundance of intriguing research that could be carried out here.

Although Einstein proposed the theory of special relativity in 1905, its impact on chemistry was not recognized until the 1970s. The characteristic properties of gold, the liquid nature of mercury, and the unusually low valence of thallium and lead compounds are all relativistic effects in chemistry. As a vivid illustration, the glitter of gold can be ascribed to relativity.[31,32] The 6s valence orbital of gold experiences a strong and preferential relativistic contraction. Accordingly, the 4d-5s band gap of silver is 3.5 eV, whereas the analogous 5d-6s band gap in gold is only 2.4 eV, moving the electronic excitation energy from the ultraviolet region into the visible spectrum. The characteristic yellow glitter of gold results from the 5d-6s absorptions in the blue visible region. If the fine structure constant were larger, such remarkable relativistic effects would occur for much lighter elements.

Once again, let us return to the hydrogen atom for some instruction, this time concerning the influence of α on electronic structure. The full four-component Dirac equation can be solved exactly for the hydrogen atom.[31,33] If the rest mass energy of the electron is removed from the solution by referencing the system to the infinitely separated particles, then

$$E(\text{H atom}) = \frac{\sqrt{1-\alpha^2}-1}{\alpha^2(\beta+1)} \quad E_h = -627.51\frac{\left(1-\sqrt{1-\alpha^2}\right)}{\alpha^2(\beta+1)} \ \text{kcal mol}^{-1} \tag{8.8}$$

is the energy of the quantum ground state. In the limit $\alpha \to 0$, we recover the nonrelativistic energy of Equation (8.5), while in the limit $\alpha \to 1$, we obtain a binding energy twice as large. Therefore, the atom itself becomes more stable against ionization as α increases, an effect in the opposite direction from that encountered above for increasing β. In essence, the increase in α enhances relativistic contraction of the electron around the nucleus.

For molecular studies of sensitivity to the fine structure constant, we can adapt existing methods[31,32] of relativistic quantum chemistry to allow for variation in α. The approximate Breit–Pauli Hamiltonian provides scalar corrections for relativistic effects and avoids the need for multi-component wavefunctions for modest values of $Z\alpha$. There are three terms in this Hamiltonian that are commonly computed in electronic structure theory: (1) the well-known mass–velocity effect resulting from the variation of the mass of the electron with its speed, as described by

$$\hat{H}_{mv} = -\frac{\alpha^2}{8} \sum_i \nabla_i^4, \tag{8.9}$$

(2) the Darwin term arising from the smearing of the charge of the electron due to its relativistic motion, quantified by

$$\hat{H}_D = \frac{\alpha^2}{8} \sum_i \nabla_i^2 V, \tag{8.10}$$

where V is the total Coulombic potential of electon–electron repulsion and electron–nuclear attraction; and (3) the spin–orbit coupling involving the interaction of the spin magnetic moment of the electron with the magnetic field generated by its orbital motion. Equations (8.9) and (8.10) are written in atomic units.

In a first-order perturbation theory approach, the mass–velocity and Darwin relativistic energy corrections (mv+D) are evaluated by taking the expectation values of $\hat{H}_{mv}$ and $\hat{H}_D$ over the nonrelativistic wavefunction Ψ_e. Scattered results for these corrections can be found in the literature for prototypical chemical species. As in our BODC analysis above, the first-order relativistic corrections can be used in rough, preliminary explorations of counterfactual chemistry by scaling them upward from the observed value of α. One difference is that the mass–velocity and Darwin corrections scale with α^2, rather than linearly as in the β case.

For the O–H dissociation energy of water, the mv+D correction is -0.14 kcal mol^{-1},[22] which serves to *reduce* the chemical binding energy in proportion to α^2. By extrapolation we deduce that $D_0(\text{H–OH})$ would become 0 at $\alpha = 0.21$, i.e., there would be no O–H chemical bond at all if the fine-structure constant were this large. Of course, this rough prediction may extend the first-order mv+D analysis outside its region of validity. However, it is noteworthy in this regard that the next higher (α^4) correction in the expansion of the exact Dirac energy formula for the hydrogen atom [Equation (8.8)] only amounts to 3 kcal mol^{-1} for $\alpha = 0.2$.

Among the limited data available in the literature, the relativistic mv+D contributions to atomization energies of [H, C, N, O] compounds are universally negative.[22] The first-order prediction of a rough $\alpha \approx 0.2$ threshold for the existence of chemical bonds seems typical for such molecules. For example, the enormous 381 kcal mol^{-1} nonrelativistic atomization energy of CO_2 is entirely canceled by the first-order mv+D correction if $\alpha = 0.20$. We should emphasize that these tentative predictions give $\alpha \approx 0.2$ as an upper limit for *any chemistry at all*; drastic changes from existing chemistry would occur well before this limit is reached. One final example highlights this point. The nonrelativistic bond energies of the CH and OH diatomic molecules are 80 and 102 kcal mol^{-1}, respectively, whereas the mv+D corrections to these quantities are -0.043 and -0.129 kcal mol^{-1}, in order.[22] Because the relativistic reduction of the bond energy is much greater for OH than for CH, there will be a critical α value above which the strength of these bonds is reversed. The first-order mv+D treatment predicts this critical point to be $\alpha = 0.12$. The consequences on organic chemistry of O–H bonds being inherently weaker than (or even comparable to) C–H bonds would be profound. If the question of fine-tuning in chemistry is to be satisfactorily explored, then some serious computational research is warranted on the dependence of chemistry on the fine-structure constant, as well as the electron–proton mass ratio.

8.5 SPAWNING COUNTERFACTUAL CHEMISTRY FROM THE BORN–OPPENHEIMER SEPARATION

In the classic fine-tuning arguments of physics,[12] the forms of the physical laws are assumed to be unchanged, and the investigation centers on the consequences of changing dimensionless ratios among the fundamental constants therein. As Barrow[34] nicely explains, "The fact that we can conceive of so many alternative universes, defined by other values of the constants of Nature, may be simply a reflection of our ignorance about the strait-jacket of logical consistency that a Theory of Everything demands." In searching strenuously for such a unified theory, one of Einstein's motivations was to discover "whether God could have made the world in a different way; that is, whether the necessity of logical simplicity leaves any freedom at all."[35]

In addition to varying fundamental constants, an obvious means of generating counterfactual physics is to change the exponents in the gravitational and Coulombic inverse-square laws. However, going all the way back to Immanuel Kant (1724–1804),[36] it has been understood that there is an inextricable connection between the exact, integral exponent of two in these laws and the existence of three (unfolded) dimensions of space. Ehrenfest[37,38] showed that in worlds with more than three such spatial dimensions, no stable atoms exist. Conversely, if three spatial dimensions are present, then an inverse-square law is required for stable classical orbits of an object around a central mass. In brief, three-dimensional worlds with inverse-square laws and a single arrow of time have a unique structure for the existence of chemistry and complex life.[39] This type of fine-tuning would be destroyed by changing the exponent of the Coulombic potential energy term in the Schrödinger equation.

In this essay we have already discussed how quantum mechanics with only electromagnetic forces dictates the individual molecular and thermodynamic properties of chemical substances. Without significantly restructuring the physical laws, the only parameters available for fine-tuning chemistry are the fine structure constant and the electron–proton mass ratio. How can we logically escape this "strait-jacket" to conceive of counterfactual universes with a vast richness of chemical possibilities? The best answer lies in the Born–Oppenheimer separation of nuclear and electronic motion, as embodied in Equations (8.3) and (8.4).

Once the potential energy surfaces $\varepsilon_e(\mathbf{\rho}_{nuc})$ are provided in Equation (8.4), the nuclear Schrödinger equation governs molecular properties and chemical reactivities without explicit reference to electrons. Here we might break the link and release the existing constraints of electronic quantum mechanics on $\varepsilon_e(\mathbf{\rho}_{nuc})$. By invoking perturbations in $\varepsilon_e(\mathbf{\rho}_{nuc})$, we are implicitly assuming: (a) unspecified, alternative (probably more complicated) laws of electronic motion can be constructed that are logically consistent with the perturbation; or (b) electrons are not present in the counterfactual universe, and the $\varepsilon_e(\mathbf{\rho}_{nuc})$ functions are elevated to the status of fundamental constants. To retain photochemistry, we would have to independently specify not only the $\varepsilon_e(\mathbf{\rho}_{nuc})$ surfaces but also the transition probabilities among them. In actuality, we would not care about arbitrary variations of the $\varepsilon_e(\mathbf{\rho}_{nuc})$ surfaces, but would only focus on the salient features such as the positions and energy separations of local minima and saddle points.

To illustrate the concept, consider a mathematical form that is currently employed to represent the ground-state potential energy surface of H_2O in terms of its bond distances (r_1, r_2) and bond angle (θ)[40]:

$$V(r_1,r_2,\theta) = V_1(r_1) + V_1(r_2) + V_2[r_{HH}(r_1,r_2,\theta)] + V_3(r_1,r_2,\theta), \tag{8.11}$$

where

$$V_1(r) = De^{-a(r-r_0)}[e^{-a(r-r_0)} - 1], \tag{8.12a}$$

$$V_2(r) = Ae^{-br}, \tag{8.12b}$$

and $V_3(r_1,r_2,\theta)$ is an exponentially damped polynomial expansion with numerous adjustable coefficients. The $V_1(r)$ function is a simple Morse potential, whose parameters (D, r_0, a) are the (dissociation energy, equilibrium distance, curvature) for stretching an individual O–H bond, the types of parameters that might serve as unconstrained, "fundamental constants" in counterfactual chemistry made possible by the Born–Oppenheimer separation. The model potential energy functions for water clusters contain similar parameters, such as the hydrogen-bond strength for two-body interactions.

The countless possibilities for counterfactual chemistry of this type could be explored by large-scale computer simulations in which the key parameters in the potential energy surfaces for the interaction of water molecules with themselves and with biochemical macromolecules are independent fine-tuning variables. This effort would restore the fitness inquiries of Henderson,[10,11] but now with a logical basis and computational means for probing the space of counterfactual water chemistry and quantitatively assessing the fine-tuning issue. Representative topics for investigation would be: (a) What is the precise functional dependence of the heat capacity and surface tension of bulk water on the intermolecular hydrogen-bond strength? (b) What changes in water cluster potentials would prevent water from expanding upon freezing? (c) How much can the dipole moment of water be altered before there is erosion of the hydrophobic cores of proteins, the cellular reaction chambers for organic synthesis? Questions of this kind have been addressed previously by statistical mechanics and molecular dynamics investigations; however, the possibilities for further research are virtually unlimited.

The chapter by Dr. Lynden-Bell in this volume describes interesting preliminary research on quasi-water that we would categorize as spawning counterfactual chemistry from the Born–Oppenheimer separation. Therein the complexities of intermolecular water interaction potentials are stripped down to the most essential physics by investigating a rudimentary model. This instructive model has a spherical hard-wall repulsion component added to an orientational term built from simple point charges on the hydrogen and oxygen positions. Computer simulations were used to map the behavior of liquid quasi-water as a function of density, temperature, and hydrogen bond strength, revealing boundaries between tetrahedral structure and "normal" liquids. In our view, much more research of this type is warranted to enrich fine-tuning discussions.

8.6   PERTURBING THE PAULI PRINCIPLE

Before 1927, there was no satisfactory explanation for the formation of strong chemical bonds between atoms, and thus the existence of stable molecules was an enigma. In that year Heitler and London[41] solved this fundamental problem by formulating a quantum mechanical valence bond (VB) theory for the paradigmatic hydrogen molecule. It is worth reviewing the key elements of the VB treatment of H_2,[42] because the analysis reveals the striking consequences of the Pauli principle for all of chemistry.

Within the Born–Oppenheimer approximation, a trial wavefunction for the two-electron H_2 system can be constructed by multiplying one-electron functions centered on the individual hydrogen atoms (labeled H_A and H_B). A trenchant choice for the one-electron functions is the familiar $1s$ orbitals s_A and s_B centered on H_A and H_B, respectively, comprising the ground-state quantum mechanical wavefunctions of the isolated hydrogen atoms. Without the Pauli principle, one would naively write the (unnormalized) trial H$_2$ wavefunction simply as

$$\Psi_e(1,2) = s_A(1)s_B(2). \tag{8.13}$$

However, Equation (8.13) violates the indistinguishability of electrons. To satisfy the Pauli principle, one must instead write for the ground-state wavefunction of H$_2$

$$\Psi_e(1,2) = [s_A(1)s_B(2) + s_A(2)s_B(1)][\alpha(1)\beta(2) - \alpha(2)\beta(1)], \tag{8.14}$$

where $\alpha(k)$ and $\beta(k)$ are the usual spin eigenfunctions for electron k. In the ansatz of Equation (8.14), the Pauli principle is satisfied by multiplying a spatial factor $s_A(1)s_B(2) + s_A(2)s_B(1)$, that is symmetric with respect to interchange of the electrons, by a spin factor $\alpha(1)\beta(2) - \alpha(2)\beta(1)$, that is antisymmetric with respect to such an interchange.

In the classic Heitler–London VB treatment, the trial wavefunction of Equation (8.14) is employed to compute the electronic energy of the system in the usual manner:

$$\varepsilon_e(R) = \frac{\left\langle \Psi_e \left| \hat{H} \right| \Psi_e \right\rangle}{\left\langle \Psi_e \middle| \Psi_e \right\rangle}, \tag{8.15}$$

where $\hat{H}$ is the electronic Hamiltonian of Equation (8.3), and $\varepsilon_e(R)$ is the resulting Born–Oppenheimer potential energy curve of H$_2$, which is a function only of the internuclear distance R. The eight-dimensional integrations in Equation (8.15) are over the spatial and spin variables of both electrons. The cylindrical symmetry of the problem allows one to perform these integrations analytically in elliptic coordinates, yielding

$$\varepsilon_e(R) = -1 + E_J + E_K, \tag{8.16}$$

where

$$E_J(R) = \frac{e^{-2R}}{1+S^2}\left(\frac{5}{8} + \frac{1}{R} - \frac{3R}{4} - \frac{R^2}{6} \right), \tag{8.17}$$

$$E_K(R) = \frac{1}{1+S^2}\left[K(R) - 2Se^{-R}(R+1) + \frac{S^2}{R} \right], \tag{8.18}$$

$$S(R) = e^{-R}\left(1 + R + \frac{R^2}{3} \right), \tag{8.19}$$

and $K(R)$ is a rather complicated analytic form involving special functions not widely known.[42] $S(R)$ is the overlap integral between the 1s orbitals centered on H$_A$ and H$_B$. Atomic units are assumed in Equations (8.16) through (8.19).

The key physical insights of the VB treatment come from the interpretation of the E_J and E_K contributions to the potential energy function $\varepsilon_e(R)$. The E_J term comprises the classical Coulombic interactions, specifically, the average electron–nuclear attraction, the average electron–electron repulsion, and the simple nuclear–nuclear repulsion. In contrast, E_K is the exchange term that is a direct consequence of the Pauli principle; it cannot be interpreted via classical electrostatic concepts and is purely quantum mechanical in nature. If $E_J(R)$ is plotted vs. R, one finds a single minimum at $R = 1.0$ Å with a depth of only 9 kcal mol^{-1} compared to the separated atom limit. We conclude that the classical electrostatic terms contribute less than 10 kcal mol^{-1} to the binding energy of the H_2 molecule! By adding the exchange contribution $E_K(R)$ to $E_J(R)$, the equilibrium distance (r_e) moves inward to 0.87 Å, and more importantly, the binding energy increases enormously to 73 kcal mol^{-1}. If the effective nuclear charge in the hydrogenic orbitals is made a variational parameter, then the simple VB treatment including exchange terms yields $r_e = 0.74$ Å and a binding energy of 87 kcal mol^{-1}.[42,43] Further improvement of the trial wavefunction, particularly the incorporation of dynamical electron correlation, is necessary to recover the full $D_e = 109$ kcal mol^{-1} in the H_2 covalent bond. However, the qualitative picture arising from this analysis is indisputable: *the strong covalent bond in the H_2 prototype would not exist without the dictates of the Pauli principle.*

The conclusions of the classic valence-bond treatment of the hydrogen molecule have largely been forgotten by the computational chemistry community. Decades ago it was established that the inclusion of exchange terms is essential in describing chemistry via molecular quantum mechanics, and in modern chemical research the philosophical and cosmological ramifications of the Pauli principle are not given a moment's thought. It seems to us that there would be merit in revisiting the Pauli principle vis-à-vis fine-tuning concepts. A fundamental question would be: In some space of counterfactual universes, does the Pauli principle in its current form represent a kind of extremum on the life-fitness landscape?

Naive considerations of a universe without the familiar particle symmetry restrictions lead to amazing conclusions. In the case of water, the omission of the exchange terms from the quantum mechanical treatment should produce an unstable chemical species without strong covalent bonds, although we are not aware of a state-of-the-art computational study on the properties of "exchangeless" quasi-water. If instead the electrons were treated as spin 3/2 rather than spin 1/2 fermions, then water would be a free radical, because four electrons would be required to fill each spatial molecular orbital, and the number of electrons (10) in H_2O is not divisible by four. While these types of musings can give an enlightening context regarding the structure of our current existence, the fundamental question raised above requires us to conceive of the consequences of perturbing the Pauli principle in some continuous manner.

In elementary algebra the technique of decomposing an arbitrary function into a sum of even and odd functions is well known. The same idea can be applied to many-electron wavefunctions via the elegant algebra of the symmetric group (S_n) of $n!$ permutations of n electrons.[44] From a (Hartree) product of molecular spin-orbitals $\chi_1(1)\chi_2(2)\chi_3(3) \ldots \chi_n(n)$, projection operators P_μ can be applied to generate many-electron functions transforming as the μth irreducible representation (irrep) of S_n. It is the totally antisymmetric irrep (a) that satisfies the Pauli principle for electrons, and the corresponding wavefunction is the Slater determinant $|\chi_1(1)\chi_2(2)\chi_3(3) \ldots \chi_n(n)|$. One means of perturbing the Pauli principle in a continuous fashion is to mix functions of different permutational irreps into the usual antisymmetric electronic wavefunction:

$$\Psi_e = \left|\chi_1(1)\chi_2(2)\ldots\chi_n(n)\right| + \sum_{\mu \neq a}\lambda_\mu P_\mu\left[\chi_1(1)\chi_2(2)\ldots\chi_n(n)\right], \qquad (8.20)$$

where the λ_μ are perturbation parameters. The energy of such a wavefunction would be evaluated in the usual way as an expectation value of the nonrelativistic Hamiltonian operator. In this construction one could perform a type of stability analysis of the Pauli principle by tracking molecular properties as a function of the λ_μ parameters. In the simple H_2 paradigm, the antisymmetric VB wavefunction of Equation (8.14) would be augmented only by a symmetric analog with a variable mixing coefficient. In the case of water, the number of irreps to explore would be much larger, and the possible λ_μ parameters would provide an enormous range of flexibility. A search over the space of λ_μ variables could be executed to determine whether the chemical bond strengths attained when the Pauli principle is rigorously obeyed could possibly be exceeded. We prefer not to speculate on what conclusions may be derived from these ideas for future research, but we find them appealing for opening up the Pauli principle to fine-tuning considerations.

8.7 SUMMARY

In our universe, [H, C, N, O] chemistry is an inescapable consequence of the nonrelativistic Schrödinger equation of quantum mechanics with electromagnetic forces alone, as well as the inherent indistinguishability of all electrons as expressed in the Pauli principle, and the Born–Oppenheimer separation of nuclear and electronic motion. The microscopic and macroscopic properties of water, although unusual in comparison with other substances, are not independent fine-tuning variables in our world; they are simply remarkable consequences of an underlying quantum mechanics that displays no real anomalies. The two dimensionless fundamental parameters on which chemistry depends are the fine structure constant ($\alpha \approx 1/137$) and the electron-proton mass ratio ($\beta \approx 1/1836$). Mere perturbations of α and β will not appreciably affect chemistry, because we exist in a relatively insensitive regime near the limit in which both values vanish. However, *increasing* both ratios more than a factor of 10 will have dramatic consequences, and we have outlined a rigorous theoretical approach that would elucidate the ensuing counterfactual chemistry. Of particular intrigue is the route to molecular structure catastrophe as β approaches unity and the marked decrease in covalent bond energies with increasing α. We have also discussed a foundation for additional research in which the constraints on counterfactual chemistry imposed by the existing quantum mechanics of electrons are released by means of the Born–Oppenheimer separation. This program would restore the fitness inquiries of Henderson[10,11] on the biocentricity of water, except in a modern form with a more rigorous epistemology built on large-scale molecular simulations. Finally, we have examined the far-reaching consequences of the Pauli principle for chemistry. Without the exchange terms arising from the indistinguishability and fermionic character of electrons, the covalent bond in the H_2 paradigm would be an order of magnitude weaker than observed in nature. Such a bond strength is much too weak for a stable molecule to persist under physiological conditions. An abstract formalism has been presented by which the Pauli principle can be mathematically perturbed and attendant changes in molecular properties can be tracked. These ideas for exploring states of altered electronic permutational symmetry might be effective in opening up a third research initiative into counterfactual chemistry and biochemical fine-tuning. It is our hope that the analyses of molecular quantum mechanics presented here will help move fine-tuning discussions out of the realm of qualitative argument into the world of quantitative mathematical physics and large-scale computations.

REFERENCES

1. Barrow, J. D., and F. J. Tipler. 1986. *The Anthropic Cosmological Principle.* Oxford: Oxford University Press.
2. Carr, B. J., and M. J. Rees. 1979. The anthropic principle and the structure of the physical world. *Nature* 278:605.
3. Hogan, C. J. 2000. Why the universe is just so. *Rev. Mod. Phys.* 72:1149.
4. Ball, P. 2001. Life's matrix: Water in the cell. *Cell. Mol. Biol.* 47:717.
5. Ball, P. 2003. How to keep dry in water. *Nature* 423:25.
6. Ball, P. 2004. Astrobiology: Water, water, everywhere? *Nature* 427:19.
7. Benner, S. A., A. Ricardo, and M. A. Carrigan. 2004. Is there a common chemical model for life in the universe? *Curr. Opin. Chem. Biol.* 8:672.
8. Dirac, P. A. M. 1929. Quantum mechanics for many-electron systems. *Proc. R. Soc. (London)* 123:714.
9. Herbert, N. 1985. *Quantum Reality.* New York, NY: Anchor Books.

10. Henderson, L. J. 1970 (Reprint). *The Fitness of the Environment.* Cambridge, MA: Harvard University Press.

11. Henderson, L. J. 1917. *The Order of Nature.* Cambridge, MA: Harvard University Press.

12. Barrow, J. D., and F. J. Tipler. 1986. *The Anthropic Cosmological Principle.* Oxford: Oxford University Press, p. 524.

13. Curtiss, L. A., D. J. Frurip, and M. Blander. 1979. Studies of molecular association in H_2O and D_2O vapors by measurement of thermal conductivity. *J. Chem. Phys.* 71:2703.

14. Klopper, W., J. G. C. M. van Duijneveldt-van de Rijdt, and F. B. van Duijneveldt. 2000. Computational determination of equilibrium geometry and dissociation energy of the water dimer. *Phys. Chem. Chem. Phys.* 2:2227.

15. Shirin, S. V., O. L. Polyansky, N. F. Zobov, P. Barletta, and J. Tennyson. 2003. Spectroscopically determined potential energy surface of $H_2{}^{16}O$ up to 25,000 cm^{-1}. *J. Chem. Phys.* 118:2124.

16. Shostak, S. L., W. L. Ebenstein, and J. S. Muenter. 1991. The dipole moment of water: I. Dipole moments and hyperfine properties of H_2O and HDO in the ground and excited vibrational states. *J. Chem. Phys.* 94:5875.

17. Tschumper, G. S., M. L. Leininger, B. C. Hoffman, E. F. Valeev, H. F. Schaefer, and M. Quack. 2002. Anchoring the water dimer potential energy surface with explicitly correlated computations and focal point analyses. *J. Chem. Phys.* 116:690.

18. Polyansky, O. L., A. G. Császár, S. V. Shirin, N. F. Zobov, P. Barletta, J. Tennyson, D. W. Schwenke, and P. J. Knowles. 2003. High-accuracy ab initio rotation-vibration transitions for water. *Science* 299:539.

19. Császár, A. G., G. Czakó, T. Furtenbacher, J. Tennyson, V. Szalay, S. V. Shirin, N. F. Zobov, and O. L. Polyansky. 2005. On equilibrium structures of the water molecule. *J. Chem. Phys.* 122:214305.

20. Ruscic, B., D. Feller, D. A. Dixon, K. A. Peterson, L. B. Harding, R. L. Asher, and A. F. Wagner. 2001. Evidence for a lower enthalpy of formation of hydroxyl radical and a lower gas phase bond dissociation energy of water. *J. Phys. Chem. A* 105:1.

21. Ruscic, B., A. F. Wagner, L. B. Harding, R. L. Asher, D. Feller, D. A. Dixon, K. A. Peterson, Y. Song, X. Qian, C.-Y. Ng, J. Liu, W. Chen, and D. W. Schwenke. 2002. On the enthalpy of formation of hydroxyl radical and gas-phase bond dissociation energy of water and hydroxyl. *J. Phys. Chem. A* 106:2727.

22. Tajti, A., P. G. Szalay, A. G. Császár, M. Kállay, J. Gauss, E. F. Valeev, B. A. Flowers, J. Vázquez, and J. F. Stanton. 2004. HEAT: high accuracy extrapolated ab initio thermochemistry. *J. Chem. Phys.* 121:11599.

23. Barrow, J. D. 2002. *The Constants of Nature. From Alpha to Omega—The Numbers That Encode the Deepest Secrets of the Universe.* New York, NY: Pantheon Books, p. 49.

24. Tarczay, G., A. G. Császár, W. Klopper, V. Szalay, W. D. Allen, and H. F. Schaefer. 1999. The barrier to linearity of water. *J. Chem. Phys.* 110:11971.

25. Barrow, J. D. 2002. *The Constants of Nature.* New York, NY: Pantheon Books, p. 167.

26. Barrow, J. D. 2002. Ibid., p. 166.

27. Valeev, E. F., and C. D. Sherrill. 2003. The diagonal Born–Oppenheimer correction beyond the Hartree-Fock approximation. *J. Chem. Phys.* 118:3921.

28. Schwenke, D. W. 2001. Beyond the potential energy surface: *Ab initio* corrections to the Born–Oppenheimer approximation for H_2O. *J. Phys. Chem. A* 105:2352.

29. Bunker, P. R., and R. E. Moss. 1980. The effect of the breakdown of the Born–Oppenheimer approximation on the rotation-vibration Hamiltonian of a triatomic molecule. *J. Mol. Spectrosc.* 80:217.

30. Handy, N. C., Y. Yamaguchi, and H. F. Schaefer. 1986. The diagonal correction to the Born–Oppenheimer approximation: Its effect on the singlet-triplet splitting of CH_2 and other molecular effects. *J. Chem. Phys.* 84:4481.

31. Balasubramanian, K. 1997. *Relativistic Effects in Chemistry: Part A, Theory and Techniques.* New York, NY: Wiley.

32. Balasubramanian, K. 1997. *Relativistic Effects in Chemistry: Part B, Applications.* New York, NY: Wiley.

33. Merzbacher, E. 1970. *Quantum Mechanics*, 2nd ed. New York, NY: Wiley.

34. Barrow, J. D. 2002. *The Constants of Nature.* New York, NY: Pantheon Books, p. 277.

35. Hawking, S. W., and W. Israel. In *Einstein: A Centenary Volume.* Cambridge: Cambridge University Press, p. 128.

36. Kant, I. 1929. In *Kant's Inaugural Dissertation and Early Writings.* ed. J. Handyside. Chicago, IL: University of Chicago Press.

37. Ehrenfest, P. 1917. In what way does it become manifest in the fundamental laws of physics that space has three dimensions? *Proc. Amsterdam Acad.* 20:200.

38. Ehrenfest, P. 1920. Welche Rolle spielt die Dreidimensionalität des Raumes in den Grundgesetzen der Physik. *Ann. Phys.* 61:440.

39. Barrow, J. D. 2002. *The Constants of Nature,* Chap. 10. New York, NY: Pantheon Books.

40. Partridge, H., and D. W. Schwenke. 1997. The determination of an accurate isotope dependent potential energy surface for water from extensive *ab initio* calculations and experimental data. *J. Chem. Phys.* 106:4618.

41. Heitler, W., and F. London. 1927. Wechselwirkung neutraler Atome und Homopolare Bindung nach der Quantenmechanik. *Z. Phys.* 44:455.

42. McQuarrie, D. A. 1983. *Quantum Chemistry.* Mill Valley, CA: University Science Books.

43. Wang, S. C. 1928. The problem of the normal hydrogen molecule in the new quantum mechanics. *Phys. Rev.* 31:579.

44. R. Pauncz. 1979. *Spin Eigenfunctions.* New York, NY: Plenum Press.

9 Properties of Nanoconfined Water

Branka M. Ladanyi

CONTENTS

9.1 INTRODUCTION

It is perhaps not surprising that water molecules are strongly affected by their surroundings. Water properties near interfaces differ significantly from the bulk, given that interactions with interface components affect the structural features associated with the hydrogen bond network and impact molecular mobility. Further restrictions on mobility arise because of confinement: In various self-assembled structures, such as folded proteins, vesicles, and reverse micelles, water droplets are confined to a nanoscopic volume over periods that are long on the molecular scale. For many practical purposes then, the overall scale of molecular motion is bounded by the size and shape of the confining region.

There has been considerable interest in studying confined water through a number of experimental methods. Theoretical and computational approaches have also been applied to this problem. Most of these studies have focused on systems in which properties such as the confinement geometry and the chemical structure of the interface can be experimentally controlled. The main types of such systems are porous hydrophilic glasses (Crupi et al., 2000a) and reverse micelles (Luisi and Straub, 1984; Luisi et al., 1988). In glasses, the nanopores are approximately cylindrical in shape, while in reverse micelles, the confining geometry is approximately spherical. The pore diameter and the size of the aqueous droplet in reverse micelles are under experimental control, thus allowing systematic studies of water properties as a function of the extent of confinement. Given that reverse micelles are liquid self-assembled structures, they resemble more closely the type of confinement that is biologically relevant (Luisi et al., 1988). As in membranes, vesicles, and atmospheric aerosols, water droplets in reverse micelles are confined by a surfactant layer. Therefore, this chapter will focus primarily on reverse micelle properties and especially on water confined in these systems. Readers interested in confinement in nanopores are referred to the special issue on confined liquids published in 2003 (volume 12, no. 1) in *The European Journal of Physics* and the feature article by Crupi et al. (2000b) in *The Journal of Physical Chemistry*.

The remainder of this chapter is organized as follows: a general overview of reverse micelle properties; a review of the results of experiments and computer simulations on the water structure and dynamics; a discussion of issues related to solvation in confined water; a discussion of reverse micelles as reaction media; and, finally, conclusions.

9.2 REVERSE MICELLES

Reverse micelles or water-in-oil microemulsions, shown schematically in Figure 9.1, are self-assembled structures in which a surfactant layer encloses water dispersed in a continuous nonpolar phase. Generally, the tendency to form reverse micelles is reinforced by the steric bulk of the surfactant, depicted in Figure 9.1 as a double-chain molecule. Additional branching enhances this tendency (Nave et al., 2000). This is why the most frequently used, by far, surfactant for reverse micelle formation is sodium bis(2-ethylhexyl) sulfosuccinate with a common name aerosol-OT or AOT (De and Maitra, 1995; Nave et al., 2000), which, as Figure 9.2 indicates, has additional branches. Experiments comparing properties of water/surfactant/oil systems with different degrees of branching of the two chains indicate that additional branching aids in reverse micelle formation (Nave et al., 2000, 2002), most likely by reinforcing the tendency to form inwardly curving surfaces as schematically depicted in Figure 9.2. Surfactants with less bulky tails usually require a chain alcohol or amine cosurfactant to form reverse micelles (De Gennes and Taupin, 1982). For a given water/surfactant/oil system, the reverse micelle size can be controlled by varying the water/surfactant concentration ratio. The diameter of approximately spherical reverse micelles increases linearly with water content (Luisi and Straub, 1984; De and Maitra, 1995),

$$w_0 = [H_2O]/[Surfactant]. \qquad (9.1)$$

Values of w_0 in the range 1–15 produce reverse micelles in the nanometer size range.

Many aspects of reverse micelle properties have been studied as a function of w_0, temperature, surfactant, and the nonpolar phase. Some of these studies are aimed at characterizing the mesoscopic aspects of the system such as the sizes and shapes of the reverse micelles (De Gennes and Taupin, 1982). Dynamic light scattering (Eicke and Rehak, 1976) and small-angle neutron (Giordano et al., 1993), and x-ray (Fulton et al., 1995) scattering are the techniques most frequently used for this purpose. Information on how w_0 varies with system composition and temperature comes from such studies. Figure 9.3 illustrates the results for the water/AOT/isooctane system for different [H$_2$O] and

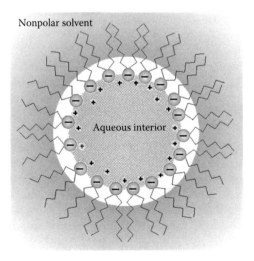

FIGURE 9.1 A schematic drawing of an aqueous reverse micelle.

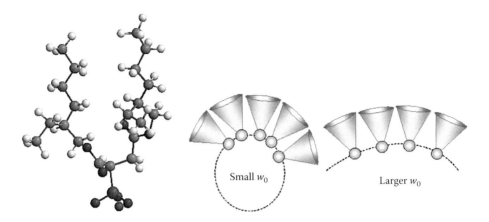

FIGURE 9.2 Left: Molecular structure of AOT, the surfactant most frequently used in formation of reverse micelles. Middle and right: A schematic illustration of the effects of bulky surfactant tails on the surface area per headgroup in reverse micelles at low and high water content w_0.

temperature, obtained from dynamic light scattering. As can be seen from this figure, the relation between the hydrodynamic radius of the reverse micelle and w_0 remains approximately linear up to $w_0 \simeq 30$. This indicates that the water/AOT/isooctane reverse micelle is approximately spherical in shape.

The hydrodynamic or Stokes radius corresponds to the size measured by monitoring translational diffusion of the reverse micelle and thus includes the surfactant layer. The inner water pool radius can be estimated, for example, by combining dynamic light scattering data with an estimate of the surfactant chain length. Table 9.1 includes values estimated in this way by Eicke and

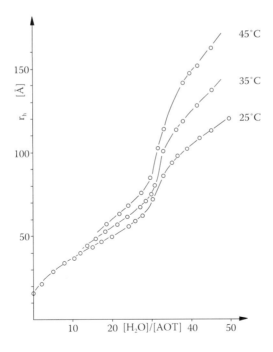

FIGURE 9.3 Stokes radius of the microemulsion formed by water and AOT in isooctane as a function of water content for three temperatures. The AOT concentration is 0.025 M. (Reprinted with permission from Zulauf, M., and Eicke, H. F., *J. Phys. Chem.* 83, 480, 1979. Copyright 1979 by American Chemical Society.)

TABLE 9.1
Composition and Water Pool Radius for Model Water/AOT/Isooctane Reverse Micelles

w_0	n_{water}	$n_{surfactant}$	Radius (Å)	A_{surf} (Å²)
1.0	21	21	10.25	35.9
2.0	52	26	11.6	40.0
4.0	140	35	14.1	48.3
7.5	525	70	19.4	51.3
10.0	980	98	22.9	53.4

Rehak (1976) for H₂O/AOT/isooctane reverse micelles, with small adjustments made by Faeder and Ladanyi (2000). The numbers of water and surfactant molecules are also included in the table. Note that the pool includes the surfactant headgroups and the counterions.

Table 9.1 also lists the values of A_{surf}, the areas per surfactant headgroup. As already noted in connection with Figure 9.2, because of steric effects of the surfactant tails, A_{surf} increases with increasing water content. At large w_0 values, beyond those listed in Table 9.1, A_{surf} eventually levels off. For the system considered in Table 9.1, this occurs at around $w_0 = 20$ and experimental estimate for this limiting value of A_{surf} is 55 Å² at 298 K.

Even though reverse micelles are often approximately spherical, they are liquid-phase self-assembled structures. Because of this, their shape and, to a smaller extent, their size, fluctuate (De Gennes and Taupin, 1982). In computer simulations, the amplitude of shape fluctuations has been measured using the eccentricity parameter (Tobias and Klein, 1996):

$$e = 1 - I_{min}/I_{avg}, \qquad (9.2)$$

where I_{min} is the smallest principal moment of inertia of the reverse micelle and I_{avg} is the average over the three principal moments.

The amplitude and the time scale of fluctuations in e vary with the surfactant and the nonpolar phase (Tobias and Klein, 1996; Allen et al., 2000; Salaniwal et al., 2001; Senapati and Berkowitz, 2003a, 2004). A typical result, from molecular dynamics (MD) simulation of a reverse micelle formed by perfluoropolyether ammonium carboxylate (PFPECOO⁻NH₄⁺) surfactant in supercritical CO₂ is shown in Figure 9.4.

Other studies of the physical properties of reverse micelles focus on a smaller length scale, dealing with the structural features of the reverse micelle components and their interactions. A variety

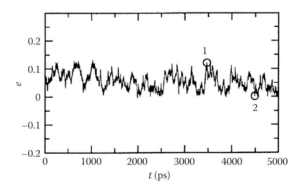

FIGURE 9.4 Time-evolution of the eccentricity of a $w_0 = 8.4$ aqueous reverse micelle formed by the PFPECOO⁻NH₄⁺ surfactant in supercritical CO₂. Labels 1 and 2 indicate maximum and minimum values of e in the 5 ns time interval. (Reprinted with permission from Senapati, S., and Berkowitz, M. L., *J. Phys. Chem. A*, 107, 12906, 2003. Copyright 2003 by American Chemical Society.)

of spectroscopic techniques, including nuclear magnetic resonance (NMR), infrared, and optical spectroscopies have been used to investigate these aspects of reverse micelle properties. In this chapter, I will concentrate on the studies of the properties of the water pool.

In addition to studying the properties of the water pool directly, research has also focused on the effects of confinement on solvation and reactivity. Information on solvation free energies and dynamics has been obtained through several spectroscopic methods. A number of different reactions have been carried out in reverse micelles and the effects of confinement on their rates and equilibria have been studied. These have included reactions carried out entirely within the water pool and reactions in which reactants are brought together across the surfactant interface.

9.3 PROPERTIES OF THE WATER POOL IN REVERSE MICELLES

The properties of the water pool in a given type of reverse micelle depend on two key factors. The first is the fact that the system is heterogeneous. The water molecules near the interface experience strong electrostatic interactions with the surfactant headgroups and, in the case of ionic surfactants, the counterions, while water molecules away from the interface are in an environment that is more bulk-like. Second, due in part to steric effects associated with surfactant tails, the surface area per headgroup increases with w_0, so that the properties of water in a given range distances from the interface vary with reverse micelle size. This effect is illustrated in Figure 9.2.

A considerable effort has been devoted to investigating the properties of the reverse micelle water pool. Many experimental techniques have been devoted to this problem as have a number of theoretical and computational studies. Some experimental studies have focused directly on water, while others have used solutes as probe molecules. This section concentrates on the results that have been obtained by focusing directly on water.

A key structural feature of liquid water is its hydrogen bond network. Studying how this network is perturbed when water is confined can provide a great deal of information on how nanodroplets differ from the bulk. The main experimental method that has been used for this purpose is infrared (IR) spectroscopy. The focus has been on two different spectroscopic signatures of hydrogen (H) bonding. Near-IR spectroscopy measures absorption spectra of intramolecular modes, while far-IR spectroscopy covers a lower-frequency range in which the intermolecular modes dominate. In liquid water, H-bonding lowers the intramolecular OH stretch frequency. This spectral region (3000–3800 cm^{-1}) is considerably broadened due to distortions of the H-bond network and fluctuations of the number of H-bonds per molecule. H-bonding also impacts the spectral region below 1000 cm^{-1} where there are no intramolecular vibrational modes of water molecules. The most prominent feature in this spectral region of liquid water is a peak centered at around 650 cm^{-1} that corresponds to H-bond libration, i.e., the oscillatory motion of hydrogen in the O–H–O H-bond–covalent bond state.

In reverse micelles, proximity to the surfactant interface perturbs the H-bond network. Given that water coordinates to the surfactant headgroups and, if the surfactant is ionic, to the counterions, water molecules near the interface make fewer H-bonds to other water molecules (Faeder and Ladanyi, 2000). This has an impact on the distributions of OH stretch and O–H–O librational frequencies. Both have been investigated experimentally (Onori and Santucci, 1993; D'Angelo et al., 1994; Amico et al., 1995; Fioretto et al., 1999a; Venables et al., 2001). Representative near-IR spectroscopy results for water in reverse micelles are shown in Figure 9.5. The results illustrate the strong w_0 dependence of the OH stretch band and the seemingly complicated progression with increasing w_0 of the IR spectra. The complicated appearance of the spectra occurs because, in all cases, water molecules in different H-bonding arrangements contribute to the band. The spectra of pure water (top left panel) and of reverse micelles are all fit to three Gaussian functions, whose peak frequencies remain essentially constant as w_0 is varied. The peak locations are associated with the different H-bonding states of the water molecules, the lowest frequency peak with fully H-bonded molecules, the highest with the free O–H stretch and the middle with intermediate H-bond arrangements (Onori and Santucci, 1993). The relative intensities of the peaks change when water

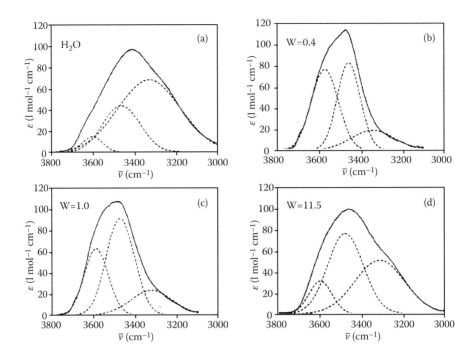

FIGURE 9.5 OH stretch band (full line) of bulk water (a) and of water in $H_2O/AOT/CCl_4$ system at different water content (denoted here as W): (b) 0.4, (c) 1.0, and (d) 11.5. The fits of the band to three Gaussian components are depicted by dashed lines. (Reprinted with permission from Onori, G., and Santucci, A., *J. Phys. Chem.* 97, 5430, 1993. Copyright 1993 by American Chemical Society.)

is confined in reverse micelles of different size. A comparison of the bulk water spectrum to the spectrum of water in the smallest reverse micelle (top right panel) shows increased intensity of the high-frequency peak, which corresponds to water that is weakly H-bonded, while the spectrum of water in the largest reverse micelle (bottom right panel) shows higher intensity of lower frequency peaks corresponding to molecules with larger numbers of H-bonds.

Despite the apparent complexity of the spectra, a fairly simple interpretation of the data is possible. Specifically, the spectra can be explained by assuming that they represent contributions from two distinct water populations, "bound" (to the surfactant) and "free," i.e., bulk-like water. At each value of w_0, the molar extinction coefficient ε is then a sum of these two contributions:

$$\varepsilon = x\varepsilon_{bound} + (1 - x)\varepsilon_{bulk}, \qquad (9.3)$$

where x is the mole fraction of bound water. The extinction coefficient ε_{bulk} is assumed to be the same as in bulk liquid water, whereas ε_{bound} is deduced from the spectra at small w_0. As expected, x decreases with increasing w_0 (Onori and Santucci, 1993).

The far-IR spectroscopy results provide further support for this interpretation of the water pool properties (Venables et al., 2001). Figure 9.6 shows the progression of the librational band shape as a function of w_0. An isosbestic point (i.e., a single frequency value at which the curves for different w_0 cross), which is evident in the figure, is a signature of spectra arising from two distinct populations of molecules. The estimates of x vs. w_0 from OH stretch (D'Angelo et al., 1994) and librational (Venables et al., 2001) band spectra agree reasonably well with each other for AOT surfactant reverse micelles.

More detailed information about the structural features of water in the reverse micelle interior can be obtained from computer simulation studies (Linse, 1989; Allen et al., 2000; Faeder

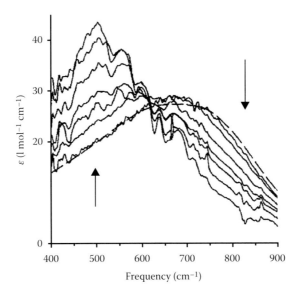

FIGURE 9.6 Water librational band in H_2O/AOT/isooctane at different values of water content w_0. The progression of sizes from large to small w_0 values is indicated with the arrows. The high-frequency portion of the spectrum decreases, while the low-frequency portion increases with decreasing w_0. Bulk water is shown as a smooth dashed curve, followed by reverse micelles with $w_0 = 40, 20, 10, 6, 4, 2$, and 1. (Reprinted with permission from Venables et al., *J. Phys. Chem. B*, 105, 9132, 2001. Copyright 2001 by American Chemical Society.)

and Ladanyi, 2000; Senapati and Berkowitz, 2003b; Abel et al., 2004). The physical picture that emerges from these studies is that the structure is strongly perturbed in a few interfacial layers, but then becomes bulk-like. Of course, in small to moderate size reverse micelles, i.e., for $w_0 < 10$, only a few interfacial layers are present, so a relatively large fraction of the molecules belongs to the perturbed set. For example, from the librational band data, Venables et al. (2001) estimate that $x \cong 0.3$ at $w_0 = 10$.

Information on layering of water relative to the interface can be obtained in computer simulation in the form of the density profiles. Typical results for a model of water/AOT/oil reverse micelle (Faeder and Ladanyi, 2000) are shown in Figure 9.7.

The water density profiles show several peaks, the most prominent one corresponding to water trapped between the headgroups. The intensity of this peak grows with w_0, given that the surface area per headgroup increases (see Figure 9.2). The Na^+ counterion density profile is concentrated near the interface. It has a double peak structure.

The inner peak, which corresponds to Na^+ partially solvated by water, grows in intensity as the water content w_0 increases. The onset of uniform water density, after about three layers, signals bulk-like water structure. However, stronger perturbations are found in water dynamics.

Several experimental techniques have been applied to the study of water mobility in the reverse micelle interior. Among the earliest ones was proton NMR (Wong et al., 1977), which showed a dramatic change in the single-molecule rotational relaxation time τ_c (for the intramolecular H–H vector) with increasing w_0. In the smallest reverse micelles studied, τ_c turned out to be only about a factor of 5 smaller than the time estimated for tumbling of the whole reverse micelle and about 50 times longer than in bulk water, while bulk-like time scales are approached as w_0 increases to about 15. These results are depicted in Figure 9.8.

NMR provides only a value of the relaxation time τ_c, i.e., the total time associated with the loss of memory of molecular orientation. Other techniques can access more detailed information on water rotational mobility. For example, recent experiments using time-resolved vibrational photon echo

spectroscopy (Tan et al., 2005) have measured the time correlation $C_2(t)$ for the reorientation of the O–D bond of HOD present in low concentration in the AOT reverse micelle water pool:

$$C_2(t) = \langle P_2[\cos\theta_{OD}(t)]\rangle. \tag{9.4}$$

Here $P_2(x) = \dfrac{3}{2}x^2 - \dfrac{1}{2}$ is the second-order Legendre polynomial, $\theta_{OD}(t)$ is the angle by which the orientation of the O–D vector changes in time t, and $\langle\cdots\rangle$ denotes an equilibrium ensemble average.

These experiments show that water reorientation is quite different from this process in bulk water. Specifically, in the bulk, $C_2(t)$ decays approximately exponentially except at very short times (<50 fs), while in reverse micelles, the decay rate is highly nonexponential. Tan et al. (2005) were able to fit their results to a reorientation model (Lipari and Szabo, 1980), which involves fast restricted (in angular range) rotation and slow tumbling. MD computer simulation results (Harpham et al., 2004), which agree well with these experiments, indicate that the short-time "wobbling in a cone" motion is due to water molecules coordinated to the Na^+ counterions. This coordination is caused primarily by ion–dipole interactions, so the wobbling corresponds to motion of the O–D vector around a fixed orientation of the molecular dipole. The behavior of $C_2(t)$ for two reverse micelle sizes and bulk water is compared in Figure 9.9.

As can be seen from the figure, there is a strong dependence of $C_2(t)$ on the reverse micelle size, with slower long-time decay of orientational correlations at smaller w_0 values. The nonexponential nature of the decay in reverse micelles, but not in the bulk, is evident.

Information on translational mobility of water in reverse micelles is also accessible experimentally, for example, through quasielastic neutron scattering (QENS) (Harpham et al., 2004), which takes advantage of the fact that the ^1H isotope has a much larger incoherent scattering cross section than any other nucleus, including deuterium, D (Bée, 1988). If the reverse micelle surfactant and the nonpolar phase are deuterated, the QENS spectrum, $S(\mathbf{Q},\omega)$, will be primarily because of water hydrogens. At small values of the momentum transfer $Q = |\mathbf{Q}|$, the spectrum is due mainly to translational motion of water molecules. In the bulk, the Lorentzian width of the translational band, $T(\mathbf{Q},\omega)$, can be simply related to the translational diffusion constant D_T (Teixeira et al., 1985). In reverse micelles, this connection is somewhat less straightforward because $T(\mathbf{Q},\omega)$ is not Lorentzian (Harpham et al., 2004). Nevertheless, information obtained from low-Q QENS data indicates that

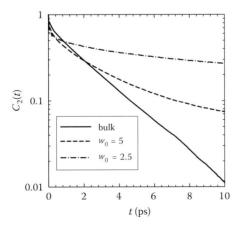

FIGURE 9.9 The orientational time correlation $C_2(t)$ for the H_2O molecule center-of-mass to H vector (similar to the O–H bond vector) in the bulk liquid, and in model H_2O/AOT/isooctane reverse micelles corresponding to water content $w_0 = 2.5$ and 5. (Data from Harpham et al., *J. Chem. Phys. B*, 121, 7855, 2004.)

water translational mobility in reverse micelles in the size range $1 \leq w_0 \leq 5$ is much lower than in the bulk (Harpham et al., 2004).

MD simulation studies of mean squared displacements,

$$\left\langle \left| \Delta \mathbf{r}(t) \right|^2 \right\rangle = \left\langle \left| \mathbf{r}(t) - \mathbf{r}(0) \right|^2 \right\rangle. \tag{9.5}$$

where $\mathbf{r}(t)$ is the position of the center-of-mass of a water molecule at time t, in reverse micelles show generally lower translational mobility than in the bulk (Allen et al., 2000; Senapati and Berkowitz, 2003b) and a strong dependence on w_0 (Faeder and Ladanyi, 2000; Abel et al., 2004). In the bulk, the slope of $\left\langle \left| \Delta \mathbf{r}(t) \right|^2 \right\rangle$ becomes time-independent at long times and equal to $6D_T$. In confined systems, the range of translational displacements is restricted by the dimensions of the confining region. A local D_T in reverse micelles may be defined for $\left\langle \left| \Delta \mathbf{r}(t) \right|^2 \right\rangle$ values significantly smaller than the water droplet size.

Figure 9.10 illustrates the behavior of $\left\langle \left| \Delta \mathbf{r}(t) \right|^2 \right\rangle$ as a function of w_0, obtained by Abel et al. (2004) from MD simulation of water/AOT/isooctane. It is clear from these results that translational mobility of water molecules is strongly suppressed in small reverse micelles and that even in a fairly large reverse micelle, $w_0 = 7$, which in this simulation contains 574 water molecules, $\left\langle \left| \Delta \mathbf{r}(t) \right|^2 \right\rangle$ is significantly lower than in bulk water.

In addition to single-molecule relaxation processes, collective water relaxation has been investigated experimentally by THz (Boyd et al., 2002) and dielectric spectroscopies (Fioretto et al., 1999a; Freda et al., 2001). Both sets of results have shown strong w_0 dependence. The latter results have also shown strong dependence on surfactant counterion (Fioretto et al., 1999a, 1999b). It should be noted that because the dielectric permittivity is a collective property, in reverse micelle systems it cannot be attributed solely to water but also to the other system components, mainly the surfactant. Some of the low-frequency dynamics that are observed are most likely due to the reverse micelle shape fluctuations (Boyd et al., 2002).

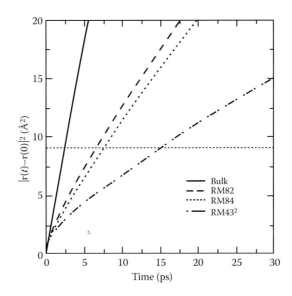

FIGURE 9.10 Mean squared displacement of the water oxygen atoms vs. time. Different curves represent bulk water, and water/AOT/isooctane reverse micelles: reverse micelle82, RM64 and RM43^a correspond to water content $w_0 = 7$, 5, and 3, respectively. (Reprinted with permission from Abel et al., *J. Phys. Chem. B*, 108, 19458, 2004. Copyright 2004 by American Chemical Society.)

Experimental results on water mobility, such as those of Figure 9.8, represent an average over all the molecules within the reverse micelle water pool. However, the mobility of water molecules for a given reverse micelle composition varies with distance from the surfactant interface. This aspect of reverse micelle properties is accessible experimentally through the use of probe molecules, as will be discussed in the next section. MD simulation data can be used to extract such information in the absence of the probe. This can be done by tagging molecules in different interfacial layers and analyzing their trajectories. The water density profiles, such as those shown in Figure 9.7, can be used to define the boundaries of different interfacial layers (Linse, 1989; Faeder and Ladanyi, 2000; Senapati and Berkowitz, 2003b). So far, only ionic surfactant reverse micelles have been studied by MD simulation. In these systems, water mobility in the vicinity of the interface is found to be considerably lower than in the core (Linse, 1989; Faeder and Ladanyi, 2000; Faeder et al., 2003; Senapati and Berkowitz, 2003b, 2004). In each region, the mobility increases with w_0. This is due both to a decrease of the surfactant interfacial density (see Figure 9.2) and to an increase in the droplet size (see Figure 9.3). Typical results illustrating the w_0 dependence of rotational mobility in interfacial and core regions are shown in Figure 9.11. The time correlation plotted is the first-order orientational autocorrelation of the molecular dipole moment μ:

$$C_{1\mu}(t) = \left\langle \cos\theta_\mu(t) \right\rangle = \left\langle \mu(0) \cdot \mu(t) \right\rangle \big/ \left\langle \mu^2 \right\rangle. \qquad (9.6)$$

It is evident from Figure 9.11 that water dipole reorientation is considerably slower near the surfactant interface than in the core of the reverse micelle. Comparison of the results for $w_0 = 2$ and 7.5 illustrates the effects of increasing reverse micelle size on water mobility in each region. In each region, water is considerably more mobile in the larger reverse micelle.

Given that water–surfactant interactions play an important role in decreasing water mobility in reverse micelles, changing surfactant properties should have an appreciable impact on water interfacial structure and mobility. One of the ways of doing this for ionic surfactants is by varying the counterion (Eastoe et al., 1993; Fioretto et al., 1999a, 1999b; Faeder et al., 2003; Senapati and Berkowitz, 2004; Harpham et al., 2005). Counterions are found predominantly in the interfacial region (see Figure 9.7), so the effects of counterion substitution can be expected to have a stronger effect on the interfacial mobility than on the core mobility of water molecules. This is indeed the case (Faeder et al., 2003; Senapati and Berkowitz, 2004; Harpham et al., 2005).

Figure 9.12 illustrates the effects of counterion substitution on $C_{1\mu}(t)$ of water molecules in reverse micelles formed by PFPECOO$^-$ salts in CO_2. The counterions are Ca^{2+}, Na^+, and NH_4^+. Of these,

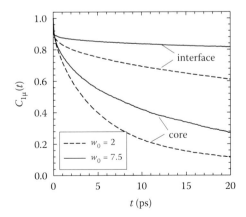

FIGURE 9.11 Water dipole orientational time correlation for molecules at the interface ($D < 2$ A in Figure 9.7) and the core ($D > 4$ Å) of model water/AOT/isooctane reverse micelles for $w_0 = 2$ and 7.5. (Data from Faeder, J., and Ladanyi, B. M., *J. Phys. Chem. B*, 104, 1033, 2000.)

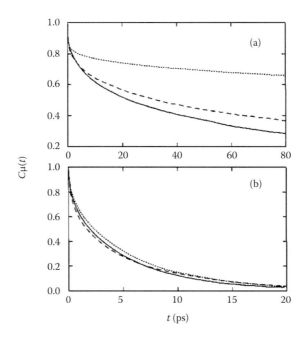

FIGURE 9.12 Molecular dipole time autocorrelation, $C_{1\mu}(t)$ for water in reverse micelles formed by PFPECOO⁻ salts of Ca^{2+} (dotted line), Na^+ (dashed line), and NH_4^+ (full line) in CO_2 for the composition [H₂O]/[PFPECOO⁻] = 8.4. (a) Interface; (b) core. (Reprinted with permission from Senapati, S., and Berkowitz, M. L., *J. Phys. Chem. A*, 108, 9768, 2004. Copyright 2004 by American Chemical Society.)

Ca^{2+} has the strongest electrostatic interactions with water and NH_4^+ the weakest. The rate of relaxation of $C_{1\mu}(t)$ in the interfacial region reflects this ordering, with considerably slower decay in the presence of Ca^{2+} than for the two singly charged counterions.

The counterion dependence of $C_{1\mu}(t)$ becomes almost negligible in the core. Similar behavior of $C_{1\mu}(t)$ is observed in model reverse micelles in which the usual AOT counterion Na^+ was substituted by larger alkali ions, K^+ and Cs^+ (Faeder et al., 2003; Harpham et al., 2005).

Note that the time intervals for the (a) and (b) panels in Figure 9.12 differ by a factor of 4, illustrating the large difference in mobility between core and interfacial water. Senapati and Berkowitz (2004) showed that interfacial water in reverse micelles has significantly lower rotational mobility than water in the first coordination shell of the same counterions in bulk aqueous solution. This is partly because, in reverse micelles, the counterions are close to the surfactant headgroups and some water molecules occupy bridging sites between them.

9.4 SOLVATION IN REVERSE MICELLES

In computer simulation, water molecules can be tagged and information on the impact of the heterogeneity of the reverse micelle environment on water mobility can be studied by monitoring the dynamics of water molecules in different interfacial regions. In experiments, one can obtain related information by using probe molecules. Ionic and highly polar solutes partition into the aqueous reverse micelle phase. Solute locations in surfactant assemblies can be determined experimentally using techniques that are sensitive to interatomic distances such as the NMR nuclear Overhauser effect (Bachofer et al., 1991; Vermathen et al., 2002). They can also be deduced from the way that solute spectra depend on w_0 and on the surfactant type.

In reverse micelle systems, different probe molecules reside in different environments and their spectra are affected to different degrees by the heterogeneity of the solvent environment. The location of the probe relative to the interface depends on its interactions with the surfactant head-groups (Behera et al., 1999; Sando et al., 2004, 2005). Attraction to the headgroups leads to a stronger tendency for the probe to be near the interface and its spectra are then strongly dependent on w_0, while much weaker dependence is found for probes that are repelled by the headgroups.

Electronic (Zhang and Bright, 1991, 1992; Heitz and Bright, 1996; Sarkar et al., 1996; Levinger et al., 1998; Pant et al., 1998; Riter et al., 1998a, 1998b; Behera et al., 1999; Bhattacharyya et al., 2002) and infrared (Zhong et al., 2003; Sando et al., 2004, 2005) spectroscopies of probe molecules have been used to study solvation in reverse micelles. The probe molecules for electronic spectroscopy are typically fairly large compared to water molecules and, therefore, to the thickness of interfacial layers in reverse micelles, so they are sensitive to the water mobility averaged over several layers. Steady-state and time-resolved spectra typically show strong dependence on w_0. Figure 9.13 illustrates the solvation dynamics observed by monitoring the time-evolution of the Stokes shift in the fluorescence spectrum of coumarin 343 (C343) in water/AOT/isooctane reverse micelles. These results are compared with the response of a concentrated aqueous electrolyte solution.

The quantity plotted is the difference between the Stokes shift Δv at time t and the extrapolated value of the $t = 0$ shift (Riter et al., 1998b):

$$S(t) = \Delta v(t) - \Delta v(0). \tag{9.7}$$

The Stokes shift is approximately proportional to the change in the solute–solvent interaction energy resulting from solute electronic excitation. As can be seen in Figure 9.13, the total change in the magnitude of $S(t)$ within 10 ns and its rate of decay are strongly dependent on reverse micelle size, w_0. A much larger total shift and faster decay of $S(t)$ is observed in electrolyte solution than even in the largest reverse micelle, $w_0 = 40$.

The C343 chromophore is incompletely solvated by water in the smallest reverse micelle, but in $w_0 \geq 2$ the number of water molecules is sufficient for at least one solvation shell. However, for $w_0 = 5$, for which several solvation shells can surround the chromophore, the reorganization of the shell is still much slower than for $w_0 \geq 7.5$.

In the previous section, the effects of counterion substitution on water mobility in reverse micelles were described. As one might expect, solvation dynamics is affected by these changes (Riter et al., 1998a). Levinger and coworkers (Pant et al., 1998; Riter et al., 1998a) found a faster decay of $S(t)$

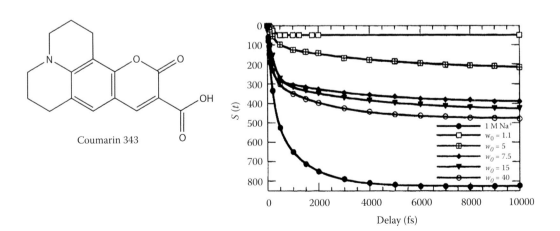

FIGURE 9.13 Left panel: The solvation dynamics probe molecule, C343. Right panel: Fluorescence Stokes shift response in reverse micelles composed of H2O/AOT/isooctane and in 1 M aqueous Na^+ (Na_2SO_4) solution. (Reprinted with permission from Riter et al., *J. Phys. Chem. B*, 102, 2705, 1998b. Copyright 1998 by American Chemical Society.)

when K$^+$ and NH$_4^+$ ions were substituted and slower decay when Ca^{2+} was substituted for Na$^+$ in AOT surfactant reverse micelles.

A number of interesting results on solvation in reverse micelles were obtained recently by Owrutsky and coworkers (Zhong et al., 2003; Sando et al., 2004, 2005) using time-resolved IR spectroscopy of ions dissolved these systems. The probes in these studies were considerably smaller than the chromophores used in solvation studies via electronic spectroscopy, and therefore appeared to be quite sensitive to the length scale of heterogeneity in the reverse micelle environment. For example, in the case of the azide, N$_3^-$ ion, different results for the w_0 dependence of the vibrational excitation responses were found in anionic, cationic, and nonionic surfactant reverse micelles (Sando et al., 2004). These differences are very likely due to the different locations of the azide relative to the reverse micelle water/surfactant interface. The trends that are observed are consistent with the expectation that an anion such as azide will be attracted to positively charged headgroups and repelled by the negatively charged ones. Thus the ion would be more likely to be located in the interfacial layer in cationic than in anionic surfactant reverse micelles.

Figure 9.14 illustrates these trends for the peak, ν_{max}, of the asymmetric stretch band of N$_3^-$ in reverse micelles composed of H$_2$O/surfactant/oil or D$_2$O/surfactant/oil compared with bulk H$_2$O, D$_2$O, and H$_2$O/tri(ethylene glycol) monomethyl ether (TGE) mixtures. The surfactants used were

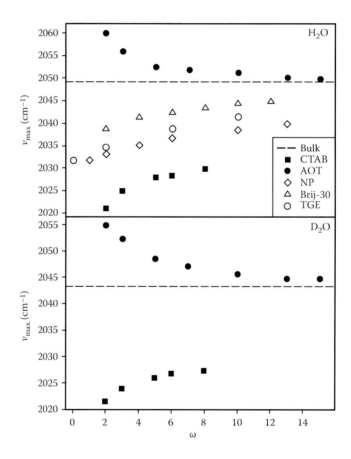

FIGURE 9.14 Peak of antisymmetric stretching band of azide as a function of w_0 (here denoted as ω) for H$_2$O (top) and D$_2$O (bottom) containing reverse micelle and TGE mixture solutions. Circles are AOT/n-heptane, squares are CTAB/CH$_2$Cl$_2$, open triangles are Brij-30/cyclohexane, open diamonds are NP/cyclohexane, and open circles are H$_2$O/TGE mixtures. (Reprinted with permission from Sando et al., *J. Phys. Chem. A*, 108, 11209, 2004. Copyright 2004 by American Chemical Society.)

AOT, which is anionic, cetyl trimethylammonium bromide (CTAB), which is cationic, and two non-ionic surfactants, NP (nonylphenyl polyoxyethylene) and Brij-30 (tetraoxyethylene dodecyl ether), with the nonpolar oil phases specified in the figure caption.

Lowering of ν_{max} occurs because of favorable interactions between the probe ion and the surrounding medium. As can be seen from Figure 9.14, ν_{max} is strongly influenced by w_0. For the reverse micelles, this quantity increases with reverse micelle size. For the water–TGE mixture, w_0 = [H$_2$O]/[TGE] and its increase represents an increase in the water mole fraction.

For AOT in which azide-headgroup electrostatic interactions are repulsive, ν_{max} is higher than in bulk water (or D$_2$O), while for CTAB, in which there is a Coulomb attraction between azide and the headgroup, ν_{max} is considerably lower. In both cases, differences between the reverse micelle and bulk water results diminish with increasing w_0, although in the case of CTAB they do not disappear, indicating that aide remains in the interfacial layer.

In reverse micelles formed by nonionic surfactants, one does not expect a strong preference of the azide for the interfacial relative to the core location. However, the red shift of ν_{max} in reverse micelles relative to bulk H$_2$O and the trends with w_0 indicate a more favorable interaction with the surfactant headgroup than with water. A very similar trend is seen for the H$_2$O/TGE mixture, in which it is likely to be due to preferential solvation by the TGE mixture component. TGE has a similar molecular structure as the hydrophilic portions of the two nonionic surfactants, which further supports the idea that azide resides preferentially near the interface of these systems.

While the ν_{max} results reflect the trends related to salvation-free energies, time-resolved IR spectroscopy also produces data on the rates of solute reorientation and vibrational energy relaxation. The results on rotational relaxation indicate reduced rotational mobility of probe ions in small reverse micelles, as one might expect based on the reduced water mobility in these systems.

Solvation in confined systems, including reverse micelles, has also been studied by simulation (Faeder and Ladanyi, 2001, 2005; Thompson, 2002, 2004). Generally, these results show that a significant portion of the effect of confinement comes from the likely location of the solute relative to the interface (Faeder and Ladanyi, 2001, 2005; Thompson, 2002, 2004). In the case of solvation dynamics in which the relaxation of the system is monitored after solute electronic excitation, a further important factor is the change in the solute location relative to the interface (Thompson, 2004; Faeder and Ladanyi, 2005). This arises because solute electrostatic interactions with the surrounding medium change as a result of electronic excitation. MD simulations of solvation dynamics for a solute ion oppositely charged to the surfactant headgroups show that this effect can be quite pronounced in reverse micelles (Faeder and Ladanyi, 2005). Thus, unlike in systems such as bulk liquids, which are isotropic, in heterogeneous systems solute motion relative to the interface can contribute significantly to solvation and reaction dynamics.

9.5   REVERSE MICELLE AS MEDIA FOR BIOLOGICALLY RELEVANT REACTIONS

There has been considerable interest in investigating and exploiting reverse micelles as reaction media. One of their technologically significant uses has been as "nanobeakers" (Luisi et al., 1988; Pileni, 1989, 1993). For example, the fact that the size and shape of reverse micelles can be controlled by varying w_0 and counterion charge has been exploited to use them as templates to produce size- and shape-selected semiconductor and metal nanoparticles (Pileni, 1993; Pileni et al., 1999). More generally, there has been a considerable amount of interest in solvation and reactivity in reverse micelles, because they can serve as media for aqueous chemistry as well as for heterogeneous reactions between species that are soluble in different reverse micelle phases (Luisi and Straub, 1984; Luisi et al., 1988). One of the directions of this research has been to look upon the interior region as an environment that mimics biologically relevant conditions. There is a considerable amount of evidence that biological macromolecules retain more of their *in vivo* structure and function when they are enclosed in surfactant assemblies than when they are in bulk aqueous solution (Griffiths and Tawfik, 2000; Savelli et al., 2000).

Given that reverse micelles resemble biological membranes, they have been used to solubilize proteins, DNA fragments, and enzymes. Solubilization of these species and the effects of reverse micelle environment on biologically relevant reactions have been very active research areas for many years (Luisi and Straub, 1984; Luisi et al., 1988; Pileni, 1993; Savelli et al., 2000). Reviewing the results of these investigations is well beyond the scope of this chapter. However, it is worth noting that the heterogeneity of the reverse micelle environment plays a significant role in enzyme-catalyzed reactions (Luisi et al., 1988; Savelli et al., 2000). It facilitates reactions of enzymes that partition predominantly into the aqueous phase and substrates that are only slightly water-soluble, given that such species can be brought into close proximity across the reverse micelle surfactant layer. In some cases, confinement in reverse micelles has been shown to play a direct role in the enzymatic reaction mechanism. For example, for enzymes such as recombinant horseradish peroxidase, which have different catalytic activities in monomeric and dimeric forms, smaller reverse micelles can accommodate a single monomer, whereas as w_0 increases, the dimeric form can also be accommodated (Gazaryan et al., 1997). This gives rise to a bimodal distribution of the catalytic activity vs. w_0.

Even though their linear dimensions are typically a couple of orders of magnitude smaller than those of natural cells (Pohorille and Deamer, 2002), reverse micelles and liposomes have featured prominently in attempts to design artificial cells, i.e., nanoscopic structures that mimic some functions of cells as reaction media (Luisi et al., 1994; Walde et al., 1994; Pohorille and Deamer, 2002). While confinement and interactions of enzymes and substrates with the surfactant layer are important ingredients in the design strategy, other ingredients and processes are also needed. They include the presence of a template polymer such as RNA, small molecules, and ions that can be transported across the membrane and used as reactants, enzymes that can catalyze the reactions of interest, the ability of the membrane to grow and reform into separate compartments, and an external source of energy to drive the chemical reactions. Readers are referred to, for example, a review by Pohorille and Deamer (2002) on recent progress in creating artificial cells.

9.6   CONCLUDING REMARKS

This chapter deals with water in nanoscopic confinement. The main focus is on water in reverse micelles, which are self-assembled structures in which a surfactant coats liquid droplets dispersed in a continuous nonpolar phase. Because the size of the water droplet can be easily controlled by varying the water content, w_0, a great deal of information has been obtained by studying how the properties of water and of the solutes partitioned into the aqueous phase vary with confinement. Some of the key methods for studying confined water and their results are described. The overall picture that emerges is that water–surfactant interactions lead to disruption of the water H-bond network and to a reduction in the mobility of the molecules near the interface. While the structural effects propagate for only a few layers, the effects on the mobility seem longer-ranged, due, in part, to the restrictions imposed by the confinement geometry. As w_0 increases, the surface density of the surfactant headgroups decreases and the droplet volume increases, both of which contribute to decreasing the overall effects of confinement.

Properties of solutes partitioned into the aqueous phase are affected by confinement and by their location relative to the interface. The rate of solvent reorganization in response to a change in solute properties, measured in time-dependent fluorescence experiments and relevant to the dynamics of chemical reactions, is strongly affected by reverse micelle size and is much slower in small reverse micelles than it is in bulk aqueous electrolyte solution. Small-size solutes are sensitive to the heterogeneity of the reverse micelle environment. Their likely location is determined by their affinity to the surfactant headgroups. Solutes that remain in the interfacial region are strongly affected by the reverse micelle size and surfactant properties.

In living systems water is often enclosed in surfactant assemblies. This has stimulated research into the effects of reverse micelle environment on the structure and function of biological macromolecules. It has often been found that enzyme activity and stability can be controlled by dissolving them in aqueous surfactant assemblies. Reverse micelles and vesicles have also been used in the design of artificial cells. It is likely that confinement and proximity to a surfactant interface, and not just the presence of the aqueous solvent, play important roles in the rates and mechanisms of chemical reactions occurring in living systems.

REFERENCES

Abel, S., F. Sterpone, S. Bandyopadhyay, et al. 2004. Molecular modeling and simulations of AOT-water reverse micelles in isooctane: Structural and dynamic properties. *J. Phys. Chem. B* 108:19458–19466.

Allen, R., S. Bandyopadhyay, and M. L. Klein. 2000. C12E2 reverse micelle: A molecular dynamics study. *Langmuir* 16:10547–10552.

Amico, P., M. D'Angelo, G. Onori, et al. 1995. Infrared absorption and water structure in aerosol OT reverse micelles. *Nuovo Cimento* 17:1053–1065.

Bachofer, S. J., U. Simonis, and T. A. Nowicki. 1991. Orientational binding of substituted naphthoate counterions to the tetradecyltrimethylammonium bromide micellar interface. *J. Phys. Chem.* 95:480–488.

Bée, M. 1988. *Quasielastic Neutron Scattering*. Bristol: Hilger.

Behera, G. B., B. K. Mishra, P. K. Behera, et al. 1999. Fluorescent probes for structural and distance effect studies in micelles, reversed micelles and microemulsions. *Adv. Colloid Interface Sci.* 82:1–42.

Bhattacharyya, K., K. Hara, N. Kometani, et al. 2002. Solvation dynamics in a microemulsion in near-critical propane. *Chem. Phys. Lett.* 361:136–142.

Boyd, J. E., A. Briskman, C. M. Sayes, et al. 2002. Terahertz vibrational modes of inverse micelles. *J. Phys. Chem. B* 106:6346–6353.

Crupi, V., S. Magazu, D. Majolino, et al. 2000a. Confinement influence in liquid water studied by Raman and neutron scattering. *J. Phys.: Condens. Matter* 12:3625–3630.

Crupi, V., D. Majolino, P. Migliardo, et al. 2000b. Diffusive relaxations and vibrational properties of water and H-bonded systems in confined state by neutrons and light scattering: State of the art. *J. Phys. Chem. A* 104:11000–11012.

D'Angelo, M., G. Onori, and A. Santucci. 1994. Study of aerosol-OT reverse micelle formation by infrared-spectroscopy. *J. Phys. Chem.* 98:3189–3193.

De Gennes, P. G., and C. Taupin. 1982. Microemulsions and the flexibility of oil/water interfaces. *J. Phys. Chem.* 86:2294–2304.

De, T., and A. Maitra. 1995. Solution behavior of aerosol OT in non-polar solvents. *Adv. Colloid Interface Sci.* 59:95–193.

Eastoe, J., T. F. Towey, B. H. Robinson, et al. 1993. Structures of metal bis(2-ethylhexyl) sulfosuccinate aggregates in cyclohexane. *J. Phys. Chem.* 97:1459–1463.

Eicke, H. F., and J. Rehak. 1976. On the formation of water/oil microemulsions. *Helv. Chim. Acta* 59: 2883–2891.

Faeder, J., M. V. Albert, and B. M. Ladanyi. 2003. Molecular dynamics simulations of the interior of aqueous reverse micelles: A comparison between sodium and potassium counterions. *Langmuir* 19:2514–2520.

Faeder, J., and B. M. Ladanyi. 2000. Molecular dynamics simulations of the interior of aqueous reverse micelles. *J. Phys. Chem. B* 104:1033–1046.

Faeder, J., and B. M. Ladanyi. 2001. Solvation dynamics in aqueous reverse micelles. *J. Phys. Chem. B* 105:11148–11158.

Faeder, J., and B. M. Ladanyi. 2005. Solvation dynamics in reverse micelles: The role of headgroup-solute interactions. *J. Phys. Chem. B* 109:6732–6740.

Fioretto, D., M. Freda, S. Mannaioli, et al. 1999a. Infrared and dielectric study of Ca(AOT)(2) reverse micelles. *J. Phys. Chem. B* 103:2631–2635.

Fioretto, D., M. Freda, G. Onori, et al. 1999b. Effect of counterion substitution on AOT-based micellar systems: Dielectric study of Cu(AOT)(2) reverse micelles in CCl4. *J. Phys. Chem. B* 103:8216–8220.

Freda, M., G. Onori, and A. Santucci. 2001. Infrared and dielectric spectroscopy study of the water perturbation induced by two small organic solutes. *J. Mol. Struct.* 565:153–157.

Fulton, J. L., D. M. Pfund, J. B. McClain, et al. 1995. Aggregation of amphiphilic molecules in supercritical carbon-dioxide: A small-angle x-ray-scattering study. *Langmuir* 11:4241–4249.

Gazaryan, I. G., N. L. Klyachko, Y. K. Dulkis, et al. 1997. Formation and properties of dimeric recombinant horseradish peroxidase in a system of reversed micelles. *Biochem. J.* 328:643–647.

Giordano, R., P. Migliardo, U. Wanderlingh, et al. 1993. Structural properties of micellar solutions. *J. Mol. Struct.* 296:265.

Griffiths, A. D., and D. S. Tawfik. 2000. Manmade enzymes: From design to in vitro compartmentalisation. *Curr. Opin. Biotechnol.* 11:338.

Harpham, M. R., B. M. Ladanyi, N. E. Levinger, et al. 2004. Water motion in reverse micelles studied by quasielastic neutron scattering and molecular dynamics simulations. *J. Chem. Phys.* 121:7855–7869.

Harpham, M. R., B. M. Ladanyi, and N. E. Levinger. 2005. The effect of the counterion on water mobility in reverse micelles studied by molecular dynamics simulations. *J. Phys. Chem. B* 109:16891–16900.

Heitz, M. P., and F. V. Bright. 1996. Rotational reorientation dynamics of aerosol-OT reverse micelles formed in near-critical propane. *Appl. Spectrosc.* 50:732–739.

Levinger, N. E., R. E. Riter, D. M. Willard, et al. 1998. Solvation dynamics in restricted environments: Solvent immobilization in reverse micelles. *Springer Ser. Chem. Phys.* 63:553–535.

Linse, P. 1989. Molecular dynamics study of the aqueous core of a reversed ionic micelle. *J. Chem. Phys.* 90:4992–5004.

Lipari, G., and A. Szabo. 1980. Effect of librational motion on fluorescence depolarization and nuclear magnetic-resonance relaxation in macromolecules and membranes. *Biophys. J.* 30:489–506.

Luisi, P. L., M. Giomini, M. P. Pileni, et al. 1988. Reverse micelles as hosts for proteins and small molecules. *Biochim. Biophys. Acta* 947:209–246.

Luisi, P. L., and B. E. Straub. 1984. *Reverse Micelles: Biological and Technological Relevance of Amphiphilic Structures in Apolar Media.* New York, NY: Plenum.

Luisi, P. L., P. Walde, and T. Oberholzer. 1994. Enzymatic RNA-synthesis in self-reproducing vesicles: An approach to the construction of a minimal synthetic cell. *Phys. Chem. Chem. Phys.* 98:1160–1165.

Nave, S., J. Eastoe, R. K. Heenan, et al. 2000. What is so special about aerosol-OT? Part II—Microemulsion systems. *Langmuir* 16:8741–8748.

Nave, S., J. Eastoe, R. K. Heenan, et al. 2002. What is so special about aerosol-OT? Part III—Glutaconate versus sulfosuccinate headgroups and oil-water interfacial tensions. *Langmuir* 18:1505–1510.

Onori, G., and A. Santucci. 1993. IR investigations of water-structure in aerosol OT reverse micellar aggregates. *J. Phys. Chem.* 97:5430–5434.

Pant, D., R. E. Riter, and N. E. Levinger. 1998. Influence of restricted environment and ionic interactions on water solvation dynamics. *J. Chem. Phys.* 109:9995–10003.

Pileni, M. P., Ed. 1989. Structure and reactivity in reverse micelles. *Stud. Phys. Theor. Chem.* Amsterdam: Elsevier.

Pileni, M. P. 1993. Reverse micelles as microreactors. *J. Phys. Chem.* 97:6961–6973.

Pileni, M. P., B. W. Ninham, T. Gulik-Krzywicki, et al. 1999. Direct relationship between shape and size of template and synthesis of copper metal particles. *Adv. Mater.* 11:1358–1362.

Pohorille, A., and D. Deamer. 2002. Artificial cells: Prospects for biotechnology. *Trends Biotechnol.* 20:123–128.

Riter, R. E., E. P. Undiks, and N. E. Levinger. 1998a. Impact of counterion on water motion in aerosol OT reverse micelles. *J. Am. Chem. Soc.* 120:6062–6067.

Riter, R. E., D. M. Willard, and N. E. Levinger. 1998b. Water immobilization at surfactant interfaces in reverse micelles. *J. Phys. Chem. B* 102:2705–2714.

Salaniwal, S., S. T. Cui, H. D. Cochran, et al. 2001. Molecular simulation of a dichain surfactant water carbon dioxide system. 1. Structural properties of aggregates. *Langmuir* 17:1773–1783.

Sando, G. M., K. Dahl, and J. C. Owrutsky. 2004. Vibrational relaxation dynamics of azide in ionic and nonionic reverse micelles. *J. Phys. Chem. A* 108:11209–11217.

Sando, G. M., K. Dahl, and J. C. Owrutsky. 2005. Surfactant charge effects on the location, vibrational spectra, and relaxation dynamics of cyanoferrates in reverse micelles. *J. Phys. Chem. B* 109:4084–4095.

Sarkar, N., K. Das, A. Datta, et al. 1996. Solvation dynamics of coumarin 480 in reverse micelles. Slow relaxation of water molecules. *J. Phys. Chem.* 100:10523–10527.

Savelli, G., N. Spreti, and P. Di Profio. 2000. Enzyme activity and stability control by amphiphilic self-organizing systems in aqueous solutions. *Curr. Opin. Colloid Interface Sci.* 5:111–117.

Senapati, S., and M. L. Berkowitz. 2003a. Molecular dynamics simulation studies of polyether and perfluoropolyether surfactant based reverse micelles in supercritical carbon dioxide. *J. Phys. Chem. B* 107:12906–12916.

Senapati, S., and M. L. Berkowitz. 2003b. Water structure and dynamics in phosphate fluorosurfactant based reverse micelle: A computer simulation study. *J. Chem. Phys.* 118:1937–1944.

Senapati, S., and M. L. Berkowitz. 2004. Computer simulation studies of water states in perfluoro polyether reverse micelles: Effects of changing the counterion. *J. Phys. Chem. A* 108:9768–9776.

Tan, H. S., I. R. Piletic, and M. D. Fayer. 2005. Orientational dynamics of water confined on a nanometer length scale in reverse micelles. *J. Chem. Phys.* 122: Art. No. 174501.

Teixeira, J., M. C. Bellissent-Funel, S. H. Chen, et al. 1985. Experimental-determination of the nature of diffusive motions of water-molecules at low-temperatures. *Phys. Rev. A* 31: 1913–1917.

Thompson, W. H. 2002. A Monte Carlo study of spectroscopy in nanoconfined solvents. *J. Chem. Phys.* 117:6618–6628.

Thompson, W. H. 2004. Simulations of time-dependent fluorescence in nano-confined solvents. *J. Chem. Phys.* 120:8125–8133.

Tobias, D. J., and M. L. Klein. 1996. Molecular dynamics simulations of a calcium carbonate/calcium sulfonate reverse micelle. *J. Phys. Chem.* 100:6637–6648.

Venables, D. S., K. Huang, and C. A. Schmuttenmaer. 2001. Effect of reverse micelle size on the librational band of confined water and methanol. *J. Phys. Chem. B* 105:9132–9138.

Vermathen, M., P. Stiles, S. J. Bachofer, et al. 2002. Investigations of monofluoro-substituted benzoates at the tetradecyltrimethylammonium micellar interface. *Langmuir* 18:1030–1042.

Walde, P., A. Goto, P. A. Monnard, et al. 1994. Oparin's reactions revisited—Enzymatic-synthesis of poly(adenylic acid) in micelles and self-reproducing vesicles. *J. Am. Chem. Soc.* 116:7541–7547.

Wong, M., J. K.Thomas, and T. Nowak. 1977. Structure and state of H_2O in reversed micelles. 3. *J. Am. Chem. Soc.* 99: 4730–4736.

Zhang, J., and F. V. Bright. 1991. Nanosecond reorganization of water within the interior of reversed micelles revealed by frequency-domain fluorescence spectroscopy. *J. Phys. Chem.* 95:7900–7907.

Zhang, J., and Bright, F. V. 1992. Steady-state and time-resolved fluorescence studies of bis(2-ethylhexyl) sodium succinate (AOT) reserve micelles in supercritical ethane. *J. Phys. Chem.* 96:5633–5641.

Zhong, Q., A. P. Baronavski, and J. C. Owrutsky. 2003. Reorientation and vibrational energy relaxation of pseudohalide ions confined in reverse micelle water pools. *J. Chem. Phys.* 119:9171–9177.

Zulauf, M., and H. F. Eicke. 1979. Inverted micelles and microemulsions in the ternary system water/aerosol-OT/isooctane as studied by photon correlation spectroscopy. *J. Phys. Chem.* 83:480.

Part III

Water in Biochemistry

10 Water
Constraining Biological Chemistry and the Origin of Life

Steven A. Benner

CONTENTS

10.1   INTRODUCTION

Why are things in biology the way they are? Such questions are not new to humankind. A century of work in biological chemistry now allows us, however, to expand their scope to the molecular level. We can therefore ask such questions about the chemical details of biomolecules ranging from genes to proteins to metabolism. Why are our genes built from four nucleotide letters (Benner, 2004)? Why are our proteins built from 20 directly encoded amino acids? Why is citric acid a key metabolite in the oxidative degradation of acetate? Why does terran life exploit water as its principal solvent?

A simple "We don't know" may be the most correct answer for such questions. It is, however, an uninteresting answer. Further, attempts to answer such questions, even at the level of speculative hypothesis, can drive the generation of new ideas and new experiments. These, in turn, can force the development of our understanding of the natural world (Benner et al., 2002). This chapter covers some of these attempts, and includes a brief discussion of the intellectual processes that might stand behind them (Benner, 2009).

10.2 ADAPTATION, ACCIDENT, AND VESTIGIALITY

Within the Darwinian paradigm, it is axiomatic that the life we know is a product of natural selection superimposed upon random variation in the structure of biological molecules. This suggests three categories of explanation for individual features of the life that we know (Benner and Ellington, 1988). The first considers function, and hypothesizes that a terran biomolecular structure is found because it offers the best solution to a particular biological problem. Under the Darwinian paradigm, "best" means "confers most fitness," or (at least) more fitness than available (but unused) alternatives. This explanation implies that the living system had access to alternative solutions, individuals chose among these, and those who chose poorly failed to leave progeny.

The second category of explanation considers historical accident. The universe of chemical possibilities is huge. For example, the number of different proteins 100 amino acids long, built from the standard set of 20 amino acids, is 20^{100}, a number larger than the number of atoms in the universe. Life on Earth certainly did not have time to sample all possible sequences to find the best protein for any particular biological role. Rather, the protein sequences that exist in modern terran life must reflect contingencies, chance events in history that led to one choice over another, whether or not the choice was optimal.

In the case of proteins as a biomolecule, balancing explanations of the first and second categories is difficult because we know little about the hypersurface that relates the sequence of a protein to its ability to confer fitness (Benner and Ellington, 1988). If that hypersurface is relatively smooth, then optimal solutions might be easily reached via Darwinian search processes, regardless of where contingencies began the search. If the surface is rugged, however, then many structures of terran life are locally optimal, and suboptimal with respect to the universe of possibilities. Darwinian search strategies cannot easily find a globally optimal solution on a rugged hypersurface.

A third category of explanation draws on the concept of vestigiality. Here, it is recognized that some features of contemporary terran biochemistry may reflect ancient constraints and selective pressures that no longer exist. Contemporary features may therefore not be optimized for the modern environment, but rather be vestiges of constraints or optimization from times past. Vestigiality presumably explains, for example, the human appendix, and many of the biochemical details found in modern life (such as the RNA component of many cofactors used in contemporary terran metabolism) (Benner et al., 1989; Benner, 2009).

10.3 ASKING "WHY?" QUESTIONS ABOUT THE STRUCTURE OF DNA

These categories of explanations are illustrated by possible answers to a question asked by many biochemistry students: Why does DNA use adenine, which forms only two hydrogen bonds with thymine (Figure 10.1)? Aminoadenine, an analog of adenine with one additional amino group, can form three hydrogen bonds to thymine, forming a stronger nucleobase pair leading to a higher "melting temperature" for the DNA duplexes that contain it (here, "melting temperature" refers to the temperature at which two DNA strands separate) (Geyer et al., 2003). Are double helices not better if they are more stable? Would not aminoadenine be better than adenine as a component of DNA?

Consider three ways to explain the presence of adenine, rather than aminoadenine, in terran DNA. A functional explanation might argue that genomes are optimal when they have access to *both* a weakly bound nucleobase pair (A–T) *and* a strongly bound (G–C) nucleobase pair. According to this model, biological systems need to modulate the melting temperature of its DNA to achieve an optimum, which may not be the maximum melting temperature. If DNA has both a strong and weak base pair, it can do so by adjusting the ratio of AT and GC nucleobase pairs. An organism that has both a strong base pair (GC) and a weak base pair (AT) has, according to this explanation, a competitive advantage over an organism that uses aminoadenine, does not have a weak base pair, and therefore does not have the opportunity to modulate the melting temperature of its DNA by choosing between a strong and weak pair. This explanation implies that at some point in life's

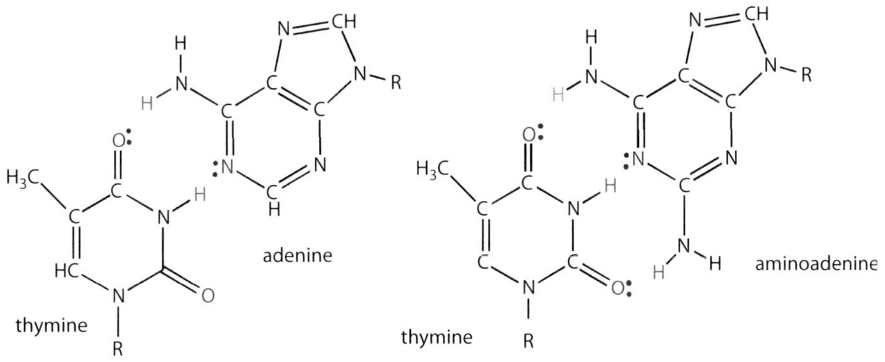

FIGURE 10.1 The thymine–adenine T–A base pair is joined by two hydrogen bonds. Why is aminoadenine (right) not used instead of adenine (left), to give a stronger base pair joined by three hydrogen bonds instead of two? Function, history, and vestigiality might be separately invoked. R is the 2′-deoxyribose ring of the DNA backbone.

history, the biosphere had access to both aminoadenine and adenine, and rejected the former as it accepted the latter.

A historical model can also be constructed to explain the structure of adenine. This model suggests that adenine was arbitrarily chosen over aminoadenine by accident. To explain the conservation of this historical accident, the models suggest that a metabolically complex environment becomes committed in many ways in a particular genetic structure. Once DNA is built with adenine, for example, DNA polymerases must evolve to accept it (and not aminoadenine). Biosynthetic pathways then arise to give adenine (not aminoadenine). In these and other ways, the adenine originally chosen by accident soon becomes difficult to replace by aminoadenine, even if it conferred greater fitness.

The nature of the accident may not be determinable, especially if it was very ancient. Of course, the accident must have been consistent with physical law as constrained by the environment. Beyond that, however, the choice was made for no particular reason. This makes historical models for contemporary biomolecular structure often the most difficult to test.

An explanation focused on vestigiality might notice that adenine, and perhaps not aminoadenine, is made from ammonium cyanide under conditions presumed to have been present in the cosmos before life emerged (Oró, 1960). According to this explanation, adenine was then preferentially suited for starting life in preference to aminoadenine in times past. In the modern world, metabolic pathways provide access to many molecules that are not accessible prebiotically. Thus, the constraints on modern genetic structures are in no sense the same as those on the first genetic structures; if modern life wanted aminoadenine, it could biosynthesize it. Therefore, it is possible that in the modern world, adenine may no longer have an advantage over aminoadenine (i.e., a biological system making the choice today would be indifferent with respect to the two). Indeed, aminoadenine might be preferable today over adenine, but not exploited because it is too difficult to replace.

It is useful to construct these alternative explanations as hypotheses, even if no experiments can presently distinguish them. The act of constructing the trialectic allows scientists to appreciate how accessible alternative explanations are. This, in turn, provides a level of discipline to a discussion where the "We don't know" answer is probably most appropriate (Benner, 2009).

10.4 SYNTHESIS AS A WAY OF TESTING ANSWERS TO "WHY?" QUESTIONS IN BIOLOGY

The development of synthetic biology as a discipline (Benner, 2003; Ball, 2004; Gibbs, 2004) has provided a process to explore such "Why?" questions. This process goes beyond simply constructing the function–history–vestige trialectic on paper. Rather, it challenges the scientist to prepare,

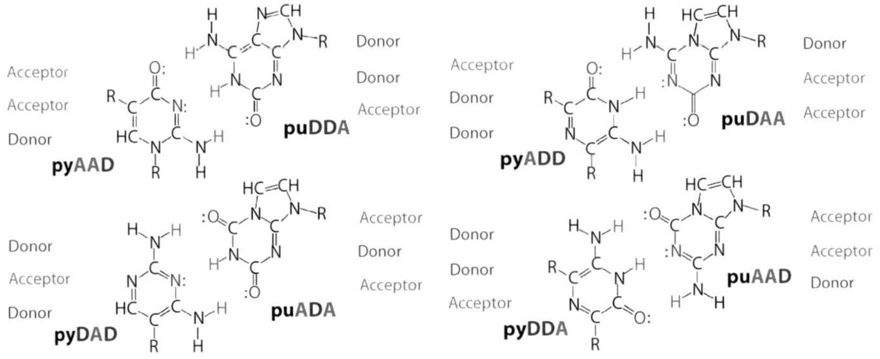

FIGURE 10.2 Synthetic biology can create DNA that is different from that found in contemporary terran life. The small pyrimidines (left) pair with large purines (right) to achieve size complementarity. Hydrogen bonding complementarity is achieved by pairing hydrogen bond donors (D) with acceptors (A). Shuffling these donor and acceptor groups creates new nucleobase pairs. The fact that these compounds can be prepared by synthetic biologists, and once prepared, have been shown to behave in a way acceptable as part of a genetic system, makes it clear that the answer to the counterfactual question—Can DNA have a different structure than it has naturally?—is "yes."

in the laboratory, structures that are not natural (e.g., DNA that contains aminoadenine) and to see how well they work. These experiments have been done with aminoadenine. Aminoadenine indeed binds more tightly to thymidine than does adenine (Geyer et al., 2003). For that reason, many experiments seeking to develop artificially evolving systems in the laboratory use it instead of adenine.

Synthetic biology permits us to be more ambitious than nature has been when constructing genetic molecules. For example, the Watson–Crick base pair follows two rules of complementarity; under the first (size complementarity), large purines pair with small pyrimidines. Under the second, hydrogen bond donors pair with hydrogen bond acceptors. Given these constraints, it is clear that six mutually exclusive nucleobase pairs are possible within the Watson–Crick pairing geometry (Figure 10.2). Why are only two of these present in contemporary terran DNA?

Synthetic biology allowed us to rule out one conceivable answer to this question: "Because the other four do not form duplexes." The Benner laboratory prepared representative samples of each of the nonstandard base pairs shown in Figure 10.2, and found that they did indeed support duplex function. Indeed, it was possible to show that expanded genetic alphabets could be replicated and replicated again, in principle forming a type of artificial life capable of Darwinian evolution (Sismour et al., 2004; Sismour and Benner, 2005). This implies that DNA could have been different from how we find it on Earth. Why today's terran DNA does not exploit more genetic letters must be explained using the trialectic outlined above (Benner, 2009, Chapter 6).

This exemplifies synthesis as a paradigm for doing "counterfactual research" to answer "Why?" and "Why not?" questions. Synthesis allows the "counterfact" to become a "fact," at least in the laboratory. The approach is strongly rooted in chemistry, where alternatives to biomolecules are often synthesized by preparative organic chemistry. With nucleic acids, synthesis has delivered to us a series of molecules that are *not* natural DNA, but behave like it. The products of such syntheses also have practical value. Today, some 400,000 patients each year have their health care improved using diagnostics systems based on compounds in Figure 10.2 (Elbeik et al., 2004).

10.5 SYNTHESIS AS A TOOL TO PREVENT SELF DECEPTION

Synthesis of counterfactuals also has a broader intellectual value as we attempt to address questions where the best answer might be "We do not know." As imagination becomes increasingly more distant from the reality that we know, intellectual process (how we think about thinking) becomes increasingly important. The need for process reflects a theme that recurs in the history of science: It is easy for humans to convince themselves that data contain patterns that they do not, that patterns compel models when they need not, and that models are reality, which they are not. The desire to believe is strong, especially when the object of belief is a theory or model that the believer himself/herself developed.

These comments are not pejorative. They simply reflect how human minds work. Further, human scientists can be creative and productive *because* human minds work this way. Thus, while the scientific method taught in middle school emphasizes the importance of unfiltered observations, prejudice-free data analysis, and value-neutral experiments, the productivity of scientists does not depend on the extent to which they meet this largely fictitious ideal, but rather how they manage the "closed-ness" of mind, the values, and the filters that come naturally with human cognition. Nevertheless, "science" can be defined as an intellectual process that embodies a mechanism to prevent the scientists from believing what they want to believe.

Synthesis is a way to manage these. Synthesis sets a goal, a large "put a man on the moon" type of challenge. By forcing scientists to meet the challenge, synthesis forces scientists across uncharted terrain. There, scientists must solve problems that they would not normally encounter in a purely analytical exercise. In the process of encountering and solving these unexpected problems, scientists are forced to make discoveries. Not the least of these discoveries might be that their model for reality is inadequate to guide the synthesis. In this event, the discovery can neither be ignored nor rationalized away, as the synthesis then fails, the challenge is not met, and this is obvious to everyone.

The value of synthesis as part of an intellectual process arises because the challenge is unforgiving. Nowhere is this better exemplified than in "rocket science." For example, in the Mars Climate Orbiter, the guidance system software employed the metric system of measurement, while the hardware (mistakenly) employed the English system. Throughout the transit flight to Mars, this incompatibility caused the craft to misbehave. But, as humans do, this misbehavior was rationalized away by much of the mission staff. Only the unforgiving goal, placing the craft in orbit, intervened to require the scientists to recognize that their model of reality was false. The orbiter crashed.

Synthesis has frequently provided the same kind of drive to change our view of biomolecular structure. For example, starting in 1990, various laboratories attempted to make "antisense" nucleic acid analogs that lacked the repeating negative charge on the DNA backbone. They hoped that this counterfactual DNA analog would be soluble in lipids, passively enter cells, and still retain the Watson–Crick pairing rules of standard DNA. The Watson–Crick model did not propose any particular reason to think otherwise; it dogmatically proposed no role for the repeating charges in the DNA backbone. As a consequence, many organizations (and many venture capitalists) invested huge sums of money in the challenge of getting a membrane-permeable nucleic acid analog based on a model for reality that argued that the backbone charges were dispensable.

This model proved to be wrong. For very short sequences, nonionic analogs of DNA were found that could pair according to Watson–Crick rules. For longer analogs, however, this proved not to be possible. Long nonionic DNA analogs were found to fold, precipitate, and fail to bind to their complements (Hutter et al., 2002). The failure to achieve robust Watson–Crick pairing in long nonionic DNA analogs forced scientists to reconsider their view that the backbone was not important to duplex formation. They began to ask whether the repeating charge in the backbone of terran genetic molecules might play a role.

These questions generated the discovery that the charges were important to Watson–Crick pairing. This prompted the further thought that they might be critical for Darwinian processes that support life. In general, even modest changes in the structure of an organic molecule lead to significant changes in its physical properties. To support Darwinian evolution, however, a living system needs a genetic molecule that can suffer small changes in chemical structure *without* significantly changing its physical properties.

A repeating charge in its backbone generates this feature in a polymer. A repeating charge dominates the physical properties of any molecule that has it. Therefore, changing the information-encoding parts of the molecule (in DNA and RNA, the nucleobases), while leaving the repeating charge unchanged, can change the information content of a genetic molecule without changing its physical properties. Indeed, it is difficult to conceive of any other way to achieve this result. Thus, a repeating charge might be necessary for *any* genetic molecule acting in water. This conclusion, different from that proposed in the Watson–Crick first generation model for DNA, was forced on scientists through their failure to synthesize molecules based on models that ignored it.

Other exercises in synthetic biology have tested alternative genetic systems that lack the ribose and deoxyribose sugars that are found in natural RNA and DNA (Benner, 2004). One of the most striking conclusions of this work has been that ribose and 2′-deoxyribose are outstanding for supporting Watson–Crick pairing, and better than almost all other scaffolding molecules. This too was not anticipated from the first generation Watson–Crick model for DNA structure. As we shall mention below, ribose also appears to be a particularly easy carbohydrate to synthesize prebiotically (Ricardo et al., 2004), a virtue that appears to be unrelated to its ability to support Watson–Crick pairing.

As a consequence of these studies, our understanding of the rationale behind the structure of genetic material has become quite broad. This rationale includes historical, functional, and vestigial features, and makes broad hypotheses about the structure of the genetic matter that might, for example, be encountered in nonterran life elsewhere in the galaxy (Benner and Hutter, 2002).

10.6   THE STRUCTURES OF TERRAN BIOMOLECULES APPEAR TO BE GENERALLY ADAPTED FOR WATER

The application of strategies from synthetic biology to ask "Why?" and "Why not?" questions in genetics brought those constructing artificial genetic systems face to face with water as a solvent. A solute like DNA that has a repeating charge is "friendly" with water. It dissolves well in water, and works well in water, as DNA and RNA show. It does not work well in less polar solvents, such as ethanol, dioxane, or hexane.

Many other molecules found in standard terran biology are also friendly to water. These include many metabolites, including the citric acid mentioned in the introduction. Citric acid has three negative charges. Many other intermediates in metabolism are also polyelectrolytes (molecules bearing multiple charges) because they are phosphorylated. These dissolve well in water. Conversely, such molecules also do not dissolve well in less polar solvents, such as ethanol, dioxane, or hexane.

There is little doubt that the compatibility of contemporary terran biomolecules with water can be explained by a model that holds that the biomolecules themselves have been adapted via Darwinian mechanisms to water. This leaves open, however, the question as to whether (presumably different) sets of molecules could have evolved to work equally well in a solvent that is very different from water (such as the methane–ethane oceans on Titan, for example).

The discussion above suggests that the answer to this question might be "no." As molecules containing a repeating charge would not work in methane–ethane oceans, genetic systems on Titan could not use a repeating charge to enable mutation in the way that DNA uses its repeating charge to enable mutation in water. Thus, it is difficult to conceive of a molecular structure that has the capability of changing its encoding parts while leaving its physical properties largely unchanged in methane–ethane mixtures, as DNA does in water.

10.7   IS WATER ADAPTED TO BIOMOLECULES?

Given the manifest matching of the properties of terran biomolecules to the demands of water, it is interesting to ask the reverse question, whether the properties of water are adapted to biomolecules. This question is not truly a reversal. A protein or DNA sequence has access to a mechanism for introducing changes in structure that are modest with respect to the overall structure. These small changes can, in many cases, permit a slightly discrete search of "function space" (Benner and Ellington, 1988). Changing one amino acid need not (although it may) change dramatically the behavior of a protein containing 400 amino acids (hence "slightly discrete"). Thus, one can search a fitness landscape in small steps with a biopolymer. Similar mechanisms are accessible to change the structures of metabolites, less discretely, but still within a largely stable environment.

No such slightly discrete stages are possible to evolve water. Other molecules exist, of course, that might replace water. The simplest arise from combinations of second row elements with hydrogen (the "ices," as they are known to planetary scientists, including BH_3, CH_4, NH_3, OH_2, and FH).

These ices do not, however, provide a set of molecules that are slightly different in their physical properties. These mutant forms of water are extremely different from water in their properties, even by naive metrics. These different properties include melting points, boiling points, and intrinsic reactivities. Because the ices cannot change, even slightly, their structure and properties, the function–structure landscape that includes them as points must be rugged. A rugged structure–function landscape defeats the Darwinian search for optimality. This implies that ices could not have evolved in a Darwinian sense to serve themselves, or their biological contents.

10.8   IS WATER WELL SUITED FOR LIFE?

This discussion, of course, assumes immutable physical law. This assumption can be relaxed, much as it is being relaxed in physics. If we assume that the properties could have been different, we can ask whether a different *kind* of water might be better suited to life.

In 1986, Barrow and Tipler (1986) addressed this question as part of their larger discussion about whether physical laws in general are well suited to support life. These might include physical laws that determine, *inter alia*, the properties of water. Thus, the universe, and more specifically the chemistry in the universe, might be adapted to facilitate life.

Barrow and Tipler (1986) discuss two general ways to view a match between the physical properties of water and the requirements for a solvent to support life. A match of some sort is clearly necessary for a sentient being to exist, and therefore be able to perceive that the match exists. Humans would not be alive to comment on the suitability of water as a solvent for life if it were not suited (in some sense) for life.

This need not be cast as a Darwinian model, but it might be. To be Darwinian, the physical laws that we know might have been chosen from a variety of law possibilities from a variety of universes. The variant universes must have progeny that have slightly different physical laws, the differences must themselves be heritable, and the fitter variant universes must survive to the exclusion of less fit universes. Under this particular version of the anthropic hypothesis, a large number of alternative universes exist, some whose ices behave in ways conducive for the emergence of life, and others whose ices do not.

The second view falls within the category of intelligent design. This version of the hypothesis requires a chooser, a supernatural entity (using this adjective in the literal sense) that arranges to have a universe in which life might emerge, and therefore chooses the physical laws (including properly behaved ices) that would permit life to emerge.

Synthesis would be a useful paradigm here as well. If we could synthesize alternative universes with different physical laws, we might run experiments to learn whether the alternatives are indeed more or less suited for life, just as we experimented with alternative synthetic DNA. Unless reality is very different from how we perceive it, this is not possible. To test these hypotheses, we do not have access to synthesis to manage our prejudices and to drive discovery. We can, however, ask simpler questions. For example, if water had different properties, how would the molecular structures of life need to be adjusted?

10.9   THE PROPERTIES OF WATER ARE NOT *VERY* FINELY TUNED TO LIFE

Discussions in physics have suggested that physical laws are very finely tuned to support life. In some cases, if physical parameters were different by a few percent, the cosmos would be very different from the way it is now.

We know that this is not true for water as a biosolvent. We can change (and scientists have changed) slightly the properties of water by replacing the protium isotope of hydrogen with deuterium, creating heavy water (a synthetic water, in a sense). Deuterated molecules have different properties; in some cases they react seven times slower than protiated analogs. Thus, the perturbation arising through the replacement of a protium by a deuterium is not negligible.

We can (and scientists have) put living systems adapted to protiated water into deuterated water to see what happens (Katz et al., 1962, 1964). Living systems often do not do too well at first. But if the life is microbial and has access to adaptive responses (epigenetic adaptation seems to be sufficient), the system changes to perform as well as it did in protiated water. There is little doubt that life, even at the level of the mouse, would become fully adapted to the tweaked water, as well as it is to undeuterated water, if given full access to Darwinian processes.

The many studies by Katz and his coworkers of life in deuterated water were so detailed that the issue has not been revisited in recent years. We might (in principle) repeat the experiment with carrier-free tritiated water, another synthetic water, as a solvent. Replacing protium by tritium creates a larger perturbation, with tritiated molecules reacting typically 10 to 20 times slower than their protiated analogs. The experiment has not been done. But there is little doubt that concerning the chemical properties of the modified solvent, life could adapt perfectly well.

At this point, the experiments must stop. No more slightly discrete variations are available in water to do any more analyses of this type. The experiments with deuterated water make indisputable, however, the view that the properties of water are not as finely tuned as has been proposed for some fundamental physical parameters, such as the fine structure constant (Barrow and Tipler, 1986).

10.10   MORE INTELLECTUAL PROCESS

The inability to synthesize alternative solvents with modestly changed properties leaves those asking about the intrinsic suitability of water for life with only a few intellectual tools; analysis and modeling are among them. Pure analysis (in ancient Greece, "metaphysics") relies heavily on concepts of "elegance" and "intuition." These are themselves largely reflections of the perspective of the analyzer.

The limitations of pure analysis are well known in physics. For example, Ginsparg and Glashow pointed out two decades ago that the intimate interconnection of space and time would not likely have been discovered had Maxwell's equations not emerged from experiment (Ginsparg and Glashow, 1986). We certainly would not have generated the idea of the quantum without the preceding century of chemistry, which relied on synthesis and experiment, to identify the elements from molecules to enable spectroscopy. Experiment is needed to drive ideas away from prejudice and toward novelty.

Two analytical approaches are clearly unable to address this question. First, the observation that all of the life that we know uses water is not a reliable guide. As noted above, life on Earth has almost certainly evolved in water. In doing so, it undoubtedly exploited opportunities to change its genetically encoded molecules and metabolites, slightly discretely, to become optimal for life in water. Thus, it is not surprising, given the Darwinian paradigm, to find that terran life is well adapted to water, at least to the extent to which various chemical constraints permit it.

Furthermore, terran life is the only life that we know. Therefore, given the anthropocentric tendencies of human thought processes, our models for life are influenced (to the point of being determined) by our experience with terran life. Indeed, given the general lack of creativity of the human mind, it is hard to find a model for life that does not resemble the terran life that we know; the aliens on *Star Trek* generally resemble Hollywood actors with prostheses. It is natural to slip from the Darwinian axiom (that life has adapted to water) to the assumption that water must be adapted to life.

10.11   THE LOGICAL DIALECTIC AS AN INTELLECTUAL PROCESS: CONSIDER WATER

For many years, scientists have played a dialectical game as part of an intellectual process to manage pure analysis (Levins and Lewontin, 1985). Here, the game involves the construction of a hypothesis and the anti-hypothesis. An argument between the two then follows, often in the literature. We can play this game to ask the question: Is water finely tuned to support life?

It is not difficult to find properties of water that, given our view of life based on our experience with terran life, appear to make water well suited to support life in general. For example, frozen water floats. Water is an excellent solvent for salts. Water has a large range of temperatures where it is liquid.

To the human mind, these are compelling, even to the point of demanding water as a requirement for life. Indeed, NASA has chosen to "follow the water" as it searches for life in the solar system. Given very real limits on NASA budgets, this seems to be a sensible strategy. But being pragmatic as we search for life that is different does not answer the question above. Let us try to manage our terracentricity by playing a game with a single rule: For every example of a property of water that we think is optimal to support life, we must find a property of water that is *not* optimal for life.

For example, we might (and do) say that water has a large temperature range over which it is liquid, and this is proposed to be good. But formamide ($HCONH_2$) has a larger range at atmospheric pressure (255–480 K). So, by this criterion, water is worse than formamide as a biosolvent. Ammonia has a larger range, if one is not tied to a linear view of the Celsius scale (the temperature range 1–2 K is much more significant than 273–274 K). Thus, by this criterion, ammonia might be viewed as a better solvent than water.

Indeed, we may view the temperature range over which water is a liquid as *not* a good choice to support life broadly. Water is a liquid in only a very small fraction of the environments that the cosmos offers; throughout most of the cosmos, water is frozen solid. Assuming that a fluid is necessary for life, and accepting the premise that water was given properties that make it optimally suited to support life in the cosmos, would water not be better as a biosolvent if its freezing and boiling points allowed it to be fluid over a larger range of the space in the galaxy? This point has been made recently by Ward and Brownlee in their book *Rare Earth* (Ward and Brownlee, 2000). The book argues that life is distributed sparsely, perhaps extremely sparsely, in part because it requires water as a liquid, and liquid water is sparsely distributed in the cosmos.

Barrow and Tipler (1986) discuss at many points the broad range at which water is a liquid, and argue that this makes water uniquely suited for life. We live on Earth, of course, and find this argument compelling. If we lived on Mars, however, the argument would be less compelling. Liquid water cannot exist on the surface of Mars; water ice sublimes directly to water vapor without going through an intermediate liquid phase at the low pressures of the Martian atmosphere (about 1% the pressure at sea level on Earth). Why do many other authors write that water has a large liquid range and ignore, for example, formamide? Most likely, it is because we live on Earth, not Mars.

Further, considering the low temperature of much of the cosmos, we might actually prefer water–ammonia mixtures as a biosolvent. Ammonia is an excellent antifreeze. Water–ammonia eutectics may even be liquid even at the low temperature of Titan. Water–ammonia eutectics are certainly more abundant in the cosmos, they have a wider temperature range of liquidity, and they are not bad as solvents.

Water is, as noted above, an excellent solvent for salts, and this is proposed to be good for supporting life. But formamide is also an excellent solvent for salts. Formamide has a broader temperature range of fluidity. Formamide wins again.

We might also, dialectically, argue that the ability to dissolve salt is *not* desirable for life. Life is made of hydrocarbons. Water is distinctly unable to dissolve many of these (as is, incidentally, formamide, although it is a modestly better solvent than water for some). In this view, in examining terran metabolism, we might observe that the structures of metabolic intermediates (citrate, oxaloacetate, malate, etc.) are unfortunately constrained to have enough charged groups as to make them soluble in water, which can dissolve well only such species.

As a consequence, some of these metabolic intermediates have decidedly disadvantageous properties. Oxaloacetate, for example, decomposes at 340 K with a half life measured in seconds; thermophiles (and these might have been the earlier forms of life on Earth (Gaucher et al., 2003) have a hard time keeping enough oxaloacetate around to support the biosynthesis of amino acids—all to manage the unfortunate limitations of water as a solvent.

The dialectic game can be carried further. For example, water supports protein folding by creating a hydrophobic effect that stabilizes the fold. To play the game, we might argue that water does *not* support protein folding, because it disrupts hydrogen bonds that stabilize the fold. Indeed, if one searches the chemical literature for examples where chemists consciously try to obtain molecules that self-organize, one observes that water is rarely used as a solvent, precisely because it disrupts noncovalent directional bonding such as hydrogen bonding. Chloroform, for example, is much preferred as a solvent by organic chemists seeking to get intermolecular assembly.

Likewise, water dissolves functionalized carbon containing molecules, and carbon is viewed as being essential for life. Dialectically, however, we might view this as a constraint. Water reacts with silanes. Maybe if this were not an intrinsic property of water, we would think that silicon is essential for life (because we would be made of it) (Bains, 2004).

The parlor game can continue indefinitely, generating as many reasons why water is a bad solvent for life as reasons why water is a good solvent for life. Many make reference to the fact that water itself is a very reactive compound. In principle, one might like a solvent *not* to react with dissolved solutes. Water is certainly not such a solvent.

Water presents both a nucleophilic oxygen and an acidic hydrogen at 55 molar concentration. As a consequence, chemists avoid water when they seek an inert solvent. Indeed, if one searches a recent issue of *Journal of Organic Chemistry* and tabulates the solvents selected by organic chemists seeking to do chemistry, one discovers that chemists chose a solvent other than water to run their reactions more than 80% of the time (Figure 10.3). Thus, in many senses, hydrocarbon solvents are better than water for managing complex organic chemical reactivity.

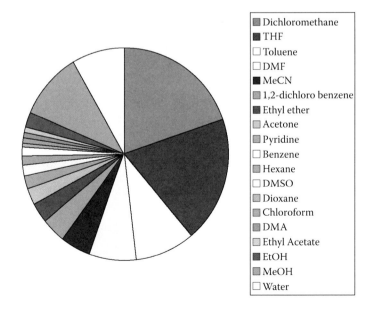

FIGURE 10.3 A pie chart showing the relative frequency of use of different solvents by chemists who wish to avoid the undesirable reactivity of water, and therefore control the outcome of chemical reactions. Water is only rarely in the solvent of choice. Data are from a single issue of the *Journal of Organic Chemistry*.

The fact that the intrinsic reactivity of water is undesirable for life is especially obvious when considering RNA and DNA. Cytosine, one of the four standard nucleobases in both DNA and RNA, hydrolytically deaminates to give uracil with a half-life of ca. 70 years in water at 300 K (Frick et al., 1987). Adenosine hydrolytically deaminates (to inosine), and guanosine hydrolytically deaminates to xanthosine at only slightly slower rates. As a consequence, water is constantly reacting with DNA in a way that causes it to lose its information; terran DNA in water must therefore be continuously repaired.

Indeed, the reactivity of water, and the consequent destruction of genetic material, creates the biggest obstacle in understanding how life arose on Earth. Creative chemists have generated prebiotic syntheses of nucleobases (Oró, 1960; Ferris et al., 1968). Expedient use of borate-containing minerals gets ribose from prebiotic precursors (Ricardo et al., 2004). Reaction of iron phosphides arriving via meteorites generates reactive phosphites that lead to the nucleotides (Pasek and Lauretta, 2005). Yet all is for naught if the RNA molecule is placed in water; it falls apart by hydrolysis. Even in laboratory cases as developed by David Bartel and Jack Szostak (Bartel and Szostak, 1993; Lawrence and Bartel, 2003), where *in vitro* evolution was used to select an RNA molecule that catalyzes the template-directed synthesis of RNA (Figure 10.4), the outcome is incompletely successful because the RNA catalyst spontaneously hydrolyzes . . . *in water!*

Again, a less reactive solvent such as formamide would be better. The synthesis of phosphate ester bonds (such as those found in RNA) is reported under prebiotic conditions in formamide (Schoffstall, 1976; Schoffstall et al., 1982; Schoffstall and Liang, 1985). Nucleobases are known to be synthesizable from formamide as well (Yamada et al., 1978). We might imagine the synthesis of longer RNA molecules in formamide. Indeed, if one wished to fine-tune the properties of water to make it better suited as a solvent for the origin of life, one might tune its reactivity to make it more like the reactivity of formamide.

Even the fact that water ice floats on water liquid is frequently cited as a benefit for water as a biological solvent. According to this view, floating ice insulates a body of liquid water that lies below, and prevents it from freezing. This feature becomes less compelling when one plays the dialectic game. Water ice does float in liquid water, but only if it is ice 1, the best known (on Earth) phase of solid ice. In contrast, ice 2 (another phase of solid water, with different crystal packing) does not float, nor do other solid forms of ice having a higher number. So why do we consider ice 1 to be the relevant form of solid water? Ice 1 is the thermodynamically most stable form of ice at

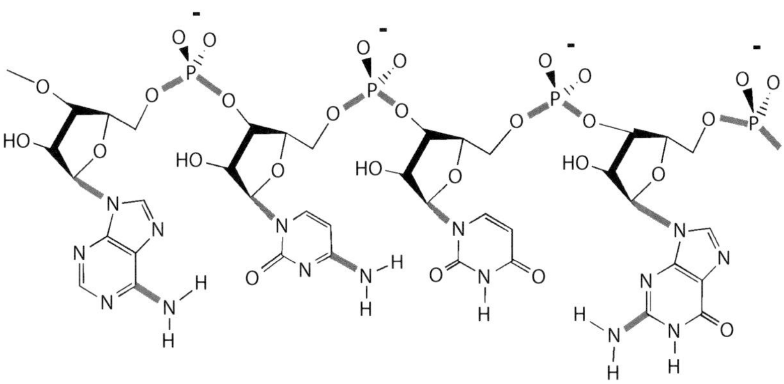

FIGURE 10.4 A generic structure for RNA. The red bonds are thermodynamic unstable with respect to hydrolysis in water. Each of these bonds represents a problem for prebiotic synthesis. In modern life, the aggressive reactivity of water with respect to molecules like RNA and DNA is managed using sophisticated repair systems. Such repair systems were presumably not present at the dawn of life. This creates a paradox in the structure of genetic matter at the dawn of life. On one hand, the repeating backbone charge suggests that RNA must work in a hydrophilic solvent such as water. But the number of hydrolyzable bonds suggests that RNA was difficult to synthesize prebiotically in water.

atmospheric pressure at sea level on Earth. Why should we consider terran atmospheric pressure to be the standard? Only because many of us live near (terran) sea level, which is itself under one terran atmosphere of pressure.

Likewise, Barrow and Tipler (1986) prefer water over ammonia based on a single criterion (water ice floats); they acknowledge that ammonia is in fact better by many of their other criteria. It is mentioned, however, that water ice has a higher albedo than water liquid. Thus, when water ice floats, it reflects more light from the sun, which leads to more cooling, more ice on the surface, a higher albedo, and still more cooling. Thus, the fact that water ice floats sets the stage for runaway glaciation. Indeed, it appears to have done just this on Earth, most recently in the Ice Ages of the past few million years. Thus, the fact that water ice floats causes water to amplify, not damp, perturbations in energy flux coming to a planet. If one believes that a useful property in a bulk solvent is one that supports a stable environment, the dialectic position is that *it is bad that water ice 1 floats*. Indeed, the floating ice causing global ice ages might be viewed to be a far worse disadvantage for this property of water than the advantage of floating ice in insulating sub-surface liquid (which is at best a kinetic factor).

An analogous dialectic process can be used to match Barrow and Tipler's (1986) comments on the desirability of the behavior of carbon dioxide as a component of life. Carbon dioxide is, as they note, a gas with poor solubility in water; this permits it to be distributed easily on a planetary scale. But water and carbon dioxide prove to be a problematic pair. The carbon of carbon dioxide is a good electrophilic center. But carbon dioxide itself is poorly soluble in water (0.88 v/v at 293 K and 1 atm). Therefore, at pH 7, carbon dioxide is present primarily in the form of the bicarbonate anion. Bicarbonate, however, has its electrophilic center shielding by the anionic carboxylate group. This means that bicarbonate is intrinsically unreactive as an electrophilic. Thus, the metabolism of carbon dioxide is caught in a conundrum. The reactive form is insoluble; the soluble form is unreactive.

Terran metabolism has worked hard to manage this conundrum. The reactivity of the biotin vitamin was discussed nearly three decades ago in this context (Visser and Kellogg, 1978). Biotin is metabolically expensive, however, and cannot be used to manage carbon dioxide and its problematic reactivity in large amounts. The enzyme ribulose bisphosphate carboxylase attempts to manage the problem without biotin. But here, the problematic reactivity of carbon dioxide competes with the problematic reactivity of dioxygen. Even in highly advanced plants, a sizable fraction of the substrate intended to capture carbon dioxide is destroyed through reaction with dioxygen (Ogren and Bowes, 1972). Terran life has not found a compelling solution to this problem in the billion years that dioxygen has been present in the atmosphere in abundance. Indeed, if we encounter nonterran carbon-based life, it will be interesting to see how they have come to manage the unfortunate properties of carbon dioxide. Thus, if the properties of carbon dioxide were fine-tuned to support life, why did not "someone" do something about this conundrum?

This disadvantage can, of course, be construed as an advantage. Barrow and Tipler (1986), for example, note that the low solubility of carbon dioxide gas in water may help recycle the material; a gas can easily move from place to place. Playing the dialectic game, however, we may note the problem of carbon dioxide as a greenhouse gas, and the possibility of runaway global warming (as has happened on Venus).

The dialectic game can be enormously productive as a way of constraining the human predilection toward "terracentricity." If one plays the game enough, one comes to realize that the reason why we think that water is optimally suited for life is because it is the solvent that supports the life familiar to us, and for no other reason.

10.12   IF NOT WATER, WHAT?

There are strong reasons based on common knowledge from chemistry to suspect that some solvent is desirable as a matrix to hold life. A liquid phase facilitates chemical reactions by allowing dissolved reactants to encounter each other at rates higher than the rates of encounter between species in a solid. Chemical reactions can take place in the gas and solid phases as well, of course. But each of these has disadvantages relative to the liquid phase.

This does not rule out entirely, of course, the possibility of life in the gas or solid phases. In the gas phase, chemistry is limited to molecules that are sufficiently volatile to deliver adequate amounts of material at moderate temperatures, and/or to molecules sufficiently stable to survive higher temperatures where vapor pressures are higher. Obviously, if volume is abundant, pressures are low, and time scales are long, low concentrations of biomolecules might support a biosystem in the gas phase. It is even conceivable that in the near-vacuum of interstellar space, life might be supported by a molecular metabolism involving compounds present at high dilution reacting in the gas phase. This entirely hypothetical form of life would need to manage challenges to its existence, including the high radiation flux and the difficulty of holding its components together. On the other hand, such life need not be encumbered by the lifetime of a planetary system. Indeed, one can imagine life in the gas phase associated with a galaxy and its energy flux for nearly the age of the universe.

Likewise, species can diffuse through solids to give chemical reactions (Huang and Walsh, 1998). Solid phase diffusion is slow, however. Nevertheless, given cosmic lengths of time, and the input of energy via high-energy particles, a biochemistry able to support Darwinian evolution can be conceived in the solid phase (Goldanskii, 1996). For example, an unusual life form might reside in solids of the Oort cloud living in deeply frozen water, obtaining energy occasionally from the trail of free radicals left behind by ionizing radiation, and carrying out only a few metabolic transformations per millennium.

These comments aside, it is best at present, however, to consider a liquid to be an essential substrate for life. What liquid might be best? It is hard to say. Liquid ammonia is a possible solvent for life, and is analogous to water in many of its properties. Ammonia, like water, dissolves many organic compounds, including many polyelectrolytes. Preparative organic reactions are done in ammonia in the laboratory. Ammonia, like water, is liquid over a wide range of temperatures (195–240 K at 1 atm). The liquid range is even broader at higher pressure. For example, when the pressure is 60 times that of the terran atmosphere at sea level, ammonia is liquid from 196 to 371 K. Further, liquid ammonia may be abundant in the solar system. A large amount of the inventory of liquid ammonia in the solar system exists, for example, in clouds in the Jovian atmosphere.

As in water, hydrophobic phase separation is possible in ammonia, although at lower temperatures. For example, Brunner reported that liquid ammonia and hydrocarbons form two phases, where the hydrocarbon chain contains from 1 to 36 CH_2 units (Brunner, 1988). Different hydrocarbons become miscible with ammonia at different temperatures and pressures. Thus, phase separation useful for isolation would be conceivable in liquid ammonia at temperatures well below its boiling point at standard pressures. Nevertheless, the increased ability of ammonia to dissolve hydrophobic organic molecules (again compared with water) suggests an increased difficulty in using the hydrophobic effect to generate compartmentalization in ammonia, relative to water.

The greater basicity of liquid ammonia must also be considered. The species that serve as acid and base in pure water are H_3O^+ and HO^-. In ammonia, NH_4^+ and NH_2^- are the acid and base, respectively. H_3O^+, with a pK_a of –1.7, is ca. 11 orders of magnitude stronger (in water) as an acid than NH_4^+, with a pK_a of 9.2 (in water). Likewise, NH_2^- is about 15 orders of magnitude stronger as a base than HO^-.

The increased strength of the dominant base in ammonia, as well as the corresponding enhanced aggressivity of ammonia as a nucleophile, implies that ammonia would not support the metabolic chemistry found in terran life. Terran life exploits compounds containing the C=O carbonyl unit. In ammonia, carbonyl compounds are (at the very least) converted to compounds containing the corresponding C=N unit. Nevertheless, hypothetical reactions that exploit a C=N unit in ammonia can be proposed in analogy to the metabolic biochemistry that exploits the C=O unit in terran metabolism in water (Figure 10.5) (Haldane, 1954). Given this adjustment, metabolism in liquid ammonia is easily conceivable.

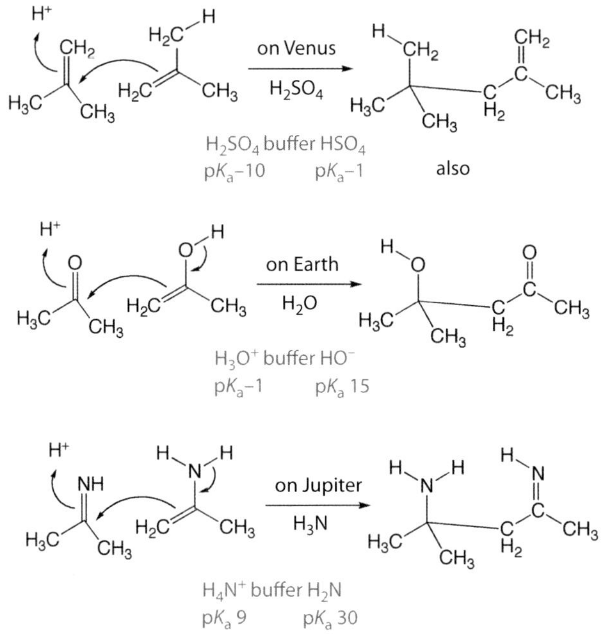

FIGURE 10.5 Analogous mechanisms comparing a simple carbon–carbon bond forming reaction involving C=O species (in water), C=N species (in ammonia), and C=C species (in sulfuric acid).

Further, ammonia is a potent antifreeze for water. Recently recovered data from Titan suggest that the moon is periodically being resurfaced by a liquid having a viscosity comparable to that of a water–ammonia eutectic, which is liquid even in an environment that experiences methane rain. This means that ammonia–water mixtures are an interesting alternative to pure water as a liquid that might serve as a biosolvent in a variety of environments.

Ammonia is not the only polar solvent that might serve as an alternative to water. For example, sulfuric acid dissolves many species, including polyelectrolytes. Sulfuric acid in liquid form is known above Venus (Kolodner and Steffes, 1998). Here, three cloud layers at 40–70 km are composed mostly of aerosols of sulfuric acid, ca. 80% in the upper layer and 98% in the lower layer (Schulze-Makuch et al., 2004). The temperature (ca. 310 K at ca. 50 km altitude, at ca. 1.5 atm) is consistent with stable carbon–carbon covalent bonds.

Metabolic hypotheses are available for a hypothetical life in acidic aerosols (Olah et al., 1978). In strong acid, the C=C bond is reactive as a base, and can support a metabolism as an analog of the C=O unit (Figure 10.5). This type of chemical reactivity is exemplified in some terran biochemistry. For example, acid-based reactions of the C=C unit is used by plants as they synthesized fragrant molecules (Kreuzwieser et al., 1999; Weyerstah, 2000).

Formamide is another potential biosolvent. Formamide is formed by the reaction of hydrogen cyanide with water; both are abundant in the cosmos. Like water, formamide has a large dipole moment and is an excellent solvent for almost anything that dissolves in water, including polyelectrolytes. Formamide is not reactive like water, as noted above. In formamide, many species that are thermodynamically unstable in water with respect to hydrolysis, are stable. Indeed, some are spontaneously synthesized. This includes ATP (from ADP and inorganic phosphate), nucleosides (from ribose borates and nucleobases), peptides (from amino acids), and others (Schoffstall, 1976; Schoffstall et al., 1982; Schoffstall and Liang, 1985).

Formamide is itself hydrolyzed by water, meaning that it persists only in a relatively dry environment, such as a desert. As desert environments have recently been proposed as being potential sites for the prebiotic synthesis of ribose (Ricardo et al., 2004), they may hold formamide as well.

We might also consider nonpolar solvents. Hydrocarbons, ranging from the smallest (methane) to higher homologs (ethane, propane, butane, etc.) are abundant in the solar system. Methane, ethane, propane, butane, pentane, and hexane have boiling points of ca. 109, 184, 231, 273, 309, and 349 K, respectively, at standard terran pressure. Thus, oceans of ethane may exist on Titan, the largest moon of Saturn. At a mean surface temperature estimated to be 95 K, methane (which freezes at 90 K) would be liquid, implying that oceans of methane could cover the surface of Titan. These oceans would undoubtedly contain large amounts of ethane and other dissolved hydrocarbons.

Many discussions of life on Titan have considered the possibility that water, normally frozen at the ambient temperature, might remain liquid following heating by impacts (Sagan et al., 1992). Life in this aqueous environment would be subject to the same constraints and opportunities as life in water. Water droplets in hydrocarbon solvents are, in addition, convenient cellular compartments for evolution, as Tawfik and Griffiths (1998) have shown in the laboratory. An emulsion of water droplets in oil is obtainable by simple shaking. This could easily be a model for how life on Titan achieves Darwinian isolation.

But why not use the hydrocarbons that are naturally liquid on Titan as a solvent for life directly? Broad empirical experience shows that organic reactivity in hydrocarbon solvents is no less versatile than in water. Indeed, many terran enzymes are believed to catalyze reactions by having an active site that is not like water. Furthermore, hydrogen bonding is difficult to use in the assembly of supramolecular structures in water. In ethane as a solvent, a hypothetical form of life would be able to use hydrogen bonding more than in water, as hydrogen bonding to water would not compete with the hydrogen bonding needed to form supramolecular structures. In addition, hydrogen bonds would have the strength appropriate for stable (but not too stable) structures. Indeed, at low temperatures, one might imagine that hydrogen bonds to be the kinetic analog of covalent bonds in liquid water at terran temperatures. Further, hydrocarbons with polar groups can be hydrocarbon-phobic; acetonitrile and hexane, for example, form two phases. One can conceive of liquid/liquid phase separation in bulk hydrocarbon that could achieve Darwinian isolation.

The reactivity of water means that it destroys hydrolytically unstable organic species. Thus, a hypothetical form of life living in a Titan hydrocarbon ocean would not need to worry as much about the hydrolytic deamination of its nucleobases, and would be able to guide reactivity more easily than life in water. Other liquids are still more abundant. For example, the most abundant compound in the solar system is dihydrogen, the principal component (86%) of the upper regions of the gas giants, Jupiter, Saturn, Uranus, and Neptune. The other principal component of the outer regions of the giant planets is helium (0.14). Minor components, including methane (fractional composition 2×10^{-3}, or 0.2%), water (6×10^{-4}, or 0.06%), ammonia (2.5×10^{-4}, or 0.025%), and hydrogen sulfide (7×10^{-5}, or 0.007%) make up the rest.

But is dihydrogen a liquid? The physical properties of a substance are described by a phase diagram that relates the state of a material (solid of various types, liquid, or gas) to temperature and pressure. A line typically extends across the phase diagram. Above this line, the substance is a gas; below the line, the substance is a liquid. Typically, however, the line ends at a critical point. Above the critical point, the substance is a supercritical fluid, neither liquid nor gas. Table 10.1 shows the critical temperatures and pressures for some substances common in the solar system.

The properties of supercritical fluids are generally different than those of the regular fluids. For example, supercritical water is relatively nonpolar and acidic. Further, the properties of a supercritical fluid, such as its density and viscosity, change with changing pressure and temperature, and especially dramatically as the critical point is approached.

The changing solvation properties of supercritical fluids near their critical points are widely used in industry (Lu et al., 1990). For example, supercritical carbon dioxide, having a critical temperature of 304.2 K and pressure of 73.8 atm, is used to decaffeinate coffee. Supercritical water below the earth's surface leads to the formation of many of the attractive crystals that are used in jewelry.

Little is known about the behavior of organic molecules in supercritical dihydrogen as a solvent. In the 1950s and 1960s, various laboratories studied the solubility of organic molecules (e.g.,

TABLE 10.1
Critical Temperature and Pressure for Selected Substances

Liquid	Critical Temperature (K)	Critical Pressure (atm)
Hydrogen	33.3	12.8
Neon	44.4	26.3
Nitrogen	126	33.5
Argon	151	48.5
Methane	191	45.8
Ethane	305	48.2
Carbon dioxide	305	72.9
Ammonia	406	112
Water	647	218

naphthalene) in compressed gases, including dihydrogen and helium (Robertson and Reynolds, 1958; King and Robertson, 1962). None of the environments examined in the laboratory, however, explored high pressures and temperatures.

Throughout most of the volume of gas giants where molecular dihydrogen is stable, it is a supercritical fluid. For most of the volume, however, the temperature is too high for stable carbon–carbon covalent bonding. We may, however, define two radii for each of the gas giants. The first is the radius where dihydrogen becomes supercritical. The second is where the temperature rises to a point where organic molecules are no longer stable; for this discussion, this is chosen to be 500 K. If the second radius is smaller than the first, then the gas giant has a "habitable zone" for life in supercritical dihydrogen. If the second radius is larger than the first, however, then the planet has no habitable zone.

If such a zone exists on Jupiter, it is narrow (Sagan and Salpeter, 1976). Where the temperature is 300 K (clearly suitable for organic molecules), the pressure (ca. 8 atm) is still subcritical. At about 200 km down, the temperature rises above 500 K, approaching the upper limit, where carbon–carbon bonds are stable (West, 1999). For Saturn, Uranus, and Neptune, however, the habitable zone appears to be thicker (relative to the planetary radius). On Saturn, the temperature is ca. 300 K when dihydrogen becomes supercritical. On Uranus and Neptune, the temperature when dihydrogen becomes supercritical is only 160 K; organic molecules are stable at this temperature.

The atmospheres of these planets convect. To survive on Jupiter, hypothetical life based on carbon–carbon covalent bonds would need to avoid being moved by convection to positions in the atmosphere where they can no longer survive. This is, of course, true for life in any fluid environment, even in terran oceans. Sagan and Salpeter (1976) presented a detailed discussion of what might be necessary for a floater to remain stable in the Jovian atmosphere. Last, Bains (2004) has recently discussed dinitrogen as a solvent. The molecule is again abundant in the cosmos, but underexplored as a solvent for organic reactions.

10.13   CONCLUSION

The discussion above summarizes only some of the issues that might be addressed as we attempt to combine the constraints of physical and chemical laws, the consequent behavior of solvents (like water), and the structure of biological molecules. Synthesis is an experimental approach to better understanding of this interaction, and especially for how underlying physical law has constrained biology at the molecular level. It also provides intellectual processes that manage the natural tendency for scientists to become advocates for their individual prejudices.

Historical accident and natural selection have almost certainly shaped the structures of terran living systems to be adapted to water. Synthesis shows, however, that the terran solution is not unique for DNA molecules. Synthesis has driven our models for DNA forward in unexpected ways, however. From synthesis, we can now say that water, because it dissolves polyelectrolytes (e.g., DNA and RNA), can serve as a solvent for a life form (such as ours) that uses a repeating charge in its genetic biopolymer to support Darwinian evolution. This property is not shared by many nonpolar solvents in the cosmos (e.g., methane, ethane, and dinitrogen), but it is shared by certain dipolar solvents, including ammonia, sulfuric acid, and formamide. If a biopolymer requires a repeating charge to support Darwinian evolution, and if Darwinian evolution must stand behind life, we can conclude that a solvent like water is the only solvent that can support life.

Matching this feature of water as a good solvent for life are many of its properties that appear to be inimical to life, especially near its origin. Foremost among these are its own intrinsic reactivity. Water damages DNA and RNA, changing their information content as it does so. Modern biology handles the damage that water does to DNA and RNA through the use of sophisticated repair mechanisms. These would presumably not have been available to prebiotic systems and to early life. Indeed, even though possibly prebiotic routes to make nucleobases, sugars, and phosphates are all known, these pieces cannot be assembled into RNA-like genetic molecules without solving the "water problem."

The availability of isotopically substituted variants of water with different properties excludes the possibility that the properties of water have been as finely tuned to support life as certain physical constant appear to be. Nevertheless, we can see a remarkable range of good fortune that appears to stand behind the history of the biomolecules that we know on Earth. Ribose appears to be a particularly good carbohydrate backbone to support the rule-based molecular recognition seen in RNA; it also, apparently coincidentally, is a carbohydrate that is easy to get via prebiotic processes in the presence of borate minerals (Ricardo et al., 2004). Phosphate appears conveniently to support a backbone with a repeating charge, at the same time as having the kinetic stability needed to support a genetic biopolymer.

Synthesis, historical reconstruction, laboratory evolution, and other tools useful to dissect chance and necessity in the structure of contemporary terran biomolecules are all increasing in power. We can expect in the next 5 years to have in hand artificial chemical systems capable of Darwinian evolution, partly by design and partly by laboratory selection (Szostak et al., 2001). Obtaining these is truly a "put-a-man-on-the-moon" class of challenge. Our efforts to attain it cannot help but deepen our understanding of the intimate connection between physics, chemistry, and biology, and allow us to obtain ever deeper answers to the "Why?" questions that motivate so much of human inquiry.

ACKNOWLEDGMENTS

The author is indebted to NASA for supporting research related to this review, and to the John Templeton Foundation for encouraging the discussions of these questions at a conference in Varenna in 2005, and elsewhere.

REFERENCES

Bains, W. 2004. Many chemistries could be used to build living systems. *Astrobiology* 4:137–167.

Ball, P. 2004. Starting from scratch. *Nature* 431, 624–626.

Barrow, J. D., and F. J. Tipler. 1986. *The Anthropic Cosmological Principle*. New York, NY: Oxford University Press.

Bartel, D. P., and J. W. Szostak. 1993. Isolation of new ribozymes from a large pool of random sequences. *Science* 261:1411–1418.

Benner, S. A., M. D. Caraco, J. M. Thomson, and E. A. Gaucher. 2002. Planetary biology. Paleontological, geological, and molecular histories of life. *Science* 293:864–868.

Benner, S. A. 2003. Synthetic biology. *Nature*. 421:118.

Benner, S. A. 2004. Understanding nucleic acids using synthetic chemistry. *Acc. Chem. Res.* 37:784–797.

Benner, S. A. 2009. *Life, the Universe, and the Scientific Method*. Gainesville, FL: The FfAME Press. www. ffame.org.

Benner, S. A., A. D. Ellington, and A. Tauer. 1989. Modern metabolism as a palimpsest of the RNA world. *Proc. Natl. Acad. Sci.* 86:7054–7058.

Benner, S. A., and A. D. Ellington. 1988. Interpreting the behavior of enzymes. Purpose or pedigree? *CRC Crit. Rev. Biochem.* 23:369–426.

Benner, S. A., and D. Hutter. 2002. Phosphates, DNA, and the search for nonterran life. A second generation model for genetic molecules. *Bioorg. Chem.* 30:62–80.

Benner, S. A., A. Ricardo, and M. A. Carrigan. 2004. Is there a common chemical model for life in the universe? *Curr. Opin. Chem. Biol.* 8:672–689.

Brunner, E. 1988. Fluid mixtures at high pressures: 7. Phase separation and critical phenomena in 18 (*n*-alkane + ammonia) and 4 (*n*-alkane + methanol) mixtures. *J. Chem. Thermodyn.* 20:1397–1409.

Elbeik, T., J. Surtihadi, M. Destree, J. Gorlin, M. Holodniy, S. A. Jortani, K. Kuramoto, V. Ng, R. Valdes, A. Valsamakis, and N. A. Terrault. 2004. Multicenter evaluation of the performance characteristics of the Bayer Versant HCV RNA 3.0 assay (bDNA). *J. Clin. Microbiol.* 42:563–569.

Ferris J. P., R. A. Sanchez, and L. E. Orgel. 1968. Studies in prebiotic synthesis: III. Synthesis of pyrimidines from cyanoacetylene and cyanate. *J. Mol. Biol.* 33:693–704.

Frick, L., J. P. MacNeela, and R. Wolfenden. 1987. Transition state stabilization by deaminases. Rates of non-enzymic hydrolysis of adenosine and cytidine. *Bioorg. Chem.* 15:100–108.

Gaucher, E. A., J. M. Thomson, M. F. Burgan, and S. A. Benner. 2003. Inferring the palaeoenvironment during the origins of bacteria based on resurrected ancestral proteins. *Nature* 425:285–288.

Geyer, C. R., T. R. Battersby, and S. A. Benner. 2003. Nucleobase pairing in expanded Watson–Crick like genetic information systems. The nucleobases. *Structure* 11:1485–1498.

Gibbs, W. W. 2004. Synthetic life. *Sci. Am.* 290(5):74–81.

Ginsparg, P., and S. Glashow. 1986. Desperately seeking superstrings. *Phys. Today* 39:7.

Goldanskii, V. I. 1996. Nontraditional mechanisms of solid-phase astrochemical reactions. *Kinet. Catal.* 37:608–614.

Haldane, J. B. S. 1954. The origins of life. *New Biol.* 16:12–27.

Huang, B., and J. J. Walsh. 1998. Solid-phase polymerization mechanism of poly(ethyleneterephthalate) affected by gas flow velocity and particle size. *Polymer* 39:6991–6999.

Hutter, D., M. O. Blaettler, and S. A. Benner. 2002. From phosphate to dimethylenesulfone: non-ionic backbone linkers in DNA. *Helv. Chim. Acta* 85:2777–2806.

Katz, J. J., H. L. Crespi, D. M. Czajka, and A. J. Finkel. 1962. Course of deuteration and some physiological effects of deuterium in mice. *Am. J. Physiol.* 203:907–913.

Katz, J. J., H. L. Crespi, and A. J. Finkel. 1964. Isotope effects in fully deuterated hexoses, proteins and nucleic acids. *Pure Appl. Chem.* 8:471–481.

King, Jr., A. D., and W. W. Robertson. 1962. Solubility of naphthalene in compressed gases. *J. Chem. Phys.* 37:1453–1455.

Kolodner, M. A., and P. G. Steffes. 1988. The microwave absorption and abundance of sulfuric acid vapor in the Venus atmosphere based on new laboratory measurements. *Icarus* 132:151–169.

Kreuzwieser, J., J. P. Schnitzler, and R. Steinbrecher. 1999. Biosynthesis of organic compounds emitted by plants. *Plant Biol.* 1:149–159.

Lawrence, M. S., and D. P. Bartel. 2003. Processivity of ribozyme-catalyzed RNA polymerization. *Biochemistry* 42:8748–8755.

Levins, R., and R. Lewontin. 1985. *The Dialectical Biologist*. Cambridge, MA: Harvard University Press.

Lu, B. C. Y., D. Zhang, and W. Sheng. 1990. Solubility enhancement in supercritical solvents. *Pure Appl. Chem.* 62:2277–2285.

Ogren, W. L., and G. Bowes. 1972. Oxygen inhibition and other properties of soybean ribulose 1,5-diphosphate carboxylase. *J. Biol. Chem.* 247:2171–2176.

Olah, G. A., G. Salem, J. S. Staral, and T. L. Ho. 1978. Preparative carbocation chemistry: 13. Preparation of carbocations from hydrocarbons via hydrogen abstraction with nitrosodium hexafluorophosphate and sodium nitrite trifluoromethanesulfonic acid. *J. Org. Chem.* 43:173–175.

Oró, J. 1960. Synthesis of adenine from ammonium cyanide. *Biochem. Biophys. Res. Commun.* 2:407–412.

Pasek, M. A., and D. S. Lauretta. 2005. Aqueous corrosion of phosphide minerals from iron meteorites. A highly reactive source of prebiotic phosphorus on the surface of the early Earth. *Astrobiology* 5:515–535.

Ricardo, A., M. A. Carrigan, A. N. Olcott, and S. A. Benner. 2004. Borate minerals stabilize ribose. *Science* 303:196.

Robertson, W. W., and R. E. Reynolds. 1958. Effects of hydrostatic pressure on the intensity of the singlet-triplet transition of 1-chloronaphthalene in ethyl iodide. *J. Chem. Phys.* 29:138–141.

Sagan, C., and E. E. Salpeter. 1976. Particles, environments, and possible ecologies in the Jovian atmosphere. *Astrophys. J.* 32:737–755.

Sagan, C., W. R. Thompson, and B. N. Khare. 1992. Titan. A laboratory for prebiological organic chemistry. *Acc. Chem. Res.* 25:286–292.

Schoffstall, A. M. 1976. Prebiotic phosphorylation of nucleosides in formamide. *Orig. Life Evol. Biosph.* 7:399–412.

Schoffstall, A. M., R. J. Barto, and D. L. Ramo. 1982. Nucleoside and deoxynucleoside phosphorylation in formamide solutions. *Orig. Life Evol. Biosph.* 12:143–151.

Schoffstall, A. M., and E. M. Liang. 1985. Phosphorylation mechanisms in chemical evolution. *Orig. Life Evol. Biosph.* 15:141–150.

Schulze-Makuch, D., D. H. Grinspoon, O. Abbibas, L. N. Irwin, and M. A. Bullock. 2004. A sulfur-based survival strategy for putative phototropic life in the Venusian atmosphere. *Astrobiology* 4:1–8.

Sismour, A. M., S. Lutz, J.-H. Park, M. J. Lutz, P. L. Boyer, S. H. Hughes, and S. A. Benner. 2004. PCR amplification of DNA containing non-standard base pairs by variants of reverse transcriptase from human immunodeficiency virus-1. *Nucleic Acids Res.* 32:728–735.

Sismour, A. M., and S. A. Benner. 2005. The use of thymidine analogs to improve the replication of an extra DNA base pair: a synthetic biological system. *Nucleic Acids Res.* 33:5640–5646.

Szostak, J. W., D. P. Bartel, and P. L. Luisi. 2001. Synthesizing life. *Nature* 409:387–390.

Tawfik, D. S., and A. D. Griffiths. 1998. Man-made cell-like compartments for molecular evolution. *Nat. Biotechnol.* 16:652–656.

Visser, C. M., and R. M. Kellogg. 1978. Biotin. Its place in evolution. *J. Mol. Evol.* 11:171–178.

Ward, P. D., and D. Brownlee. 2000. *Rare Earth. Why Complex Life Is Uncommon in the Universe.* New York, NY: Springer Verlag.

West, R. A. 1999. Atmospheres of the giant planets. In *Encyclopedia of the Solar System* 1999. Chapter 14, ed. P. Weissman, L. McFadden, and T. Johnson. New York, NY: Academic Press.

Weyerstah, P. 2000. Synthesis of fragrance compounds isolated from essential oils. *Proc. Phytochem. Soc. Europe.* 46, *Flavour and Fragrance Chemistry.* Chapter 6, ed. V. Lanzotti and O. Taglialatela-Scafati. Dordrecht: Kluwer Academic Publishing.

Yamada, H., M. Hirobe, K. Higashiyama, H. Takahashi, and T. Suzuki. 1978. Detection of C-13-N-15 coupled units in adenine derived from doubly labeled hydrogen cyanide or formamide. *J. Am. Chem. Soc.* 100:4617–4618.

11 Fine-Tuning and Small Differences between Large Numbers

John L. Finney

CONTENTS

11.1 INTRODUCTION

This chapter will discuss an analog between an aspect of two fine-tuning examples from cosmology and an aspect of biological function in which water is a major actor. The biological example is the stability of a folded, native protein against its unfolded state in its normal aqueous environment. This comparison suggests a property of water for which numerical values are critical to this aspect of biological functionality in a way that has a parallel in the cosmological cases. This will not demonstrate that no other molecule could perform in a similar way to water in this situation, but it does underline how critical the small difference between two large numbers can be in the biological as well as in the cosmological case. Whether these values are fine-tuned, in the case of water, or merely aspects that evolution has been able to take advantage of, remains a matter of speculation.

11.2 TWO COSMOLOGICAL EXAMPLES

11.2.1 THE TRIPLE ALPHA REACTION

It is this process that leads to the formation of carbon from helium (Salpeter, 1952; Hoyle, 1954). Without this process, it is difficult to see how elements heavier than helium could be formed in stars. The process is a two stage one:

$$^4He + {}^4He \rightarrow {}^8Be \tag{11.1}$$

$$^8\text{Be} + {}^4\text{He} \rightarrow {}^{12}\text{C}. \tag{11.2}$$

^{8}Be is unstable, so it needs to live long enough for the second reaction to occur. Moreover, the possible further reaction:

$$^{12}\text{C} + {}^4\text{He} \rightarrow {}^{16}\text{O} \tag{11.3}$$

must not occur, or the carbon would not be stable.

For reaction (11.2), the excited resonance level of ^{12}C* is at 7.65 MeV (Hogan, 2000). This is only slightly (0.28 MeV) higher than the energy of ^{8}Be + ^{4}He (7.37 MeV). This small difference of less than 4% is sufficient to allow reaction (11.2) to occur before the unstable ^{8}Be decays. With respect to the possibility of reaction (11.3), the precise value of the difference is critical: the level of ^{16}O is 7.1187 MeV, which is just (0.043 MeV) below that of ^{12}C + ^{4}He (7.1616 MeV). Were the ^{16}O level a mere 0.6% higher, reaction (11.3) would destroy the carbon and we would have to look for a form of life other than one based on carbon. These two small differences between much larger numbers are critical to the ability of stars to produce the carbon essential to our life form. Perturb these energy levels by only a few percent or less and the universe would not contain the essential elements that are needed for the evolution of life as we know it anywhere within it. More detailed recent re-examinations have been made of the possible consequences of energy level perturbations in stars of different masses using modern stellar evolution codes. The results suggest that changes to the carbon energy level of up to a few percent can change the ratio of carbon to oxygen produced over the full range from a regime that produces all oxygen to one that produces all carbon (Oberhummer et al., 2000; Schlattl et al., 2004).

11.2.2   QUARK MASSES AND THE STABILITY OF THE PROTON

The proton is composed of two up and one down quarks (*uud*), while the neutron is composed of one up and two down quarks (*udd*). The total masses of the proton and neutron are 938.272 and 939.566 MeV, respectively, the mass difference between them being only approximately 1.4% (Hogan, 2000). The stability of the proton depends on the difference between the masses of the up and down quark, $m_u - m_d$. The down quark must be just enough heavier than the up quark to overcome the difference in electromagnetic energy to make the proton sufficiently lighter than the neutron and, hence, ensure its stability. However, this difference must not be too large, or the deuteron would become unstable, thus making it difficult to make nuclei that are heavier than hydrogen.

Here again, we have a difference whose value is critical to enabling processes in stars that lead to the building blocks that are essential for the creation of the elements. If this difference in quark masses varied by only a small fraction either way, nuclear astrophysics would change in a major way. So, we can hypothesize that these quantities might be fine tuned to make the subsequent development of the universe possible.

11.3   A POSSIBLE BIOLOGICAL EXAMPLE: THE STABILITY OF A NATIVE PROTEIN

The tertiary structure of a globular protein is essential for its activity. This native structure is often contrasted with that of a notional (in that its structure is not well defined experimentally) unfolded or denatured state which is inactive. Once synthesized, the polypeptide chain that forms the protein must fold to the native state before it can function. We note here that the free energy of stability of the folded protein is very low—between approximately 10 to 20 kcal mol^{-1}. This is equivalent to the energies of only 2 to 4 hydrogen bonds, of which there are several hundred in a typical protein–water system. In the context of the different contributions that combine to give this result, this free energy of stability is again a small difference between two much larger numbers (the free energies of the unfolded and the native structures in solvent). We might note here the consequence that even very small changes in the individual free energy contributions could either make the protein unstable and therefore nonfunctional, or perhaps even too stable to function effectively.

Assume for the sake of this discussion that the unfolded structure is one in which the majority of chemical groups on the protein are exposed to the solvent, which for simplicity we assume to be just water. Thus, the polar groups of the unfolded polypeptide will be making hydrogen bonding links to the surrounding water, while the nonpolar groups will have hydration shells in which the waters do not directly bond to the water. When the molecule folds, a large number of changes take place, both to the structure of the protein and to the interacting waters. Without going into these changes in detail (see for example Finney et al., 1980, for a discussion of the free energy contributions to folding and stability and also the chapter in this book by Pace et al.), the long chain polypeptide, when folded, will make a number of intramolecular interactions between parts of the molecule that end up in contact with each other. Such interactions will include contacts between nonpolar groups, as well as hydrogen bonds between polar groups. This process is shown schematically in Figure 11.1.

We focus on the latter process and make the following very simplistic but illustrative, and perhaps illuminating, argument. In making direct hydrogen bonding interactions, the polar groups concerned must first remove themselves from interaction with the water molecules that were present in the unfolded state. These groups will then form hydrogen bonds with other similarly released polar groups on the same molecule, while the released waters will be expelled to the bulk solvent where their hydrogen bond donor or acceptor functionalities will, to first order, hydrogen bond to other water molecules. Furthermore, some of the polar groups—those that end up on the surface of the native structure—will continue to make hydrogen bonds to solvent water molecules. The enthalpic contribution to the free energy of stability of the folded state will therefore depend on the number of each of such interactions and their energies *in both the unfolded and native structures.*

Let P represent a polar group on the protein, W a polar group on a water molecule, and P–W a hydrogen bond between a polar group on the protein and a water molecule. We can then represent the hydrogen bond accounting during the folding process as:

$$[n_{P-W} + m_{W-W}]_{unfolded} \Rightarrow [p_{P-P} + q_{P-W} + r_{W-W}]_{native}, \tag{11.4}$$

where n, m, p, q, and r are integers.

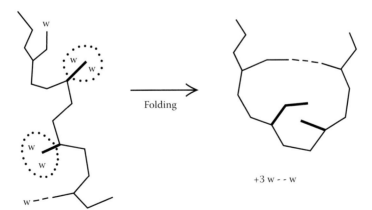

FIGURE 11.1 Schematic showing part of a hypothetical unfolded protein (left) and the same part after folding (right). The two thicker lines represent nonpolar groups, dashed lines represent hydrogen bonds, and w denotes a water molecule. After folding, the six water molecules shown are expelled to the solvent and make hydrogen bonds with other bulk water molecules.

We consider two cases. In Case 1, we assume:

(1) The total number of hydrogen bonds is the same in both the unfolded and folded structures. Thus, a polar group making a hydrogen bond to a water molecule in the unfolded state will in the folded state make either a similar hydrogen bond or one to another polar group on the protein. With respect to Equation (11.4), this is equivalent to $n + m = p + q + r$. This assumption is reasonable and is broadly consistent with experiments on simpler model systems (Dixit et al., 2002);

(2) The energies of all hydrogen bonds are identical. In the terms of Equation (11.4), denoting the energy of a hydrogen bond between two kinds of polar groups as $E_{group1\text{-}group2}$, this means that $E_{P–W} = E_{P–P} = E_{W–W}$. This assumption is not quantitatively correct, but it enables us to set a base line for the purpose of the discussion.

The conclusion from Case 1 is obvious: hydrogen bonding will make absolutely no contribution to the stability of the folded protein. The total number of hydrogen bonds does not change on folding, and the energies of all hydrogen bonds are assumed equal.

For Case 2, we retain assumption a, but make the simplest nontrivial assumption about hydrogen bond energies, namely that $E_{P–W} \neq E_{P–P} \neq E_{W–W}$. In this case, we are assuming that all hydrogen bonds between protein groups have the same energy and that all hydrogen bonds between water molecules are of the same energy, which is different from that between protein polar groups. Moreover, the energy of the protein–water hydrogen bonds are also all the same, but again different from the energies of the two like–like hydrogen bonds.

The conclusion here is obvious: whether the hydrogen bonding contributes to stability in a positive or negative way depends on the actual numerical values of the different hydrogen bond energies. For example, if we choose a P–P interaction to be stronger than a W–W interaction, with the P–W hydrogen bond strength in between, there will be a favorable hydrogen bonding contribution to the enthalpy of stability. If we choose the values to be the other way round, then the hydrogen bonding contribution will be unfavorable. Whatever the resultant number for the enthalpy difference is, it will depend—not just in magnitude, but also in sign—on these differences between the energies of the different kinds of hydrogen bond. And considering that the system we are considering is likely to contain several hundred hydrogen bonds, only very small shifts in relative hydrogen bonding strengths can change the final situation dramatically. The protein may not fold at all—or it may when folded become functionally inactive through being too stable.

Is this an echo of the energy levels in the triple alpha interaction? Or of the difference between the masses of the up and down quarks? In those cases, small variations in small differences between two larger numbers would have prevented the universe from developing in the way that it has. In this case also, the balance between the average values of the three different kinds of interactions needs to resolve in a particular way or the protein would not fold.

11.4   ILLUSTRATIONS FROM MODEL LIQUID SYSTEMS

It would be an obvious next step to look at the actual values for the different strengths of these different hydrogen bonding interactions in a real protein in solution. Unfortunately, such information is not available. Several attempts have been made to analyze the different contributions to protein stability (e.g., Finney et al., 1980), but these generally make the simple assumption that all hydrogen bonds, on average, contribute the same enthalpy to the system—i.e., assumption (b) of Case 1 above. We do not have sufficiently good experimental information on the differences between the strengths of different kinds of hydrogen bonds to make any other realistic assumption and, if we did, there would be variations between different examples of the same kind of hydrogen bond—e.g., N–H...O—because of distortion from the minimum energy configuration. We are also oversimplifying the situation by ignoring dynamics and entropic contributions to the free energy difference (although attempts are made to estimate these also in the folding analyses that have been made—see again Finney et al., 1980 and Pace et al., this volume).

On the other hand, we do have computer simulations of this type of system, and these use model potential functions for each of these types of hydrogen bonding interaction. However, it should be clear from the above discussion that the actual parameters used in such simulations are likely to be critical to the ability of a simulation to give the marginally stable folded molecule we obtain experimentally. Although, in principle, a computer simulation analysis might be made of the effects on folding of varying the different contributions in a way similar to varying, for example, the carbon and oxygen resonance levels in the astrophysical example, there are some technical problems in doing this. Although these problems might be overcome, no such variational work appears to have been attempted.

However, we do have good structural data on a number of simpler liquid systems that have some useful similarity to the protein case exampled. In these cases, the solute is an amphiphile and we will focus on one of these: tertiary butanol. Being an amphiphile, like the amino acids that make up a protein, it contains polar and nonpolar chemical groups. The polar groups can interact through hydrogen bonding either with polar groups on other solute molecules (as between two hydrogen bonded amino acids in a folded protein) or with solvent water molecules (as they would in an unfolded protein).* These experimental systems can enable us to probe the degree of sensitivity of structures to the potential functions that control the system. We therefore now consider two liquid systems in this context. As background to this discussion, we summarize briefly how the required structural information is accessed.

11.4.1 DETERMINING LIQUID STRUCTURES

Using neutron scattering with appropriate substitution of isotopes, it is possible to obtain detailed atomic-level structures of not only single component liquids such as water (Soper, 2000) but also of more complex solutions (see, e.g., Enderby and Neilson, 1979; Finney and Soper, 1994; Finney and Bowron, 2002). Without going into the full methodological details, by taking neutron scattering measurements on a number of isotopically distinct samples of a given solution (most usually substituting deuterium for hydrogen), several sets of scattering data are obtained for each chemically similar, but isotopically different, solution. These data can then be used to obtain information on the average environments of the different atoms in the system, information that is contained in the set of partial radial distribution functions, each of which tells us how a particular kind of atom is surrounded by another particular kind of atom. For example, for liquid water, which contains two chemically distinct kinds of atom (oxygen and hydrogen), we require three such distribution functions. The oxygen–oxygen partial radial distribution function tells us, in essence, the probability of finding neighboring oxygen atoms at different distances from any other oxygen, whereas the oxygen–hydrogen function tells us the distribution of hydrogens around oxygens (and vice versa). The third function, the hydrogen–hydrogen partial, similarly tells us about the distribution of hydrogen atoms around other hydrogens. We thus have three unknowns: the three partial radial distribution functions—and three sets of data, three experimental diffraction measurements using different isotope substitutions, which in this case means essentially using samples of H_2O, D_2O, and a 50:50 mixture of these two.

However, for a complete structural characterization of the example tertiary butanol–water system, we actually need many more than three partials to describe the structure: there are seven chemically distinct types of atom, which means we need 28 partial radial distribution functions to describe this structure. Formally, therefore, we need diffraction experiments on 28 isotopically distinct samples

---

* This is of course an oversimplification in that, in the folded protein, there are still polar groups that are exposed to the solvent and interact with the neighboring water through hydrogen bonding. For the purposes of this discussion, however, the model used is a useful and instructive one.

to obtain these structural functions. This is not, at least yet, a practical proposition. The best we can do is to take neutron diffraction measurements on seven isotopically distinct samples.

Despite the fact that we have performed only seven different experiments, we can, in fact, use these data together with other known chemical information to extract the partial radial distribution functions we require. This is not a case of getting something for nothing; it is, in effect, doing a similar sort of thing that a crystallographer does when refining a crystal structure in which the number of independent reflections that can be measured is—as is usually the case for very large molecules—less than the number of structural parameters that need to be refined. Such crystallographic refinements are therefore constrained by what we know of the chemistry of the system, for example standard bond lengths and angles, and the nonoverlap of atoms.

We do something similar in the liquid case, although the actual refinement process used is a little different. The way we do this is relevant to the point we are discussing, so it is useful to summarize the procedure, known as Empirical Potential Structure Refinement (Soper, 1996).

We start by running a Monte Carlo computer simulation of the system we have measured using a set of standard potential functions. Obtained from the literature, these are functions that are used in simulations of aqueous solutions, sometimes of proteins, so that we should expect them to predict reality reasonably well. After an initial equilibration time, the same partial structure factors that were experimentally measured are calculated for the simulated system. These in general (in all cases we are aware of so far) fail to give adequate agreement with experiment, indicating in passing that the starting potentials are inadequate to reproduce the experimental data. This is a point that is relevant to the current discussion and we will come back to it a little later. We then continue the simulation after making small modifications to the interaction potentials in a way that is designed to drive the simulated system toward agreement with experiment (see Soper, 1996, for details). This procedure is repeated many times until adequate agreement is obtained between the scattering observed and that predicted by the simulated system. The resulting experimentally consistent computer simulated ensembles can then be interrogated to obtain atom–atom partial radial distributions. These can then be examined to explore the detailed structure of the liquid system.

To explore the sensitivity of structure to the potential functions that control an amphiphilic liquid system in the context of the biological analogy we are pursuing, we consider now two experimental systems in which the liquid structures have been determined: pure liquid *t*-butanol and a *t*-butanol–water solution.

11.4.2 LIQUID TERTIARY BUTANOL

Although only a single component system, the way in which amphiphilic molecules organize themselves in their liquid phase does provide an illustration of the sensitivity of the precise way that amphiphiles interact with each other (remember that proteins are also amphiphiles) to the details of the intermolecular interactions that we have argued above are critical in processes like protein folding.

We take here the example of liquid tertiary butanol, $(CH_3)_3COH$. This molecule, shown in Figure 11.2, has a bulky but essentially spherical nonpolar head group consisting of the three methyl (CH_3) groups (the top of the molecule in the orientation shown in Figure 11.2). This is attached via a fourth carbon atom to a polar alcohol OH 'tail' group (the group at the bottom of the molecule in Figure 11.2).

Looking at this molecule, we can suggest that there are three main kinds of intermolecular interactions we need to consider. First, there is the possibility of hydrogen bonding between the polar (alcoholic) ends of the molecules. These are the interactions that are indicated between the three molecules shown in Figure 11.3a. Second, there are possible interactions between the nonpolar methyl groups on neighboring molecules, such as in the orientation shown in Figure 11.3b. Being a van der Waals interaction, this interaction between the head groups would be expected to be weaker than the hydrogen bonding between the tails of the molecules. Third, there are possible mixed polar–

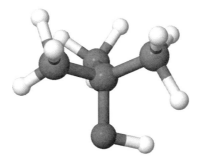

FIGURE 11.2 **(See color insert following page 302.)** A tertiary butanol molecule oriented with its polar alcohol tail group pointing downward and its nonpolar head group pointing upward. (With acknowledgment to Dr. Daniel Bowron, ISIS Facility, Rutherford Appleton Laboratory, UK.)

nonpolar interactions between a head of one molecule and a tail of another to consider. We would again expect this latter interaction to be weak, although with one of the two partners having a dipolar character and the other (the nonpolar head) having a significant polarizability, we might expect the average polar–nonpolar interaction (the like–unlike interaction in the protein folding discussion above) to be stronger than the like–like nonpolar–nonpolar one.

A simple intuitive assessment of this liquid might, therefore, lead us to expect the major intermolecular interaction to be hydrogen bonding between the tails of the molecules. Provided there are enough polar groups present (and there are), we would intuitively expect the polar tails to maximize their hydrogen bonding, giving us local hydrogen bonded structures such as that shown schematically in Figure 11.3a.

This kind of structure is indeed what we get when we start the EPSR procedure on the experimental neutron data (Bowron et al., 1998a): the standard (literature) potential functions that are used in the initial stage of the refinement give an average number of intermolecular hydrogen bonds of 1.8 ± 0.2. This is entirely consistent with intuitive expectation and with Figure 11.3a. Unfortunately (or fortunately for this discussion), the simulated experimental data using this model do not agree with the actual measured data. Continuing the refinement, which, as stated above, effectively involves a perturbation of the potential functions operating in the system, we eventually obtain agreement between the experimental and simulated data. When we now look at the experimentally consistent structure, we find an average of only 1.1 ± 0.2, a situation consistent with the kind of hydrogen bonded dimer shown in Figure 11.4.

We conclude from this that the standard potential functions that were fed in as starting potentials for the refinement probably had a hydrogen bonding term that was too strong in relation to the other interactions in the system. This led to an over-hydrogen-bonded model structure that was not consistent with the experiment. In effect, the relative strengths of the different competing intermolecular interactions were incorrectly balanced, leading to local molecular aggregates that were chemically incorrect. During the refinement process that brought the computer simulated structure into agreement with the experimental data, the potential functions themselves were perturbed; as a result of these small changes to the potentials, the interaction chemistry of the local molecular aggregates of solute molecules was corrected.

We now refer back to the earlier discussion of protein stability, where we argued that small perturbations in the relative strengths of different kinds of hydrogen bonding interactions—between E_{P-P}, E_{P-W}, and E_{W-W} in the terms of Equation (11.4)—could result in incorrect folding, or perhaps no folding at all. The simple amphiphile case discussed here illustrates something similar in a much simpler system: that a small perturbation of the potential functions that gave an incorrect hydrogen bonded structure resulted in a structural change that regained the experimentally observed structure.

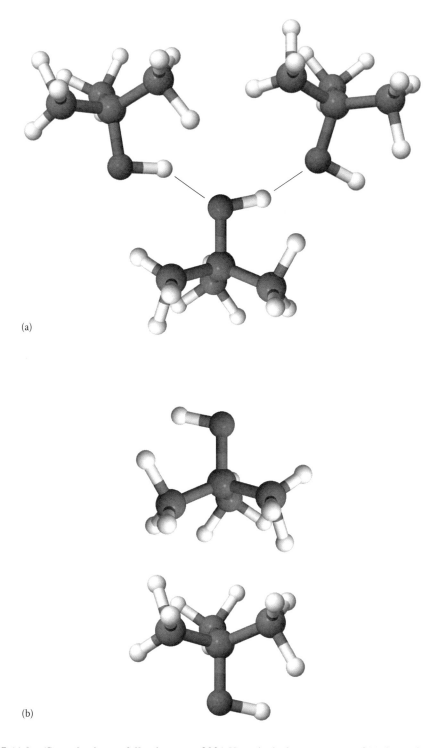

(a)

(b)

FIGURE 11.3 **(See color insert following page 302.)** Hypothetical arrangement of (a) three *t*-butanol molecules in the liquid interacting with each other through hydrogen bonding (indicated by the two linking lines) of their alcohol tails and (b) two *t*-butanol molecules in the liquid interacting with each other through nonpolar head group interactions. (With acknowledgment to Dr. Daniel Bowron, ISIS Facility, Rutherford Appleton Laboratory, UK.)

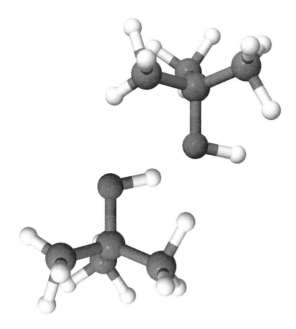

FIGURE 11.4 **(See color insert following page 302.)** Schematic of the actual average intermolecular hydrogen bonding situation found in liquid *t*-butanol at room temperature. (With acknowledgment to Dr. Daniel Bowron, ISIS Facility, Rutherford Appleton Laboratory, UK.)

11.4.3 Aqueous Solutions of Amphiphiles

We can extend the above experiments to look at detailed structures of amphiphiles in water. Whereas the hydrogen bonding in the pure liquid case discussed above is limited to a single type, between the polar alcohol tails of neighboring molecules, adding water brings in three different kinds of hydrogen bonding interactions. This additional complexity enables us to consider the effects on solute aggregation of differences in strengths of hydrogen bonds between solute and solute, solute and solvent, and solvent and solvent. In this way, we can explore the sensitivity of the solute assembly to changes in the balance between these three interactions, echoing the sensitivity of protein folding and stability to the relative strengths of these three kinds of hydrogen bonding interactions that are at the center of the biological analogy being discussed.

Again, intuitively, we might expect significant hydrogen bonding between the alcohol molecules to be retained as we add water. This is what I was led to expect in school chemistry, and it has not been possible to verify or otherwise this expectation until recently. Knowing what we now know about the pure liquid from the above experiments, full hydrogen bonding such as that shown in Figure 11.3a would not be expected, but perhaps quite a lot of the type of configuration in Figure 11.4 might be retained.

But yet again, the experimental results on *t*-butanol–water solutions show something different (Bowron et al., 1998b, 2001). As soon as we add water, these added molecules appear to go straight for the polar alcohol and hydrogen bond with it. Thus, at all concentrations that have been looked at so far for which there are enough waters to hydrate every polar alcohol group (up to 0.16 mol fraction alcohol), there is virtually no intermolecular solute–solute hydrogen bonding at all. There are, on average, essentially no alcohol molecules directly hydrogen bonded to another *t*-butanol in the way shown in Figure 11.4. Rather, at these concentrations, each alcohol OH group is hydrogen bonded to approximately between 2 and 2.5 water molecules, depending on the concentration. The kind of structure shown in Figure 11.5 is what is found experimentally. It is only when we get to very low water concentrations where there is insufficient water to satisfy the hydrogen bonding requirements

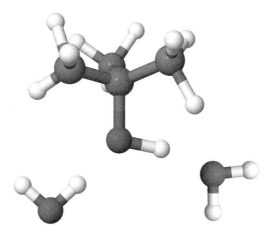

FIGURE 11.5 **(See color insert following page 302.)** Schematic picture of the actual average hydrogen bonding situation observed in all but the most concentrated aqueous solutions of *t*-butanol in which there is no hydrogen bonded interaction between solute molecules. The water molecules take up the hydrogen bonding capability of the alcohol's hydroxyl group. (With acknowledgment to Dr. Daniel Bowron, ISIS Facility, Rutherford Appleton Laboratory, UK.)

of the alcohol that we see some degree of alcohol–alcohol hydrogen bonding, for example in the 0.86 mol fraction solution (Bowron et al., 2002), which has, on average, only one-sixth of a water molecule for every molecule of *t*-butanol.

Although entropic effects have not been considered here, and may materially affect the overall argument, there is an implication that the interaction between water and alcohol is stronger than that between either alcohol and alcohol or water and water. Although the calculations have not specifically been made, we could imagine performing simulations of this type of system, but making variations in the relative strengths of the water–water, water–alcohol, and alcohol–alcohol hydrogen bonds to explore in more detail the sensitivity of the structure to such small changes (see the variational work on the effects on liquid water structure of changes in the water–water potential function discussed in the chapter in this book by Lynden-Bell and Debenedetti). Whether the standard potentials that are used to computer simulate systems such as these would reproduce the experimentally observed "no dimer" structure or produce significant polar dimerization is not clear, although there are indications from unpublished work on methanol–water solutions that the latter—incorrect— behavior may be obtained by at least some of the potential functions that are used (Réat and Finney, Bowron and Finney, unpublished work).

It should perhaps be emphasized that the central point emerging from the two examples discussed above of pure liquid *t*-butanol and aqueous solutions of this amphiphile is not to comment on the adequacy or otherwise of the potential functions generally used in computer simulations of both simple solutions and proteins. The point is rather that they provide an experimental demonstration of the high sensitivity of the way the amphiphile in the system self-assembles to the potential functions controlling the intermolecular interactions. The model amphiphile system has useful similarities with the protein folding and stability phenomena that are the focus of the biological fine-tuning analogy that we are exploring. The way in which these systems self-assemble locally depends on how the system resolves the competition between competing interactions that themselves are of similar strengths.

The protein argument encapsulated in Equation 11.4 is that the enthalpic contribution to the stability of the native (active) structure can depend on the small differences between the three different kinds of hydrogen bonds in the system: between protein and protein, protein and water, and water and water. The *t*-butanol examples show that indeed small perturbations in the strengths of competing interactions can result in qualitative changes in the geometry of the self-assembled amphiphiles that echo the protein folding situation. Indeed, the standard literature potentials that were used as starting points for the fitting of the simulated liquid structure to the experimental results fail to reproduce the local structures that are observed experimentally. That only apparently small perturbations need to be made to the potential functions in the corresponding simulated systems to reproduce the experimental structures underlines the sensitivity of important structures to the relative strengths of the different—but quantitatively very similar—intermolecular interactions.

11.5 CONCLUDING REMARKS

The actual protein folding situation is, of course, much more complex than suggested here (Finney et al., 1980). For example, there are many other contributions to the free energy of folding which we have not even begun to address, for example entropic terms with respect to both the protein and the solvent. These can have major effects on the structure. Moreover, we don't have any real experimental pointer as to values that we should use for the energies of these different hydrogen bonds. But it does seem an interesting and informative biologically relevant scenario to discuss in the fine-tuning context, involving as it does sensitivity of active biological structures to the details of water-involved interactions.

Just as small differences between certain fundamental constants need to have precisely defined values in order for the universe to have developed the way it has, so in this biologically relevant case we need to pay attention to small differences between large numbers when considering the viability of processes such as protein folding. This is particularly critical in that the stability of a native protein is in itself marginal (which it may well need to be to function efficiently), being quantitatively of the order of only 2 to 4 hydrogen bonds in a system that contains several hundred of such bonds.

The simple analogy and the illustrations discussed argue that we need to pay attention also to possible small differences in energies of hydrogen bonds between different molecules—in the simplest case considering all water–water hydrogen bonds to be the same strength, all protein–protein polar hydrogen bonds to be equal in strength (but different from the water–water value), with the water–protein interactions similarly all the same but again different from the other two. The precise values of these average energies—and the relatively small differences between them—could well be critical to the stability of the protein in a way that echoes the criticality of the values of the cosmological constants discussed in the introduction. It is perhaps difficult to think of these differences in the biological case being *a priori* fine-tuned, but perfectly reasonable to accept that evolution has taken advantage of them in building life on Earth. And the discussion does, though incidentally, underline the importance of obtaining and refining potential functions for computer simulations that can reproduce the right chemistry in the solute association that is observed experimentally. Otherwise, they cannot hope to realistically simulate a folded protein with its marginal stability.

REFERENCES

Bowron, D. T., Finney, J. L., and Soper, A. K. 1998a. The structure of pure tertiary butanol. *Mol. Phys.* 93: 531–543.

Bowron, D. T., Finney, J. L., and Soper, A. K. 1998b. Structural investigation of solute-solute interactions in aqueous solutions of tertiary butanol. *J. Phys. Chem. B* 102: 3551–3563.

Bowron, D. T., Soper, A. K., and Finney, J. L. 2001. Temperature dependence of the structure of a 0.06 mole fraction tertiary butanol-water solution. *J. Chem. Phys.* 114: 6203–6219.

Bowron, D. T., and Díaz Moreno, S. 2002. The structure of a concentrated aqueous solution of tertiary butanol: Water pockets and resulting perturbations. *J. Chem. Phys.* 117: 3753–3762.

Dixit, S., Crain, J., Poon, W. C. K., Finney, J. L., and Soper, A. K. 2002. Molecular segregation observed in a concentrated alcohol–water solution. *Nature* 416: 829–832.

Enderby, J. E., and Nielson, G. W. X-ray and neutron scattering by aqueous solutions of electrolytes. In *Water. A Comprehensive Treatise*. Vol. 6. Edited by F. Franks. New York, NY: Plenum Press, 1–46.

Finney, J. L., and Bowron, D. T. 2002. Experimental determination of the structures of complex liquids: Beyond the PDF. In *From Semiconductors to Proteins: Beyond the Average Structure*. Edited by S. H. Billinge and M. F. Thorpe. Norwell, MA: Kluwer Academic/Plenum, pp. 219–244.

Finney, J. L., Golton, I. C., Gellatly, B. J., and Goodfellow, J. M. 1980. Solvent effects and polar interactions in the structural stability and dynamics of globular proteins. *Biophys. J.* 32: 17–33.

Finney, J. L., and Soper, A. K. 1994. Solvent structure and perturbations in solutions of chemical and biological importance. *Chem. Soc. Rev.* 1994: 1–10.

Hogan, C. J. 2000. Why the universe is just so. *Rev. Mod. Phys.* 72: 1149–1161.

Hoyle, F. 1954. The nuclear reactions occurring in very hot stars. *Astrophys. J. Suppl.* 1: 121.

Oberhummer, H., Csótó, A., and Schlattl, H. 2000. Stellar production rates of carbon and its abundance in the universe. *Science* 289: 88–90.

Salpeter, E. E. 1952. Nuclear reactions in stars without hydrogen. *Astrophys. J.* 115: 326.

Schattl, H., Heger, A., Oberhummer, H., Rauscher, T., and Csótó, A. 2004. Sensitivity of the C and O production on the 3α rate. *Astrophys. Space Sci.* 291: 27–56.

Soper, A. K. 1996. Empirical potential Monte Carlo simulation of fluid structure. *Chem. Phys.* 202: 295–306.

Soper, A. K. 2000. The radial distribution functions of water and ice from 220 to 673 K and at pressures up to 400 MPa. *Chem. Phys.* 258: 121–137.

12 Fine-Tuning Protein Stability

Carlos Warnick Pace, Abbas Razvi, and J. Martin Scholtz

CONTENTS

12.1 INTRODUCTION

In all living systems, proteins perform the tasks that make life possible. To carry out their biological functions, the chain of amino acids making up the protein must fold, in the presence of water, to a unique, three-dimensional structure called the native state. These proteins are called globular proteins and they are, by far, the most abundant class of proteins. Sometimes, proteins have more than one globular unit in a single polypeptide chain and these units are then called domains. Other times, these globular units associate noncovalently to form larger proteins and this assembly and its architecture is called quaternary structure. Here, we will focus our attention on two proteins that differ markedly in size but fold to give a single globular unit. We are most interested in the forces that determine the stability of this tertiary structure of globular proteins (Petsko and Ringe, 2004).

The conformational stability of a globular protein is defined as the free energy change, ΔG, for this reaction:

$$\text{Folded (Native)} \leftrightarrow \text{Unfolded (Denatured).} \tag{12.1}$$

The stability is remarkably low: the folded native state is generally only 10 to 40 kJ/mol more stable than the large ensemble of unfolded states that make up the denatured state of a protein (Pace,

1990). This is what evolution has chosen to do. We now know that proteins could have become more stable, but did not. Why not? One possibility is that the biological function of the protein requires some flexibility and this is more difficult if the protein is too stable (Fields, 2001; Beadle and Shoichet, 2002). Another possibility is that the stability is low so that the protein can be degraded quickly when this is required by the cell or organism. There is evidence that supports both of these possibilities (Petsko and Ringe, 2004).

Based on what we have learned about protein stability over the past 50 years, large proteins should be more stable than small proteins (Pace et al., 1996). However, they are not, and it is not clear why. This is the main question we will consider in this chapter. Why are small proteins just as stable as large proteins? Or, put another way, what forces are used to destabilize larger proteins so that they do not become too stable for their biological function? Or, more generally, how is protein stability fine tuned?

12.2 PROTEIN FOLDING

The constituent groups of a protein are much more accessible to water in the denatured state than in the native state. When larger (>100 residues) proteins fold, more than 80% of the nonpolar side chains and the polar peptide groups are buried in the interior of the protein, out of contact with water (Lesser and Rose, 1990). The protein interior is tightly packed and more closely resembles a solid than a liquid. Because of this tight packing (≅20% tighter than alkane liquids and ≅5% tighter than alkane crystals), the van der Waals interactions will be much stronger in the native states of proteins than they are in the denatured states where the interactions will be mainly with water molecules (Pace et al., 2004).

When a protein folds, most of the nonpolar side chains are removed from contact with water and form a hydrophobic core. This burial of the nonpolar groups is the result of the hydrophobic effect, an interaction that is one of the major forces stabilizing the native state of proteins (Dill, 1990). When a protein folds, most of the polar peptide groups will form secondary structures called α-helices and β-sheets in which the polar groups are almost completely hydrogen bonded to other polar groups (Stickle et al., 1992). Hydrogen bonding is the other major force stabilizing the native state of proteins (Myers and Pace, 1996). The van der Waals interactions resulting from the tight packing of the nonpolar and polar groups in the protein interior makes a third major contribution to protein stability (Pace, 2001; Takano et al., 2003).

To attempt to answer the question posed in the Introduction, we will compare the structure and stability of one of the smallest globular proteins with 36 amino acids, with one of the largest, single-domain globular proteins with 341 amino acids. The small protein will be called VHP36 [villin head piece (VHP)]. This small protein is a subdomain in the headpiece of a protein called villin from chickens (McKnight et al., 1996). The large protein will be called BBSP341 (BBSP = *Borrelia burgdorferi* surface protein). The large protein is a surface protein found on a microorganism called *B. burgdorferi* that causes Lyme disease (Eicken et al., 2002).

Both VHP36 and BBSP341 closely approach two-state folding mechanisms and have similar stabilities, but they fold at remarkably different rates. VHP36 folds in 5 μs (Kubelka et al., 2003), but BBSP341 requires 5 s (Jones and Wittung-Stafshede, 2003). The function, structure, and stability of these proteins will be discussed in the next two sections.

12.3 VHP36

Villin is a protein whose function is to crosslink actin filaments in the cytoskeleton inside cells. Villin contains an 84 kDa N-terminal domain that binds actin and an 8 kDa C-terminal domain called "headpiece" that also binds actin. Thus, villin is able to crosslink actin filaments into structural bundles that are key components of the cytoskeleton. These filaments help control cell shape and the movement of the cell surface. For example, they are important in cytokinesis, the separation

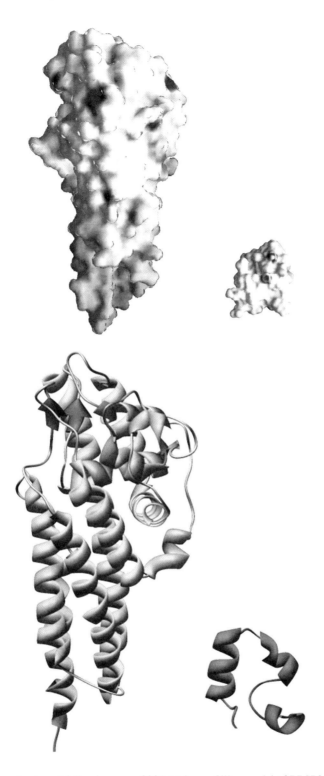

FIGURE 12.1 **(See color insert following page 302.)** (a) Space filling model of BBSP341 (left) and VHP36 (right). (b) Chimera ribbon diagram of BBSP341 (left) and VHP36 (right) showing the α-helices and β-sheets (Pettersen et al., 2004). The arrows show the β-sheets. The structures are from the Protein Data Bank files 1L8W for BBSP341 and 1VII for VHP36.

of daughter cells in the last stage of mitosis. Actin filaments are found in all eukaryotic cells, generally at high concentrations, and there are many other actin-binding proteins in addition to villin.

Both NMR structures (McKnight et al., 1997) and high-resolution x-ray crystal structures (Chiu et al., 2005) of VHP36 have been determined. The average NMR structure is shown as a ribbon diagram and a space filling model in Figure 12.1. The protein folds to give a cluster of three α-helices with a compact central core that is dominated by three Phe side chains that are ~95% buried and make a large contribution to the stability (Frank et al., 2002). The α-helices contain 8, 10, and 13 residues, giving the protein an 86% α-helical content. The protein has no disulfide bonds and its folding has been studied in the absence of all ligands that it might bind. Thus, the folding is determined by the intramolecular interactions of the polypeptide chain and intermolecular interactions of the polypeptide chain and water.

Because VHP36 is one of the smallest polypeptides that is both fast folding and very thermostable, its structure and stability have been studied in detail both experimentally (Kubelka et al., 2003 and references therein) and theoretically (Ripoll et al., 2004 and references therein). At pH 7 in water, the melting temperature of VHP36 is ~70°C and the conformational stability is 15 kJ/mol (McKnight et al., 1996).

12.4   BBSP341

B. burgdorferi is a spirochete that expresses large amounts of BBSP341 on its surface. This protein is thought to play a major role in the immune response to Lyme disease (Zhang et al., 1997). In vivo, the N-terminal cysteine residue of the protein is covalently attached to a lipid in the outer half of the lipid bilayer of the membrane to link the protein to the surface of the spirochete. In this environment, BBSP341 will be surrounded by other proteins and one suggested function is to present a variable region to the host's immune system. For this function, low stability will allow rapid turnover and contribute to high antigenic variation and, therefore, to the survival of the spirochete (Jones and Wittung-Stafshede, 2003).

The crystal structure of BBSP341 shows that the molecule is not quite globular but has dimensions of 31 × 37 × 80 Å, as shown in Figure 12.1 (Eicken et al., 2002). In the crystal structure, some residues at the N and C termini, and residues 93 to 112 in a loop on the surface are not visible. This suggests that they do not have a fixed conformation in the crystal and this is also likely to be the case in solution. We will assume that they do not contribute to the stability since they will be exposed to solvent in both the native and denatured states of the protein. Consequently, only 296 of the 341 residues will be analyzed in estimating the contribution of the various forces to the stability. BBSP341 has 11 α-helices with lengths ranging from 6 to 36 residues and 4 short strands of β-sheet. Consequently, 44% of the residues in BBSP341 are present as α-helices and 4% are present as β-sheets. BBSP341 has no disulfide bonds and binds no cofactors that might contribute to the stability. So, as with VHP36, the folding will be determined by just intramolecular interactions of the polypeptide chain and intermolecular interactions of the polypeptide chain with water.

We chose BBSP341 for this study because it is one of the largest single domain proteins whose stability has been studied. At pH 7 in water, the melting temperature is ~55°C and the conformational stability is 20 kJ/mol (Jones and Wittung-Stafshede, 2003).

12.5   COMPARISON OF THE STRUCTURES OF VHP36 AND BBSP341

The structures of VHP36 and BBSP341 are shown in Figure 12.1. Despite the big difference in size, the stability and the residue composition of the proteins is similar: nonpolar residues, 53% for VHP36 and 56% for BBSP341; polar residues, 19% for VHP36 and 15% for BBSP341; and charged residues, 28% for VHP36 and 29% for BBSP431. However, as the size of single domain proteins becomes larger, it is obvious that a greater fraction of the side chains and peptide groups will be buried on folding. This is shown in Table 12.1. Only 53% of the nonpolar side chains are

TABLE 12.1
Burial of Nonpolar Side Chains, Peptide Groups, Polar Groups, and Ionizable Groups in the Folding of VHP36 and BBSP341[a]

Nonpolar Side Chain	VHP36		BBSP341	
	Number Present	Number Buried	Number Present	Number Buried
Ale	3	1.2	58	45.6
Ile	0	0	15	14.2
Leu	5	3.2	16	14.4
Met	2	0.2	0	0
Phe	4	2.8	10	7.6
Pro	1	0.5	5	3.8
Trp	1	0.3	0	0
Tyr	0	0	2	1.1
Val	1	0.8	24	22.0
Total	17	9 (53%)	130	109 (84%)
Peptide groups	36	24.7 (68%)	295	252.7 (86%)
Side chain polar groups[b]	11	4.2 (38%)	57	33.1 (58%)
Ionizable groups[c]	18	3.8 (21%)	131	65.5 (50%)

[a] The solvent accessibility of all of the atoms in the proteins was estimated using the program pfis (Hebert et al., 1998), and the structures from the Protein Data Bank: file 1VII for VHP36 and file 1L8W for BBSP341.

[b] The accessibility of the following polar groups was determined: the –OH group for the Ser, Thr, and Tyr side chains, and the O and NH_2 groups for the Asn and Gln side chains.

[c] The accessibility of the following charged groups was determined: both O atoms for the carboxyl groups, both NH_2 groups in the Arg side chain, both NH groups in His side chains, and the NH_2 groups in the Lys side chain and at the amino terminus.

buried when VHP36 folds, but 84% are buried when BBSP341 folds. Similarly, 68% of the peptide groups are buried when VHP36 folds, but 86% are buried when BBSP341 folds. For reasons that will be discussed below, we also compared the burial of the uncharged polar groups and charged groups of the side chains in the two proteins. For the uncharged polar groups, 38% are buried in VHP36, but 58% are buried in BBSP341. For the charged groups, 21% are buried in VHP36, but 50% in BBSP341. So, as expected, more groups are buried when a large protein folds than when a small protein folds and the difference is greatest for the nonpolar side chains (31% increase) and the charged groups (29% increase).

12.6   MAJOR FORCES CONTRIBUTING TO PROTEIN STABILITY

For this discussion, we will use VHP36 and BBSP341 as examples to illustrate the magnitude of the contributing forces. The results are summarized in Table 12.2.

12.6.1   CONFORMATIONAL ENTROPY

The major force favoring the denatured state of a protein is conformational entropy. Rotation around the many bonds in a protein increases when the protein unfolds, and this provides a strong entropic driving force for unfolding. Experimental and theoretical studies agree that conformational entropy should favor the unfolded state by $\cong 7$ kJ/mol per residue (Spolar and Record, 1994; D'Aquino et al.,

TABLE 12.2
Estimates of the Major Forces Contributing to the Stability of VHP36 and BBSP341

Factor	VHP36 (kJmol)	BBSP431 (kJ/mol)
Hydrophobic effect[a]	+225	+2185
Hydrogen bonding[b]	+96	+1060
Conformational entropy[c]	−250	−2070
Net estimated stability[d]	+71	+1175
Measured stability	+15	+20

[a] The free energy of transfer, ΔG_{tr}, from water to cyclohexane was used to estimate the contribution of the burial of the nonpolar side chains to the conformational stability of the proteins. The ΔG_{tr} values are from Radzicka et al. (1988) and are summarized by Pace (1995). The number of buried nonpolar side chains is given in Table 12.1. The contribution of the hydrophobic effect to the stability is the sum of the contributions of the individual side chains. (See Pace et al., 1998, for a more complete description of this and the following contributions.)

[b] The number of hydrogen bonds formed in the folded proteins was estimated by the program pfis (Hebert et al., 1998) to be 24 hydrogen bonds for VHP36, and 265 hydrogen bonds for BBSP341. We assume that each hydrogen bond contributes 4 kJ/mol to the stability.

[c] The conformational entropy was calculated assuming 7 kJ/mol per residue (Spolar and Record, 1994).

[d] Net estimated stability = sum of the contribution of the hydrophobic effect, hydrogen bonding, and conformational entropy.

[e] The measured stabilities are from McKnight et al. (1996) for VHP36 and from Jones and Wittung-Stafshede (2003) for BBSP341.

1996). We theorize that this estimate is more likely to be too high than too low and there is considerable uncertainty in the estimate. Based on this, conformational entropy will favor unfolding by 250 kJ/mol for VHP36 and 2070 kJ/mol for BBSP341. What are the forces that overcome this instability?

12.6.2 HYDROPHOBIC EFFECT

Bernal (1939) was the first to suggest that the hydrophobic effect might be the dominant stabilizing force in protein folding. Kauzmann (1959) and Tanford (1962) popularized this idea and Tanford concluded in 1962 that ". . . the stability of the native conformation in water can be explained. . . entirely on the basis of the hydrophobic interactions of the nonpolar parts of the molecule." In an influential review, Dill (1990) concluded: "More than 30 years after Kauzmann's insightful hypothesis, there is now strong accumulated evidence that hydrophobicity is the dominant force of proteins folding. . . . There is evidence that hydrogen bonding or van der Waals interactions among polar amino acids may be important, but their magnitude remains poorly understood." Most biochemists share this view, as illustrated by some of the chapters in this book. However, more recent studies using several different approaches suggest that the burial and hydrogen bonding of the polar groups may make a contribution to the stability comparable to that of the hydrophobic effect (Pace et al., 2004). This evidence will be summarized in the next section.

What do the experimental data tell us? On the basis of studies of hydrophobic mutants of proteins, each $-CH_2-$ buried contributes ~5 kJ/mol to the stability of the protein (Pace, 1995). For comparison, the ΔG_{tr} for transfer of a $-CH_2-$ from water to the vapor phase is −0.8 kJ/mol, to n-octanol is −2.5 kJ/mol, and to cyclohexane is −4.2 kcal/mol (Pace, 2001). These differences reflect the more favorable van der Waals interactions between the $-CH_2-$ group and the nonpolar phase. Why is the value based on studies of proteins higher? Several recent analyses suggest that it is due to the more favorable van der Waals interactions that result from the tight packing of the hydrophobic groups in the protein interior (see, e.g., Jain and Ranganathan, 2004). To illustrate, the fraction of space occupied by atoms is 36% in liquid water, 44% in cyclohexane, but 75% in the protein interior (Pace et al., 2004). Thus, the classical explanation of the hydrophobic effect used to explain model compound data underestimates the contribution of the hydrophobic effect to protein stability because the protein interior is more tightly packed than solvents (such as n-octanol and cyclohexane) that are used to model the interior of a protein.

Because the ΔG_{tr} values for cyclohexane are closest to the experimental values, we will use these to estimate the contribution of the hydrophobic effect to the stability of the native state. The results suggest that the contribution of the hydrophobic effect to the stability would be 225 kJ/mol for VHP36 and 2185 kJ/mol for BBSP341 (Table 12.2). Thus, it is clear that the hydrophobic effect will make a large contribution to the stability of each protein, enough to overcome the unfavorable conformational entropy by itself for BBSP341, supporting Tanford's (1962) conclusion.

12.6.3 Hydrogen Bonding

Table 12.3 compares the thermodynamics of hydration of a peptide group and a nonpolar leucine side chain. The leucine side chain prefers a vacuum over water and this is entirely due to an unfavorable entropic effect, the classical explanation of the hydrophobic effect. In contrast, the peptide group strongly prefers water, and this is entirely due to a favorable enthalpic effect. Thus, it is not surprising to find opinions like this (Honig and Cohen, 1996): ". . . a crucial property of the polypeptide backbone is that it contains polar NH and CO groups whose removal from water involves a significant energetic penalty." Is it possible that burying a peptide group in the interior of a protein might make a favorable contribution to stability? Several experimental studies of proteins suggest the answer is yes (Hebert et al., 1998; Deechongkit et al., 2004). A buried, hydrogen bonded peptide group can overcome the large dehydration penalty and still make a favorable contribution of 8 to 12 kJ/mol to the stability of the native protein (Pace, 2001; Takano et al., 2003). This means that the hydrogen bonding and van der Waals interactions of a peptide group in the interior of a folded protein can be more favorable than the interactions with water in the unfolded protein.

The majority of the hydrogen bonds that proteins form on folding are hydrogen bonds between buried peptide groups. Stickle et al. (1992) studied a sample of 42 globular proteins containing an average of 160 amino acids and found that proteins form an average of 1.1 intramolecular hydrogen bonds per residue on folding. Of these, 68% of the hydrogen bonds are between the carbonyl oxygen and the amide hydrogen of the peptide group. Later, McDonald and Thornton (1994) estimated that ". . . 9.5% and 5.1% of the buried main-chain nitrogen and oxygen atoms, respectively, fail to hydrogen bond under our standard criteria." However, when they relaxed the hydrogen bonding criteria they found ". . . there remain some buried atoms (1.3% NH and 1.8% CO) that fail to hydrogen bond without any immediately obvious compensating interactions." More recently, Fleming and Rose (2005) published a paper titled "Do all backbone polar groups in proteins form hydrogen bonds?" and concluded that they almost all do. BBSP341 forms 265 hydrogen bonds when it folds, and 204 (77%) are between peptide groups, 44 (17%) are between a peptide group and a side chains, and 17 (6%) are between side chains. VHP36 forms 24 hydrogen bonds on folding and all but one are between peptide groups.

TABLE 12.3
Thermodynamics of Hydration of a Peptide Group and a Leucine Side Chain

	ΔG^a (kJ/mol)	ΔH^a (kJ/mol)	$T\Delta S^a$ (kJ/mol)
Peptide group	−50	−58	−8
Leu	+8	−17	−25

[a] These are free energy (ΔG), enthalpy (ΔH), and entropy ($T\Delta S$), changes for the transfer of a peptide group and a leucine side chain from the vapor phase to water as reported by Privalov and Makhatadze (1993).

Studies of Asn to Ala mutants at largely buried residues where the amide group of the Asn was originally hydrogen bonded to a peptide group can be used to estimate the contribution of peptide hydrogen bonds to protein stability. These studies suggest that each hydrogen bond formed by a peptide group contributes 4.5 ± 1.5 kJ/mol to the stability (Hebert et al., 1998). These and many other mutational studies of proteins and a variety of other studies support the idea that each intramolecular hydrogen bond contributes 4 to 8 kJ/mol to the stability of the native states of proteins (Myers and Pace, 1996).

We will make the conservative estimate that each intramolecular hydrogen bond contributes 4 kJ/mol to the stability of the native state of proteins. This leads to a contribution to the stability of VHP36 of 96 kJ/mol and of BBSP341 of 1060 kJ/mol (Table 12.2).

12.6.4 SUMMING UP

So, based mainly on experimental studies of the three major forces contributing to protein stability, we estimate the stability of VHP36 to be: 225 + 96 − 250 = 71 kJ/mol and the stability of BBSP341 to be: 2185 + 1060 − 2070 = 1175 kJ/mol. These results are summarized in Table 12.2. Note that the estimated stability of VHP36 is 56 kJ/mol greater than the measured value and the estimated stability of BBSP341 is 1155 kJ/mol greater than the measured value. One interesting result is that just the hydrophobic effect and hydrogen bonding can contribute enough free energy to stabilize the globular conformation of even a small 36 residue protein such as VHP36. The second interesting finding is that the estimated stability of the large protein is more than 1100 kJ/mol greater than the measured stability. This, of course, illustrates the question we posed in the Introduction. We will now consider the forces that might be important in destabilizing large proteins.

12.7   POSSIBLE FORCES CONTRIBUTING TO PROTEIN INSTABILITY

12.7.1   BURYING CHARGED GROUPS

A few buried charged groups can make large favorable contributions to protein stability. To do so, they must form some good hydrogen bonds like Asp76 in RNase T1 (Giletto and Pace, 1999), or a strong ion pair like an Asp and a His form in T4 lysozyme (Anderson et al., 1990). More often, burying a charged group makes an unfavorable contribution to the stability due to the Born self-energy (Braun-Sand and Warshel, 2004). For example, Asp 79 in RNase Sa is buried and forms no hydrogen bonds. When it is replaced by Ala, the stability increases by ~14 kJ/mol (Trevino et al., 2005). To see if charged group burial might contribute to the stability of the two proteins considered here, we estimated the extent of burial of the charged groups in both proteins. Of the 18 charged groups in VHP36, 21% are buried, but of the 131 charged groups in BBSP341, 50% are buried (Table 12.1). This is a striking difference and is probably the major factor that destabilizes BBSP341. This possibility was first suggested by Kajander et al. (2000). They showed that, as proteins get larger, they bury more charged groups and suggested that this might be a strategy used to keep large proteins from becoming too stable. They estimated that, on average, 37% of the charged groups would be buried in a 100 residue protein, but 61% would be buried in a 700 residue protein. The increase in % burial for different types of groups in going from a 100 to a 700 residue protein is 14% for aromatic residues, 35% for aliphatic residues, 37% for polar, uncharged residues, and an astonishing 65% for charged residues. For our two proteins, the fractional burial of charged groups is more than doubled in the larger protein (Table 12.1).

12.7.2   COULOMBIC INTERACTIONS AMONG THE CHARGED GROUPS

At neutral pH, the side chains of Asp and Glu will have a negative charge, the side chains of Lys and Arg will have a positive charge, and the side chains of His will sometimes have a positive charge. Thus, both attractive and repulsive Coulombic interactions between the charged groups are possible. Most of these charged groups are located on the surface of proteins. At pH 7, VHP36 will have 7 positive charges and 5 negative charges. For a rough estimate of the contribution of these charge–charge interactions to the stability, we used Coulomb's law with a dielectric constant of 80 to sum up the charge–charge interactions and we find that they destabilize the protein by 14 kJ/mol (Grimsley et al., 1999). We should probably use a lower dielectric constant than 80. In a previous study, we estimated the pKs of groups on the surface of RNase Sa most accurately when we used a dielectric constant of 45 (Laurents et al., 2003). If we use the same value with VHP36, the decrease in stability would be 25 kJ/mol. This brings the estimated stability closer to the measured stability for VHP36 (Table 12.2).

At pH = 7, BBSP341 will have ~43 positive charges and ~42 negative charges. Because some parts of the structure seem to be flexible and do not have a localized position in the native state, we can not estimate the contribution of charge–charge interactions to the stability. However, because the charges are evenly balanced, it is less likely that it will make a large contribution to the stability.

12.7.3 BURYING POLAR SIDE CHAIN GROUPS

We also considered the burial of the uncharged, polar groups in the side chains of the amino acids. We analyzed the burial of the –OH groups for the Ser, Thr, and Tyr residues, and the burial of the O and –NH_2 groups in the amide groups of the Asn and Gln residues. In VHP36, there are only 11 polar groups from the 2 Asn, 2 Gln, 2 Ser, and 1 Thr residues and none form intramolecular hydrogen bonds. Of these 11 polar groups, 38% are buried in VHP36. In BBSP341, 58% of the 57 polar groups are buried. In this case, they form a total of 33 intramolecular hydrogen bonds, which is considerably less than the 145 hydrogen bonds these groups could form. Hydrogen bonding will be discussed further below. Thus, BBSP341 also buries a much higher percentage of its uncharged polar groups than VHP36 and this may also be a strategy that large proteins use to destabilize their structure. This depends in part on whether the polar group forms hydrogen bonds and this will be discussed in the next section.

12.7.4 CONTRIBUTION OF BURIED, POLAR, SIDE CHAIN GROUPS TO PROTEIN STABILITY

Table 12.4 summarizes studies of the stability of a large number of Tyr to Phe, Thr to Val, and Val to Thr mutants. These studies were designed to gain a better understanding of the contribution of hydrogen bonds to protein stability (Takano et al., 2003). The results for the Tyr to Phe mutants show that the hydrogen bonding and van der Waals interactions of an –OH group in folded proteins

---

TABLE 12.4
$\Delta(\Delta G)$ Values for 52 Tyr $\Rightarrow$ Phe Mutants, 40 Thr $\Rightarrow$ Val Mutants, and 40 Val to Thr Mutants[a]

	Hydrogen Bonded		Not Hydrogen Bonded	
Mutation	No.	$\Delta(\Delta G)$ (kJ/mol)	No.	$\Delta(\Delta G)$ (kJ/mol)
Tyr $\Rightarrow$ Phe[b]	35	−5.9 ± 3.8	17	−0.8 ± 1.6
Thr $\Rightarrow$ Val[c]	25	−3.8 ± 4.2	15	0.0 ± 2.1
Val $\Rightarrow$ Thr[c]	40	−7.5 ± 4.2		

[a] $\Delta(\Delta G) = \Delta G°$ (mutant protein) – $\Delta G°$ (wild-type protein). See Takano et al. (2003) for a more detailed description.
[b] 52 Tyr $\Rightarrow$ Phe mutants (Pace et al., 2001).
[c] 40 Thr $\Rightarrow$ Val, and 40 Val to Thr mutants (Takano et al., 2003).

can be more favorable than the interactions with water in an unfolded protein. They also show that –OH groups that are not hydrogen bonded using the standard definition can still make a favorable contribution to protein stability. The results for the Thr to Val mutants show that when a non-hydrogen bonded –OH group is buried in an environment designed for an –OH group, it contributes about the same amount to the stability as a –CH₃ group buried at the same site. The results for the Val to Thr mutants show that replacing a buried –CH₃ group with an –OH group at a site designed for a –CH₃ group can markedly destabilize a protein. These results show that evolution can stabilize proteins with buried polar groups, especially if they are hydrogen bonded, but, more important for the present discussion, evolution can destabilize proteins by replacing nonpolar groups with polar groups as in Val to Thr mutations or by inserting buried polar groups as in Ala to Ser or Phe to Tyr mutations. Thus, proteins can be destabilized by either burying charged groups or polar groups and our comparison of VHP36 and BBSP341 suggest that both contribute to lowering the stability of BBSP341.

12.7.5 CONTRIBUTION OF BURIED, PEPTIDE GROUPS TO PROTEIN STABILITY

As discussed above, peptide groups that form intramolecular hydrogen bonds contribute to protein stability. Based on model compound studies, hydrogen bonds should be stronger when the distance and geometry are optimal. The average hydrogen bond length for the 23 main chain hydrogen bonds in VHP36 is 2.85 Å, but for the 204 main chain hydrogen bonds in BBSP341 it is 2.98 Å. In their survey of 42 high-resolution protein structures, Stickle et al. (1992) found that the hydrogen bonds in α-helices had an average length of 2.99 Å and those in the β-sheets were 2.91 Å. So the hydrogen bonding in BBSP341 is typical of that generally observed in proteins, but the hydrogen bonds in VHP36 are significantly shorter and this might make their contribution to the stability greater than those in BBSP341.

As discussed above, the hydrogen bonding of most of the main chain polar groups is satisfied, but those that are not might contribute as much as 20 to 25 kJ/mol to the instability of the protein (Fleming and Rose, 2005). Savage et al. (1993) showed an interesting correlation between lost hydrogen bonds and buried surface area of nonpolar atoms. The greater the number of unformed hydrogen bonds, the greater the amount of nonpolar surface buried. This suggests that larger proteins might fine tune their stability simply by forming fewer hydrogen bonds.

12.8 FINE-TUNING PROTEIN STABILITY

The results in Table 12.1 show that BBSP341 buries about 30% more of its nonpolar side chains than VHP36, and this is the main reason that the estimated stability of BBSP341 is so much greater than the measured stability. For both proteins, the results in Table 12.2 suggest that the hydrophobic effect makes the major contribution to the stability.

The results in Table 12.1 also show that BBSP341 buries about 20% more of its polar groups than VHP36. VHP36 has 52 buried polar groups and 39 (75%) are hydrogen bonded. BBSP341 has 540 buried polar groups and 405 (75%) are hydrogen bonded. The hydrogen bonds formed by these groups, 24 for VHP36 and 265 for BBSP341, make substantial contributions to the stability of both proteins (Table 12.2).

VHP36 has 13 buried polar groups that are not hydrogen bonded and BBSP341 has 135. It has been suggested that each buried, non-hydrogen-bonded polar group might contribute 20 to 25 kJ/mol to the instability because they will be completely hydrogen bonded by water molecules in the unfolded protein (Fleming and Rose, 2005). If we assume that each buried polar group contributes 20 kJ/mol to the instability, then the estimated stabilities become –190 kJ/mol for VHP36 and –1525 kJ/mol for BBSP341, far less than the observed stabilities. So this argues against a large energetic penalty for buried, non-hydrogen-bonded polar groups in folded proteins. The results in Table 12.4 also argue against this. First, they show that buried polar groups that are not hydrogen bonded may contribute favorably to the stability, and, second, they show that even when a buried –CH$_3$ group is replaced by a non-hydrogen–bonded –OH group, the destabilization will be only about 8 kJ/mol. In conclusion, the burial of non-hydrogen–bonded polar groups may contribute to the instability of BBSP341, but it is not likely to be the major factor.

The results in Table 12.1 show that BBSP341 buries about 30% more of its charges than VHP36. We think this is the primary mechanism that large proteins use to lower their stability. First, the energetic consequences of burying a charge will generally be greater than burying a polar group. For example, burying a charge in RNase Sa decreases the stability by ~14 kJ/mol (Trevino et al., 2005), but replacing a –CH$_3$ group with an –OH group only decreases the stability by ~7 kJ/mol (Takano et al., 2003). Second, in VHP36, the number of charged groups buried is about the same as the number of polar side chain groups buried, but in BBSP341 the number is more than doubled. We have begun an experimental and computational study that should lead to a better understanding of this important question.

We will conclude by considering the important contribution of water to the forces that determine protein stability. Everyone agrees that the hydrophobic effect is a major force stabilizing proteins, and that the process is entropy-driven (Table 12.3). However, despite sophisticated experimental (Finney et al., 2003; Bowron, 2004) and theoretical (Huang and Chandler, 2002) studies, it has proven difficult to understand the hydrophobic effect at the molecular level. Probably many of the properties of water are important: for example, the small size of water molecules, the structure of water in the vicinity of nonpolar molecules, and the greater affinity of water molecules for each other than for nonpolar molecules. One thing is clear: water is of crucial importance to the hydrophobic effect.

The contribution of polar group burial to protein stability is a more controversial subject (Takano et al., 2003; Kortemme et al., 2003; Baldwin, 2003). Wolfenden's studies (1978) of hydration of the peptide group led him to conclude: ". . . (1) that the peptide bond represents an extreme among uncharged functional groups in the degree to which it is stabilized by solvent water; (2) that the very hydrophilic character of the peptide bond may be associated mainly with hydrogen bonding of the solvent to the carbonyl oxygen (rather than the N–H group). . . ." As shown in Table 12.3, ΔG for hydration of a peptide bond is 50 kJ/mol. Thus, to gain stability from a buried peptide group, the hydrogen bonding, other electrostatic interactions, and van der Waals interactions with the protein would have to be greater than 50 kJ/mol. Before relevant experimental data became available, this seemed unlikely and most biochemists believed that the burial and hydrogen bonding of peptide groups would at best allow you to "break even." The first convincing studies indicating that hydrogen bonds were likely to make a favorable contribution to protein stability were published by the Fersht group (1987), and almost all experimental studies since then support this idea. Why is polar group burial in protein folding generally a favorable process? Our guess is that two main factors contribute: (1) the intramolecular hydrogen bonds in the folded protein interior are stronger than the hydrogen bonds to water in the unfolded protein because they are surrounded by an environment with a lower dielectric constant and this strengthens electrostatic interactions; and (2) because of the tight packing of the groups in the interior of folded proteins, the van der Waals interactions of the polar groups will be more favorable in the protein interior than they will be with water in the unfolded protein. We will not be sure about these ideas until the theoreticians gain a better understanding of forces stabilizing proteins, and this will require that they understand water and the interactions of water with the constituent groups of a protein (Kortemme et al., 2003).

Finally, how will the properties of water influence the burial of a charge in the protein interior? One important factor is the Born self-energy. Transferring a charge from a medium with a high dielectric constant such as water to a medium with a low dielectric constant such as the protein interior is unfavorable. Since the interior of a protein is heterogeneous, the dielectric constant surrounding a given charge will vary. Garcia-Moreno's group has shown that burying a carboxyl group in the hydrophobic core of staphylococcal nuclease raises the pK from 4 to 9 and makes a very unfavorable contribution to the stability (Fitch et al., 2002). However, the temperature is important. For water, the dielectric constant is 88 at 0°C, but falls to 55 at 100°C. Thus, the cost of burying a charge will decrease at higher temperatures, but the stability gained from the coulombic interactions between charges will increase. The strengthening of charge–charge interactions at higher temperatures may be the most important strategy used by thermophillic organisms to stabilize their proteins to function at higher temperatures (Pace, 2000).

Makhatadze and Privalov (1995) wrote: "... water is a mysterious liquid. All attempts regarding its quantitative theoretical description have so far failed. As for hydration effects, they are doubly mysterious." We have made some progress since 1995 (Ren and Ponder, 2004), but we still do not have a good understanding of water and its interactions with other molecules. Tanford (1980) wrote: "Life as we know it originated in water and could not exist in its absence." A symposium was recently held to consider this possibility (Finney, 2004). For we humans, water is clearly a uniquely important solvent because it is our solvent. With each passing year, we learn more about how incredibly complex we humans are and it seems likely that much of this complexity will be linked one way or another to water, our solvent.

REFERENCES

Anderson, D. E., W. J. Becktel, and F.W. Dahlquist. 1990. pH-induced denaturation of proteins: A single salt bridge contributes 3–5 kcal/mol to the free energy of folding of T4 lysozyme. *Biochemistry* 29:2403–2408.

Baldwin, R. L. 2003. In search of the energetic role of peptide hydrogen bonds. *J Biol Chem* 278: 17581–17588.

Beadle, B. M., and B. K. Shoichet, 2002. Structural bases of stability-function tradeoffs in enzymes. *J Mol Biol* 321:285–296.

Bernal, J. D. 1939. Structure of proteins. *Nature* 143:663–667.

Bowron, D. T. 2004. Structure and interactions in simple solutions. *Philos Trans R Soc Lond B Biol Ser* 359:1167–1179.

Braun-Sand, S., and A. Warshel. 2004. Electrostatics of proteins: principles, models, and applications. In *Protein Folding Handbook*, ed. J. Bucher and T. Kiefhaber. Weinheim: Wiley-VCH Verlag, pp. 163–200.

Chiu, T. K., J. Kubelka, R. Herbst-Irmer, W. A. Eaton, J. Hofrichter, and D. R. Davies. 2005. High-resolution x-ray crystal structures of the villin headpiece subdomain, an ultrafast folding protein. *Proc Natl Acad Sci USA* 102:7517–7522.

D'Aquino, J. A., J. Gomez, V. J. Hilser, K. H. Lee, L. M. Amzel, and E. Friere. 1996. The magnitude of the backbone conformational entropy change in protein folding. *Proteins: Struct Func Gen* 25:143–156.

Deechongkit, S., H. Nguyen, E. T. Powers, P. E. Dawson, M. Gruebele, and J. W. Kelly. 2004. Context-dependent contributions of backbone hydrogen bonding to beta-sheet folding energetics. *Nature* 430:101–105.

Dill, K. A. 1990. Dominant forces in protein folding. *Biochemistry* 29:7133–7155.

Eicken, C., V. Sharma, T. Klabunde, M. B. Lawrenz, J. M. Hardham, S. J. Norris, and J. C. Sacchettini. 2002. Crystal structure of Lyme disease variable surface antigen VlsE of *Borrelia burgdorferi*. *J Biol Chem* 277:21691–21696.

Fersht, A. R. 1987. The hydrogen bond in molecular recogition. *Trends Biochem Sci* 12:301–304.

Fields, P. A. 2001. Protein function at thermal extremes: Balancing stability and flexibility. *Comp Biochem Physiol A Mol Integr Physiol* 129:417–431.

Finney, J. L. 2004. Water? What's so special about it? *Philos Trans R Soc Lond B Biol Sci* 359:1145–1163; 1163–1145, 1323–1148.

Finney, J. L., D. T. Bowron, R. M. Daniel, P. A. Timmins, and M. A. Roberts. 2003. Molecular and mesoscale structures in hydrophobically driven aqueous solutions. *Biophys Chem* 105:391–409.

Fitch, C. A., D. A. Karp, K. K., Lee, W. E. Stites, L. E. Lattman, and E. B. Garcia-Moreno. 2002. Experimental pK(a) values of buried residues: analysis with continuum methods and role of water penetration. *Biophys J* 82:3289–3304.

Fleming, P. J., and G. D. Rose. 2005. Do all backbone polar groups in proteins form hydrogen bonds? *Protein Sci* 14:1911–1917.

Frank, B. S., D. Vardar, D. A. Buckley, and C. J. McKnight. 2002. The role of aromatic residues in the hydrophobic core of the villin headpiece subdomain. *Protein Sci* 11:680–687.

Giletto, A., and C. N. Pace. 1999. Buried, charged, non-ion–paired aspartic acid 76 contributes favorably to the conformational stability of ribonuclease T1. *Biochemistry* 38:13379–13384.

Grimsley, G. R., K. L. Shaw, L. R. Fee, R. W. Alston, B. M. P. Huyghues-Despointes, R. L. Thurlkill, J. M. Scholtz, and C. N. Pace. 1999. Increasing protein stability by altering long-range coulombic interactions. *Protein Sci* 8:1843–1849.

Hebert, E. J., A. Giletto, J. Sevcik, L. Urbanikova, K. S. Wilson, Z. Dauter, and C. N. Pace. 1998. Contribution of a conserved asparagine to the conformational stability of ribonucleases Sa, Ba, and T1. *Biochemistry* 37:16192–16200.

Honig, B., and F. E. Cohen. 1996. Adding backbone to protein folding: Why proteins are polypeptides. *Fold Des* 1:R17–R20.

Huang, C., and D. Chandler. 2002. The hydrophobic effect and the influence of solute-solvent attractions. *J Phys Chem B* 106:2047–2053.

Jain, R. K., and R. Ranganathan. 2004. Local complexity of amino acid interactions in a protein core. *Proc Natl Acad Sci USA* 101:111–116.

Jones, K., and P. Wittung-Stafshede. 2003. The largest protein observed to fold by two-state kinetic mechanism does not obey contact-order correlation. *J Am Chem Soc* 125:9606–9607.

Kajander, T., P. C. Kahn, S. H. Passila, D. C. Cohen, L. Lehtio, W. Adolfsen, J. Warwicker, U. Schell, U., and A. Goldman. 2000. Buried charged surface in proteins. *Struct Fold Des* 8:1203–1214.

Kauzmann, W. 1959. Some factors in the interpretation of protein denaturation. *Adv Protein Chem* 14:1–63.

Kortemme, T., A. V. Morozov, and D. Baker. 2003. An orientation-dependent hydrogen bonding potential improves prediction of specificity and structure for proteins and protein-protein complexes. *J Mol Biol* 326:1239–1259.

Kubelka, J., W. A. Eaton, and J. Hofrichter. 2003. Experimental tests of villin subdomain folding simulations. *J Mol Biol* 329:625–630.

Laurents, D. V., B. M. Huyghues-Despointes, M. Bruix, R. L. Thurlkill, D. Schell, S. Newsom, G. R. Grimsley, K. L. Shaw, S. Trevino, M. Rico, et al. 2003. Charge-charge interactions are key determinants of the pK values of ionizable groups in ribonuclease Sa (pI=3.5) and a basic variant (pI=10.2). *J Mol Biol* 325:1077–1092.

Lesser, G. J., and G. D. Rose. 1990. Hydrophobicity of amino acid subgroups in proteins. *Proteins: Struct Func Gen* 8:6–13.

Makhatadze, G. I., and P. L. Privalov. 1995. Energetics of protein structure. *Adv Protein Chem* 47:307–425.

McDonald, I. K., and J. M. Thornton. 1994. Satisfying hydrogen bonding in globular proteins. *J Mol Biol* 238:777–793.

McKnight, C. J., D. S. Doering, P. T. Matsudaira, and P. S. Kim. 1996. A thermostable 35-residue subdomain within villin headpiece. *J Mol Biol* 260:126–134.

McKnight, C. J., P. T. Matsudaira, and P. S. Kim. 1997. NMR structure of the 35-residue villin headpiece subdomain. *Nat Struct Biol* 4:180–184.

Myers, J. K., and C. N. Pace. 1996. Hydrogen bonding stabilizes globular proteins. *Biophys J* 71:2033–2039.

Pace, C. N. 1990. Conformational stability of globular proteins. *Trends Biochem. Sci.* 15:14–17.

Pace, C. N. 1995. Evaluating contribution of hydrogen bonding and hydrophobic bonding to protein folding. *Methods Enzymol* 259:538–554.

Pace, C. N. 2000. Single surface stabilizer. *Nat Struct Biol* 7:345–346.

Pace, C. N. 2001. Polar group burial contributes more to protein stability than nonpolar group burial. *Biochemistry* 40:310–313.

Pace, C. N., E. J. Hebert, K. L. Shaw, D. Schell, V. Both, D. Krajcikova, J. Sevcik, K. S. Wilson, Z. Dauter, R. W. Hartley, et al. 1998. Conformational stability and thermodynamics of folding of ribonucleases Sa, Sa2 and Sa3. *J Mol Biol* 279:271–286.

Pace, C. N., G. Horn, E. J. Hebert, J. Bechert, K. Shaw, L. Urbanikova, J. M. Scholtz, and J. Sevcik. 2001. Tyrosine hydrogen bonds make a large contribution to protein stability. *J Mol Biol* 312:393–404.

Pace, C. N., B. A. Shirley, M. McNutt, and K. Gajiwala. 1996. Forces contributing to the conformation stability of proteins. *FASEB J* 76:75–83.

Pace, C. N., S. Trevino, E. Prabhakaran, and J. M. Scholtz. 2004. Protein structure, stability and solubility in water and other solvents. *Philos Trans R Soc Lond B Biol Sci* 359:1225–1234.

Petsko, G. A., and D. Ringe. 2004. *Protein Structure and Function*. Abingdon: New Science Press, 195.

Pettersen, E. F., T. D. Goddard, C. C. Huang, G. S. Couch, D. M. Greenblatt, E. C. Meng, and T. E. Ferrin. 2004. UCSF Chimera—A visualization system for exploratory research and analysis. *J Comput Chem* 25:1605–1612.

Privalov, P. L., and G. I. Makhatadze. 1993. Contribution of hydration to protein folding thermodynamics: II. The entropy and Gibbs energy of hydration. *J Mol Biol* 232:660–679.

Radzicka, A., L. Pederson, and R. Wolfenden. 1988. Influences of solvent water on protein folding: Free energies of solvation of *cis* and *trans* peptides are nearly identical. *Biochemistry* 27:4538–4541.

Ren, P. Y., and J. W. Ponder. 2004. Temperature and pressure dependence of the AMOEBA water model. *J Phys Chem B* 108:13427–13437.

Ripoll, D. R., J. A. Vila, and H. A. Scheraga. 2004. Folding of the villin headpiece subdomain from random structures. Analysis of the charge distribution as a function of pH. *J Mol Biol* 339:915–925.

Savage, H. J., C. J. Elliot, C. M. Freeman, and J. L. Finney. 1993. Lost hydrogen bonds and buried surface area: Rationalising stability in globular proteins. *J Chem Soc Faraday Trans* 89:2609–2617.

Spolar, R. S., and M. T. J. Record. 1994. Coupling of local folding to site-specific binding of proteins to DNA. *Science* 263:777–784.

Stickle, D. F., L. G. Presta, K. A. Dill, and G. D. Rose. 1992. Hydrogen bonding in globular proteins. *J Mol Biol* 226:1143–1159.

Takano, K., J. M. Scholtz, J. C. Sacchettini, and C. N. Pace. 2003. The contribution of polar group burial to protein stability is strongly context-dependent. *J Biol Chem* 278:31790–31795.

Tanford, C. 1962. Contribution of hydrophobic interactions to the stability of the globular conformation. *J Am Chem Soc* 84:4240–4247.

Tanford, C. 1980. *The Hydrophobic Effect*. New York, NY: John Wiley and Sons.

Trevino, S. R., K. Gokulan, S. Newsom, R. L. Thurlkill, K. L. Shaw, V. A. Mitkevich, A. A. Makarov, J. C. Sacchettini, J. M. Scholtz, and C. N. Pace. 2005. Asp79 makes a large, unfavorable contribution to the stability of RNase Sa. *J Mol Biol* 354:967–978.

Wolfenden, R. 1978. Interaction of the peptide bond with solvent water: A vapor phase analysis. *Biochemistry* 17:201–204.

Zhang, J. R., J. M. Hardham, A. G. Barbour, and S. J. Norris. 1997. Antigenic variation in Lyme disease borreliae by promiscuous recombination of VMP-like sequence cassettes. *Cell* 89:275–285.

13 Water and Information

Thomas C. B. McLeish

CONTENTS

13.1 INTRODUCTION

This chapter has two aims. The first is to raise an issue that underpins complex structure and function at all length scales and that is especially relevant to the origin and development of life. The second is to articulate potential pitfalls of projects such as this one in the context of the special physical features of water that this and other contributions to this volume are tackling. These dangers are, I think, the cause of an uneasiness that rises quite generally in the background of discussions about special structure, optimization, and design.

The physical aspect I would like to urge for discussion is that of the embodiment of information in matter and, in particular, in aqueous matter. It has long been understood that the development and maintenance of order in the material world carries as much significance as the existence of matter itself. The early Greek developers of the idea of *logos,* as much as modern considerations of statistical mechanics in cosmology, suggest that the creation of matter itself is not as challenging or as implausible an idea as the creation of *ordered* matter (or alternatively information-containing matter). We are accustomed to thinking of particular examples of designed matter such as computers or microwave links as processors of information, and yet it may seem strange to generalize this property to all matter. Yet this is essentially where the science of thermodynamics directs the way we think about complex matter. Early in the nineteenth century came the "top down" identification by Clausius and others of a quantity created from the heat flow into and out of a system and the temperature at which it flowed, called the "entropy." The total entropy may increase or be kept constant but not decrease, endowing it with the same character of "nonconservation" as information, which may be transferred or lost, but not created. The upward march of total entropy is enshrined in the second law of thermodynamics. As the properties of heat and matter were tackled from the bottom up, atomic point of view later in the century by Boltzmann, Gibbs, and Maxwell, so molecular interpretations of the thermodynamic quantities clarified. In the same way that heat now became the random motion of the particulate constituents of matter, so the internal structure of entropy became visible as the disorder of a system. So, a gas has greater entropy when it has greater volume because there is more space for its molecules to move in: they can access a greater number of states. Compressing the gas decreases its entropy because fewer configurations are available to the molecules. The tendency of a system together with its environment to maximize its disorder and, therefore, entropy, became the key to understanding the commonplace but deeply mysterious changes of state exhibited by matter, such as freezing. Atoms or molecules in a crystalline array are much more ordered than in a liquid, but if by freezing they are able to increase the entropy of their environment by more than the entropy they lose by adopting the crystalline pattern, then this process becomes spontaneous. Delivering energy to that environment in the form of heat given up on freezing is a way of doing exactly this—and at the freezing temperature the entropy budgets of liquid and environment just balance.

If the loss of entropy within a system can be read as loss of disorder, then it is a short conceptual step to see it as equivalently a gain of information. That it is possible to quantify this is the result of the program by Shannon (1948) and others: matter embodies information, and statistical mechanics can be read as a theory of information transfer and loss. The disorder of entropy just becomes negative information and the insight that information can be passed on, degraded but not invented, is a restatement of the second law. Information flow raises challenges for cosmological accounts of the universe as a whole: the very high order, or low entropy, that characterized the early universe asks for an explanation. The operation of the second law of thermodynamics requires a continuous global loss of information (at the classical level at least; information loss in quantum mechanics is linked to the act of observation). The current structure of the world is the consequence of information processing and loss at a global level. The remarkable, and highly significant, local increase in order pertinent to the formation of complex structures and the operation of life, is made possible at the expense of vastly greater information loss at the global level.

The centrality of information processing in the physical world appears at all levels and is by no means confined to the largest length scales. The most striking example is found in life itself: developed forms of matter that are not only able to represent information but also to process it. It is often loosely explained that metabolism constitutes a consumption of energy, yet the energy sum itself is zero. All of the chemical energy consumed by an animal is passed on as other forms, including heat, to its environment. Eating is more accurately a consumption of information: low entropy intake is converted to high entropy output of matter and heat, while a small fraction of the entropy gain is moderated by the synthesis of new tissue and biomolecules. The formation of new lipids and their membranes, the transport of organelles, the constriction of muscle in higher organisms as much of the growth and retraction of microtubules in single cells are all driven by thermodynamic forces, or in other words, by the flow of information. The remarkable properties of biological matter to process information follow in part from the organic chemistry of carbon, oxygen, and nitrogen. It requires a range of interactions both bonded and nonbonded to sustain a wealth of molecules that are stable enough to survive their thermal environment, but not so stable that they may not be broken down by enzymes and reconstituted. Yet all biochemistry operates in an aqueous medium. One of the roles of water is to provide an environment in which subtly differentiated forms of information flow may exist. To what extent does water require a special structure to constitute a theater in which a rich structure of information processing may exist? Does it rely on the unusual properties of water as a liquid, or is there a broad parameter-space of molecular properties that would sustain something like a *systems biology*?

It is worth reiterating that this discussion should by no means be confined to the operation of brains and nervous systems. Examples of advanced information processing in molecular biology abound, long before one enters the world of synapses, neuronal networks, and brains. Advanced signaling and decision-making abounds in protein and gene-regulatory networks within single-celled life. Bacterial chemotaxis furnishes a classic example that we will now turn to in more detail (Bray, 2003). Binding of external stimulants to receptors on transmembrane proteins modify methylation and phosphorylation of sites within the cytosol. This in turn affects the charge state of groups of mobile proteins (CheA, CheR, CheB, CheX, and CheY) that shift the balance of driven cyclical biochemical reactions responsible for intermittency of molecular motors. The special and delicate environment of water permits and informs the information flow in at least four ways in this scenario: First, water endows proteins with the possibility to act as logic gates, switches, and transmitters of information about binding to states of catalysis. In particular the hydrophobic interaction permits entropy exchange between regions of solvent and protein central to the delicate process of the folding of constituent proteins in the first place. Second, the ability to dissolve salts of different valency and concentration turns electrostatics from a strong, nonlocal, and fixed interaction into a weak (comparable to thermal energy scales), local, and tuneable one. Third, it provides the diffusive transport medium in which the signaling proteins diffuse from receptors to motors. The same hydrogen-bonded structure that sets the environment for energetic and entropic contributions to binding events also sets the timescales for diffusive searches for substrates, and so the kinetic constraints on response times of the information processing network. Fourth, perhaps trivially, it constitutes the prior for the entire data structure of the system (it is all about a micron-sized creature swimming to a purpose through water).

The first of these water-based information processes bears more attention in detail, including a little coarse-grained statistical mechanics. At the heart of the protein biochemistry of information processing is the phenomenon of *allostery* (Monod et al., 1965). The term describes the class of controlled-binding events in which the binding of a protein to a substrate B depends on whether the same protein is also bound to another substrate A, frequently at a site spatially separated from the binding site for B. Classic examples are furnished by the prokaryotic gene repressor proteins that control gene expression by binding to DNA conditionally on the presence of other coeffectors in the cytosol. The bacterial chemotaxis transmembrane proteins are also allosteric, transferring information on binding events exterior to the cell membrane many tens of nanometers to the interior of the cell (Kim et al., 1999). Theories of allosteric response were originally dominated by the idea of switching between alternative static structures (Monod et al., 1965; Koshland et al., 1966), but more recently the more subtle role of the internal dynamics of allosteric proteins has been recognized as essential to allosteric function (Weber, 1972; Hawkins and McLeish, 2004). Remarkably, modifications to the binding interface between monomers in a dimeric protein, or between protein domains, induced by the binding of an effector molecule, can alter the amplitudes of slow, global thermal motions of the entire protein in such a way that the equilibrium binding constant at distant sites can be modified significantly. So thermally excited global modes of the protein become *channels* for information flow, even though their dynamics is itself in thermal equilibrium, and therefore represents no information content. A very simple, toy model of a dimeric allosteric protein serves to make the point, and to calculate the order of magnitude of allosteric free energy that this mechanism may, in principle, supply. We suppose that two domains pivot around a central, strong binding region, and that other regions at the extremities possess mutual interactions that depend on the binding of their respective substrates (see Figure 13.1). On binding substrate A, we assume, within a harmonic approximation, that the effective stiffness of the attractive region between the protein

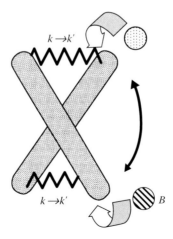

FIGURE 13.1 A dimeric allosteric protein is modelled as two rodlike regions pivoting around a central point, with binding regions at their extremities for substrates A and B. On binding, the effective spring constants of the attractive interactions between the protein domains change.

domains local to the binding site changes from k to k'. For simplicity, we assume the same local change on binding B at the distant binding site.

The quality measure for allosteric signaling is the difference in the free-energy changes of binding B for the cases when A is bound and unbound, $\Delta\Delta F$. In this very simple case, the free energy is calculated from the entropy of the mode corresponding to the mutual rotation of the domains (indicated by the double-headed arrow in the figure):

$$F = -kT \ln \left\{ \int_{-\infty}^{\infty} e^{-(k_A + k_B)x^2/2kT} \, \mathrm{d}x \right\}. \tag{13.1}$$

Performing the Gaussian integrals in the cases of both substrates bound, just one, and neither, then gives the allosteric free energy as a simple function of the dimensionless change in the effective spring constant on binding $\tilde{k} = k'/k$:

$$\Delta\Delta F = kT \ln \left(\frac{4\tilde{k}}{\left(1 + \tilde{k}^2\right)} \right). \tag{13.2}$$

Note that this allosteric signal is entirely entropic in character (it is proportional to the absolute temperature T). Typically, at this coarse-grained level, the order of magnitude of the free energy signal will be no more than a few multiples of the thermal energy unit kT. However, this can be amplified very significantly if the local vibrational modes of side-groups on the surfaces of the dynamic domains are coupled to the global mode itself. This in turn is possible if hydrophobic patches and groups on the surfaces of the domains and side groups interact. When the global mode is far from its mean position, the environment for local side chains is less native-like, and their local motion will be less constrained than when the global mode brings the two principal domains close to their mean position. In this case, the side chains are generically more likely to be presented with hydrophobic or electrostatic interactions that can be satisfied by binding tightly to the surfaces of the principal domains. In so doing, their configurational entropy is reduced. Although this switching of side groups from high entropy to low entropy states is an entirely local process, so not initially a candidate for a contribution to the global signaling entropy, it does contribute to $\Delta\Delta F$ by virtue of their "enslaving" to the global mode. It is possible to calculate the quantitative contribution exactly within various models of the coupling interactions, but the generic result for the new allosteric free energy, when N local modes are enslaved to the global one, takes the form

$$\Delta\Delta F = kT \ln \left(\frac{4\tilde{k}}{\left(1 + \tilde{k}^2\right)} \right) \gamma N, \tag{13.3}$$

for some dimensionless coupling constant, γ, that depends on the details of the interactions.* This rather subtle dynamic effect within such proteins allows the "signal strength" for information on binding to be amplified by the (large) factor N. A good example of an allosteric protein of this kind (although in this case containing large and almost compensating enthalpic and entropic contributions to the allosteric free energy), able to transmit signals of many times the thermal energy yet without conformational change, is the *met* repressor (Cooper et al., 1994). There is considerable evidence that signal amplification also occurs in the allosteric transmembrane signaling of the

---

* I am indebted to Dr. Rhoda Hawkins for this result.

bacterial chemotaxis receptor cluster (Kim et al., 1999). In this case, an additional level of interprotein interaction seems to be able to couple the internal state and dynamics of the transmembrane proteins with their assembly into clusters within the membrane.

Nothing explicit concerning the properties of the solvent of the protein and substrates appears in this coarse-grained model for allosteric information flow; however, a great deal has been implied. Systems such as this are only possible in environments in which the energetic interactions involved in the signaling are comparable to the thermal energy and somewhat stronger, yet coexist with much more stable interactions within the protein domains themselves. Screened electrostatics, or hydrophobic interactions, are both candidates for the essential weak, tuneable interactions at the binding regions and between the principle domains and the side groups. Both of these are essential properties of liquid water at the nanoscale. An example of the explicit structural dynamics of water in the allosteric binding of oxygen in hemoglobin has been pointed out by Ball in this volume. Although collapse-transitions of nonaqueous peptoids have been engineered in nonaqueous solvents (Ball, this volume), it is not clear that alternative solvents present the range of sensitivity of interactions between groups of different chemistry that nonetheless remain in solution.

At a smaller scale, strong information processing occurs at the level of single biomolecules. Let us extend some of these ideas to the first information processing role we identified for water in the chemotaxis example, that of the folding of functional proteins in the first place. Protein folding constitutes an extremely rich case of information flow. It has increasingly been realized that the primary sequence actually contains two libraries of data: one determines the native folded state itself, but equally important is the information that guides the folding toward the native state. The analogy of a funneled landscape has been invoked to explain how the physical interactions (not necessarily just the "native" ones) between a protein's residues constrain its internal Brownian motion to a set of pathways that represent a tiny fraction of the possible configurational searches that it might conceivably explore (Dinner et al., 2000). The structured energy landscape that guides a folding protein away from unfruitful configurations has been termed the "folding funnel," although the space in which the energy funnel exists is actually of very high dimension indeed, corresponding to the very large number of degrees of freedom that must simultaneously find their native values in the folded state. The funnel picture can be misleading: it only arises when the very high dimensionality of the true configurational space of a polymer is projected onto a very low dimensional subspace (in analogy with the projection within statistical mechanics that produces the free energy from the energy). The full structure of the search requires the implicit presence of an "Ariadne's thread" of information that guides the molecule into lower and lower subspaces, within which the random search is contained. The field of candidate coarse-grained structures of this very high-dimensional funneled landscape has recently been extended: energetically favorable and sequential formation of native-like regions is one route to the successive reduction in the dimensionality of the random search through configuration space that is the only general way of overcoming Levinthal's paradox. But there are many more ways of doing this if the same type of coarse-grained global modes are appealed to, as they were above in the case of dynamic allostery (McLeish, 2005a). Carefully tailored hydrophobic interactions between partially folded domains may restrict searches to those combinations of the active degrees of freedom that satisfy accessible low-energy states. The dimensional reduction is a very remarkable design feature of the sequence: to attain reasonable folding rates, it is essential that no more than about 10 degrees of freedom are ever required to engage in a simultaneous random search. This, in turn, implies that the search path through the several hundred initial degrees of freedom of even a moderately sized protein must be implicitly ordered, by the set of attractive and repulsive interactions responsible for folding, in a highly structured way. When the protein is partially folded or collapsed into a molten globule state, these interactions may be nonnative, in that actively attracting residues at a key stage in the folding process may not interact strongly or at all in the native state. In this way, the protein uses a potentially much richer store of information than is present in the native contacts alone. For a protein of N amino acids, there are $O(N)$ native interactions that may be engineered to contain the information on the structure of the native state. It is possible that these interactions may also contain information on the folding pathway, but the $O(N^2)$ nonnative interactions represent an independent data set in which information on the folding pathway may be stored. Evidence from both experiments (Capaldi et al., 2002) and numerical simulations (Paci et al., 2002) that nonnative interactions are important in defining the folding funnel is increasing. But if this is so, then it also implies that the very subtle range of weak, local interactions between biomolecules in an aqueous medium that we identified before as playing an essential role in information flow of signaling proteins may play a similar part in transmitting the information on the folding space stored within the primary sequence to the dynamical search of folding proteins themselves.

Whether the navigation of the complex folding space of proteins may be achieved from the deposit of information stored in the native interactions alone, or by processing information contained in the nonnative interactions, there is a quite generic aspect of water that greatly reduces the change in enthalpy that a naive estimate would make of the folding process. Furthermore, this can be interpreted directly as a type of information exchange between the solvent and the protein. It emerges from the highly solvent-mediated changes in entropy during folding. From the point of view of information theory, finding the single native state of a protein of N residues is equivalent to the transmission of qN bits of information to the molecule, where q is a dimensionless number of order unity. The entropy reduction of the protein is therefore $\Delta S_{\text{pro}} \sim Nk_{\text{B}}$, where k_{B} is Boltzmann's constant. Given this enormous restriction of configurational entropy implicit in the folding of a protein, one might expect the folding itself to be a highly exothermic process in general (i.e., requiring an enthalpy of order $Nk_{\text{B}}T$ at the point at which the folded and unfolded states are in equilibrium). The fact that it is not reflects another remarkable property of the aqueous solvent: it is a potential repository of information. There is an entropy change associated with the water in the environment of the protein so that the total free energy change involves a moderated entropic contribution:

$$\Delta G_{\text{tot}} = \Delta H_{\text{pro}} - T(\Delta S_{\text{water}} + \Delta S_{\text{pro}}). \tag{13.4}$$

It is essential to be clear that this does not invoke any particular model for the many possible cases of the hydrophobic interaction (see Ball, this volume, for a succinct survey). The structured-cage concept of Kauzmann (1969) is just one way in which the entropy content of water might be modified. The more recently simulated restriction of the orientational distribution of water dipoles parallel to hydrophobic surfaces is another (Blokzijl and Engberts, 1993). But independent of mechanism, the temperature dependence of hydrophobic effects indicates that the burial of hydrophobic surface on protein folding is associated with an increase in the entropy of the solvent, so that the entropic terms in Equation (13.4) generally compensate. The ability of water to exchange entropy with a folding protein is one of its more astonishing properties. It contributes in no small way to the headaches of modelers that would simulate the folding of proteins numerically, as the solvent entropy is held nonlocally, whether the source resides in the degrees of freedom explored by the hydrogen-bonded network, or by a more subtle restriction of the orientational order parameters in the vicinity of hydrophobic groups on the protein.

The current state of uncertainty surrounding the hydrophobic interaction suggests that it involves more, rather than less, content than that of current models. It constitutes an effective potential between elements of partially formed protein structure that has finite (but not trivial) range, depends in a detailed way on the surface structure, effective size and configuration of those elements, and their relative orientation, and is modified by the presence of other elements. These properties are recruited in supporting the flow of information in processes of self-assembly within the cytosol in general; protein-folding is just one example.

A third intriguing arena for information processing is the lipid membrane. Poon (this volume) points out that the nontrivial existence of a liquid phase has, as one consequence, the natural proliferation of interfaces of all kinds. Water has the ability to sustain these flexible and rich structures that crucially partition (from the point of view of living matter) its own volume. But, again, it is instructive to focus on the role of water in the information processing of the membranes. We have met one implicit example already in the transmembrane signaling of chemotaxis. It is worth drawing attention to the many unresolved mysteries of this phenomenon in general. Single helical domains of transmembrane proteins are capable of allosteric signaling from one side of the membrane to the other. It is hard, if not impossible, to conceive of mechanisms for this process in so simple a protein structure that do not engage the membrane itself, and the surrounding solvent, in the description and action of the switchable state. Examples of transmembrane allostery that are already understood include complexation of proteins within the membrane (e.g., external binding induces dimerization; it is only the protein dimers that catalyze a reaction in the cell interior). As in the case of DNA-binding proteins, there is a growing realization that allosteric information may be transmitted through membranes dynamically (Hawkins and McLeish, 2004). The common coiled-coil motif may be exploited in transmembrane signaling for this reason: thermal fluctuations in the mutual motion of the two coils may be moderation by binding events at either the extracellular or intracellular extremity, so in turn affecting binding constants nonlocally.

The delicate balance that provides a membrane robust enough to constitute a permanent barrier to ions, proteins, and peptides, yet fluid enough to support internal and transmembrane transport, is a remarkable but necessary prerequisite to its emergent property as another arena for information processing and storage. Membrane information content is now identifiably not simply resident in the type and configuration of the proteins it hosts, but in the subtle variation in the local lipid composition (out of ~10^2) species, typically. Its internal degrees of freedom and their resulting entropy constitute a third channel for information. These properties rely on the same structuring and information exchange property of water that we identified in the context of protein folding, since the lipid–lipid interactions that determine local concentrations are highly mediated by the water immediately adjacent to the membrane.

Poon (this volume) raises the question of evolvability in the context of interfaces. Naturally, the issue of information storage and flow arises in this discussion as much as in the functions of metabolism. There is one aspect of information processing that strongly suggests advantages in an interfacial context for the origin of life: the advantage of low dimensionality. We have already appealed to the need for strongly guided searches if the dimension of the search-space is large and the underlying dynamics thermally driven, as in the case of protein folding. Diffusive searches are, however, ubiquitous in molecular biology, determining the rates of enzyme–substrate binding, DNA-binding for control of gene expression, and the self-assembly of cytoskeletal components such as microtubules and filamentous actin. Max Delbrück pointed out long ago (Adam and Delbrück, 1968) that the rates for high-dimensional searches are qualitatively slower than for low dimension, and that for molecular biology to work (to process information) situating biochemistry in an effective low-dimensional space is necessary. The relevance for membrane-bound biology is that the boundary between high-dimensional and low-dimensional, in this case, occurs at $d = 2$. To make this clearer we recall the mean search times τ_d (or reaction times) for a diffusing reactant of diffusion constant D to find and bind to a reactive site of size r within a d-dimensional space of size R.

$$\tau_1 = \frac{4R^2}{\pi D}$$

$$\tau_2 = \frac{R^2}{2D}\left[\ln\left(\frac{R}{r}\right) - \frac{1}{2}\right] + \frac{r^2}{D}. \tag{13.5}$$

$$\tau_d = \frac{R^2}{D}\frac{1}{d(d-2)}\left(\frac{R}{r}\right)^{(d-2)} \quad (d \geq 3)$$

The low-dimensional search time is controlled only by the size of the search space in $d = 1$, with only logarithmic correction in $d = 2$. But for all higher dimensions the ratio (R/r) appears (when d becomes very much larger than one, the third expression of Equation (13.5) embodies the Levinthal paradox). Especially in the case of early life, when strategies for effective dimensional reduction within three (and higher) dimensional worlds have yet to evolve, the naturally fast search times in two dimensions offer a compact environment within which random processes can sample intermolecular interactions much more rapidly than in bulk solutions.

The counterfactual issues here are rich: liquid hydrocarbons are as capable of supporting flexible self-assembled surfactant membranes (naturally inverted) as water, so the issue of two-dimensional structures that support diffusion is not dependent on the special properties of water. However, several features of membrane-based molecular information processing do depend on the aqueous, polar phase constituting the external boundary. Control of local curvature, for example, important for transmembrane signaling and for the topological transitions required in endocytosis and budding, requires the electrostatic and steric interactions of the proteins that drive them to operate at the membrane surface rather than its interior. The modification of membrane-bound proteins by sugars similarly exploits the high dielectric and hydrogen-bonding water environment to switch and modify function.

We have considered water-mediated information flow at the levels of intraprotein signaling (allostery) in protein folding, and across membranes, but as the example of chemotaxis illustrates, information as well as structure pervades biology at all length scales. Moving up one level brings us to the protein-signaling networks responsible for processing the information received by clusters of receptor proteins into behavior of the rotary motor responsible for bacterial motion (Bray, 2002). A highly schematic representation of the pathway in *Escherichia coli* is shown in Figure 13.2.

The binding of a repellent (in this case) modifies the phosphorylation of the protein CheA, which, in turn, modulates an active phosphorylation cycle between the proteins CheY and CheZ such that the former binds to the cellular domain of the flagellar motor, causing it to increase the proportion of time it spends in clockwise rotation. This, in turn, induces tumbling, rather than swimming, behavior of the bacterium. If the rate of tumbling is correlated to the rate of change of repellent-binding, it can be shown that the net bacterial motion is indeed down the concentration gradient of the repellent. So, the protein signaling network needs to retain an element of recent memory. This seems

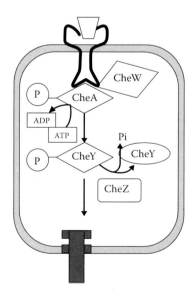

FIGURE 13.2   Schematic representation of a bacterium showing the chemotaxis pathway.

to reside within the transmembrane proteins themselves: the charge state of several side chains on their coiled coils adjusts as a function of the mean background concentration of the affector, either attractant or repellent, so that the signaling to the protein network is always controlled by changes in extracellular binding.

Very complex processing indeed can be achieved by protein networks of this kind. Perhaps the canonical example is the *Stentor* (Tartar, 1961). This single-celled funnel-feeder modulates its strategy for dealing with toxic environments in a highly contingent way, acquiring a detailed memory of its previous actions and their effects. Again, the information is contained and processed within the methylation and acetylation states of proteins and their complexation and enzymatic biochemical reactions. Memory of past environments and reactions is contained within metastable driven states of different lifetimes.

The counterfactual questions at this level are even more challenging than those at the level of single molecule processes. Currently there are several research programs addressing the synthesis of "chemical logic gates" and their effective circuits (Uchiyama et al., 2005). There is no reason to suppose that chemical information processing at the level of networks of reactions and their catalytic control is not possible in nonaqueous solvents. The more promising issue is, therefore, the second necessary condition of evolvability. Are the relatively harsh chemical conditions imposed by water, together with its rich coexistence of the steric, van der Waals, hydrophobic, hydrogen bonding, and electrostatic interactions necessary for the evolution of information processing networks? The aqueous environment certainly provides alternative chemical and self-assembling strategies to provide emergent order, rather than the more restricted chemical possibilities offered by other liquid solvents. Perhaps it is the availability of different options that enhances evolution of complex systems. But until experimental programs on evolving systems address this question, we cannot form any firm conclusions at this higher level of logic networks.

The counterfactual agenda implicit in this project may be more timely now at the more molecular spatial scales we have addressed. What fundamental properties are required of a solvent with the information-containing and processing properties of water? How special is all this? Can one envisage other chemistries of A-proteins and A-lipids that would behave similarly in liquid ammonia? A minimal complexity would seem to be the presence of a polarity tuned in energy scale to the temperature of the environment, and a consequent ability to create partially ordered structures of intermediate range. Can one create an implicit folding code for globular macromolecules that rely on other chemistries of solvent? Can self-assembled membranes in hydrophobic solvents process information in the same or analogous ways as cell membranes?

We have begun to explore possible routes to answering these questions in this chapter, but we conclude with some discussion of what would follow from firm answers to them. Suppose the answer is "no"—that we have discovered another remarkable specificity in the physical structure of the world that goes right to the heart of living matter. The route from information processing matter to an analysis of the "software" of biology may conceivably be the first steps of a route to understanding what meaning we might assign to words like "consciousness" and "soul." I have drawn attention elsewhere (McLeish, 2005b) to the way that the Church Fathers use "soul" to talk about the observing, deducing, and debating entity that we would call "mind" (see Gregory of Nyssa's *On the Soul and the Resurrection* for a clear example that, incidentally, contains a discussion of the special physical properties of water). What we would then have to decide is whether this adds qualitatively (as well as quantitatively) to the sum total fine-tuning of the universe, and the consequences we can draw from it. The extraordinary specificity of the parameters of a universe containing nontrivial structure has already been recorded time and again in professional and public literature. From a skeptical viewpoint, it initially seems difficult to see how more can be added to a purpose other than to build the (increasingly ornate) sandcastle a little higher. At first it does not seem to offer any fresh way out of the impasse of the anthropic principle (in its weak form). Can anything be deduced from the delicate fine-tuning at this, or any other level, given that we are present and attempting the deduction?

Again, the answer may well be, "Nothing": a Bayesian may rightfully point out that this is a debate about priors and their probabilities, which we have no ground to estimate. Yet there is a different flavor to a discussion about such emergent specificity as the properties of water (even the existence of water itself is emergent), as compared to the flatness of space–time or the value of the fine-structure constant. Those parameters must exist as soon as it is possible for them to do so. Water remains contingent, even in worlds where it is possible, at least in the form of lakes and oceans. Its information processing properties require more still to be true in any realization. All the essential auxiliary functions we have discussed that water performs at the four levels of coarse-graining remain completely latent until other molecular species arise to exploit them. This returns us to the central question of evolvability: the highest level of information processing that biology supports. Water waits for life to emerge and talk to itself through the subtle structure of the solvent, even exchanging information with it, but it also constitutes the environment in which life began. We have made very few inroads into a vast ignorance over the question of the origin of life, in spite of considerable efforts spent evaluating different molecular candidates. This discussion suggests that we might do better by looking for sources and conduits of software (information) rather than hardware (molecules) even at the level of origins.

REFERENCES

Adam, G., and M. Delbrück. 1968. Reduction of dimensionality in biological diffusion processes. In *Structural Chemistry and Molecular Biology*, ed. A. Rich and N. Davidson. New York, NY: W. H. Freeman and Co., pp. 198–215.

Blokzijl, W., and J. B. F. N. Engberts. 1993. Hydrophobic effects. Opinions and facts. *Angew. Chem. Int. Ed.* 32:1545–1579.

Bray, D. 2002. Bacterial chemotaxis and the question of gain. *Proc. Natl. Acad. Sci. USA* 99:7–9.

Bray, D. 2003. Molecular networks: the top-down view. *Science* 301:1864–1865.

Capaldi, A. P., C. Kleanthous, and S. E. Radford. 2002. Im7 folding mechanism: Misfolding on a path to the native state. *Nat. Struct. Biol.* 9:209–216.

Cooper, A., A. McAlpine, and P. G. Stockley. 1994. Calorimetric studies of the energetics of protein-DNA interactions in the *Escherichia-coli* methionine repressor (Metj) system. *FEBS Lett.* 348:41–45.

Dinner, A. R., A. Šali, L. J. Smith, C. M. Dobson, and M. Karplus. 2000. Understanding protein folding via free-energy surfaces from theory and experiment. *TIBS*. 25:331–339.

Hawkins, R. J., and T. C. B. McLeish. 2004. Coarse-grained model of entropic allostery. *Phys. Rev. Lett.* 93:098104.

Hawkins, R. J., and T. C. B. McLeish. 2006. Coupling of global and local vibrational modes in dynamic allostery of proteins. *Biophys. J.* 91:2055–2062.

Kim, K. K., H. Yokota, and S. H. Kim. 1999. Four helical-bundle structure of the cytoplasmic domain of a serine chemotaxis receptor. *Nature* 400:787–792.

Kauzmann, W. 1969. Some factors in the interpretation of protein denaturation. *Adv. Protein Chem.* 14:1–63.

Koshland, D., G. Nemethy, and D. Filmer. 1966. Comparison of experimental binding data and theoretical models in proteins containing subunits. *Biochemistry* 5:365–385.

McLeish, T. C. B. 2005a. Protein folding in high-dimensional spaces: Hypergutters and the role of nonnative interactions. *Biophys. J.* 88:172–183.

McLeish, T. C. B. 2005b. Values in science research. In *Values in Higher Education,* ed. S. Robinson and C. Katulushi. Leeds: Aureus Publishing.

Monod, J., J. Wyman, and J.-P. Changeux. 1965. On the nature of allosteric transitions: A plausible model. *J. Mol. Biol.* 12:88.

Paci, E., M. Vendruscolo, and M. Karplus. 2002. Native and non-native interactions along protein folding and unfolding pathways. *Proteins* 47:379–392.

Shannon, C. E. 1948. A mathematical theory of communication. *Bell Syst. Technol. J.* 27: 379–423, 623–656.

Tartar, V. 1961. *The Biology of Stentor.* New York, NY: Pergamon Press.

Uchiyama, S., G. D. McClean, K. Iwai, and A. Prasama de Silva. 2005. Membrane media create small nanospaces for molecular computation. *J. Am. Chem.* 127:8920–8921.

Weber, G. 1972. Ligand binding and internal equilibriums in proteins. *Biochemistry* 11(5):864–878.

14 Counterfactual Biomolecular Physics

Protein Folding and Molecular Recognition in Water and Other Fluid Environments

Peter G. Wolynes

CONTENTS

14.1 INTRODUCTION

Counterfactual biochemistry, in more recent times, has captured many youthful imaginations. Silicon-based life forms have been a staple of science fiction. In one form, the computer network, silicon-based life now seems to be on the verge of becoming a reality. Naturally occurring life, on other worlds, may have biochemistry similar to our own, but it would be foolhardy to rule out alternatives. Life, as we see it, on this planet seems to use chemical mechanisms in a manner quite distinct from the now emerging silicon-based life forms. It is very difficult to see how the elegantly efficient large-scale deterministic mechanisms of electronic life could have arisen without preexisting intelligences. Natural terrestrial life, as befits its molecular nature, in contrast exhibits elements of haphazardness and chance. Given modern DNA sequence analysis, to doubt the apparent role of chance in the evolution of species would be as contrarian and insensitive to masses of observations. Nevertheless the precise way terrestrial life arose in the first place must be counted a contemporary mystery. What if things had been different? What if key molecular ingredients of modern living things were unavailable, would life have been impossible? Can we, ourselves, create synthetic life with only marginal design input?

Modern life forms are largely made up of water. Under unusual conditions, living things can persist with much lower water contents (spores, hibernating insects) but they do not do much in these dry states. The physicochemical peculiarities of water feature in many biophysical mechanisms: water's properties are crucial in the self-organization processes of nucleic acids and proteins to form precisely fitting shapes. Without spontaneous self-organization into three-dimensional structures (called folding) terrestrial-style molecular life is impossible. Is the existence of a milieu for life a problem for alternative forms of molecular life on other worlds? I will first review astrophysically possible fluid milieu. I will then review what we know about the minimal requirements for folding, and what is known about the role of the terrestrial fluid, water, in the folding processes. These surveys will reveal that water's specialness is probably not an indispensable prerequisite for folding. Yet, so far, no alternative milieu has been demonstrated to be a complete alternative. Progress toward nonaqueous based self-assembly is being made and achieving it may actually be of some practical value.

14.2   ASTROPHYSICAL ENVIRONMENTS FOR FLUID BIOCHEMISTRY

An astronomer or nuclear physicist would not find the ubiquity of water in biochemistry a total surprise. Our understanding of the origin of the elements is consistent with their observed abundances (Suess and Urey, 1956). Clearly, hydrogen (the most abundant element) and oxygen (the third most abundant) are major candidates for chemical combination. The second most abundant element, helium, doesn't enter into chemical combinations. Hydrogen peroxide is rather unstable (Cotton et al., 1988). Water, H$_2$O, must be present on planets (Williams and da Silva, 1996).

Under some thermodynamic conditions, hydrogen oxide is fluid. Fluidity seems essential for active life. Completely solid-phase life, if it exists, must be very slow in its actions. Modern silicon-based entities that might eventually become living would rely on the fluidity of their electrons to function: the molecular structures of silicon devices are quite solid.

Other fluids, however, may be planetologically plausible milieu for life. Elemental hydrogen itself is fluid at low temperatures and, as a plasma, is fluid at high temperatures. It exists on the larger planets. In addition to H$_2$O, other hydrides such as CH$_4$ and higher hydrocarbons, H$_2$S, NH$_3$, and other oxides, CO$_2$, SO$_2$, etc., have been found on planets and moons in our solar system. Finally, molten silicates or metals are candidate media for life that may occur in known planetological settings, as are also molten sulfur and sulfuric acid. Most other liquids familiar to us (ethanol, methanol, benzene, etc.) are usually metastable with respect to simple combinations and are unlikely to be found in sufficient amounts on planets.

Which of these available liquids are compatible with the existence of complex molecules that can maintain their integrity sufficiently to bear information and yet self-organize? The most honest answer is, "we don't know." But some media seem to present fundamental problems from a physical point of view: for molecular life, we need a fluid where chemical bonds are stable. Plasma hydrogen-based life forms, if they were to exist (Hoyle, 1957), would have to rely on some as yet unobserved self-organization of electric currents—they cannot take advantage of the quantum mechanical-based mechanisms of structural integrity provided by the covalent bond. On similar grounds, liquid metals are tricky growth media for life. By the way, stellar life would face similar problems—both in hot, ordinary stars and in the cores of neutron stars where we have a neutron fluid. The crusts of neutron stars, on the other hand, have been hypothesized to have in them complex, relatively stable, organized structures (resembling in many ways biomembranes) (Lorenz, 1993) and are not so easily dismissed as homes for lifelike entities (Hoyle, 1957). The still-high temperatures of the somewhat cooler but still hot liquids like sulfur and silicate likewise strongly restrict the chemical possibilities for long-lived information-bearing molecules.

Very cold matter is more promising. The low temperature of liquid hydrogen presents no problems for stable molecules. Sure, things slow down in a cryogenic environment but even complex chemical reactions can occur at ultralow temperatures by tunneling. The big barrier here to life would seem to be solubility: any complex molecule is likely to interact with itself by van der Waals forces whose magnitude greatly exceeds the thermal energies in liquid hydrogen. Hydrogen, not being a very polarizable entity, will poorly compensate for these van der Waals forces through intermolecular interaction. Supercritical molecular hydrogen, on the other hand, is not so easily discarded as a possibility. Fluid water, supercritical hydrogen, the oxides, and hydrides are then compatible with dissolved information-bearing molecules. As media, are they compatible with self-organization and folding? Our knowledge of protein folding can help us sharpen the discussion.

14.3 FOLDING—WHAT IS REQUIRED?

Not all information-bearing molecules can self-organize into three-dimensional structures (Wolynes, 1995). In fact, our current understanding of the statistical mechanics of folding suggests that only a small fraction of polymer molecules have foldability. Before one leaps to invoke intelligent design as the origin of foldable molecules, one should be aware that theory suggests there are rather rapid selection routes to finding foldable sequences. The problem of finding foldable sequences has been called Hoyle's paradox, as he worried about the problem of life's origin in the finite time allowed by a Big Bang universe.

What are the requirements for a foldable sequence? The thermodynamic and kinetic aspects of this question have been addressed by the energy landscape theory of protein folding (Bryngelson et al., 1995). The sequence must possess a funneled energy landscape. By this, we mean that, out of the myriads of possible conformations, one particular organized three-dimensional structure must be separated from deepest alternate local free energy minima by many multiples of the standard deviations of the free energy of such structures. In contrast, completely random heteropolymers possess a rugged energy landscape in which many conflicting interactions between pairs of amino acids can only be poorly satisfied in conflicting ways. Such compromises always yield a modest number of candidate low energy structures. The kinetics of searching through these states is slow and inefficient. The Brownian search involves successive escape from structurally distinct traps. In addition, if a low energy structure of such a random heteropolymer were to be useful to the organism (for binding or catalysis, say) under one set of thermodynamic conditions (pH, solvent, etc.), the thermal occupancy of this structure can entirely be erased and ceded to a functionally worthless competitor structure when those thermodynamic conditions are changed, even slightly. Such a molecule would not confer fitness on the organism except under a narrow range of conditions (fine-tuning at its worst). Natural proteins are not like this. By virtue of evolved protein landscapes being funneled (see Figure 14.1, which contrasts the landscapes of evolved proteins and random heteropolymers) significant robustness is conferred on their low energy structures. When the landscape has a funnel topography, at a sufficiently low temperature, correct geometrical juxtapositions can form in different parts of the molecule with relative independence. It is no longer necessary to undo energetically favorable but incorrect contacts that would trap the molecule in a wildly different structure. In general, homopolymers do not satisfy the conditions needed for a funnel landscape: the compact structures of a homopolymer differ energetically by just small surface terms, giving little driving force to a native structure. A few magic number homopolymers might possess the necessary free energy gap to guide their folding process but magic cluster polymers do not have the sufficiently

Highly evolved protein   Random heteropolymer

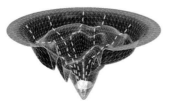   

Structure robust to mutation   Structures very sensitive to mutation

FIGURE 14.1   (See color insert following page 302.) The energy landscape of an evolved protein is funneled (left). Random sequences have rugged surface with structurally disparate minima (right).

large diversity of structures needed for setting up biochemical networks. They usually correspond to simple polyhedral folds (Wolynes, 1996).

Having a funnel-like landscape thus confers foldability. Foldability requires a sufficient diversity of interactions with the solvent—a heteropolymer with interactions of sufficiently varying strengths can fold. Water allows this diversity of interaction—charged groups that dissolve readily, aliphatic groups that are not very soluble, and polar but noncharged groups that are somewhere in between. The greater the range of possible solvent-induced interactions the larger the diversity of structures into which a polymer molecule can self-organize, just through the interactions mediated by the solvent. Attempts in the laboratory using random combinatorial synthesis suggest that two kinds of different subunits would suffice to build the very simplest structures such as 4-helix bundles, but for more elaborate structures often found in nature five different kinds of amino acids are needed (Wolynes, 1997).

Achieving foldability via funneling may seem to employ some sort of intelligent design, but this is not true. A molecule whose active site region, alone, is stably folded already can function as a mediocre catalyst. Presumably the earliest proteins only successfully folded a small part of the molecule needed for catalytic function and were acceptable, but still poor, enzymes. Because only a small part of such a primordial protein needs to be organized for catalysis, such a globally floppy molecule can have been synthesized by chance. Fixing ever longer fragments of the protein successively improves catalytic ability and the best catalysts would eventually have funnel-like landscapes guiding all parts of the molecule to an organized structure. Recently, Sasai and coworkers have shown the viability of this scenario by carrying out simulated evolution and dynamical simulations. Evolution can design a funnel landscape without much difficulty (Nagao et al., 2005).

Superficially, therefore, it appears that the diversity of interactions of simple organic molecules with several of the candidate, astrophysically available fluids—even supercritical hydrogen, etc.— would be sufficient to construct a diverse set of foldable molecules in all these environments.

14.4   FOLDING AND BIOLOGY'S EXPLORATION OF WATER'S PECULIARITIES

While energy landscape theory suggests several of the simpler fluids could provide at least minimally appropriate folding environments, the use of energy landscape theory of folding has also revealed that, in fact, today's macromolecules have evolved to exploit the peculiarities of water's physical chemistry in somewhat unexpected ways.

Energy landscape theory provides the mathematical tools to infer effective solvent-mediated interactions from the known macromolecular structures, i.e., it provides a physical basis for bioinformatics (Goldstein et al., 1992, 1996). The foldability requirement translates this inverse problem of finding potentials into an optimization problem—what potentials give the most funneled energy landscapes for the existing proteins? The mathematical form of this optimum requires maximizing the ratio of the folding temperatures to the glass transition temperature. This optimization problem, i.e., minimizing the effects of frustration, from a model energy function, like the sequence design optimization, is completely tractable. This tractability resolves the Hoyle Paradox. When optimized short-range pair potentials are inferred in this way from known protein structures, they are found to be quite consistent with the biochemical intuitions that have been built up anecdotally by structural biologists. The optimized energy functions primarily depending on whether the hydrophobic/hydrophilic character of the amino acids' interactions is complete. Nevertheless, the optimization shows that additional characteristics are needed to encode the energetic diversity needed to fold molecules into their existing shapes. The resulting bioinformatically optimized short-range pair interactions are still not the whole story, however. Such potentials are found to do a somewhat limited job in actually predicting protein structure from sequence. They work best on small structures of simple α-helical topology. They perform very poorly when specific associations between larger proteins are studied (Papoian et al., 2003, 2004). In studying protein dimers by energy landscape analysis, a pattern emerged: while these contact potentials do fine in guiding some specific associations that are dominated by hydrophobic interactions, i.e., in creating interfaces that resemble protein interiors, some interfaces observed crystallographically (about 50% of them in fact) are not as hydrophobic or well packed as protein interiors. Instead these more hydrophilic interfaces contain interpolated water molecules. These interpolated molecules bridge hydrophobic groups giving an effective attraction between those groups. When a nonadditive potential model was optimized over the protein–protein interface database, it was found this specific solvent-mediated attraction was, in fact, quantitatively stronger than the (screened) Coulomb interaction: like charges apparently attract rather than repel. Including these nonadditive interactions in structure prediction algorithms greatly improves their performance on larger protein molecules.

Why might this solvent-separated, nonadditive interaction have come to be exploited by biological macromolecules? It would seem that the answer lies in the frustration between the requirements of good folding and good binding. Such a frustration is an intrinsic feature of contact potentials. If the binding interface is to be entirely hydrophobic, allowing conventional binding, when the monomer is formed, the hydrophobic patch could alternatively bind to the interior of the protein, i.e., it opens up other possibilities for the monomer to form incorrect contacts. This ambiguity allows traps that slow the folding process. A conventional interface, therefore, cannot be too big. This size limitation reduces the amount of specificity that can be encoded in a hydrophobic interface. Exploiting the next nearest neighbor, water mediated interactions allow both of the optimization problems, folding and binding, to be solved with less conflict between them. From an evolutionary perspective, the solvent mediated interaction opens up much more sequence space for life to utilize.

Is this more complex sort of solvent mediated interaction likely to occur in other nonaqueous solvents? It seems to rely on the strongly polar and amphoteric nature of H_2O. It is difficult to imagine it occurring in supercritical H_2 or CH_4, though detailed calculations would be needed to check this. It could certainly occur in alcohols but these would be planetologically rare.

Although it is exciting that Nature has exploited this peculiarity of water as a solvent, it clearly is not an absolute requirement for folding and binding to coexist: 50% of the binding interfaces do not utilize the solvent interpolated interaction.

14.5 CONCLUSION

On Earth, protein folding and molecular recognition exploit water's physical chemistry to the fullest. This is a natural consequence of molecular evolution. The diversity of ways that water's interactions have been used in molecular biology is strongly suggestive of an opportunistic evolutionary process. Multiple solutions of the need for structural specificity exist that do not contradict our emerging understanding of how specific biological structures form. This understanding suggests foldability and bindability are collective emergent properties, and could arise in vastly different physical forms. Our understanding of folding and recognition suggest, however, that water is not unique in its ability to provide a fluid environment for molecular self-organization and biomolecular recognition—even among the fluids available in large quantities on planets.

REFERENCES

Bryngelson, J., Onuchic, J., Socci, N., and Wolynes, P. G. 1995. Funnels, pathways, and the energy landscape of protein folding: A synthesis. *Proteins: Struct. Funct. Genet.* 21:167–195.

Cotton, F. A., Wilkinson, G., Murillo, C. A., and Bachman, M. 1988. *Advanced Inorganic Chemistry.* New York, NY: Wiley Press.

Goldstein, R., Luthey-Schulten, Z., and Wolynes, P. G. 1992. Optimal protein-folding codes from spin-glass theory. *Proc. Natl. Acad. Sci. USA* 89:4918–4922.

Goldstein, R. A., Luthey-Schulten, Z., and Wolynes, P. G. 1996. The statistical mechanical basis of sequence alignment algorithms for protein structure recognition. In *New Developments in Theoretical Studies of Proteins,* Advanced Series in Physical Chemistry 7. Edited by R. Elber. Singapore: World Scientific, pp. 359–388.

Hoyle, F. J. 1957. *The Black Cloud*. New York, NY: Harper.

Lorenz, C. P., Ravenhall, D. G., and Pethick, C. J. 1993. Neutron star crusts. *Phys. Rev. Lett.* 70:379–382.

Nagao, C., Terada, T. P., Yomo, T., and Sasai, M. 2005. Correlation between evolutionary structural development and protein folding. *Proc. Natl. Acad. Sci. USA* 102:18950–18955.

Nelson, J. C., Saven, J. G., Moore, J. S., and Wolynes, P. G. 1997. Solvophobically driven folding of nonbiological oligomers. *Science* 277, 1793–1796.

Papoian, G. A., Ulander, J., and Wolynes, P. G. 2003. The role of water mediated interactions in protein-protein recognition landscapes. *J. Amer. Chem. Soc.* 125:9170–9178.

Papoian, G. A., Ulander, J., Eastwood, M. P., Luthey-Schulten, Z., and Wolynes, P. G. 2004. Water in protein structure prediction. *Proc. Natl. Acad. Sci. USA* 101:3352–3357.

Suess, H. E., and Urey, H. C. 1956. Abundances of the elements. *Rev. Mod. Phys.* 28:53–74.

Williams, R. J. P., and Frausto da Silva, J. J. R. 1996. *The Natural Selection of the Elements*. Oxford: Oxford University Press.

Wolynes, P. 1995. Three paradoxes of protein folding. In *Proceedings on Symposium on Protein Folds: A Distance-Based Approach*. Edited by H. Bohr, S. Brunak. Boca Raton, FL: CRC Press, 3–17.

Wolynes, P. G. 1996. Symmetry and the energy landscapes of biomolecules. *Proc. Natl. Acad. Sci. USA* 93:14249–14255.

Wolynes, P. G. 1997. As simple as can be? *Nat. Struct. Biol.* 4 (11):871–874.

Part IV

Water, the Solar System,
and the Origin of Life

15 Sources of Terrestrial and Martian Water

Humberto Campins and Michael J. Drake

CONTENTS

15.1 INTRODUCTION

Are Earth-like planets with water oceans rare or common among planetary systems? Although we do not yet have an answer to this question from direct observations, the study of the origin of water on Earth and Mars can provide important insights. Current evidence points toward a variety of mechanisms for the delivery of water to Earth and Mars. Most, if not all, of these delivery mechanisms are expected to be common to the formation of other planetary systems. Hence, the presence of water on Earth, and its role in the origin of life, does not appear to be unique.

Water is a common chemical compound in our galaxy and in our solar system. In addition to Earth, it has been identified in asteroids, comets, meteorites, Mars, the moon, and in the atmospheres, rings, and moons of giant planets. There is indirect evidence for it in the poles of Mercury. The high deuterium to hydrogen (D/H) ratio of Venus's atmosphere has been interpreted as evidence for Venus having had far more water in the past than is present today. Here, we address the origin of terrestrial and Martian water. There is no agreement on the origin of water on Earth and Mars.

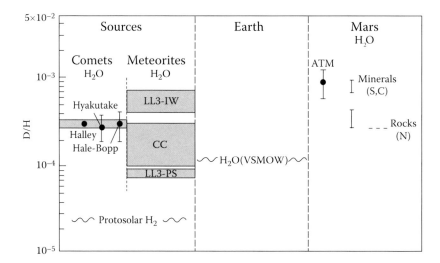

FIGURE 15.1 The D/H ratios in H$_2$O in three comets, meteorites, Earth (Vienna standard mean ocean water - VSMOW), protosolar H$_2$, and Mars. CC, carbonaceous chondrite meteorites; LL3-IW, interstellar water in Semarkona meteorite; LL3-PS, protostellar water in Semarkona meteorite. (After Drake, M. J., and Righter, K., *Nature*, 416, 39–44, 2002.)

Possible sources include capture of solar nebula gas, adsorption of water from gas onto grains in the accretion regions of these planets, accretion of hydrous minerals forming in the inner solar system, hydrous silicates migrating from the asteroid belt, and impacts with comets and hydrous asteroids. Other reviews that discuss this topic include Mottl et al. (2007), Pepin and Porcelli (2002, 2006), Righter et al. (2006), Robert (2003, 2006), Drake (2005), Campins et al. (2004), Lunine et al. (2003), Drake and Righter (2002), and Robert et al. (2000).

We have information on Martian water mainly through studies of the Martian meteorites and in situ measurements by spacecraft. However, Mars presents unique challenges because compositional information is more limited than for Earth. An important example is the global value of the D/H ratio for Martian water. It is possible that we do not know Mars' intrinsic D/H isotopic ratio. Unlike Earth, Mars lacks plate tectonics and, hence, has no means of cycling water between mantle and crust. The Martian meteorites we measure on Earth could be sampling water delivered to Mars by cometary and asteroidal impacts subsequent to planetary formation. Note, however, that the contrary has been argued (Watson [Leshin] et al., 1994), i.e., that Martian meteorites provide evidence for an intrinsic Martian mantle D/H ratio up to 50% higher than Earth's mantle (see Figure 15.1). Because we know most about water on Earth, the main focus of this paper will be examining the origin of such water, and whenever possible we address the case of Martian water.

15.2 WATER ON EARTH

15.2.1 ACCRETION HISTORY

The accretion of the Earth was dominated by a violent series of events. Dynamical theory of planetary accretion points to a hierarchy of accreting planetesimals, with the largest object accreting at any given time during the growth of a planet being one tenth to one third of the mass of the growing planet (Wetherill, 1985; Chambers, 2001). During the later stages of accretion, these collisions deposited enough energy to at least partly (and possibly completely) melt the Earth. Earth probably experienced serial magma ocean events. The most massive impact ejected material into Earth orbit; this disk of ejected material is thought to have subsequently accreted to form the Moon (Canup and

Asphaug, 2001). Even after the formation of Earth's moon, an intense bombardment (of smaller yet substantial objects) continued; this period is commonly known as the "late heavy bombardment" and ended about 3.9 Ga ago (Tsiganis et al., 2005).

Metal delivered by accreting planetesimals sank through these serial magma oceans and ponded at their bases for some period before transiting diapirically through the lower mantle to the center of the planet. Metal appears to have equilibrated with silicate at the base of the magma ocean. The mean depth calculated corresponds to at least the depth of the current upper mantle/lower mantle boundary of the Earth (Drake, 2000). That depth probably represents some ensemble average memory of metal-silicate equilibrium in a series of magma oceans and should not be taken as the literal depth of the last magma ocean. Comparable conclusions about magma oceans can be drawn for the Moon, Mars, and the asteroid Vesta (Righter and Drake, 1996). As discussed below, a primitive atmosphere and water ocean appear to have formed very early in Earth's history. Core formation, magma ocean solidification, water ocean, and atmospheric outgassing were essentially complete by 4.45 Ga ago.

15.2.2 EVIDENCE FOR AN EARLY WATER OCEAN

It had long been thought that, in the solar nebula, the accretion disk at one astronomical unit (AU) from the Sun was too hot for hydrous phases to be stable (Boss, 1998), and that water was delivered by impacts of asteroids and/or comets with Earth after it formed. However, recent geochemical evidence increasingly argues against asteroids and comets being the main sources of Earth's water and the evidence points to the Earth accreting "wet" throughout its growth (Drake and Righter, 2002; Drake, 2005; Mottl et al., 2007). In either case, the Earth may have had water oceans very early in its history. Several arguments support the existence of these early oceans on Earth.

The different reservoirs of $^{129}Xe/^{132}Xe$ in the atmosphere, and the mid-ocean ridge basalt (MORB) source support the presence of an early ocean. The isotope ^{129}I decays to ^{129}Xe with a half-life of about 16 Ma and is produced only by stellar nucleosynthesis in precursor astrophysical environments. Distinct $^{129}Xe/^{132}Xe$ reservoirs on Earth (and Mars) must have formed within 100 Ma of the last event of astrophysical nucleosynthesis of ^{129}I. It is difficult, although not impossible, to fractionate I from Xe by purely magmatic processes (Musselwhite and Drake, 2000); the problem is that I and Xe are both volatile and incompatible (they both have low vaporization temperatures and both prefer magmas to solid mantle), although Xe is a little less incompatible. However, water is extremely effective at fractionating I from Xe (Musselwhite et al., 1991). This is because I dissolves in liquid water (it is a halogen) while Xe bubbles through as a gas. If accretion ceased while ^{129}I still existed and any magma ocean solidified, liquid water could become stable at the Earth's surface. Consequently, as Musselwhite (1995) showed, outgassed ^{129}I could be recycled hydrothermally into the oceanic crust and subducted back into a mantle reservoir, by then partially degassed of Xe. Subsequent decay of ^{129}I into ^{129}Xe would give a MORB source an elevated $^{129}Xe/^{132}Xe$ ratio relative to the earlier outgassed atmosphere (Figure 15.2). A related conclusion was drawn for Mars (Musselwhite et al., 1991). Therefore, based on the observed elevated $^{129}Xe/^{132}Xe$ in MORB, it can be inferred that the Earth had a primitive atmosphere, large bodies of water, and a tectonic subduction zone by 4.45 Ga ago.

Further evidence for the existence of an early water ocean on Earth comes from detrital zircons. Wilde et al. (2001) and Mojzsis et al. (2001) independently reported zircons of 4.4 Ga age and 4.3 Ga age respectively. On the basis of magmatic oxygen isotope ratios and micro-inclusions of SiO_2, these authors concluded that the zircons formed from magmas that interacted with liquid water. Wilde et al. (2001) specifically suggest that the 4.4 Ga zircons represent the earliest evidence of both continental crust and water oceans on Earth. Support for this argument was provided by Watson and Harrison (2005), who used a geothermometer based on the Ti-content of zircons to conclude that they crystallized at ~700°C, a temperature indistinguishable from granitoid zircon growth today. In other words, water-bearing, evolved magmas appear to have been present on Earth within 200 Ma

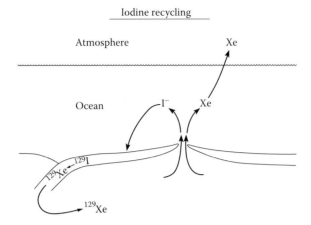

FIGURE 15.2   Illustration of outgassing of I and Xe while ^{129}I is still available. Xenon bubbles through the water ocean into the atmosphere. ^{129}I dissolves in water and is recycled hydrothermally into the crust and subducted back into the mantle. Subsequent decay of ^{129}I leads to mid-ocean ridge basalts (MORB) being erupted with elevated ^{129}Xe/^{132}Xe, compared to the atmosphere. Separate Xe reservoirs imply fractionation of I from Xe within 100 Ma after nucleosynthesis of ^{129}I, possibly by water oceans. (After Musselwhite, D. S., Experimental geochemistry of iodine, argon, and xenon: Implications for the early outgassing histories of the Earth and Mars. Ph.D. dissertation, University of Arizona, 1995.)

of solar system formation. The implication is that modern patterns of crust formation, erosion, and sediment formation had been established prior to 4.4 Ga, implying liquid water oceans already existed.

15.3   PROPOSED SOURCES OF WATER AND METHODS OF DISCRIMINATION

In principle, it should be possible to determine the main sources of and relative contributions to Earth's water, if they have distinct chemical and isotopic signatures. In practice this task can be made more difficult by the complex formation and evolution histories of Earth and its likely water sources.

15.3.1   Discriminators

Chemical and isotopic signatures that are used as discriminators include the following:

- The D/H ratio of water in Earth, Mars, comets, meteorites, and the solar nebula;
- The relative abundances and isotopic ratios for noble gases on Earth, Mars meteorites, comets, and the Sun;
- The ratio of noble gases to water on Earth, meteorites, and comets;
- The isotopic composition of the highly siderophile (very strongly metal-seeking) element Os in Earth's primitive upper mantle (PUM), in the Martian mantle, and in meteorites.

15.3.2   D/H Ratios

Deuterium is a primordial product of Big Bang nucleosynthesis, formed in the early universe. Nuclear reactions in stars convert D into He. Cycling of primordial matter through stars tends to lower the D/H ratio of the interstellar medium over cosmological time. However, chemical and physical mass fractionation processes can produce local enhancements in the D/H ratio. For example, low-temperature, ion–molecule reactions in the cores of molecular clouds can enhance the D/H ratio in icy grains by as much as 2 orders of magnitude above that observed in the interstellar medium (Gensheimer et al., 1996). In our solar system, there is evidence for more than one primitive reservoir of hydrogen (Drouart et al., 1999; Mousis et al., 2000; Robert, 2001; Hersant et al., 2001, and references therein). The solar nebula gas D/H ratio is estimated from observations of CH_4 in Jupiter and Saturn to be $2.1 \pm 0.4 \times 10^{-5}$ (Lellouch et al., 2001). Jupiter and Saturn likely obtained most of their hydrogen directly from solar nebula gas. This estimate is also consistent with the protosolar D/H value inferred from the solar wind implanted into lunar soils (Geiss and Gloecker, 1998). A second reservoir, enriched in D compared with the solar nebula gas, contributed to bodies that accreted from solid grains, including comets and meteorites.

We have high D/H ratios (relative to Earth's water) measured spectroscopically from water in three comets (all from the Oort cloud): Halley ($3.2 \pm 0.1 \times 10^{-4}$) (Eberhardt et al., 1995); Hyakutake ($2.9 \pm 1.0 \times 10^{-4}$) (Bockelée-Morvan et al., 1998); and Hale–Bopp ($3.3 \pm 0.8 \times 10^{-4}$) (Meier et al., 1998). These are all about twice the D/H ratio for terrestrial water (1.49×10^{-4}) (Lecuyer et al., 1998), and about 15 times the value for the solar nebula gas ($2.1 \pm 0.4 \times 10^{-5}$) (Lellouch et al., 2001). These comet D/H ratios are consistent with the range of values for "hot cores" of dense molecular clouds (2 to 6×10^{-4}) (Gensheimer et al., 1996). Carbonaceous chondrites have the highest water abundance of all meteorites (up to 17 wt% H_2O) (Jarosewich, 1990). Their D/H ratios range from 1.20×10^{-4} to 3.2×10^{-4} (Lecuyer et al., 1998). The largest D-enrichment in a water-bearing mineral in a meteorite was measured at $7.3 \pm 1.2 \times 10^{-4}$ in the LL3 chondrite Semarkona (Deloule and Robert, 1995). These results are illustrated in Figure 15.1.

The measurement of the D/H ratio of water in three comets is significant. Different authors, however, interpret these ratios in very different ways. Some (e.g., Dauphas et al., 2000; Morbidelli et al., 2000; Robert, 2001; Drake and Righter, 2002) consider the high D/H ratio in these comets as evidence against a cometary origin of most of the terrestrial water. Others (e.g., Delsemme, 2000; Owen and Bar-Nun, 2001) argue that comets are the main reservoir of deuterium-rich water that raised the terrestrial D/H a factor of 6 above the protosolar value. Complicating the matter further, laboratory measurements of the D/H ratio in sublimating ices have shown that D/H isotope fractionation can occur during sublimation (Weirich et al., 2004; Moores et al., 2005). During sublimation of water ice samples in a vacuum, the D/H ratio of the evolved gas varies with time; this ratio can increase or decrease relative to the initial D/H ratio, depending on the nature of the sample. The root cause is interpreted to be differential diffusion and sublimation of HDO and H_2O (HDO is a water molecule with one deuterium and one hydrogen atom instead of two hydrogen atoms). This result implies that spectroscopically measured D/H ratios from cometary comae may not be representative of the bulk cometary values.

15.3.3   NOBLE GASES

Noble gases are chemically inert and highly volatile. Hence they probably arrive at a planet along with other volatiles, quickly move to the planet's atmosphere, and in bulk tend to avoid the chemical complications of planetary evolution. Thus, the noble gas characteristics of a planetary atmosphere can provide tracing information to the source of the planet's volatiles. Figure 15.3 shows the abundances of noble gases in Venus, Earth, and Mars compared with solar abundances as well as those from two kinds of meteorites. Note that the proportions of Ar, Kr, and Xe in the atmospheres of Earth and Mars are remarkably similar, and also significantly different from the relative abundance patterns found in meteorites or in the Sun (determined from solar wind).

We do not know much about noble gases in comets. Some of the measurements we do have appear contradictory. Krasnopolsky et al. (1997) reported an upper limit in comet Hale–Bopp of 0.5% of the solar Ne/O ratio. Stern et al. (2000) reported a tentative Ar detection in comet Hale–Bopp at a roughly solar Ar/O ratio. Weaver et al. (2002) reported upper limits for Ar/O of <10% and <8%, respectively of the solar value in comets LINEAR 2001 A2 and LINEAR 2000 WM1. All

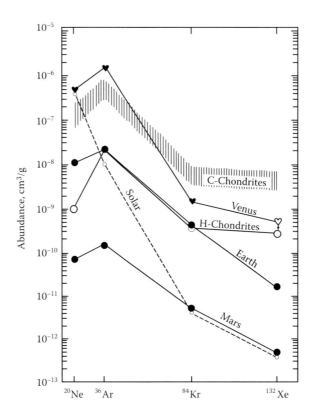

FIGURE 15.3 Noble gases in Venus, Earth, Mars, and meteorites. (After Owen, T., and Bar-Nun, A., *Icarus*, 116, 215–226, 1995.)

these observations have been made of comets from the Oort cloud. The more sensitive upper limits for Ar in the LINEAR comets are not consistent with the detection reported by Stern et al. (2000) in comet Hale–Bopp. At this point, it is not clear if comet Hale–Bopp was unusually rich in Ar, or if the tentative detection is somehow flawed. We return to this issue in the following sections.

15.3.4 SIDEROPHILE ELEMENTS IN EARTH'S MANTLE

The relatively high abundances of highly siderophile elements (HSEs), at 0.003 × CI, in Earth's primitive upper mantle and their roughly chondritic element ratios suggest that these elements arrived after Earth's core formation had ceased (Drake and Righter, 2002). Had these elements arrived sooner, they would have been quantitatively extracted into Earth's core. This material is commonly termed the "late veneer." Drake and Righter (2002) and Righter et al. (2006) argue that Earth-building materials shared some but not all properties with extant meteorites. In other words, no primitive material similar to Earth's mantle is currently in our meteorite collections.* More specifically, Os isotopes can be used to constrain the origin of the late veneer. Carbonaceous chondrites, the only abundant water-bearing meteorites, have a significantly lower ^{187}Os/^{188}Os ratio of 0.1265 than Earth's primitive upper mantle value of 0.1295, essentially ruling out their parent bodies as the main source of the late veneer (Figure 15.4). The Earth's mantle ^{187}Os/^{188}Os ratio

---

* This result is consistent with dynamical models that suggest Earth accreted about half of its current mass from material that formed near Earth's location, and the other half from embryos scattered from both smaller and larger heliocentric distances (Morbidelli et al., 2000). Because most meteorites come from the asteroid belt (between 2 and 3.2 AU), it is not surprising that the building blocks of Earth are not properly sampled by meteorites.

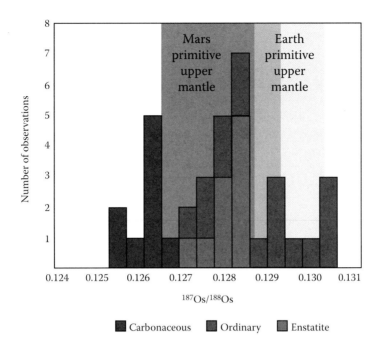

FIGURE 15.4 **(See color insert following page 302.)** $^{187}Os/^{188}Os$ ratios in carbonaceous, ordinary, and enstatite chondrites, and in the Earth's primitive upper mantle (PUM), are distinct and are diagnostic of the nature of the Earth's "late veneer." In particular, Earth's PUM is different from water-bearing carbonaceous chondrites. The $^{187}Os/^{188}Os$ ratios in Mars's PUM have been recently revised down (Muralidharan et al., 2008) and, although still uncertain, now overlap with those for Earth's PUM (gray area of the figure). Earlier estimates of the Martian mantle were probably compromised by incomplete dissolution of sample. (A. Brandon and R. Walker, personal communications, March 2008. Modified from Drake, M. J., and Righter, K., *Nature*, 416, 39–44, 2002.)

overlaps anhydrous ordinary chondrites and is distinctly higher than anhydrous enstatite chondrites. If the objects that brought Earth its late veneer also contributed a significant amount of water, their composition had to have been different from water-bearing meteorites in our collections. It has been suggested that mixtures of anhydrous ordinary chondrites and hydrous carbonaceous chondrites could yield an appropriate Os isotopic signature for Earth. This hypothesis is certainly possible, but the same mixture would have to satisfy both $^{187}Os/^{188}Os$ ratios and D/H ratios in Earth (Figures 15.1 and 15.4). Until recently, it appeared that Mars' primitive upper mantle could have had an even higher $^{187}Os/^{188}Os$ ratio of 0.132 (Brandon et al., 2005). However, new estimates (Muralidharan et al., 2008) suggest the value is more terrestrial (orange and gray areas of Figure 15.4), and therefore, more compatible with a contribution to Mars by the same source of Earth's late veneer.

15.4 SOURCES

In this section we examine the principal proposed sources of terrestrial and Martian water in the light of these discriminators. We evaluate the possible contribution from each source using the discriminators discussed in the previous section.

15.4.1 PRIMORDIAL GAS CAPTURED FROM THE SOLAR NEBULA

It has been argued that a primordial atmosphere could not have been captured directly from the solar nebula, principally because of Earth's D/H ratio in water and the noble gas abundances. However,

Campins et al. (2004) and Mottl et al. (2007) point out that the processes involved in planetary accretion, degassing, and the evolution of a hydrosphere and atmosphere are complex and may have fractionated the chemical and isotopic signatures of the water source(s). Hydrogen, for example, may be an important constituent in the outer, and possibly inner, core of Earth (Okuchi, 1997, 1998). If H and D were fractionated in that process, the residual D/H ratio in the hydrosphere may not reflect that of the original source. After accretion, the surface of the Earth continued to be modified by large impacting asteroids and comets (Sections 3.2.3 and 3.2.4). Large impact events have the capacity to completely volatilize any oceans (Zahnle and Sleep, 1997) and blow off portions of Earth's atmosphere (Melosh and Vickery, 1989; Zahnle, 2006). Fractionation of D/H and noble gas/water ratios may occur as a consequence of impact processes, which may also mask the signatures of the original source material.

15.4.2 ADSORPTION OF WATER ONTO GRAINS IN THE ACCRETION DISK

The terrestrial planets grew in an accretion disk of gas and dust grains. Hydrogen, He, and O$_2$ dominated the gas in which the dust was bathed. Some of that H$_2$ and O$_2$ combined to make water vapor. If thermodynamic equilibrium was attained, there were about two Earth masses of water vapor in the accretion disk inside of 3 AU (Drake, 2005). The mass of the Earth is 5×10^{27} g. The mass of one Earth ocean is 1.4×10^{24} g. The extreme maximum estimate for the amount of water in the Earth is about 50 Earth oceans (Abe et al., 2000), with most estimates being 10 Earth oceans or less. For example, an estimate based on the water storage potential of minerals in the silicate Earth is about 5–6 Earth oceans (Ohtani, 2005). Thus, the mass of water vapor available in the region of the terrestrial planets far exceeded the mass of water accreted. Could water vapor be adsorbed onto grains before the gas in the inner solar system was dissipated? Stimpfl et al. (2004) have examined the role of physisorption by modeling the adsorption of water onto grains at 1000 K, 700 K, and 500 K using a Monte Carlo simulation. Stimpfl et al. (2004) exploded the Earth into 0.1 μm spheres of volume equal to Earth, recognizing that grains in the accretion disk are not spherical and would be fractal in nature. If the surface area of the fractal grain was 100 times that of a sphere of corresponding volume, then one quarter of an ocean of water could be adsorbed at 1000 K, one Earth ocean could be adsorbed at 700 K, and three Earth oceans could be adsorbed at 500 K. This work is discussed in more detail by Drake (2005). There are also issues of retention of water as the grains collide and grow to make planets. However, it is clear that volatiles are not completely outgassed even in planetary scale collisions. For example, primordial ^{3}He, far more volatile than water, is still outgassing from Earth's mantle 4.5 Ga after an almost grown Earth collided with a Mars-sized body to make the Moon. Stimpfl et al. (2004) showed that the efficiency of adsorption of water increases as temperature decreases; that is, the process should have been more efficient further from the Sun than closer to the Sun. Thus, it is likely that Mars, Earth, and Venus all accreted some water by adsorption, with Mars accreting the most both because of its greater distance from the Sun and the lower energy of collisions during accretion because of its smaller final mass. The current differences in the apparent water abundances among the terrestrial planets are probably the result of both different initial inventories and subsequent geologic and atmospheric processing.

There is an interesting consequence for the evolution of planetary redox states if the terrestrial planets accreted "wet." Okuchi (1997) showed that when Fe-metal and water were compressed to 30–100 kbar and heated to between 1200°C and 1500°C, iron hydride formed. In a magma ocean environment, metal sinking to form planetary cores should contain H, and OH should be left behind in the molten silicate. As more metal was delivered to the planet as it accreted, more H would be extracted into the core and more OH liberated in the silicate mantle. Thus, planetary mantles should become progressively more oxidized with time, perhaps explaining the high redox states relative to the iron–wüstite buffer (the iron–wüstite buffer is a measure of the oxygen fugacity as a function of temperature). This process might explain the correlation of the degree of oxidation of silicate mantles with planet mass (Righter et al., 2006).

15.4.3 COMETS

Comets were long considered the most likely source of water in the terrestrial planets. A cometary source was attractive because it is widely believed that the inner solar system was too hot for hydrous phases to be thermodynamically stable (Boss, 1998). Thus, an exogenous source of water was needed. There are elemental and isotopic reasons why at best 50% and, most probably, a smaller percentage of water accreted to Earth from cometary impacts (Drake and Righter, 2002). Figure 15.1 compares the isotopic composition of hydrogen in Earth, Mars, three Oort Cloud comets, and various early solar system estimates. It is clear that 100% of Earth's water did not come from Oort Cloud comets with D/H ratios like the three comets measured so far. D/H ratios in Martian meteorites do agree with the three cometary values (Figure 15.1). That may reflect the impact of comets onto the Martian surface in a non plate-tectonics environment that precludes recycling of surface material into the Martian mantle (Drake, 2005). Conversely, there are models (Lunine et al., 2003) that have asteroids and comets from beyond 2.5 AU as the main source of Mars's water.

So what limits the cometary contribution to Earth's water? Consider, for example, that perhaps Earth accreted some hydrous phases or adsorbed water, and some amount of additional water came from comets. Indigenous Earth water could have had D/H ratios representative of the inner solar system, i.e., low values because of relatively high nebular temperatures, perhaps like protosolar hydrogen (2–3×10^{-5}) (Lecluse and Robert, 1994), in which case a cometary contribution of up to 50% is possible. Alternatively, indigenous Earth water could have had D/H ratios representative of a protosolar water component identified in meteorites ($\sim 9 \times 10^{-5}$) (Deloule and Robert, 1995), in which case there could be as little as a 10–15% cometary contribution (Owen and Bar-Nun, 2000). There are caveats to using cometary D/H ratios to limit the delivery of cometary water to Earth. First, we do not know that Oort Cloud comets Halley, Hale–Bopp, and Hyakutake are representative of all comets. Certainly they are unlikely to be representative of Kuiper Belt objects, the source of Jupiter family comets, as Oort Cloud comets formed in the region of the giant planets and were ejected, while Kuiper Belt objects have always resided beyond the orbits of the giant planets. Second, the D/H measurements available are not of the solid nucleus, but of gases emitted during sublimation. As mentioned in Section 3.1, differential diffusion and sublimation of HDO and H_2O may make such measurements unrepresentative of the bulk comet. The D/H ratio would be expected to rise in diffusion and sublimation, as has been confirmed in preliminary laboratory experiments on pure water ice (Weirich et al., 2004). Lower bulk D/H ratios would increase the allowable amount of cometary water. Intriguing experiments on mixtures of water ice and TiO_2 grains by Moores et al. (2005) suggest that D/H ratios could be lowered in sublimates. Third, the D/H ratios of organics and hydrated silicates in comets are currently unknown. Note, however, that D/H ratios up to 50 times higher than Vienna Standard Mean Ocean Water (VSMOW) have been measured in some chondritic porous interplanetary dust particles (CP-IDPs), which may have had cometary origins (Messenger, 2000). Higher aggregate D/H ratios of comets would decrease the allowable cometary contribution to Earth's water.

Delivery of water from comets can also be evaluated in light of other cometary chemical data. If one assumes 50% of Earth's water was derived from comets (the maximum amount permitted by D/H ratios), and an Ar/H_2O ratio of 1.2×10^{-7} in the bulk Earth, then comets like Hale–Bopp (with a solar ratio of Ar/H_2O) (Stern et al., 2000) would bring 2×10^4 more Ar than is presently in the Earth's atmosphere (Swindle and Kring, 2001). It is unclear if this measurement of comet Hale–Bopp is applicable to all comets. However, the true Ar/O ratios in comets would have to be at least three orders of magnitude below solar in order to be consistent with the Ar abundance of the Earth's atmosphere. Another approach to estimating the contribution of cometary materials to Earth's water budget can be made by considering the implications for the abundances of noble metals and noble gases. Dauphas and Marty (2002) show that the total mass of cometary and asteroidal material accreting to Earth after core formation is 0.7–2.7×10^{25} g and that comets contribute <0.001 by mass or $<0.7–2.7 \times 10^{22}$ g. Given that the minimum mass of water in the Earth, one Earth ocean, is 1.4×10^{24}g, this argument suggests that comets can contribute less than 1% of Earth's water.

15.4.4 ASTEROIDS

Asteroids are a plausible source of water based on dynamical arguments. Morbidelli et al. (2000) have shown that up to 15% of the mass of the Earth could be accreted late in Earth's growth by collision of one or a few asteroids originating in the Main Belt. This hypothesis is difficult to test, as it could involve a single, unique event. However, there are strong geochemical arguments against a significant contribution of water from asteroids, unless one postulates that Earth was hit by a hydrous asteroid unlike any falling to Earth today as sampled by meteorites. As discussed in Section 3.1.3, a late addition of water to Earth must be accompanied by other chemical elements such as Os. Because Earth's primitive upper mantle has a significantly higher $^{187}Os/^{188}Os$ ratio than the water-bearing meteorites (carbonaceous chondrites), this effectively rules them out as the main source of the late veneer (Figure 15.4). However, mixtures of anhydrous ordinary chondrites and hydrous carbonaceous chondrites could yield an appropriate Os isotopic signature for Earth, but the same mixture would have to satisfy both $^{187}Os/^{188}Os$ ratios and D/H ratios in Earth (Figures 15.1 and 15.4). Another caveat to consider is that thermal processing of asteroids was occurring 4.5 Ga ago. One cannot exclude the possibility that ordinary chondrites once contained water and those falling to Earth today have lost their water by metamorphism, even though it was still present 4.5 Ga ago. However, the preservation of aqueous alteration products in some carbonaceous chondrites (McSween, 1979) suggests that loss of water from initially hydrous asteroids is unlikely to proceed to the anhydrous limit.

15.4.5 EARLY ACCRETION OF WATER FROM INWARD MIGRATION OF HYDROUS SILICATES

Most solar nebula models suggested that the growth zones of the terrestrial planets were too hot for hydrous minerals to form (Cyr et al., 1998; Delsemme, 2000; Cuzzi and Zahnle, 2004). Ciesla and Lauretta (2005) suggest that hydrous minerals were formed in the outer asteroid belt region of the solar nebula and were then transported to the hotter regions of the nebula (i.e., Earth and Mars) by gas drag, where they were incorporated into the planetesimals that formed there. These hydrated minerals were able to survive for long periods in hotter regions due to sluggish dehydration kinetics. Note that this mechanism differs from the delivery of water by stochastic impacts with large planetary embryos originating in the outer asteroid belt region (Section 3.2.4). Drake (2005) points out that it seems unlikely that hydrous silicates could be decoupled from other minerals and transported into the inner solar system. Thus the proposed radial migration of hydrous minerals would be subject to the same objection involving Os isotopes (Section 3.2.4), unless the hydrous silicates arrived before the differentiation of Earth, as suggested by Ciesla and Lauretta (2005).

15.4.6 WATER AND ORGANICS

It has been postulated that water and organics were delivered from the same cometary and/or asteroidal source (Chyba, 1993; Delsemme, 2000). In light of the new evidence, such combined delivery seems less likely. In addition to comets and asteroids being inconsistent with some geochemical properties of Earth, it is unlikely that a complex organic material would have survived the magma ocean accompanying the formation of the Moon at the end of accretion of the Earth (Drake, 2000). However, it seems possible that complex organic material may have been delivered to Earth after it formed and liquid water oceans became stable. Comets, which are known to be rich in organic molecules, have been postulated to be the principal source of terrestrial amino acids (Pierazzo and Chyba, 1999; Chyba, 1993). Some meteorites are also rich in carbon and organic compounds, hence, asteroids may have also contributed significantly to Earth's organic inventory. In fact, Kring and Cohen (2002) point out that during the late heavy bombardment, asteroidal material probably delivered a large mass of organic material to Earth's surface, as much as 160 times larger than that in the total land biomass today. Even if some of the organic molecules were dissociated during the impacts, Kring and Cohen (2002) propose the formation of impact-generated hydrothermal regions with lifetimes up to 10^6 years, where complex organic molecules might reassemble.

15.5 OUTSTANDING CHALLENGES

Although a number of sources for terrestrial and Martian water have been proposed, the pieces of this puzzle do not currently fit into a coherent picture. Many issues remain to be resolved; below we list some of the most important:

1. The D/H ratio of the nebular gas is inferred from spectroscopic measurements of CH_4 in the atmospheres of Jupiter and Saturn to be $2.1 \pm 0.4 \times 10^{-5}$ (Lellouch et al., 2001), much lower than VSMOW (the D/H ratio for Vienna Standard Mean Ocean Water). If the nebular D/H ratio really is as low as implied by the Jovian and Saturnian atmospheres,[*] then mechanisms to raise the D/H ratio of nebular gas from solar to terrestrial (as VSMOW) are needed.
2. It is possible that the D/H and Ar/O ratios measured in cometary comae are not truly representative of cometary interiors. Reconnaissance experiments have shown that D/H ratios in laboratory experiments can increase or decrease with time due to differential diffusion and sublimation, depending on the physical nature of the starting material (Weirich et al., 2004; Moores et al., 2005). Further, published measurements of Ar/O ratios in comets are either upper limits or 3-sigma detection limits and seem inconsistent with each other. Depending on the location of Ar inside comets, Ar/O ratios may also be unrepresentative of the bulk cometary composition. The successful Deep Impact mission is the first attempt to expose fresh cometary interior material for spectral analysis with ground-based and space-based high spectral resolution spectrometers. Unfortunately, improvements in our understanding of cometary D/H ratios and Ar/O ratios are not expected from the Deep Impact results.
3. The key argument against an asteroidal source of Earth's water is that the Os isotopic composition of Earth's primitive upper mantle matches that of anhydrous ordinary chondrites, not hydrous carbonaceous chondrites. But are the parent bodies of the ordinary chondrites anhydrous? Could ordinary chondritic meteorites be derived from the metamorphosed outer parts of hydrous asteroids, in which case impact of a bulk asteroid could deliver water? It is probable that spacecraft spectral examination of very deep impact basins in S-type asteroids will be needed to address this question.
4. A related question is why there are any anhydrous primordial bodies, such as the parent bodies of anhydrous primitive meteorites, in the solar system if adsorption of water from gas in the accretion disk was an efficient process, as preliminary calculations suggest it might have been.
5. The timing of loss of gas from the accretion disk in the region of the terrestrial planets is unknown. For adsorption to be efficient, nebular gas must persist long enough for grains to adsorb water. Radial migration of hydrous silicates also depends on the presence of gas. The timing of loss of gas from the accretion disk will be intimately connected to the currently unknown mechanism of loss.

---

[*] Spectroscopic or direct measurements of solar D/H are not representative of the original bulk composition of the nebular source because practically all of the Sun's deuterium has been fused to make He.

ACKNOWLEDGMENTS

H. Campins gratefully acknowledges the support of the John Templeton Foundation, particularly through the "Water of Life" workshop held in Varenna, Italy. H. Campins' work was also supported by grants from NASA and from the National Science Foundation. M. J. Drake was supported by the NASA Cosmochemistry Program. Helpful comments were provided by D. A. Kring.

REFERENCES

Abe, Y., Ohtani, E. Okuchi, T., Righter, K., and Drake, M. J. 2000. Water in the early Earth. In *Origin of the Earth and Moon*, eds. R. Canup and K. Righter. Tucson, AZ: University of Arizona Press, pp. 413–434.

Bockelée-Morvan, D., Gautier, D., Lis, D. C., Young, K., Keene, J., Phillips, T. G., Owen, T., Crovisier, J., Goldsmith, P. F., Bergin, E. A., Despois, D., and Wooten, A. 1998. Deuterated water in comet P/1996 B2 (Hyakutake) and its implications for the origins of comets. *Icarus* 133:147–162.

Boss, A. P. 1998. Temperatures in protoplanetary disks. *Annu. Rev. Earth Planet. Sci.* 26:26–53.

Brandon, A. D., Humayun, M., Puchtel, I. S., and Zolensky, M. E. 2005. Re-Os isotopic systematics and platinum group element composition of the Tagish Lake carbonaceous chondrite. *Geochim. Cosmochim. Acta* 69:1619–1631.

Campins, H., Swindle, T. D., and Kring, D. A. 2004. Evaluating comets as a source of Earth's water. In *Origin, Evolution and Biodiversity of Microbial Life in the Universe*, ed. J. Seckbach. Dordrecht: Kluwer Academic Publishing, pp. 569–591.

Canup, R. M., and Asphaug, E. 2001. The Moon-forming impact. *Nature* 412:708–712.

Chambers, J. E. 2001. Making more terrestrial planets. *Icarus* 152:205–224.

Chyba, C. F. 1993. The violent environment of the origins of life: Progress and uncertainties. *Geochim. Cosmochim. Acta* 5:3351–3358.

Ciesla, F. J., and Lauretta, D. S. 2005. Radial migration and dehydration of phyllosilicates in the solar nebula. *Earth Planet. Sci. Lett.* 231:1–8.

Cuzzi, J. N., and Zahnle, K. J. 2004. Material enhancement in protoplanetary nebulae by particle drift through evaporation fronts. *Astrophys. J.* 614:490–496.

Cyr, K. E., Sears, W. D., and Lunine, J. I. 1998. Distribution and evolution of water ice in the solar nebula: Implications for solar system body formation. *Icarus* 135:537–548.

Dauphas, N., and Marty, B. 2002. Inference on the nature and mass of Earth's late veneer from noble metals and gases. *J. Geophys. Res.* 107:E12-1–E12-7.

Dauphas, N., Robert, F., and Marty, B. 2000. The late asteroidal and cometary bombardment of Earth as recorded in water deuterium to protium ratio. *Icarus* 148:508–512.

Deloule, E., and Robert, F. 1995. Interstellar water in meteorites? *Geochim. Cosmochim. Acta* 59:4695–4706.

Delsemme, A. H. 2000. 1999 Kuiper prize lecture cometary origin of the biosphere. *Icarus* 146:313–325.

Drake, M. J. 2000. Accretion and primary differentiation of the Earth: a personal journey. *Geochim. Cosmochim. Acta* 64:2363–2370.

Drake, M. J. 2005. Origin of water in the terrestrial planets. *Meteorit. Planet. Sci.* 40:519–527.

Drake, M. J., and Righter, K. 2002. Determining the composition of the Earth. *Nature* 416:39–44.

Drouart, A., Dubrulle, B., Gautier, D., and Robert, F. 1999. Structure and transport in the solar nebula from constraints on deuterium enrichment and giant planets formation. *Icarus* 140:129–155.

Eberhardt, P., Reber, M., Krankowsky, D., and Hodges, R. R. 1995. The D/H and ^{18}O/^{16}O ratios in water from comet P/Halley. *Astron. Astrophys.* 302:301.

Geiss, J., and Gloeckler, G. 1998. Abundances of deuterium and helium-3 in the protosolar cloud. *Space Sci. Rev.* 84:239–250.

Gensheimer, P. D., Mauersberger, R., and Wilson, T. L. 1996. Water in galactic hot cores. *Astron. Astrophys.* 314:281–294.

Hersant, F., Gautier, D., and Hure, J.-M. 2001. A two-dimensional model for the primordial nebula constrained by D/H measurements in the solar system: Implications for the formation of giant planets. *Astrophys. J.* 554:391–407.

Jarosewich, E. 1990. Chemical analyses of meteorites: A compilation of stony and iron meteorite analyses. *Meteoritics* 25:323–337.

Krasnopolsky, V. A., Mumma, M. J., Abbott, M., Flynn, B. C., Meech, K. J., Yeomans, D. K., Feldman, P. D., and Cosmovici, C. B. 1997. Detection of soft x-rays and a sensitive search for noble gases in comet Hale-Bopp (C/1995 01). *Science* 277:1488–1491.

Kring, D. A., and Cohen, B. A. 2002. Cataclysmic bombardment throughout the inner solar system 3.9–4.0 Ga. *J. Geophys. Res.* 107:4-1–4-6.

Lecluse, C., and Robert, F. 1994. Hydrogen isotope exchange reaction rates: Origin of water in the inner solar system. *Geochim. Cosmochim. Acta* 58:2927–2939.

Lecuyer, C., Gillet, P., and Robert, F. 1998. The hydrogen isotope composition of seawater and the global water cycle. *Chem. Geol.* 145:249–261.

Lellouch, E., Bezard, B., Fouchet, T., Feuchtgruber, H., Encrenaz, T., and de Graauw, T. 2001. The deuterium abundance of Jupiter and Saturn from ISO-SWS observations. *Astron. Astrophys.* 370:610–622.

Lunine, J. I., Chambers, J., Morbidelli, A., and Leshin, L. A. 2003. The origin of water on Mars. *Icarus* 165:1–8.

McSween, H. Y. 1979. Are carbonaceous chondrites primitive or processed? A review. *Rev. Geophys. Space Phys.* 17:1059–1078.

Meier, R., Owen, T. C., Matthews, H. E., Jewitt, D. C., Bockelée-Morvan, D., Biver, N., Crovisier, J., and Gautier, D. 1998. A determination of the HDO/H$_2$O ratio in comet C/1995 O1 (Hale-Bopp). *Science* 279:842–844.

Melosh, H. J., and Vickery, A. M. 1989. Impact erosion of the primordial atmosphere of Mars. *Nature* 338:487–489.

Messenger, S. 2000. Identification of molecular cloud material in interplanetary dust particles. *Nature* 404:968–971.

Mojzsis, S. J., Harrison, T. M., and Pidgeon, R. T. 2001. Oxygen isotope evidence from ancient zircons for liquid water at Earth's surface 4,300 Myr ago. *Nature* 409:178–181.

Moores, J. E., Brown, R. P., Lauretta, D. S., and Smith, P. H. 2005. Preliminary results of sublimations fractionation in dusty disaggregated samples (abstract). *Lunar Planet. Sci.* XXXVI: # 1973. LPI Contribution No. 1234. Houston, TX: Lunar and Planetary Institute.

Mottl, M. J., Glazer, B. T., Kaiser, R. I., and Meech, K. J. 2007. Water and astrobiology. *Chem. Erde* 67:253–282.

Mousis, O., Gautier, D., Bockelée-Morvan, D., Robert, F., Dubrulle, B., and Drouart, A. 2000. Constraints on the formation of comets from D/H ratios measured in H$_2$O and HCN. *Icarus* 148:513–525.

Morbidelli, A., Chambers, J., Lunine, J. I., Petit, J. M., Robert, F., Valsecchi, G. B., and Cyr, K. 2000. Source regions and timescales for delivery of water to the Earth. *Meteorit. Planet. Sci.* 35:1309–1320.

Muralidharan, K., Deymier, P. A., Stimpfl, M., and Drake, M. J. 2008. Adsorption as a water delivery source in the inner solar system: a kinetic Monte Carlo study. *Lunar Planet. Sci. Conf.* XXXIX: #1401. LPI Contribution No. 1391. Houston, TX: Lunar and Planetary Institute.

Musselwhite, D. S., and Drake, M. J. 2000. Early outgassing of Mars: Implications from experimentally determined solubility of iodine in silicate magmas. *Icarus* 148:160–175.

Musselwhite, D. S. 1995. Experimental geochemistry of iodine, argon, and xenon: Implications for the early outgassing histories of the Earth and Mars. Ph.D. dissertation, University of Arizona.

Musselwhite, D. S., Drake, M. J., and Swindle, T. D. 1991. Early outgassing of Mars: Inferences from the geochemistry of iodine and xenon. *Nature* 352:697–699.

Ohtani, E. 2005. Water in the mantle. *Elements* 1:25–30.

Okuchi, T. 1997. Hydrogen partitioning into molten iron at high pressure: Implications for Earth's core. *Science* 278:1781–1784.

Okuchi, T. 1998. Melting temperature of iron hydride at high pressures and its implications for temperature of the Earth's core. *J. Phys. Condens. Matter.* 10:11595–11598.

Owen, T., and Bar-Nun, A. 1995. Comets, impacts, and atmospheres. *Icarus* 116:215–226.

Owen, T., and Bar-Nun, A. 2000. Volatile contributions from icy planetesimals. In *Origin of the Earth and Moon*, ed. R. M. Canup and K. Righter. Tucson, AZ: Univ. of Arizona Press, pp. 459–471.

Owen, T., and Bar-Nun, A. 2001. Contributions of icy planetesimals to Earth's early atmosphere. *Orig. Life Evol. Biosph.* 31:435–458.

Pepin, R. O., and Porcelli, D. 2002. Origin of noble gases in the terrestrial planets. *Rev. Mineral. Geochem.* 47:191–246.

Pepin, R. O., and Porcelli, D. 2006. Xenon isotope systematics, giant impacts, and mantle degassing on the Earth. *Earth Planet. Sci. Lett.* 250:470–485.

Pierazzo, E,, and Chyba, C. F. 1999. Amino acid survival in large cometary impacts. *Meteorit. Planet. Sci.* 32:909–918.

Righter, K., and Drake, M. J. 1996. Core formation in Earth's Moon, Mars, and Vesta. *Icarus* 124:513–529.

Righter, K., Drake, M. J., and Scott, E. 2006. Compositional relationships between meteorites and terrestrial planets. In *Meteorites and the Earth Solar System II*, ed. D. S. Lauretta and H. Y. McSween Jr. Tucson, AZ: University of Arizona Press, pp. 803–828.

Robert, F. 2001. The origin of water on Earth. *Science* 293:1056–1058.

Robert, F. 2003. The D/H ratio in chondrites. *Space Sci. Rev.* 106:87–101.

Robert, F. 2006. Solar System deuterium/hydrogen ratio. In *Meteorites and the Early Solar System II*, ed. D. Lauretta and H. Y. McSween. Jr. Tucson, AZ: University of Arizona Press, pp. 341–351.

Robert, F., Gautier, D., and Dubrulle, B. 2000. The solar system D/H ratio: observations and theories. *Space Sci. Rev.* 92:201–224.

Stern, S. A., Slater, D. C., Festou, M. C., Parker, J. W., Gladstone, G. R., A'Hearn, M. F., and Wilkinson, E. 2000. The discovery of argon in Comet C/1995 01 (Hale-Bopp). *Astrophys. J.* 544:L169–L172.

Stimpfl, M., Lauretta, D. S., and Drake, M. J. 2004. Adsorption as a mechanism to deliver water to the Earth. *Meteorit. Planet. Sci.* 39:A99.

Swindle, T. D., and Kring, D. A. 2001. Implications of noble gas budgets for the origin of water in Earth and Mars. In *Eleventh Annual V.M. Goldschmidt Conference*. Abstract # 3785. LPI Contribution No. 1088. Houston, TX: Lunar and Planetary Institute.

Tsiganis, K., Gomes, R., Morbidelli, A., and Levison, H. F. 2005. Origin of the orbital architecture of the giant planets of the Solar System. *Nature* 435:459–461.

Watson, E. B., and Harrison, T. M. 2005. Zircon thermometer reveals minimum melting conditions on earliest Earth. *Science* 308:841–844.

Watson (Leshin), L., Hutcheon, I. D., Epstein, S., and Stolper, E. M. 1994. Water on Mars: Clues from deuterium/hydrogen and water contents of hydrous phases in SNC meteorites. *Science* 265:86–90.

Weaver, H. A., Feldman, P. D., Combi, M. R., Krasnopolsky, V., Lisse, C. M., and Shermansky, D. E. 2002. A search of Ar and O VI in three comets using the far ultraviolet spectroscopic explorer. *Astrophys. J.* 576:L95–L98.

Weirich, J. R., Brown, R. H., and Lauretta, D. S. 2004. Cometary D/H fractionation during sublimation. *Bull. Am. Astron Soc.* 36:1143.

Wetherill, G. W. 1985. Occurrence of giant impacts during the growth of the terrestrial planets. *Science* 228:877–879.

Wilde, S. A., Valley, J. W., Peck, W. H., and Graham, C. M. 2001. Evidence from detrital zircons for the existence of continental crust and oceans 4.4 Gyr ago. *Nature* 409:175–178.

Zahnle, K. J. 2006. Earth's earliest atmosphere. *Elements* 2:217–222.

Zahnle, K. J., and Sleep, N. H. 1997. Impacts and the early evolution of life. In *Comets and the Origin and Evolution of Life*, ed. P. J. Thomas, C. F. Chyba, and C. P. McKay. New York, NY: Springer-Verlag, pp. 175–208.

16 Water
The Tough-Love Parent of Life

Veronica Vaida and Adrian F. Tuck

CONTENTS

16.1 INTRODUCTION

In this chapter, we argue that water imparted survivability to the incipient biosphere by providing a testing, but not impossible, environment for geochemistry to become biochemistry. In Earth's atmosphere, the ability of water to sample all phases around its triple point allows it to exert major influences on climate. By forcing organic chemistry to the interface with air at the surface of the ocean and with globally distributed atmospheric aerosols, water allowed for concentration, alignment, and, ultimately, synthesis of the biomolecular building blocks of life.

We further suggest that organization can emerge from random molecular motion and that the process can be described by the fluctuation-dissipation theorem. The methods of statistical physics could be used to examine the relationship between entropy production and scale invariance, which appears to be ubiquitous. The water molecule appears to play a central role on all scales.

The parental role of water in the birth of biochemistry involved much more than providing a bulk solvent for carbon-based chemistry. Geophysics set the stage on which geochemistry was converted to biochemistry. The central role of water in this conversion is evident from its existence as a liquid present in high volume and with a large surface area at the interface with the atmosphere. This liquid presence arose on Earth because of nonlinearities in interactions between the rotational and vibrational spectral characteristics of water and convective turbulence in a gaseous atmosphere whose density profile was determined by gravity through hydrostatic balance. Temperature was and is fundamental, being determined by the balance between enthalpic and entropic processes. The beam of energetic solar photons from a black body at ~5800 K represents a low-entropy input, and the outgoing flux of low-energy infrared radiation at ~255 K over the whole 4π solid angle, mostly from the atmosphere, represents a high-entropy output. Thus, in a given planetary environment, alternative molecules to water must have the ability to not only initiate and sustain whatever biochemistry is to be engendered, but also to thermostat the space that life occupies. Water can do this

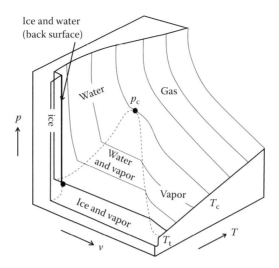

FIGURE 16.1   Phase diagram for pure water. At temperatures greater than the critical temperature, T_c, water will be a gas, unliquifiable by any pressure. At temperatures and pressures below the critical point, p_c, any two phases can exist as shown. By further application of the phase rule, all three phases can exist at only one point, T_t, which is defined by $p_t = 6.1$ mbar, $T_t = 273$ K, and specific volume $v_{t, vapor} = 2.06 \times 10^5$ m^3 kg^{-1}, $v_{t, water} = 1.00 \times 10^{-3}$ m^3 kg^{-1} and $v_{t, ice} = 1.09 \times 10^{-3}$ m^3 kg^{-1}. Note that ice floats.

because of a unique combination of several molecular properties (Bergman and Lynden-Bell, 2001), expressed macroscopically as an ability to exist in all phases near its triple point at many locales on Earth's surface, as illustrated in Figure 16.1. This ability results in water being the major stabilizing feedback in the climate system.

The emergence of biochemistry from geochemistry is inescapably an exercise in the behavior of carbonaceous molecules; it is irreducibly chemical. By confining many organic molecular species, particularly surfactants, to two dimensions rather than three, the atmosphere–ocean interface greatly enhances the opportunities for reactive counters among molecules (Adam and Delbrück, 1968, pp. 198–215). It also offers an aligning environment for linear molecules when the surface film is compressed. This latter feature is of central importance to condensation reactions between monomers when water has to be eliminated to form polypeptides and polynucleotides, which have to be linear to perform their tape-like role in transmitting the genetic code. These features suggest that the chemical ability to act as a solvent in bulk-liquid phases is not as important as surface-layer characteristics.

A geophysically inevitable feature of a liquid ocean on a rotating planet is the presence of vast populations of aerosols, small aqueous particles suspended in the atmosphere. The wind causes ocean waves, which, when they break, entrap air bubbles below the sea surface (i.e., create white-caps). On rising to the surface, these bubbles burst and form aerosols by the cap-and-jet mechanism (Blanchard, 1964; Mason, 1954). The smaller particles coagulate, while those that are several microns in diameter or larger fall rapidly under gravity, leaving a size distribution on Earth with a maximum dimension of about one micron, or the size of a bacterium (Dobson et al., 2000). While a pure water aerosol particle, being spherical, cannot divide because it has attained its minimum free energy (surface tension), the addition of an organic surfactant film does enable such fission via the energetics of film compression (Donaldson et al., 2001, 2002). For Earth, in the case of a typical n-carboxylic acid with 16 C atoms, the fission is asymmetric, with one daughter the size of a bacterium and one the size of a virus (Tuck, 2002). In prebiotic conditions, such large interacting populations arising from aerosols would provide a medium in which the widespread horizontal genetic transfer of early, primitive genetic material posited by Woese (2002) could occur. The possibility of chemical differentiation between the daughters of asymmetric fission exists (Donaldson et al., 2002; Tuck, 2002), a feature that would be absent in bulk solution.

How an organized system such as life could emerge from the random motion of the vast molecular populations in the ocean and the atmosphere is a lingering conceptual question. Earth's surface and fluid envelope is not at equilibrium, a situation arising from the enthalpic and entropic balance described earlier. A comparatively recent discovery has been the principle of maximum entropy production in Earth's atmosphere (Paltridge, 1975, 2001), which has been given a basis in nonequilibrium statistical mechanics by Dewar (2003, 2005). A fluctuation-dissipation theorem emerges in which the fluctuations represent more organized motion, while the dissipation represents random molecular motion that becomes thermalized (i.e., becomes heat). An essential adjunct is turbulence, which is ever present in a high Reynolds number fluid such as the atmosphere. While turbulence in the atmosphere has traditionally been viewed as random motion resulting from instability in organized large-scale hydrodynamic flows, this approach largely fails in specifying the detailed characteristics of the ensuing turbulence. Based on the long-known power-law decay of the molecular velocity autocorrelation function (Alder and Wainwright, 1970; Dorfman and Cohen, 1970), turbulence can also be viewed as an organized fluctuation emerging from the more random motion of the molecules (Tuck et al., 2005; Tuck, 2008) with associated scale invariance. Turbulence is intimately related to scale invariance and its associated fractality. We encounter fractality in atmospheric chemistry (Tuck et al., 2002, 2003) within sequences of amino acids in membrane proteins (Hoop and Peng, 2000; Oliviera et al., 2005; Rani and Mitra, 1995), as well as within nucleotide base sequences in nucleic acids (Mrevlishvili, 1995). It has been noted that among several reaction networks observed in astrochemistry, only Earth's atmosphere conforms to the criterion of being a scale-free network, something held to be a sign of life (Solé and Munteanu, 2004). This is yet another test that water has passed, which an alternative solvent would also have to do. Water, by providing a testing, but not impossible, fluid environment for the polymerization of amino acid and nucleotide monomers, confers survivability on the polypeptide and polynucleotide products. Tough love paid off over eons of parental supervision.

16.2 WATER AND ITS COMPLEXES IN ATMOSPHERIC CHEMISTRY AND CLIMATE

In all its phases, water is a unique contributor to the habitability of the planet. Primarily, it is of great importance in the greenhouse effect, which is responsible for trapping heat generated by the planet when it is illuminated by the solar beam. Water vapor is transparent to the incoming solar radiation with a mostly high-energy output in the visible (VIS) and ultraviolet (UV) frequencies, but it effectively absorbs the terrestrial emission of the infrared (IR) and near-infrared (NIR) frequencies. In Earth's atmosphere, water is the most significant greenhouse gas. Longwave IR radiation is absorbed when water vapor excites the symmetric and asymmetric O–H stretches and their overtone transitions as these have significant IR and NIR cross sections (Donaldson et al., 2003).

Water vapor above the liquid ocean has a pronounced concentration gradient as it decreases with increasing altitude and decreasing temperature. Consequently, an optically thick water vapor layer radiates the upwelling IR from the surface back downward. Energy is thus trapped near the surface, maintaining surface temperatures above freezing and ensuring a high water-vapor pressure. The upper, optically thinner layer emits IR radiation to space that is approximately characteristic of a black body at 255 K. This process cools the upper, optically thin layer, which overlies the heated warmer water vapor layer described above. The result is convective instability leading to cloud formation and precipitation in the troposphere, below the radiatively controlled stratosphere with its vertically stabilizing temperature inversion.

Other gases such as CO_2, CH_4, O_3, CFCs, etc. contribute to the greenhouse effect by absorbing in windows where water absorptions are sparse. Increases in anthropogenic greenhouse gases contribute to global warming (Harries et al., 2001), yet the effect of water is at present difficult to assess. As the temperature of a planet rises, water evaporates, and the additional water vapor enhances the warming effect. This simple scenario becomes very complicated as one considers water as a multiphase system where liquid (liquid aerosols, clouds, and precipitation) and ice coexist with vapor. The temperature and pressure characteristic of Earth's atmosphere sample the region around the triple point on the water-phase diagram, providing a mechanism for utilization of phase transitions and associated latent heat effects. Water is able to form highly directional hydrogen-bonded networks. Hydrogen bonding in water has been a fascinating yet elusive scientific issue explored by experimental and theoretical methods over nearly five decades.

In liquid water, tetrahedral arrangements of molecules ordered by hydrogen bonds are expected to exhibit interesting and short-time dynamics. Even so, a coherent description of the hydrogen-bonded networks in water has not yet emerged. Other molecules such as NH_3, HF, HCOOH, HNO_3, H_2SO_4, etc., also form hydrogen-bonded networks, but only water effectively condenses into liquid and ice at the temperatures and pressures characteristic of Earth. Condensation of water vapor onto atmospheric aerosols in air that has become supersaturated with water leads to cloud formation. At any moment, about 60% of Earth's surface is covered by clouds. Aerosols and clouds have a pronounced effect on the planet's radiation budget as they reflect (scatter) incoming VIS and UV solar radiation back to space. This cooling effect increases in a global warming scenario where more water evaporates from the liquid ocean, leading to enhanced aerosol and cloud formation. Cooling due to hydrogen-bonded water networks counteract, to an as yet unpredictable extent, the warming greenhouse effect of the additional water vapor, aqueous aerosols, and clouds.

At the high densities characteristic of planetary atmospheres, formation of weakly bound complexes is likely, yet relatively high temperatures characteristic of Earth's and other dense planetary atmospheres, such as that of Venus, limit the lifetime and abundance of molecular complexes. Water is believed to form molecular complexes in Earth's atmosphere, which are potentially important in absorption of solar radiation (Pfeilsticker et al., 2003; Ptashnik et al., 2004; Vaida et al., 2001). Water complexes contribute to radiative transfer, both in the mid-IR known as the "water window" (Wolynes and Roberts, 1978) and in the near-IR, where the intense and high-frequency OH overtone stretching vibrations are absorbed (Goss et al., 1999; Vaida et al., 2001, 2003).

The electronic states of water, especially the photodissociative lowest electronic state, have been studied theoretically and experimentally (Chipman, 2005; Kawasaki et al., 2004). Absorption by electronic states of water vapor is at high energy, at wavelengths below 185 nm, with liquid water, ice, and hydrates absorbing at still higher energies; water photodissociates only at photon energies that are not available in the lower atmosphere. These transitions are not accessible in contemporary Earth's atmosphere, except at very high altitude, as solar radiation at the required energies and wavelengths is absorbed by molecular oxygen. However, the prebiotic atmosphere, with a much reduced partial pressure of oxygen, could have allowed solar radiation of such high energy to penetrate the atmosphere. Excitation would have resulted in radical formation (OH + H) and subsequent processing of organic molecules in the atmosphere and at the planet's surface. Recent theoretical calculations (Harvey et al., 1998) of the UV absorption spectra of water clusters explained the shift to high energy of clusters and condensed phases of water, pointing to a low-energy-absorbing tail in the water dimer. Such a low-energy absorption into a reactive state of water would have been particularly important in the atmosphere of prebiotic Earth in promoting photochemistry through the UV absorption of the water dimer at 200–190 nm.

At present, estimating both the atmospheric abundance and the spectra of water complexes at high temperatures (200–300 K) remains a challenge (Schenter et al., 2002; Vigasin, 1983). Attempts at estimating the partial pressure, p_D, for the water dimer approximate the product of the equilibrium constant K_P and the square of the partial pressure of the water monomer, $(p_M)^2$. The partial pressure of hydrates is a nonlinear function of temperature and, subsequently, of the water vapor. In a global-warming scenario, the water vapor pressure would increase predictably with temperature, while the water dimer, trimer, and tetramer would increase nonlinearly as the square, the third, and the fourth power of the monomer partial pressure, respectively (Vaida et al., 2001). Hydrates would, therefore, provide a nonlinear amplifier to any climate change, resulting in warmer sea-surface temperatures (Vaida et al., 2001).

16.3 ATMOSPHERIC WATER–AIR INTERFACES

Atmospheric interfaces have been considered in origin-of-life scenarios (Negron-Mendoza and Ramos-Bernal, 2004; Orgel, 1998; Segrè et al., 2001). Water–air interfaces coated with organics are interesting chemical environments for prebiotic biomolecular synthesis (Dobson et al., 2000; Donaldson et al., 2004; Goldacre, 1958; Tervahattu et al., 2004; Tuck, 2002). Inputs from sources as different as comets and meteorites, sea-floor vents, volcanoes, and atmospheric and oceanic chemistry are deposited in this unique locale. In the contemporary atmosphere, the surfaces of oceans and lakes contain organic surfactants derived primarily from the biosphere. On prebiotic Earth, organics could have been produced endogenously in discharges and by UV radiation in reducing CH_4, N_2, H_2O, NH_3 atmospheres (Miller, 1953, 1998; Miller and Urey, 1959; Ourisson and Nakatani, 1994) and hydrothermal vents (Huber and Wächtershäuser, 1998). Organics also could have been exogenously brought in by meteoritic and cometary infall (Anders, 1989; Chyba and Sagan, 1992; Cronin et al., 1988).

Organic surfactants are known to reside at the water–air interface, yet even water-soluble organics have been shown to partition to the surface to some extent (Demou and Donaldson, 2002). The water–air interface has been shown to concentrate and orientate organics (Ji and Shen, 2004; Lavigne et al., 1998; Nanita et al., 2004; Scatena and Richmond, 2004; Stenstam et al., 2002; Watry and Richmond, 2002), properties that can select from the available chemical space the precursors and reactions needed for biology (Dobson, 2004). Concentration of organic molecules at an aqueous surface is responsible for formation of an "oil slick" that feeds organic material to aerosols at inception (Dobson et al., 2000; Tuck, 2002). The surface longevity of organic surfactants representative of the contemporary atmosphere has been investigated (Gilman et al., 2004) and found to be of the same order of magnitude—a residence time of a few days—as that of the aerosols in the atmosphere.

No similar experiments have yet been done with prebiotic surfactants, but the results are expected to be similar. Recent laboratory studies (Ji and Shen, 2004; Nanita et al., 2004; Scatena and Richmond, 2004; Stenstam et al., 2002; Watry and Richmond, 2002) speak to the concentration, conformation, orientation, and chiral enrichment of amino acids, polypeptides, and enzymes relevant to biology. For example, helical peptide rods appear to spontaneously arrange themselves at the air–water interface (Fukuto et al., 1999; Sjogren and Ulenlund, 2004). Long α-helical peptides have been shown to self-assemble into close-packed domains that partially cover the surface even at low surface pressure (Gillgren et al., 2002; Malcolm, 1968; Yamamoto et al., 2001). Enrichment in the α-helix over the β-sheet form of glucose oxidase monolayers on aqueous solutions was accomplished using a Langmuir–Blodgett technique (Dai et al., 1999). The relative enzyme activity intensified as the content of the α-helix increased at the surface. The properties of the aqueous interfaces are under investigation. In a few examples, control over the configuration and orientation of biomolecules has been demonstrated (Dai et al., 1999; Fukuto et al., 1999; Malcolm, 1968; Sjogren and Ulenlund, 2004; Yamamoto et al., 2001).

The surface/interface between water and air offers an auspicious environment for chemical reactions. Solar radiation and atmospheric radicals are important agents at the organic-air interface. OH is the dominant reactive radical at present; it would have been present in smaller amounts in a prebiotic atmosphere via photolysis of water. Reactions at the aqueous-organic interface involve water-soluble reagents, as well as ions whose concentration at the interface is controlled by the identity of the solute, the aqueous-solution pH, and surface thermodynamics. Reactions at the surface using sunlight to generate complex biopolymers from much simpler organics provide an analogy to metabolism. The expectation based on surface-science studies is that molecular-absorption cross sections increase at the surface. At this time, insufficient data exist for those molecules that may be found at atmospheric interfaces to allow for credible speculation.

Condensation reactions, while implicated in biopolymer formation, are handicapped on both thermodynamic and kinetic grounds. Aqueous solutions are the preferred reaction medium for biology. However, these reactions are unfavorable in bulk water as they involve eliminating H_2O. Attempts to form proteins and nucleic acids in the absence of enzymes have been successful only in water-restricted environments. Successful experiments (Kumar and Oliver, 2001) involving generation of amide bonds at water–air interfaces showed accelerated rates in the surface monolayer, rates comparable to the corresponding reactions in enzymes. While the mechanism that controls the reactivity in monolayers is not well understood, surface films are proving to be effective nonenzymatic models of ribosomal and nonribosomal peptide synthesis. In essence, the hydrophobic effect in the surface film provides an environment with low water activity, which is necessary for the condensation reactions that eliminate water to form peptide bonds and nucleoside oligomers.

On macroscopic scales, the ocean surface and the polar-sea ice exhibit statistical multifractal scale invariance, something that, we argue below, is a turbulence-associated characteristic by which organization emerges from smaller-scale randomness via the operation of a fluctuation-dissipation theorem. Water is thus intimately involved in these phenomena.

The curl of the surface wind stress is what drives the ocean circulation, along with the production of dense water under the polar pack ice, as salt is precluded from the ice on crystallization, causing the saltier, denser water to sink. The ocean buries carbon in sediments and maintains photosynthesis by plankton. These processes are important for maintaining atmospheric oxygen content. The ocean surface is the site of a large fraction of the planetary entropy production by evaporation at subtropical latitudes, a process also resulting in the hydrological cycle and rainfall, as well as the production of the most important climate regulator, water vapor and clouds.

16.4   ATMOSPHERIC AEROSOLS

The ability of water to form hydrogen bonds results in aerosol formation in the atmosphere. Under auspicious conditions of high humidity, aerosols form cloud condensation nuclei and promote cloud formation. We now consider marine aerosols because in contemporary Earth's atmosphere aqueous aerosols of continental origin are largely induced anthropogenically and naturally by emissions from land plants; therefore, they are not relevant to the present discussion.

Aerosols are small (about one micrometer in diameter) solid or liquid suspensions in air, globally distributed in Earth's atmosphere. The size and size analogy between atmospheric aerosols and single-cell bacteria have already been pointed out (Dobson et al., 2000). A spherical rotating planet will be heated differentially by the sun, giving rise to winds. In the presence of a body of water, wind action on the surface of an ocean, sea, or lake generates whitecaps, which make bubbles whose buoyancy takes them to the surface where they burst to form sea spray. The majority of the drops thus formed return to the ocean, with some becoming airborne aerosols. The life cycle of an atmospheric aerosol formed by the cap-and-jet mechanism of bubble bursting is illustrated in Figure 16.2 (Tuck, 2002). Aerosols play an important role in determining the temperature, and therefore the climate, of a planet and in promoting heterogeneous chemistry (Ellison et al., 1999). They have also been proposed as contributors in prebiotic scenarios (Dobson et al., 2000; Lerman, 1986, 1994, 1996; Lerman and Teng, 2004; Shah, 1972). The properties of atmospheric aerosols relevant to climate and chemistry are highly nonlinear. In particular, their size and number density are strongly dependent on small temperature fluctuations in the atmosphere, which themselves are associated with turbulence. Attempts to model the role of atmospheric aerosols in climate have so far been very limited, with uncomfortably large uncertainties in the magnitude and sign of aerosol effects.

To complicate this already difficult problem, atmospheric measurements have recently established that aerosols have a large organic content. To the extent that molecular speciation of collected aerosols was possible, it was established that surface-active amphiphilic organics (alcohols,

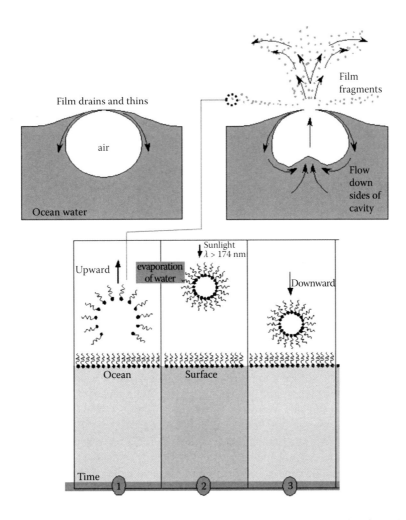

FIGURE 16.2   The air–sea–air journey, a sketch of the life cycle of an aerosol particle formed via the cap-and-jet mechanism of bubble bursting.

acids, amines, etc.) are important contributors to the organic mass found on atmospheric aerosols (Tervahattu et al., 2002, 2005). These observations led to the proposal that a significant population of organic aerosols consists of an aqueous core with an organic surface film. Chemical arguments dictate that all hydrophobic (water-repellent) organic material must be at the surface and not dissolved in the aqueous droplet. The presence of such surface films gives rise to different morphological, optical, and chemical properties. On contemporary Earth, aerosols provide coupling between the ocean, the biosphere, life, and climate. Several independent investigators (Dobson et al., 2000; Lerman and Teng, 2004; Shah, 1972; Tervahattu et al., 2004; Tuck et al., 2002) have suggested that the presence of organic films on atmospheric aqueous interfaces on the prebiotic Earth could have played a crucial role in the origin of life.

Surface-sensitive analyses of aerosols in the contemporary atmosphere have shown that surfactants do indeed concentrate at the surface of marine and terrestrial aerosols (Tervahattu et al., 2002, 2005). Using the model of the aqueous aerosol with a surrounding film of stearic acid, it can be shown that while the interior will contain no more than 25 molecules, 10^6 can be accommodated at the surface (Ellison et al., 1999). Other concentration mechanisms are possible, such as the selective evaporation of water during the aerosol's atmospheric journey in low-humidity regions (Lerman and Teng, 2004) and the coagulation of aerosols in the atmosphere (Donaldson et al., 2001); however, concentration of organics at the water–air interface is the dominant effect (Dobson et al., 2000; Tervahattu et al., 2004; Tuck et al., 2002).

Unlike their uncoated counterparts, spontaneous division of atmospheric aerosol particles covered in a compressed surface film is thermodynamically possible (Donaldson et al., 2001, 2002), as stated in the Introduction. The conclusion is that thermodynamics would allow coagulation and fission, which could have provided early forms of replication and chemical differentiation between the bacteria- and virus-sized populations of aerosols (Donaldson et al., 2004; Tervahattu et al., 2004) and a mechanism for the efficient horizontal gene transfer posited by Woese (2002).

16.5   THE EMERGENCE OF BIOCHEMISTRY FROM RANDOMNESS AT THE MOLECULAR LEVEL

The behavior of populations of molecules is described by statistical mechanics, which is particularly successful at quantitatively expressing macroscopic properties of matter at equilibrium. Nonequilibrium statistical mechanics has been far more difficult to apply, yet it is this branch of mathematical physics that concerns living systems, which do, of course, exhibit disequilibrium as one of their defining characteristics.

The media to which the framework must be applied are the planet's fluid envelope (water and air). The approaches have been molecular from the smaller scales up and fluid-mechanical from the larger scales down. The key features are the scale invariance in the macroscopic observations of dynamical variables, passive scalars, and reactive species, as well as the production of fluid-mechanical behavior (ring currents or vortices) on very short time- and space-scales in molecular-dynamic simulations of fluids subject to an anisotropic environment. It has recently become apparent that the behavior of both microscopic and macroscopic systems can be explained by fluctuation-dissipation theorems (Crooks, 1999; Dewar, 2003, 2005; Evans and Searles, 2002; Paltridge, 1975, 2001). An important feature of these treatments is that temperature remains well defined for a nonequilibrium sample of fluid, in accordance with observations. It is also true that temperature remains well defined in molecular-dynamic simulations because of the rapid equilibration of translational energy among molecules having velocities close to the mean. However, this is not true of the population in the high-velocity tail; such molecules induce higher pressure ahead of them as they collide with molecules in front, leaving lower pressure behind them. The pressures tend to equilibrate, inducing a vortical flow (Alder, 2002).

Thus, from a computer experiment on the simplest case—an ideal gas of Maxwellian atoms subject to an initial disequilibrium—an organized flow emerges, in contrast with the random molecular motion in which it originated. The velocity of the faster particles is thus fed back into the surrounding matter via the vortex in a mutually sustaining interaction that is evidently nonlinear. The result is a power-law decay of the velocity autocorrelation function (Alder and Wainwright, 1970; Dorfman and Cohen, 1970). It has been suggested in an atmospheric context that this approach could provide a microscopic bottom-up view of the turbulence in contrast to the macroscopic top-down view in which it is a disorganized state produced by hydrodynamic instabilities (Tuck, 2008; Tuck et al., 2005). Scale invariance is intimately connected with power law rather than exponential behavior and is associated with heavy-tailed probability distribution functions (Mandelbrot, 1998). Scale invariance was first noted by Richardson (1926, 1961) in the power-law growth in the separation of particles in the atmosphere and in the dependence of the length of the western coast of Great Britain on the scale of the map from which it was being measured. This was subsequently applied (Hurst, 1951) to a study of floods on the river Nile. Successful formulation of the characteristics of geophysical data as statistical multifractals resulted in the theory of generalized scale invariance (Schertzer and Lovejoy, 1987, 1991), which has been demonstrated to apply to water vapor, clouds, and rain.

The prebiotic geophysical environment had far more degrees of freedom and anisotropies than an initially isotropic, numerically equal population of Maxwellian "billiard-ball" molecules, as well as considerable chemical heterogeneity. It is not clear whether the planet has ever been, is, or ever will be at a stationary state, which would allow rigorous application of a fluctuation-dissipation theorem and, hence, of the principle of maximum entropy production to the evolution of the chemical composition of its fluid envelope. However, the attempts made under the stationarity assumption, by which is meant a fluctuating state with a stable mean, have been sufficiently illuminating (e.g., Kleidon and Lorenz, 2005) that it is worth exploring in a number of systems of varying scales, from microscopic to planetary. This is particularly true because treating it as an initial value problem in computational physics—by integrating chemical kinetic and hydrodynamic equations over time— seems to be an overwhelmingly difficult proposition.

It seems that the presence of water in liquid and vapor forms does maximize entropy production in the planet's outer, fluid (ocean and atmosphere) components, with over half the total being thus attributable (Paltridge, 1975, 2001; Paulius, 2005). A comparison has been made between the thermodynamics of the global ocean circulation and the living systems it sustains (Shimokawa and Ozawa, 2005, pp. 121–134), with the conclusion that the maximum-entropy-production principle is applicable. While a great deal more investigation is necessary on the range of systems to which applicability is possible, the progress so far should encourage further investigation. One caveat to note is the objection to Jaynes' information theory, on which maximum entropy production is based—namely, that it is intransitive and therefore not applicable to physical systems (Balescu, 2000). A resolution of the need to define entropy in a system far from equilibrium (Gallavotti, 1999) is also required, but because temperature stays well defined in molecular-dynamic simulations and in the atmosphere, this should be surmountable.

Very recently, it has been shown experimentally (Collin et al., 2005) on a single RNA molecule that the refolding and unfolding of a small hairpin obeys a fluctuation theorem (Crooks, 1999). Given that such theorems seem to apply on scales ranging from an individual biopolymer to the fluid envelope of the planet, it would be worth examining on intermediate scales, for example, an individual aerosol particle or a population of such particles that they have size distributions characterized by power laws (Pruppacher and Klett, 1998).

Sella and Hirsh (2005) have applied statistical physics to evolutionary biology, concluding that just as Gibbs free energy represents the balance between enthalpy (internal energy) and entropy in a physical system, a free-fitness function can provide an analogous expression of the balance between natural selection and stochastic drift in a genetic population and, moreover, that it will be maximized. The central point is that positive feedbacks result in skewed long-tailed probability distribution functions. These long tails represent the emergence of organization, the entropic price for which is paid by dissipation as the molecules near the most probable allow the definition of temperature.

Finally, power-law distributions and fractality characterize sequences of monomers in both proteins and nucleic acids in contemporary biopolymers and their distribution with lipids in membranes (Hoop and Peng, 2000; Mrevlishvili, 1995; Oliviera et al., 2005; Rani and Mitra, 1995). The extent to which such behavior depends on the nature of the immediate aqueous environment remains to be investigated, as does the relationship of the scale invariance to genetic function. The central point is that fluctuation-dissipation theorems and the maximum-entropy-production principle offer a statistical physics framework for examining the principles by which the interplay of scale invariance, fluctuation, dissipation, and turbulence proceed to produce macroscopic organization from microscopic randomness.

16.6 CONCLUSIONS

Water's role is central to life on Earth, on all scales, from the molecular to the global (atmosphere and ocean). This must also have been true in the prebiotic era. We see scale invariance from the power-law distributions, residue sequences in proteins and nucleic acids, and spatial arrangement of membrane species to the characteristics of meteorological dynamical variables, aerosols, clouds, and rainfall. Scale invariance seems to be associated with the operation of fluctuation-dissipation theorems working to maximize entropy production. Water is intimately involved, accounting for at least half the planetary entropy production because of its ability to thermostat the planet's fluid envelope in a range around the water triple point, its ability to form hydrogen-bonded aggregates, and its ability to concentrate organic molecules in an anhydrous film at the water–air interface. We argue that water has a greater role in generating and maintaining life than merely its imperfect ability (Ball, 2005) to operate as a bulk solvent for carbon-based chemistry. Water is an inhospitable solvent for many organic molecules and organic reactions (Benner et al., 2004): far from being a handicap, this characteristic is an essential attribute in its role as the tough-love parent of life.

The water–air interfaces that would be found at the surface of oceans, lakes, atmospheric aerosols, and clouds have the unique ability to concentrate and align organic molecules into semipermeable membranes. The rigorous yet auspicious anhydrous chemical environment at the water–air interface provided survivability via natural selection operating in a physicochemically determined milieu. Aqueous atmospheric aerosols are generated by wind action on the surface of the planet. The maximum residence time in the atmosphere of a few days selects micron-sized aerosols, similar to single-cell bacteria. Vast and globally distributed populations of aqueous aerosols with organic surface films undergo atmospheric journeys during which they are exposed to reactive radicals and radiation to constitute interesting and auspicious chemical and biochemical reactors. Natural selection would have had many opportunities to operate on this population (Dobson et al., 2000; Donaldson et al., 2004).

Any parent of life will have to perform the multiple functions of water of life and do so under the geophysical conditions relevant to the locale in the universe where such alternative life might have evolved. Water is the only example available to us, with many outstanding questions about its role in biochemistry and the maintenance of an auspicious climate for life remaining to be understood.

ACKNOWLEDGMENTS

Veronica Vaida gratefully acknowledges fellowships from the John Simon Guggenheim Memorial Foundation and the Radcliffe Institute for Advanced Study at Harvard University.

REFERENCES

Adam, G., and Delbrück, M. 1968. *Reduction of Dimensionality in Biological Diffusion Processes*. San Francisco, CA: Freeman.

Alder, B.J. 2002. Slow dynamics by molecular dynamics. *Physica A* 315: 1–4.

Alder, B.J., and Wainwright, T.E. 1970. Decay of the velocity autocorrelation function. *Phys. Rev. A* 1: 18–21.

Anders, E. 1989. Prebiotic organic matter from comets and asteroids. *Nature* 342: 255–257.

Balescu, R. 2000. *Statistical Dynamics: Matter Out of Equilibrium*. London: Imperial College Press.

Ball, P. 2005. Seeking the solution. *Nature* 436: 1084–1085.

Benner, S.A., Ricardo, A., and Carrigan, M.A. 2004. Is there a common chemical model for life in the universe? *Curr. Opin. Chem. Biol.* 8(6): 672–689.

Bergman, D.L., and Lynden-Bell, R.M. Is the hydrophobic effect unique to water? The relationship between solvation properties and network structure in water and modified water models. *Mol. Phys.* 99(12): 1011–1021.

Blanchard, D.C. 1964. Sea-to-air transport of surface active material. *Science* 146: 396–397.

Chipman, D.M. 2005. Excited electronic states of small water clusters. *J. Chem. Phys.* 122: DOI: 10.1063/1.1830438.

Chyba, C.F., and Sagan, C. 1992. Endogenous production, exogenous delivery and impact-shock synthesis of organic molecules: An inventory for the origin of life. *Nature* 355: 125–132.

Collin, D., Ritort, F., Jarzynski, C., Smith, S. B., Tinoco, I. J., and Bustamante, C. 2005. Verification of the Crooks fluctuation theorem and recovery of RNA folding free energies. *Nature* 437: 231–234.

Cronin, J.R., Pizzarello, S., and Cruickshank, D.P. 1988. *Organic Matter in Carbonaceous Chondrites, Planetary Satellites, Asteroids and Comets*. Tucson, AZ: University of Arizona Press.

Crooks, G.E. 1999. Entropy production fluctuation theorem and nonequilibrium work relation for free energy differences. *Phys. Rev. E* 60: 2721–2726.

Dai, G.L., Li, J.R., and Jiang, L. 1999. Conformation change of glucose oxidase at the water–air interface. *Colloids Surf. B* 13(2): 105–111.

Demou, E., and Donaldson, D.J. 2002. Adsorption of atmospheric gases at the air-water interface. *J. Phys. Chem. A* 106: 982–987.

Dewar, R. 2003. Information theory explanation of the fluctuation theorem, maximum entropy production and self-organized criticality in non-equilibrium stationary states. *J. Phys. A: Math. Gen.* 36: 631–641.

Dewar, R.C. 2005. Maximum entropy production and the fluctuation theorem. *J. Phys. A: Math. Gen.* 38: L371–L381.

Dobson, C.M. 2004. Chemical space and biology. *Nature* 432: 824–828.

Dobson, C.M., Ellison, G.B., Tuck, A.F., and Vaida, V. 2000. Atmospheric aerosols as prebiotic chemical reactors. *Proc. Natl. Acad. Sci.* 97: 11864–11868.

Donaldson, D.J., Tuck, A.F., and Vaida, V. 2001. Spontaneous fission of atmospheric aerosol particles. *Phys. Chem. Chem. Phys.* 3: 5270–5273.

Donaldson, D.J., Tuck, A.F., and Vaida, V. 2002. The asymmetry of organic aerosol fission and prebiotic chemistry. *Orig. Life Evol. Biosph.* 32: 237–245.

Donaldson, D.J., Tuck, A.F., and Vaida, V. 2003. Atmospheric photochemistry via vibrational overtone absorption. *Chem. Rev.* 103: 4717–4730.

Donaldson, D.J., Tervahattu, H., Tuck, A.F., and Vaida, V. 2004. Organic aerosols and the origin of life: An hypothesis. *Orig. Life Evol. Biosph.* 34: 57–67.

Dorfman, J.R., and Cohen, E.G.D. 1970. Velocity autocorrelation functions in two and three dimensions. *Phys. Rev. Lett.* 25: 1257–1260.

Ellison, G.B., Tuck, A.F., and Vaida, V. 1999. Atmospheric processing of organic aerosols. *J. Geophys. Res.-Atmos.* 104(D9): 11633–11641.

Evans, D.J., and Searles, D.J. 2002. The fluctuation theorem. *Adv. Phys.* 51: 1529–1585.

Fukuto, M., Heilmann, R.K., Pershan, P.S., Yu, S.J.M., Griffiths, J.A., and Tirrell, D.A. 1999. Structure of poly(gamma-benzyl-L-glutamate) monolayers at the gas-water interface: A Brewster angle microscopy and x-ray scattering study. *J. Chem. Phys.* 111(21): 9761–9777.

Gallavotti, G. 1999. *Statistical Mechanics: A Short Treatise.* Berlin: Springer.

Gillgren, H., Stenstam, A., Ardhammar, M., Norden, B., Sparr, E., and Ulvelund, S. 2002. Morphology and molecular conformation in thin films of poly-gamma-methyl-L-glutamate at the air-water interface. *Langmuir* 18(2): 462–469.

Gilman, J.B., Eliason, T.L., Fast, A., and Vaida, V. 2004. Selectivity and stability of organic films at the air-aqueous interface. *J. Colloid Interface Sci.* 280: 234–243.

Goldacre, R.J. 1958. *Surface Films, Their Collapse on Compression, the Shape and Size of Cells and the Origin of Life.* Oxford: Pergamon Press, 1958.

Goss, L.M., Sharpe, S.W., Blake, T.A., Brault, J.W., and Vaida, V. 1999. Direct absorption spectroscopy of water clusters. *J. Phys. Chem.* 103: 8020–8024.

Harries, J.E., Brindley, H.E., Sagoo, P.J., and Bantges, R.J. 2001. Increases in greenhouse forcing inferred from the outgoing longwave radiation spectra of the Earth in 1970 and 1997. *Nature* 410: 355–357.

Harvey, J.N., Jung, J.O., and Gerber, R.B. 1998. Ultraviolet spectroscopy of water clusters: Excited electronic states and absorption lineshapes of $(H_2O)_n$, $n = 2–6$. *J. Chem. Phys.* 109: 8747–8750.

Hoop, B., and Peng, C.-K. Fluctuations and fractal noise in biological membranes. *J. Membr. Biol.* 177: 177–185.

Huber, C., and Wächtershäuser, G. 1998. Peptides by activation by amino acids with CO on (Ni,Fe)S surfaces: Implications for the origin of life. *Science* 281: 670–672.

Hurst, H.E. 1951. Long-term storage capacity of reservoirs. *Trans. Am. Inst. Civil Eng.* 116: 519–577.

Ji, N., and Shen, Y.R. 2004. Sum frequency vibrational spectroscopy of leucine molecules adsorbed at air-water interface. *J. Chem. Phys.* 120(15): 7107–7112.

Kawasaki, M., Sugita, A., Ramos, C., Matsumi, Y., and Tachikawa, H. 2004. Photodissociation of water dimer at 205 nm. *J. Phys. Chem.* 108: 8119–8124.

Kleidon, A., and Lorenz, R.D., eds. 2005. *Non-equilibrium Thermodynamics and the Production of Entropy.* Berlin: Springer.

Kumar, J.K., and Oliver, J.S. 2001. Proximity effects in monolayer films: Kinetic analysis of amide bond formation at the air-water interface using ${}^1H$ NMR spectroscopy. *J. Am. Chem. Soc.* 124: 11307–11314.

Lavigne, P., Tancrede, P., and Lamarche, F. 1998. The monolayer technique as a tool to study the energetics of protein-protein interactions. *Biochim. Biophys. Acta* 1382: 249–256.

Lerman, L. 1986. Potential role of bubbles and droplets in primordial and planetary chemistry: Exploration of the liquid-gas interface as a reaction zone for condensation processes. *Orig. Life Evol. Biosph.* 16: 201–202.

Lerman, L. 1994. The bubble-aerosol droplet cycle as a natural reactor for prebiotic organic chemistry. *Orig. Life Evol. Biosph.* 24: 111–112.

Lerman, L. 1996. The bubble-aerosol-droplet cycle: A prebiotic geochemical reactor. *Orig. Life Evol. Biosph.* 26: 369–370.

Lerman, L., and Teng, J. 2004. In the beginning. In *Origins, Genesis, Evolution and Diversity of Life*. Edited by J. Seckbach. Dordrecht: Kluwer Academic Publishers, pp. 35–58.

Malcolm, B.R. 1968. Molecular structure and deuterium exchange in monolayers of synthetic polypeptides. *Proc. R. Soc. London Ser. A*. 305: 363–385.

Mandelbrot, B.B. 1998. *Multifractals and 1/f Noise*. New York: Springer.

Mason, B.J. 1954. Bursting of air bubbles at the surface of sea water. *Nature* 174: 470–471.

Miller, S.L. 1953. A production of amino acids under possible primitive Earth conditions. *Science* 117: 528–529.

Miller, S.L. 1998. The endogenous synthesis of organic compounds. In *The Molecular Origins of Life*. Edited by A. Brack. Cambridge: Cambridge University Press.

Miller, S.L., and Urey, H.C. 1959. Organic compound synthesis on the primitive Earth. *Science* 130: 245–251.

Mrevlishvili, G.M. 1995. Aperiodic structures, fractals, and low-temperature heat capacity of biological macromolecules. *Biofizika*. 40: 485–496.

Nanita, S.C., Takats, Z., and Cooks, R.G. 2004. Chiral enrichment of serine via formation, dissociation, and soft-landing of octameric cluster ions. *J. Am. Soc. Mass Spectrom.* 15(9): 1360–1365.

Negron-Mendoza, A., and Ramos-Bernal, S. 2004. The role of clays in the origin of life. In *Origins: Genesis, Evolution and Diversity of Life*. Edited by J. Seckbach. Dordrecht: Kluwer Academic Publishers.

Oliviera, R.G., Tanaka, M., and Maggio, M. 2005. Many length scales surface fractality in monomolecular films of whole myelin lipids and proteins. *J. Struct. Biol.* 149: 158–169.

Orgel, L.E. 1998. Polymerization on the rocks: theoretical introduction. *Orig. Life Evol. Biosph.* 28: 227–234.

Ourisson, G., and Nakatani, Y. 1994. The terpenoid theory of the origin of cellular life: The evolution of terpenoids to cholesterol. *Chem. Biol.* 1: 11–23.

Paltridge, G.W. 1975. Global dynamics and climate—A system of minimum entropy exchange. *Q. J. R. Meteorol. Soc.* 101: 475–484.

Paltridge, G.W. 2001. A physical basis for a maximum of thermodynamic dissipation of the climate system. *Q. J. R. Meteorol. Soc.* 127: 305–313.

Paulius, O.M. 2005. In *Water Vapor and Entropy Production in the Earth's Atmosphere,* Chapter 9. Edited by Kleidon and Lorenz. Berlin: Springer.

Pfeilsticker, K., Loter, A., Peters, C., and Bosch, H. 2003. Atmospheric detection of water dimers via near-infrared absorption. *Science* 300: 2078–2080.

Pruppacher, H.R., and Klett, J.D. 1998. *Microphysics of Clouds and Precipitation*. London: D. Reidel Publishing.

Ptashnik, I.V., Smith, K.M., Shine, K.P., and Newnham, D.A. 2004. Laboratory measurements of water vapor continuum absorption in spectral region 5000–5600 cm^{-1}. *Q. J. R. Meteorol. Soc.* 130: 2391–2408.

Rani, M., and Mitra, C.K. 1995. Correlation analysis of frequency distributions of residues in proteins. *J. Biosci.* 20: 7–16.

Richardson, L.F. 1926. Atmospheric diffusion shown on a distance-neighbour graph. *Proc. R. Soc. London Ser. A*. 110: 709–737.

Richardson, L.F. 1961. The problem of contiguity: An appendix of statistics of deadly quarrels. *Gen. Syst. Yearb.* 6: 139–187.

Scatena, L.F., and Richmond, G.L. 2004. Isolated molecular ion solvation at an oil/water interface investigated by vibrational sum-frequency spectroscopy. *J. Phys. Chem. B*. 108(33): 12518–12528.

Schenter, G.K., Kathmann, S.M., and Garrett, B.C. 2002. Equilibrium constant for water dimerization: Analysis of the partition function for a weakly bound system. *J. Phys. Chem. A* 106: 1557–1566.

Schertzer, D., and Lovejoy, S. 1987. Physical modeling and analysis of rain and clouds by anisotropic scaling multiplicative processes. *J. Geophys. Res.* 92: 9693–9714.

Schertzer, D., and Lovejoy, S. 1991. *Nonlinear Variability in Geophysics: Scaling and Fractals*. Chapters 4, 7, and 8. Dordrecht: Kluwer Academic Publishers.

Segrè, D., Ben-Eli, D., Deamer, D.W., and Lancet, D. 2001. The lipid world. *Orig. Life Evol. Biosph.* 31: 119–145.

Sella, G., and Hirsh, A. E. 2005. The application of statistical physics to evolutionary biology. *Proc. Natl. Acad. Sci.* 102: 9541–9546.

Shah, D.O. 1972. The origin of membranes and related surface phenomena. In *Exobiology*. Edited by C. Ponnamperuma. New York, NY: North Holland, pp. 235–265.

Shimokawa, S., and Ozawa, H. 2005. Thermodynamics of the ocean circulation: A global perspective on the ocean system and living systems. In *Non-equilibrium Thermodynamics and the Production of Entropy*. Berlin: Springer.

Sjogren, H., and Ulenlund, S. Effects of pH, ionic strength, calcium and molecular mass on the arrangement of hydrophobic peptide helices at the air-water interface. *J. Phys. Chem. B* 108: 20219–20227.

Solé, R.V., and Munteanu, A. 2004. The large-scale organization of chemical reaction networks in astrophysics. *Europhys. Lett.* 68: 170–176.

Stenstam, G.H., Ardhammar, M., Norden, B., Sparr, E., and Ulenlund, S. 2002. Morphology and molecular conformation in thin films of poly-gamma-methyl-L-glutamate at the air-water interface. *Langmuir* 18(2): 462–469.

Tervahattu, H., Juhanoja, J., and Kupiainen, K. 2002. Identification of an organic coating on marine aerosol particles by TOF-SIMS. *J. Geophys. Res. (Atmos.)*. 107(D16): doi:10.1029/2001JD001403.

Tervahattu, H., Juhanoja, J., Vaida, V., Tuck, A.F., Niemi, J.V., and Kupiainen, K. 2005. Fatty acids on continental sulfate aerosol particles. *J. Geophys. Res. (Atmos.)* 110: D06207, doi:10.1029/2004JD005400.

Tervahattu, H., Tuck, A.F., and Vaida, V. 2004. Chemistry in prebiotic aerosols: A mechanism for the origin of life. In *Origins, Genesis, Evolution and Diversity of Life*. Edited by J. Seckbach. Dordrecht: Kluwer Academic Publishers.

Tuck, A. 2002. The role of atmospheric aerosols in the origin of life. *Surv. Geophys.* 23: 379–403.

Tuck, A.F. 2008. *Atmospheric Turbulence: A Molecular Dynamics Perspective*. Oxford: Oxford University Press.

Tuck, A.F., Hovde, S.J., Gao, R.S., and Richard, E.C. 2003. The law of mass action in the Arctic lower stratospheric polar vortex January–March 2002: ClO scaling and the calculation of ozone loss rates in a turbulent fractal medium. *J. Geophys. Res.* 108:(D15). Art. No. 4451.

Tuck, A.F., Hovde, S.J., Richard, E.C., Fahey, D.W., and Gao, R.S. 2002. A scaling analysis of ER-2 data in the inner vortex during January–March 2000. *J. Geophys. Res.* 108(D5): Art. No. 8306.

Tuck, A.F., Hovde, S.J., Richard, E.C., Gao, R.S., Bui, T.P., Swartz, W.H., and Lloyd, S.A. 2005. Molecular velocity distributions and generalized scale invariance in the turbulent atmosphere. *Faraday Discuss.* 130: 181–193.

Vaida, V., Daniel, J.S., Kjaergaard, H.G., Goss, L.M., and Tuck, A.F. 2001. Atmospheric absorption of near infrared and visible solar radiation by the hydrogen bonded water dimer. *Q. J. R. Meteorol. Soc.* 127: 1627–1643.

Vaida, V., Kjaergaard, H.G., and Feierabend, K.J. 2003. Hydrated complexes: Relevance to atmospheric chemistry and climate. *Int. Rev. Phys. Chem.* 22: 203–219.

Vigasin, A.A. 1983. Structure and properties of associated water. *J. Str. Chem.* 24: 102–131.

Watry, M.R., and Richmond, G.L. 2002. Orientation and conformation of amino acids in monolayers adsorbed at an oil/water interface as determined by vibrational sum-frequency spectroscopy. *J. Phys. Chem. B.* 106: 12517–12523.

Woese, C.R. 2002. On the evolution of cells. *Proc. Natl. Acad. Sci.* 99: 8742–8747.

Wolynes, P.G., and Roberts, R.E. 1978. Molecular interpretation of the infrared water-vapor continuum. *Appl. Opt.* 17: 1484–1485.

Yamamoto, S., Tsujii, Y., and Fukuda, T. 2001. Characteristic phase-separated monolayer structure observed for blends of rodlike and flexible polymers. *Polymer* 4(5): 2007–2013.

17 What Is the Diversity of Life in the Cosmos?

Peter D. Ward

CONTENTS

17.1   INTRODUCTION

Diversity, made up of the number of taxa found in a defined area, is usually measured by counting species (species richness). In similar fashion, the diversity of genera or even families within a given area can be measured, and all of these are used as descriptors for what is called "the diversity of life." Here I will describe another kind of diversity of life—not the number of species, but the number of separate chemistries of potentially viable life. It has been taken for granted, ever since Darwin said that it was so, that there is only one kind of life on Earth, even if it is subdivided into millions of species. That is our familiar DNA-life, which uses the same genetic code and same twenty amino acids to build the proteins for all species. Thus, if the diversity of life is based on a particular chemistry of life, it can be said that the diversity of life on Earth is one. A still unanswered question is whether Earth life is the only way, or but one way, to achieve a living status and, in reality, we do not know if DNA life is the unique chemistry of life on Earth. Thus, if there is other life in the cosmos (or even undiscovered here on Earth), how similar will it be to Earth life? All Earth life found to date is united in having DNA, a specific genetic code for the same twenty amino acids, and is water based. But is our life typical of all living cells throughout the universe, or just one way to construct a living cell?

There are many definitions of life, and all use some variant on defining life as a set of chemical reactions that allow metabolism (the harvesting of energy), replication, and evolution. But this uniformity of what life does may not extend to how life can be constructed. A structural definition of life might be that it is based on a dominant structural element (in our case, carbon), it has an information storage system (in our case, DNA), and it uses a specific solvent (in our case, water). It is in these three areas that life beyond the Earth may significantly differ from Earth life. Thus while it has been argued that all life will be based on carbon chemistry (Pace, 2001), there may be many kinds of life that differ from Earth life, even in terms of using carbon (Feinnberg and Shapiro, 1980; Koerner and LeVay, 2000; Irwin and Schulze-Makuch, 2001; Crawford, 2001; Grinspoon, 2004; Benner et al., 2004; Ward, 2005), just as there may be alternatives to Earth-life DNA, and the Earth-life solvent. In this chapter, I will briefly explore alternatives in all three of these areas that have been studied by previous investigators as having the potential of being part of a viable, "life as we do not know it."

17.2 THE CHEMICAL MAKEUP OF LIFE

Until recently, any discussion of the diversity of life not at its species level, but at the other end of the taxonomic spectrum—the kind of chemistry it uses, has been handicapped by a lack of taxonomic categories at the highest end of the Linnaean taxon hierarchy (the familiar domain, kingdom, phylum, class, order, family, genus, and species). Domain, heretofore the highest category, is still composed of our familiar DNA life. Yet life based on an alternate chemistry would not fit anywhere on the Tree of Earth Life and, for this reason, Ward (2005) proposed a new taxon, called an arborea, that includes all life with a similar biochemistry. Silicon-based life, if it exists, would occur within an arborea different from that of Earth-specific DNA life.

Earth life has been designated as "CHON" life because of its preponderance of carbon, hydrogen, oxygen, and nitrogen (Benner, 2004). As noted above, the particular CHON life (the specific Earth arborea) found on Earth uses DNA for its genome, builds proteins from 20 specific amino acids, has several common metabolic systems for energy extraction, and uses a bilayer cell membrane made up of lipids. Changing any of these would create an "alien" life form. Here are ways that such life might be changed (i.e., how different arborea might be composed).

17.3 SINGLE VERSUS MULTIPLE BIOPOLYMER LIFE

Benner (2004) broke CHON life down in the following way: life with one biopolymer (which could be a nucleic acid, and conceivably protein), and life with more than one biopolymer. Life using a single biopolymer—such as hypothetical RNA life—must use its biopolymer for the necessary functions of information storage and catalysis. Earth life is a dual biopolymer system: DNA and protein. Because dual biopolymer life must evolve from the single type, Benner inferred that single biopolymer life forms might be more abundant in the universe. Thus, to start this section, we can divide CHON life using these two great categories.

There are interesting implications for single versus multiple biopolymer life. By necessity, the two will be very different kinds of life. A biopolymer evolved and used specifically as a catalyst, such as the protein used in Earth life, will be optimally built of many elemental and molecular building blocks, so as to make the widest spectrum of chemical reactions possible. But the biopolymer used for replication will want the fewest number of separate components, so as to ensure more accurate replication, since mistakes in replication are disastrous and usually lethal. Furthermore, the molecule used in catalysis, such as a protein, must be able to easily fold into multiple shapes, whereas the information-carrying molecule cannot afford to fold easily if it is to serve as template (Benner and Hutter, 2002). Thus we see that, in the dual biopolymer organisms, extreme and divergent evolution has created two classes of molecules that are radically different from one another. This affords a far wider range of possibilities in both the nature of information stored, and the diversity of chemical reactions that can be undertaken. In contrast, the single biopolymer life will be far more limited in both abilities. But there may be environments where it is advantageous, especially in cold systems. Also, they require fewer resources, and can be smaller both physically and in terms of genome size.

17.4   LIFE WITH CHEMICALLY DIFFERENT DNA

Our familiar DNA is a double helix made up of two long strands of sugar, with the steps of this twisted ladder made up of four different bases. The code is based on triplet sequences, with each triple either an order to go fetch and attach a specific amino acid floating around in the soupy, organic cell interior, or a punctuation mark—like "stop here." Within this elaborate system, there are many specific changes that could be made—at least theoretically—that would be alien and yet might still work.

Perhaps the simplest change would be to change the alphabet—there is no *a priori* reason that the current base triplet has to be specific for the amino acid it now codes for. Another simple change is to add more letters to the alphabet, and several groups have done this. The four traditional bases of the DNA ladder—adenine, thymine, cytosine, and guanine—are still used. But instead of a sequence or triplet of three being specific for each amino acid, a sequence with some other number would be used. Thus, Benner et al. (1995) created a code using 12 base pairs per letter instead of 3, whereas Sismour et al. (2004) demonstrated that nonstandard base pairs can be inserted into DNA using viruses. Chin et al. (2003) have also created an expanded genetic code, but in a eukaryote, rather than a prokaryote as in the other studies.

Another way to change the language or coding is to use new or additional bases. A DNA-like molecule containing six gene-building nucleotides instead of the traditional four has been created (Benner, 2004). This artificial molecule was able to replicate, through the photocopying-like action of the polymerase chain reaction (PCR). Benner's new DNA look-alike was able to go through five generations of replication.

A final way is to change other structural components of DNA beyond the bases. This was accomplished by Schneider and Benner (1990), while various possible configurations of alternative DNA structure were listed by Huang et al. (1991, 1993), Eschenmoser (1999), Eschgfaeller et al. (2003), and Geyer et al. (2003).

These experiments clearly show that DNA could come in many varieties of language and structure, and changing the base coding or base number would be simple. It would be interesting to know if, early in Earth history, many separate kinds of DNA with different codes competed one against each other. Is a 12-nucleotide DNA code more or less efficient than our familiar three nucleotide coding? Was there competition among a whole series of slightly different DNA, with our current version proving competitively superior, or was ours simply the first to achieve this grade of organization, suppressing a variety of equally, or even more effective, competitors through some sort of incumbency advantage?

17.5   LIFE USING DIFFERENT AMINO ACIDS

Earth life uses the same twenty amino acids. However, it has been shown that familiar polypeptides that are composed of amino acids not used by Earth life can be synthesized (Bain et al., 1989; Hohsaka and Masahiko, 2002). Thus, it is possible that there could be DNA life similar in all other respects to Earth life save for the use of amino acids not found in Earth life.

17.6   LIFE WITH A DIFFERENT SOLVENT

Cells need to be filled with liquid (because it does not appear plausible for life to exist in either an entirely solid or gas phase) and this must be a solvent. To be useful for life, a solvent must remain liquid (and thus work) under a fairly wide range of temperatures, be able to regulate internal temperatures, and be accessible to life by being abundant in the universe. Water is the solvent used by all Earth life, and it is a necessity of life as we know it. But other existing solvents might also be used by theoretical, non-Earth life.

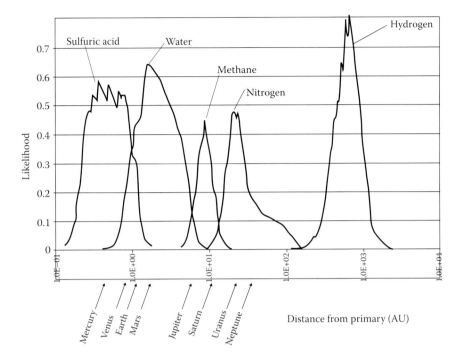

FIGURE 17.2   Potential solvents as a function of distance from the sun. (From Bains, W. 2004. *Astrobiology.* 4:137–167. With permission.)

(Dabrowska, 1984). The lower temperatures of Earth life are currently the center of intense investigation (Junge et al., 2003, 2004), but it is clear that the temperatures on Titan, Europa, and even Mars would pose severe challenges for Earth life, and other chemistries of life might do better on these planets and moons. The question thus becomes: can Earth biochemical systems function in these more exotic liquids and in the extreme cold? And if not, what kind of chemistry would do better? There is no doubt that the Earth biochemical systems are enormously adaptable, but there are limits—especially with regard to the solvent chemistry and temperature (Lu et al., 1990; Stroppolo et al., 2001). Many Earth-life biochemical molecules will not be viable in liquid such as water/ammonia solutions at very low temperatures, while the high pH found in many of these oceans (such as that of Europa) would quickly break DNA and RNA apart through hydrolysis.

Of the various kinds of potential life using different solvents, that based on ammonia has been the most studied (Firsoff, 1962, 1965; Sagan, 1973; Bains, 2004; Benner, 2004). Such life would have to have a very different kind of outer cell wall or membrane, since liposomes, a major structural part of Earth-life cell walls, dissolve in ammonia. Yet in spite of this challenge, Benner (2004) suggested that such life might be viable. Where Earth life exploits compounds using carbon–oxygen bonds in metabolic pathways (specifically the unit known as carbonyl), Benner posited a workable metabolism using carbon–nitrogen bonds.

17.7   SILICA-BASED (SILANE) LIFE

All of the life listed above is essentially carbon-based, or CHON life. In this section, I will examine potential life that is based on silica. If such life existed, it would form its own specific arborea.

As described above, cold is a challenge for carbon biochemistry because necessary chemical reactions run slowly, or not at all. At 70 K (−200°F), virtually nothing dissolves in any solvent. But at such temperatures, nitrogen, ethane, and methane exist in a liquid state. This works against carbon-based life. Organic molecules such as methane, acetylene, and carbon dioxide will dissolve in these liquids at these temperatures, but larger organic molecules will not, with one notable exception. Silicon can form an analog to carbon-based alcohol, and thus can be considered an organic molecule (Tacke and Linoh, 1989). These are called silanols (or silanes) (Lickiss, 2001), and they are soluble in a wide variety of solutes, at a huge range of temperatures, including the very cold temperatures at which nitrogen remains liquid (Harrison, 1997; McCarthy et al., 2003; Benner, 2004). When they dissolve, they populate the solute—be it ethane, methane, or liquid nitrogen—with analogs to the carbon-based organic molecules (Patai and Rappoport, 1989). They are large molecules, and once dissolved they can form polymers—the chained-together molecules that are necessary for life (Brook, 2000; West, 2002; Benner et al., 2004). And, because they are silica-based rather than carbon-based, they have weaker bonds to break—they thus have greater reactivity than carbon-based compounds (Walsh, 1981). This is a problem at Earthly temperatures, making silica a liability as a molecule central to life, but in very cold environments, this is an advantageous property for life. Because of this, less catalytic efficiency is needed to perform necessary chemical reactions by this theoretical silica. For these two reasons, the ultra-cold environments of our solar system would be a more favorable place for some type of silicon life than for carbon life.

Life is far more than simple chemical reactions. What of the structural components of silicon life? There are numerous silicate minerals that form very stable chemical compounds. Silicate rocks are composed of large chains or sheets of silicon bonded to oxygen. Silicon can also form stable polymers, a building block of life, by Si–Si molecules, as well as Si–C bonds (Hayase et al., 1989). It is, thus, a mistake to consider any possible silicon life as being composed mainly of silicon, just as it is a similar mistake to think of carbon life as mainly carbon. Even complex carbon compounds can bond to the silica chains as side branches, thus allowing silanes to form diverse and complex molecules with side chains making them analogous to carbohydrates, nucleic acids, and proteins (Figure 17.3). Silicon compounds can even form ring structures like the carbon compound benzene, and like sugars, thus illustrating silicon versatility (Sanji et al., 1999).

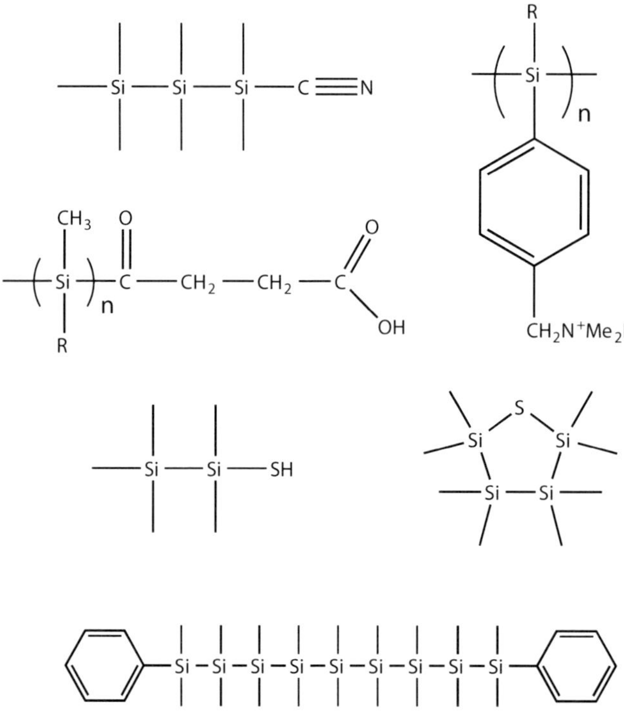

FIGURE 17.3   Kinds of silanes. (From Benner, S. A., Ricardo, A., and Carrigan, M., 2004. *Curr. Opin. Chem. Biol.* 8:672–689. With permission.)

What of metabolism—how will silicon chemistry be modified to yield energy? Where many of the metabolic systems of carbon life shuttle protons in energy harvesting systems (like the ATP–ADP system), silicon might do the same not with protons, but with electrons (Benner, 2004). There could be light activated electronic effects, mimicking photosynthesis, as well as other energy analogies.

Where might we find silicon life? Not in water, and not in water–ammonia solutions, because these solvents would quickly destroy the complex Si-organic molecules. But on Titan there are ethane/methane lakes and, on Triton, perhaps an ocean of liquid nitrogen. In such places, we might find silicon-based life.

17.8   SILICON/CARBON CLAY LIFE

Cairns-Smith (1982) proposed a second kind of silicon-based life, using crystals for structural elements. He envisioned a life form based on a flake of clay (invisible to the naked eye)—crystals of clay that actively grow—and evolve as they do so. Cairns-Smith singles out the clay known as kaolinite as a prime example of how this type of life might live. Following formation of such silicon-clay life, Cairns-Smith envisions a carbon takeover. If such life could exist, it would certainly qualify as being different from Earth life.

17.9   DISCUSSION

From the brief summary above, it seems evident that there may be a large diversity of life in the universe in the sense of the number of different chemical kinds of life (Ward, 2005), not just species (although as the chemical diversity of life goes up, so too must the species richness of the various chemical varieties of life also climb in number).

In the discussion so far, all of the kinds of life referred to relate to naturally evolved life. The actual chemical diversity of life in the cosmos is surely less than the number of chemically permissible kinds of life, for even if a specific biochemistry can work (i.e., live), it still has to be evolved. Furthermore, just as individuals and even species die out, so too with the various chemical species of life—their chemicals must fuse to form living matter. This kind of life exists for some period of time, and then it will probably go extinct, perhaps to reform through convergent evolution at some later time and place in space, but perhaps not. Because of this, the diversity of arborea in the cosmos might be constantly changing, and fascinating questions about the kinds of arborea possible during different stages of cosmos evolution can be posed. But it is becoming clear that there can be life that exists, but that has not come to its present chemistry through natural processes alone. Life chemically different from naturally evolved Earth life has now been produced on Earth.

The two methods of artificially creating life are bottom-up—essentially starting from scratch and synthesizing all the components of a cell: its wall, genome, and metabolic and replication machinery. The other is top-down—taking an existing cell and simplifying it by removing parts of its genome to minimal levels. Both kinds of work are ongoing, but the top down methodologies seem most promising.

The target organism for the top-down experimentalists is the microbe *Mycoplasma genitalium*, a bacterium with the simplest known genetic code yet discovered of only 580,000 bases in its genome, coding for about 480 genes. It has already been determined that at least 130 of these 480 genes are not essential for life, and another 85 have been identified that may not be essential either. While this is still DNA (Earth arborea) life, it shows the direction of future research toward chemically different varieties of life. Work on synthetic viruses is even further along. One group, headed by Echard Wimmer of the State University of New York at Stony Brook, has successfully synthesized the genome of the Poliovirus. The group went on to build an infectious particle with a genome length of 7500 nucleotides (Cello et al., 2002). This is not the only virus synthesized in recent months. Craig Venter and crew (the force behind the private attempt to delineate the human genome) constructed an entirely synthetic genome for another virus, a parasite on the common human-inhabiting

TABLE 17.1
Potential Alien Life Forms

Name	Scaffold Element	Gene Material	Solvent	Scaffold Element Source	Energy Source	Possible Terrestrial Habitat	Solar System Habitat	Plausibility
Earth life	Carbon	DNA	Water	CO_2, other organisms	Many	Most	Mars, Europa, Titan	It exists
RNA life	Carbon	RNA	Water	Organic molecules	?	?	Titan, Mars, Europa, Earth?	High—it once existed and may still on Earth
Protein life	Carbon	Proteins	Water	Organic molecules	?	?	Titan, Mars, Europa, Earth?	
Ammonia life	Carbon	Nucleic acids or proteins	Ammonia	CO_2? Other?	Sunlight	None		
Acid life	Carbon	Nucleic acids	Water or ammonium				Venus clouds, Jupiter clouds?	
Silicon life—Silanes	Silicon		Ethane	Crystal fluids			Titan, Triton	
Silicon/clay	Silicon		Water				Ancient Earth? Mars?	

bacterium *Escherichia coli* (Smith et al., 2003). Once again the new viral particle was infectious, demonstrating traits that many would equate with life. The genome was 5400 nucleotides in length. What was even more impressive was the rapidity with which it was synthesized: it took only 14 days to build this genome from scratch.

These two examples show that the methodology for synthesizing the genome of even the most complicated virus is now available. The same methodologies can be used to build a bacterial cell as well, or at least something that resembles a bacterium—a larger form of artificial life than a virus. An ethical review panel was commissioned to consider the implications of the Venter virus, and found no evident problem with this new line of work, concluding: "The prospect of constructing minimal and new genomes does not violate any fundamental moral precepts or boundaries, but does raise question that are essential to consider before the technology advances further."

17.10 SUMMARY

In this brief chapter I have listed a summary of potential kinds of alien life (i.e., not Earth-like life, or life as we know it), and these are summarized in Table 17.1. While these kinds of "life as we do not know it" may be theoretically plausible, they may not be present simply because there was no pathway by which they could evolve. Nevertheless, continued research into how life different from our own might be built will provide valuable insights into how our own kind of life works.

REFERENCES

Bain, J. D., Diala, E. S., Glabe, C. G., Dix, T. A., and Chamberlin, A. R. 1989. Biosynthetic site-specific incorporation of a non-natural amino acid into a polypeptide. *J. Am. Chem. Soc.* 111: 8013–8014.

Bains, W. 2001. The parts list of life. *Nat. Biotechnol.* 19: 401–402.

Bains, W. 2004. Many chemistries could be used to build living systems. *Astrobiology*. 4: 13–167.

Baross, J. A., and Deming, J. W. 1995. Growth at high temperatures: Isolation and taxonomy, physiology, and ecology. In *The Microbiology of Deep-Sea Hydrothermal Vents*. Edited by D. M. Karl. Boca Raton: CRC Press, 169–217.

Baross, J. A., and Holden, J. F. 1996. Overview of hyperthermophiles and their heat-shock proteins. *Adv. Protein Chem.* 48: 1–35.

Benner, S. A. 2004. Understanding nucleic acids using synthetic chemistry. *Acc. Chem. Res.* 37: 784–797.

Benner, S. A., and Hutter, D. 2002. Phosphates, DNA, and the search for nonterran life. A second generation model for genetic molecules. *Bioorg. Chem.* 30: 62–80.

Benner, S. A., Horlacher, J., Voegel, J., and von Krosigk, U. 1995. A new molecular encoding system. In *Supramolecular Stereochemistry*. Boston: Kluwer Academic Publishers, 47–56.

Benner, S. A., Ricardo, A., and Carrigan, M. 2004. Is there a common chemical model for life in the universe? *Curr. Opin. Chem. Biol.* 8: 672–689.

Brook, M. A. 2000. *Silicon in Organic, Organometallic and Polymer Chemistry*. Toronto: John Wiley & Sons.

Cairns-Smith, A. (1982). *Genetic Takeover and the Mineral Origins of Life*. Cambridge, UK: Cambridge University Press.

Cello, J., Paul, A. V., and Wimmer, E. 2002. Chemical synthesis of poliovirus cDNA: Generation of infectious virus in the absence of natural template, *Science* 297: 1016–1018.

Chin, J. W., Cropp, T. A., Anderson, J. C., Mukherji, M., Zhang, Z. W., and Schultz, P. G. 2003. An expanded eukaryotic genetic code. *Science* 301: 964–967.

Crawford, R. L. 2001. In search of the molecules of life. *Icarus* 154: 531–539.

Dabrowska, B. 1984. The solubility of selected halogen hydrocarbons in liquid nitrogen at 77.4 K. *Cryogenics* 24: 276–277.

Eschenmoser, A. 1999. Chemical etiology of nucleic acid structure. *Science* 284: 2118–2124.

Eschgfaeller, B., Schmidt, J., and Benner, S. A. 2003. Synthesis and properties of oligodeoxynucleotide analogs with bis(methylene) sulfone-bridges. *Helv. Chim. Acta* 86: 2957–2997.

Feinnberg, G., and Shapiro, R. 1980. *Life beyond Earth*. New York, NY: William Morrow.

Firsoff, V. 1962. An ammonia based life. *Discovery* 23: 36–42.

Firsoff, V. 1965. Possible alternative chemistries of life. *Spaceflight* 7: 132–136.

Geyer, C. R., Battersby, T. R., and Benner, S. A. 2003. Nucleobase pairing in Watson-Crick-like genetic expanded information systems. *Structure* 11: 1485–1498.

Grinspoon, D. 2004. *Lonely Planets*. New York, NY: HarperCollins.

Harrison, P. G. 1997. Silicate cages: Precursors to new materials. *J. Organomet. Chem.* 542: 141–184.

Hayase, S., Horiguchi, R., Onishi, Y., and Ushirogouchi, T. 1989. Syntheses of polysilanes with functional groups: 2. Polysilanes with carboxylic acids. *Macromolecules* 22: 2933–2938.

Hohsaka, T., and Masahiko, S. M. 2002. Incorporation of non-natural amino acids into proteins. *Curr. Opin. Chem. Biol.* 6: 809–815.

Huang, Z., Schneider, K. C., and Benner, S. A. 1991. Building blocks for analogs of ribo- and deoxyribo-nucleotides with dimethylene-sulfide, -sulfoxide and -sulfone groups replacing phosphodiester linkages. *J. Org. Chem.* 56: 3869–3882.

Huang, Z., Schneider, K. C., and Benner, S. A. 1993. Oligonucleotide analogs with dimethylene-sulfide, -sulfoxide and -sulfone groups replacing phosphodiester linkages. In *Methods in Molecular Biology,* 20: 315–353. Edited by S. Agrawal. Totowa, NJ: Humana Press.

Irwin, L. N., and Schulze-Makuch, D. 2001. Assessing the plausibility of life on other worlds. *Astrobiology* 1: 143–160.

Junge, K., Eicken, H., and Deming, J. W. 2003. Motility of *Colwellia psychrerythraea* strain 34H at subzero temperatures. *Appl. Environ. Microbiol.* 69: 4282–4284.

Junge, K., Eicken, H., and Deming, J. W. 2004. Bacterial activity at −2 to −20 degrees C in Arctic wintertime sea ice. *Appl. Environ. Microbiol.* 70: 550–557.

Koerner, D., and LeVay, S. 2000. *Here Be Dragons*. Oxford: Oxford University Press.

Lickiss, P. 2001. Polysilanols. In *The Chemistry of Organic Silicon Compounds,* Vol. 3. Edited by Z. Rappoport and Y. Apeloig. Chichester, UK: John Wiley & Sons, pp. 695–744.

Lu, B. C., Zhang, D., and Sheng, W. 1990. Solubility enhancement in supercritical solvents. *Pure Appl. Chem.* 62: 2277 –2285.

McCarthy, M. C., Gottlieb, C. A., and Thaddeus, P. 2003. Silicon molecules in space and in the laboratory. *Mol. Phys.* 101: 697–704.

Pace, N. R. 2001. The universal nature of biochemistry. *Proc. Natl. Acad. Sci. USA* 98: 805–808.

Patai, S., and Rappoport, Z., eds. 1989. *The Chemistry of Organic Silicon Compounds*. Chichester, UK: John Wiley & Sons.

Saenger, W. 1987. Structure and dynamics of water surrounding biomolecules. *Annu. Rev. Biophys. Biophys. Chem.* 16: 93–114.

Sagan, C. 1973. Extraterrestrial life. In *Communication with Extraterrestrial Intelligence (CETI)*. Edited by C. Sagan. Cambridge, MA: MIT Press, 42–67.

Sanji, T., Kitayama, F., and Sakurai, H. 1999. Self-assembled micelles of amphiphilic polysilane block copolymers. *Macromolecules* 32: 5718–5720.

Schneider, K., and Benner, S. A. 1990. Oligonucleotides containing flexible nucleoside analogs. *J. Am. Chem. Soc.* 112: 453–455.

Sismour A. M., Lutz, S., Park, J.-H., Lutz, M. J., Boyer, P. L., Hughes, S. H., and Benner, S. A. 2004. PCR amplification of DNA containing non-standard base pairs by variants of reverse transcriptase from human immunodeficiency virus-1. *Nucleic Acids Res.* 32: 728–735.

Smith H. O., Hutchison, C. A. 3rd, Pfannkoch, C., and Venter, J. C. 2003. Generating a synthetic genome by whole genome assembly: PhiX174 bacteriophage from synthetic oligonucleotides. *Proc. Natl Acad. Sci. USA* 100: 15440–15445.

Stroppolo, M. E., Falconi, M., Caccuri, A. M., and Desiderim, A. 2001. Superefficient enzymes. *Cell. Mol. Life Sci.* 58: 1451–1460.

Tacke, R., and Linoh, H. 1989. Bioorganosilicon chemistry. In *The Chemistry of Organic Silicon Compounds, Part 2.* Edited by S. Patia and Z. Rappoport. Chichester, UK: John Wiley & Sons, pp. 1143–1206.

West, R. 2002. Multiple bonds to silicon: 20 years later. *Polyhedron* 21: 467–472.

Walsh R. 1981. Bond dissociation energy values in silicon-containing compounds and some of their implications. *Acc. Chem Res.* 14: 246–252.

Ward, P. 2005. *Life as we do not know it*. New York, NY: Viking Penguin.

18 The Primordial Bubble
Water, Symmetry Breaking, and the Origin of Life

Louis Lerman

CONTENTS

18.1 A PRELUDE ON THE SYMMETRY OF ORIGINS AND THE INTERFACIAL NATURE OF SELF-ORGANIZATION

Origin problems are the most conjectural and qualitative of scientific questions. It is hardly surprising then that the origin of life, like the origin of the universe, lacks uniquely defining quantitative assumptions and initial conditions. Individually, they are Fermi questions of a functional sort, requiring the use of a first-principles conceptual approach to a situation of fundamental quantitative ignorance. Indeed, concepts useful in the exploration of one question significantly help our investigation of the other: this is true for underlying the cosmochemistry of life is a cosmophysics hinting at congruences of symmetry-breaking problems in both the origin of the universe and the origin of life. But where theorists such as Lee Smolin have sought to apply the ideas and concepts of evolution to cosmology (Smolin, 1997), in the present work we present a new concept of physical self-organization and apply it to (prebiotic chemical) evolution.*

The physicist's tool of symmetries and their breaking is applied quite naturally here, as issues of symmetry at the grandest scale are fundamental to the Anthropic Principle. This is because the Anthropic Principle can be elegantly read as a statement of symmetry breaking with regard to the "Symmetry of Probabilities" (which is itself the basis of the physical worldviews of Copernicus, Newton, and Einstein—e.g., "There does not exist a privileged position in phase space"). Hence, the numerical tuning of cosmological constants is a qualitatively similar problem to that of the bio-chemical tuning of life: How did each arise out of all the possibilities and how was this "symmetry of probabilities" broken?

We emphasize in the present work the critical role of microenvironments and their interfaces in the act of symmetry breaking and the consequent building of structure. We suggest that at every physical scale it is the boundaries of the relevant microenvironments that play an essential organizing role. Generalizing this idea further naturally incorporates the notion of symmetry breaking. In particular, we propose the following: that least-action and least-energy principles applied to symmetry breaking leads to the universal role of interfaces as the symmetry-breaking microenvironments *necessary* for self-organization to occur. The key points[†] of this thesis are arranged alliteratively as follows:

Lerman's Lemma of Self-Organizing Systems

Self-organization over all scales of nature occurs in dimensionally nested hierarchies of "first-order"[‡] microenvironments; and where the matching of the space and time scales of the microenvironments to the self-organizing process is critical.

Lerman's Law of Self-Organization

Self-organizing microenvironments are found specifically at the interfaces between phases, regimes, gradients, forces, and/or dynamic processes.   ■

By microenvironments, one means the (heterogeneous) subsets of a bulk-average environment, whatever the scale. To understand what is meant by a "nested hierarchy," one can think of nested Russian dolls. Taking the terrestrial ocean as example, microenvironments can range from the sea-surface microlayer to the region surrounding a single bubble. These two microenvironments are, in fact, intimately related; a near-surface bubble being a microcosm of the air–sea interface, which itself brings into being the unique chemistry of the sea-surface microlayer (MacIntyre, 1974a, 1974c; Sieburth, 1983). Hence, an example of two elements of a nested hierarchy of microenvironments

---

* Because of the highly interdisciplinary nature of these problems, this paper uses ideas from a variety of fields. We start with a bit of abstract conceptual physics in this section, evolve by Section 18.3 to the functional requirements of prebiotic chemistry, and starting in Section 18.5 devote the majority of this paper to explorations of planetary geochemistry and the origin of life.

† This work is part of an effort to develop a more general approach to self-organizing processes. See Lerman (2002d) on a new form of symmetry breaking in relativistic nuclear collisions, and Lerman (2003, 2005c) for initial presentations of the above phenomenologically based laws.

‡ By first-order microenvironment we mean *most probable*. It need not be a first-order phase transition, although the boundaries of these regions will often involve a phase-transition of the first-order.

that, as we will clearly demonstrate in this paper, have a set of chemistries and chemical physics unique and highly differentiated* from that of the ocean as a whole.

A simple illustration of self-organization that occurs at a boundary between dynamic processes would be "salt fingers" formed at the interface between opposing temperature and salinity gradients. The direction of a gradient-induced transport is essential, for a conservative quantity transported down its own gradient results instead in a mixed state of higher entropy. Indeed, driven by symmetry-breaking negentropic processes, self-organization at any scale *must* occur at interfaces almost by definition. The basic logic for this abstraction is as follows:

1) Symmetry and structure are inversely correlated.
   To illustrate this nonintuitive concept: Although a snowflake has greater structure, the water in a snowflake dispersed throughout a room as a vapor has much greater symmetry (the few reflection and rotation symmetries of the snowflake are dominated by the exchange symmetries of a gas).
2) Symmetry breaking → structure building
   By definition, the breaking of a symmetry changes the homogeneity of the property represented by the invariance (under the group operator) that describes the initial symmetry. The initial homogeneous state becomes heterogeneous, and almost by definition, structure is created. Structure building is the essence of a self-organizing process.
3) Symmetry breaking → boundaries
   Along with the creation of structure, a transition region (i.e., a boundary) between the now heterogeneous regions is created. Again, by definition, it is at the boundary (the interface) between these heterogeneous regimes that the describing symmetries are globally discontinuous, i.e., broken.
4) Boundaries and least energy → microenvironments
   Least-energy constraints lead to boundary conditions that, in turn, create geometrical and topological constraints on the now heterogeneous regions. At the scale of the symmetry breaking, these topologically constrained heterogeneous regions are the microenvironments that make up the structure.
5) Broken symmetries build upon (previously) broken symmetries
   By induction, the next level and scale of structure building will occur at these boundaries. One of the primary reasons for this is the decrease in entropic possibilities (in the relevant phase space) that necessarily exist at these interfaces.
6) Applying the above: Self-organization occurs at the boundaries of microenvironments.

The logic described above derives from the fundamental fact that entropy, being a quantitative measure of disorder, is equal to the number of accessible states that look the same. In other words, entropy is a measure of the symmetries of a system; so breaking symmetries reduces entropy and creates structure. Moreover, Noether's theorem (Noether, 1918) states that symmetries of a system correspond with conserved quantities; but from nonequilibrium thermodynamics we know that structure comes from systems that are far from equilibrium. This means that they involve systems in which some of the thermodynamic variables are not conserved, i.e., symmetries are broken.

It is hoped that these principles will have a philosophical component analogous to the way in which the variational principle embodies natural philosophy: and that the underlying logic and approach can be formalized and applied to many scales of phenomena. In addition, they are intended to provide critical insights into how and where (self-)organization takes place from one level of structure to the next. For example, an important consequence for astrobiology is that chemistries

---

* With respect to the world of chemistry and chemists, "This is distinct from a traditional chemical view, where heterogeneity is clear at the 'beaker scale' (things are in the beaker or not) and at the molecular scale, but at scales in between, the chemist hopes to have an averaged environment." Steve Benner (private communication, 2005).

based on different values of the fundamental physical constants, or "merely" different chemistries than our own water-soluble organics (Benner, 2004), will require analogous forms of microenvironments and their boundaries.

Alternative chemistries based on alternative physics, if they are to support the conversion of inanimate matter to structured living systems, will need to be evaluated not just with respect to bulk-state reactions but by their ability to form interfaces and microenvironments. This will restrict the number and nature of otherwise "plausible" variant chemistries (Benner, 2004). That is, if the alternative chemistries are to support the emergence of life, then a mutual incompatibility of two liquid phases would seem to be necessary. Attempts to use nonpolar solvents (such as methane on Titan) will be especially affected by this requirement. Without some form of polarity it is going to be tough going, for the polar nature of the underlying molecules is itself the symmetry-breaking template upon which the rest of the (self-)organizing process is built.

Certainly with respect to the phenomenological model described in this paper, the linkage between water, its symmetry-breaking interfaces, and its ability to support chemical self-organization in a prebiotic context is not just intimate, it is fundamental.

18.2   THE SCIENTIFIC PROBLEM OF PREBIOTIC CHEMICAL EVOLUTION

At every scale of the observable universe, the evolution from simple to complex is congruent with what we see. Thus, in this paper, we assume evolutionary self-organization regardless of the scale of the physics, chemistry, or cosmic geography. Hence, the origin of life is assumed to have occurred through the processes of prebiotic chemical evolution, presupposing successive generations of increasingly complex organic molecules combinatorially synthesized from earlier generations. Less obvious is how this combinatorial chemistry occurred or how the overall process of chemical self-organization was functionally supported at each stage of its occurrence in the prebiotic environment.

This latter point, the functional support of prebiotic chemical evolution within realistic planetary environments, is a requirement for any nook or cranny of a universe harboring ambitions to support the development of life, in any chemistry or form. It is a goal of this work to explore the mechanisms by which water, through its ability to form microenvironments, offers a fundamental functional support for the processes of prebiotic chemical self-organization.

Unfortunately, we know far more about the evolution of life on Earth, its false steps, and its successful building blocks, than we know about how it came about in the first place. But just as the biochemistry of contemporary organisms can be viewed as a "fossil" record of biogenesis, the geochemical physics of the contemporary Earth can help delineate the self-organizing processes underlying prebiotic chemistry on an early Earth, Mars, or any other terrestrial-like body. Taking the minimalist assumption of microscopic stochastic processes, we first consider the system and molecular-level requirements necessary to support a bootstrapping of molecular organization. Especially when considering alternative chemistries to the origin of life, it is critical to look at functional requirements from the standpoint of first principles, utilizing only the most basic and transparent of assumptions and imposed boundary conditions.

18.3   CHEMICAL EVOLUTION AS PROCESS: SYSTEM
AND MOLECULAR LEVEL REQUIREMENTS

Continuing on in a Fermi question-like manner, we necessarily assume ignorance of the totality of reactions involved in the creation of successive generations of increasingly complex prebiotic organic molecules. But using the stochastic nature of physics applied to the simple-to-complex nature of chemical self-organization means that each level of self-organization evolved from broadly ranging consecutive chemical trial and errors. Some pathways will work, leading to products useful for the construction of the next level of organization, but most will not. Hence, in the abstract, our presupposed processes of chemical evolution require the existence of a global chemical engineering system (Lerman, 1986, 1992, 1994a, 1994b, 1996, 2002b; Lerman and Teng, 2004a), which at the system level amplify the probabilities of stochastic self-organization at the molecular level. For this amplification to be effective at the system level, these still-to-be discovered processes must be:

- Probable ("had-to-have-been" mechanisms)
- Robust (highly efficient and rapid processes)
- Diverse (exploring many different possible chemical routes and mechanisms)
- Selective (for the good stuff)
- Semi-closed (able to retain useful materials in the total system)

At the molecular level, the race between the self-organization of increasingly complex organic structure and its dissipation due to entropy requires the following functional operations:

- Selective concentration of the desired compounds needed as reactants
- Stabilization and coordination of these reactants
- Controlled energy and "directed" synthesis
- Cycle continuity (where the products become in turn the reactants for the next stage of the cycle)
- And most importantly, in our least-assumptions approach, all must occur in a plausible, indeed likely, geophysical/geochemical environment

An additional boundary condition for the chemical evolution of water-based carbon chemistries is the existence of (micro-)environments promoting both heterogeneous and dehydration reactions. This is because the polymerization of biomolecules almost universally* requires an H^+ and an OH^- to be discarded, one each from either of the two sides of the forming bond. It is thermodynamically very difficult for this expulsion of H^+OH^- (i.e., H_2O) to occur in an aqueous environment; therefore, pure solvent (homogenous) chemistry in water seems unlikely to have been the primary process by which organic chemistry bootstrapped itself up the ladder of complexity.

Contemplation of the above-stated requirements reveals that each embodies a form of symmetry breaking in space, time, and/or thermodynamic variables. This is hardly surprising as each is an organizational (anti-entropic) step leading to some local enhanced state of structure. One of the goals of prebiotic planetary chemistry is the need to demonstrate the existence of a global chemical engineering system that *has all of the above functions, synergizes each to the other,* and that *had to have been in existence* on the early Earth or Mars. Our question of the origin of life now becomes: "On what larger organizing processes can chemical evolution piggyback itself?" Or, in the parlance of the previously stated "lemma" of self-organization: "At the scale of chemical self-organization, what 'first-order' microenvironments exist and mutually connect with each other to support increasingly complex cycles of prebiotic chemistry?"

Such a first-order system exists on the contemporary Earth: it is the bubble–aerosol–droplet cycle. Described in Section 18.5, it is the most fundamental, robust, and far-reaching of geophysical/chemical supercycles[†] involving organics in and between the ocean and atmosphere. Comparing these present-day terrestrial processes against the functional requirements listed above leads directly to the bubble–aerosol–droplet model,[‡] a phenomenologically based model with the potential to support chemical evolution in all its stages (Lerman, 1986, 1992; Lerman and Teng, 2004a). These

---

* There are a very few biopolymers that do not, e.g., terpenoids like vitamin A.
† It is a supercycle because of the many large complex hydrological cycles that are subsets of the whole.
‡ It has also been referred to by a variety of other names in the scientific literature and the popular press, including the Bubble Hypothesis, Lerman's Bubble Model, the Bubblesol Hypothesis, etc.

range from raw organic synthesis on the prebiotic Earth to the possibility of helping coordinate the transition from organic chemistry to biochemistry. Indeed, this model seems relevant to any terrestrial-like body with liquid water and simple organics. (In Section 18.15, we apply the bubble–aerosol–droplet model to an early warmer wetter Mars and consider implications for currently observed Martian geology.)

Over the many physical scales and modalities of our model, one common theme will clearly be seen: the organizing ability of the water–air interface, with these organizing properties themselves a consequence of the symmetry-breaking properties of water's heterogeneous two-phase interface.

18.4   THE AIR–WATER INTERFACE AND ADSORPTION

The keystone elements to all of the above are the symmetry-breaking surfaces of the microenvironments of the air–water interface: from the surface of a freestanding body of water (static or flowing) to whitecap-induced bubbles and their aerosol progeny. The shape of the surface of each of these microenvironments is itself a consequence of least-energy considerations, as is the topologically closed nature of the bubble or its inverse droplet.

It is for this reason that one can look at bubbles and aerosols as complementary: one is a metastabilized fluctuation of air in water, the other of water in air. The continuous layers of bubble clouds formed beneath breaking waves are the analogs of low stratus or stratocumulus clouds in the Earth's atmosphere (Kraus and Businger, 1994). Whether floating under water, bobbing at the sea surface, or drifting in the atmosphere it is the adsorption of amphiphiles that drives these self-organizing phenomena by decreasing the local surface energy and metastabilizing local fluctuations into microenvironments with organized structures capable of further organizing organics, metals, and larger scale particulate matter. Additionally, the dimensional projection from three-space to two-dimensional surfaces significantly decreases the entropic possibilities of molecular orientations and dynamics.

Central to the function and partially closed nature of the bubble–aerosol–droplet cycle is the ubiquitous bubble itself. Isolated bubbles come into existence as regimes of a gas in a liquid in which the surface of the bubble–water interface exists due to a local increase of surface free energy supported by the pressure of the vapor contained within the bubble regime. The surface energy per unit area is also termed the surface tension, but the latter is a potentially misleading term. Molecules on a bubble surface are not necessarily stretched under tension to form the surface; the term surface energy more accurately reflects the work required to bring molecules from the vapor phase (in the bubble's interior) onto the bubble surface (Adamson, 1997). Hence, a potential energy difference exists between the surface and interior of a bubble, allowing well-matched surfactants the opportunity to build more complex and stable structures through the lowering of this energy. Adsorption is thus driven by a reduction of the surface free energy as material is added. Lacking the stabilization due to adsorbed materials, a pure water bubble without any sort of a skin has a surface lifetime of less than a second. With stabilization due to adsorbed materials (organics and metal ions), the lifetime in the open terrestrial ocean can range from seconds to hours.

Adsorption is highly selective and is dependent on the details of the system, which include the following:

- The charge configurations of the molecules involved, especially with respect to hydrophobic and hydrophilic components on a single molecule (i.e., surface activity)
- The ability of the adsorbed materials to couple to previously adsorbed surfactants
- The relative sizes of the bubbles and objects to be adsorbed
- The composition of the bubble vapor, materials to be adsorbed, and various impurities

The air–water interface is an organizational foundation relatively independent of the specific chemistry of the atmosphere and ocean. Also being a function of the (curved) air–water interface, this relative independence is likewise true for the aerosol-droplet (inverted bubble) phase of the supercycle.

The robustness of the selective concentration abilities of bubbles and aerosols is underlined by the large number of industrial processes (e.g., mineral flotation and secondary and tertiary petroleum extraction) based on the use of bubbles, foams, and aerosol processes (Lemlich, 1972).

18.5   THE BUBBLE–AEROSOL–DROPLET SUPERCYCLE: A UNIVERSAL PLANETARY HYDROLOGY CYCLE

On the contemporary Earth the bubble–aerosol–droplet cycle (also known as the bubblesol cycle) include bubble formation and the adsorption of surface-active materials, bubble dissolution, and the nonequilibrium dynamics of bubble bursting (Figure 18.1). This leads to the formation of aerosols and their subsequent roles in atmospheric condensation, which in turn couples back to the bubble phase of this "supercycle" (Lerman, 1992).

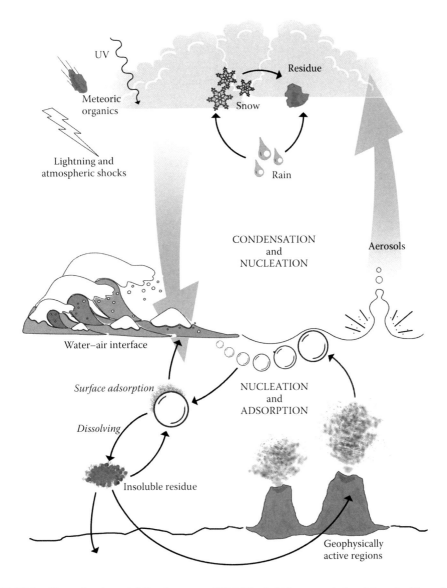

FIGURE 18.1   (See color insert following page 302.) The bubble–aerosol–droplet cycle. There is every reason to believe that this cycle also existed on the early Earth and possibly Mars.

Remarkably, this complex supercycle is a process that requires *only* the disturbance of a water-air interface metastabilized by simple amphiphilic compounds. The rest follows from the fundamentals of chemical physics, being relatively independent of specific chemistry. Specifically, bubble formation is a physicochemical process due to the Rayleigh–Taylor instability (Sharp, 1984), its initiation requiring only the existence of an air-water interface disrupted by mechanical turbulent energy (from waves, meteorites, or geophysically active regions). Its maintenance requires only a metastabilization by simple amphiphilic compounds that would have come from meteorites (carbonaceous chondrites) and comets. This latter statement is strongly supported by the fact that 70% of the organic compounds in the Murchison meteorite are polar (Cronin and Chang, 1993). Hence, it is difficult to imagine something similar to the bubble–aerosol–droplet cycle not being active in both the early Archaean on Earth and a wet Noachian on Mars.

There are two bursting mechanisms possible for an isolated bubble (MacIntyre, 1972), each leading to the formation of a different class of particulate matter injected into the atmosphere (Figure 18.2). For small bubbles (< 0.5 mm), surface tension-driven "jet drops" are formed, which in turn are the principal source of sea-salt aerosols. By some estimates this is the largest source (by mass) of particulate matter injected into the atmosphere. For larger bubbles (> 0.5 mm), an instability mechanism due to the gravity-driven draining of the bubble cap leads to the formation of large numbers of "film cap drops," particles that are much smaller than jet drops. This mechanism is possibly the largest source (by number) of particles injected into the atmosphere, and may be the major source of cloud condensation nuclei as well.

Particles injected into the atmosphere by bursting bubbles have different airborne lifetimes depending on their mass and composition. The heaviest particles injected immediately fall back to the water, forming more bubbles upon their impact. Particles swept up into the atmosphere will enter into the complex tropospheric aerosol cycle (Figure 18.3). Some will act as condensation nuclei for precipitation, and some will be scavenged by other precipitation bodies. The condensation nucleus

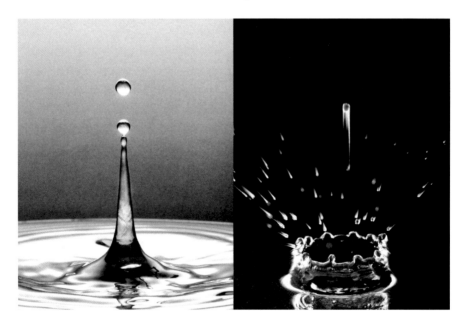

FIGURE 18.2 (See color insert following page 302.) Formation of jet drops and film cap drops from a bursting bubble. Left: For small bubbles (<0.5 mm) surface tension–driven "jet drops" are formed. The material making up successive jet drops from a single bursting bubble comes from successively microtomed layers of its air–water interface (MacIntyre, 1974c). Right: For larger bubbles a very large number of "film cap drops" are produced by a gravitationally-induced instability mediated through the bubble cap's draining. (Photographs by Prof. Andrew Davidhazy, 2009. With permission.)

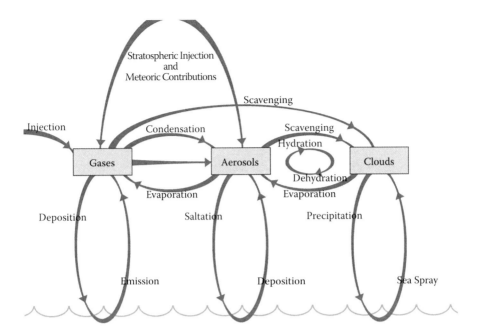

FIGURE 18.3 Atmospheric cycles of condensation, accretion, and evaporation over multiple length scales. (After Turco, R. P., O. B. Toon, R. C. Whitten, R. G. Keesee, and P. Hamill. 1982. Importance of heterogeneous processes to tropospheric chemistry: studies with a one-dimensional model. In *Heterogeneous Atmospheric Chemistry*, ed. D. Schryer. Washington D.C.: American Geophysical Union, 231–240.)

of an aerosol can undergo a number of hydration–dehydration cycles before eventually falling to the ocean surface. While in the atmosphere, these objects will be exposed to a variety of energy sources including solar radiation and the plasma and shock effects associated with lightning.

18.6   THE BUBBLE–AEROSOL–DROPLET SUPERCYCLE: A UNIVERSAL "ORGANIC WEATHER" CYCLE

Accompanying the water–air cycles are "organic weather cycles," which seem to support each of the functional requirements for chemical evolution listed in Section 18.2 at both the system and molecular levels. This is because the very existence of this first-order cycle, its nodes, and processes implicitly depends on these same functional requirements. Once a bubble forms it tends to collect surface-active surfactants, this being the first of several different concentrating phenomena associated with bubbles, aerosols, and droplets. These concentrated materials, which are predominantly organic, stabilize the bubble allowing it to follow one of several pathways (Blanchard, 1975, 1983, 1989; Blanchard and Woodcock, 1957). Each of the pathways as described below offers additional concentrating opportunities. Dovetailing between steps moreover suggests that once organic matter enters into this cycle, it tends to remain within it; at each step being joined by freshly accumulated organics, ions, and heterogeneous catalytic surfaces.

Based on the Rayleigh–Taylor instability, even short timescale disturbances of surface water (lakes, seas, or intermittent turbulent flows) can initiate this cycle, which by analogy to the current terrestrial ocean includes the following:

1. Bubble formation and the consequent adsorption of surface-active materials, bubble dissolution, and the nonequilibrium energetics of bubble bursting.

2. Organic materials and selected metals, as well as clay particles, are preferentially adsorbed onto the surface of the bubble. A bubble being a micro-version of the air–sea interface, it will adsorb and become stabilized by those dissolved materials that congregate at the air–sea interface. These include fatty acids, alcohols, proteins, and polysaccharides as primary adsorbents; these organics, in turn, adsorb metal ions, colloidal silica, and clays (kaolin and montmorillonite) (Johnson and Cooke, 1980; Lemlich, 1972).

3. This stabilizes the bubble, thus leading to a highly concentrated resultant particulate, as the now organically "dirty" bubble dissolves or bursts.

4. A dissolving bubble yields an organic-rich residue, which can then nucleate other bubble formation, or be adsorbed in turn by other bubbles.

5. All of the above leads to a significant enhancement (up to a million-fold or better compared to bulk water) of organics and metals at the millimeter-thick microlayer of the air–sea interface (MacIntyre, 1974c; Duce and Hoffman, 1976; Sieburth, 1983).

6. The bursting of bubbles injects into the atmosphere particulate matter also rich in minerals and organics, the latter being as much as 40–60% of the mass (O'Dowd and de Leeuw, 2007). See Figure 18.4. Besides organics, concentrated up to a million-fold, these processes can yield up to a 10,000-fold enhancement in phosphates (MacIntyre and Winchester, 1969) and other scarce minerals and ions (MacIntyre, 1970, 1974a, 1974b).

7. The bubble-bursting process also provides a set of energy-producing possibilities capable of driving highly nonequilibrium chemical synthesis (see Section 18.10).

8. These injected materials are then coupled to aerosol formation and the subsequent nucleation of atmospheric condensation, leading to the further heterogeneous chemistry and nonequilibrium physics associated with rain and snow (discussed in Section 18.11).

9. Precipitation concentrates organics and minerals from the atmosphere. During this scavenging process, there is evidence for the formation of an organic skin around the precipitation object (rain, snow, hail) (Gill et al., 1983). Fog particles have been found to contain mostly organic carbonates, esters, and proteins, whereas the fatty acids in aerosols have been shown to have an oceanic source. Organics and transition metals from meteoritic sources are also scavenged and concentrated (Murphy et al., 1998).

10. These take part in a variety of heterogeneous physicochemical reactions including successive hydration–dehydration reactions that may be critical in the transition of organic chemistry to biochemistry (Section 18.12).

11. Atmospheric condensation closes this supercycle since both falling raindrops and snow directly deposit their organometallic chemistry on the ocean surface, efficiently inducing new bubble formation. Precipitation-induced bubbles are an important link in merging the oceanic and atmospheric subcycles of the entire bubble–aerosol–droplet supercycle, whereas wind-induced bubbles will tend to collect the rest of the material congregating at a sea surface.

12. Once organic matter, inorganic ions, and terrestrial and meteoritic particles enter anywhere into the bubblesol cycle, they tend to be continuously recycled and are therefore available for the self-organizing chemical–physical processes discussed in the remainder of this paper.

Throughout this supercycle, coupled hydration–dehydration cycles are abundant. As indicated above, this is of unique importance to the self-organizing formation of biopolymers, for essentially all biopolymers are formed through linkages derived from a dehydration reaction. Chang and Lahav (1982) have additionally shown that such polymerizing condensation reactions are greatly enhanced if the organics undergo hydration–dehydration cycles on salt and mineral surfaces. Fortunately for our intended prebiotic purposes, such organically rich hydration–dehydration cycles naturally occur at every step—e.g., at the concentrating sites at each phase of the bubble–aerosol–droplet cycle (bubble bursting, aerosol formation, precipitation scavenging, precipitation, or wave-induced bubble formation leading to the next stage of this iterative cycle).

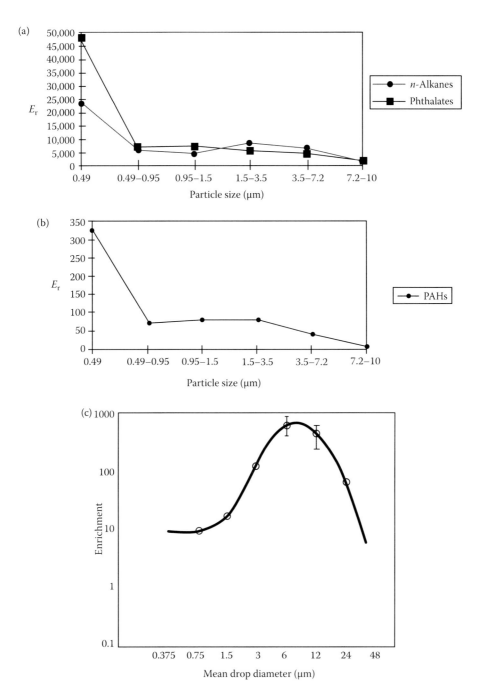

FIGURE 18.4   (a) Enrichment ratio (E_r) of *n*-alkanes and phthalates relative to Na ion as a function of particle size in an aerosol sample. (From Cincinelli, A., L. Lepri, L. Checchini, and M. Perugini. 1999. In *The Role of Sea Surface Microlayer Processes in the Biogeochemistry of the Mediterranean Sea*, ed. F. Briand CIESM Workshop Series, vol. 9, pp. 23–25. With permission.) (b) Enrichment ratio (E_r) of PAHs relative to Na ion as a function of particle size in an aerosol sample. (From Cincinelli et al., 1999. With permission.) (c) Phosphate enrichment due to bubble concentration and bursting (From MacIntyre, F., and J. W. Winchester. 1969. *J. Phys. Chem.* 73:2163–2169. With permission.)

Additionally, membrane-like phase boundaries are created at each node of the cycle, selectively concentrating organics and metal ions. These metastable boundaries may well have played an essential role in the transition from organic chemistry to biochemistry through the symmetry-breaking segregation of both materials and phases. A potentially critical example of this is discussed in Sections 18.12 and 18.13, where an arguably close-to-ideal microenvironment is created that can support early RNA activity. As an added bonus, the nested hierarchy of microenvironments naturally embodies and functionally supports in a real-world way the majority of other specialized environments and reaction possibilities postulated by other workers in the field (Sections 18.13 and 18.16). But first, we examine new experimental evidence that offers strong support for the claimed advantages of this novel approach to prebiotic chemistry.

18.7 EXPERIMENTAL SUPPORT FOR THE BUBBLE–AEROSOL–DROPLET MODEL IN CHEMICAL EVOLUTION

The first experiments using these ideas in a prebiotic context are only now beginning to be conducted, but with strikingly positive results to date. In an unpublished work, Ruiz Bermejo, Menor Salvan, Osuna Esteban, and Veintemillas Verdaguer (2005) have redone the standard Miller–Urey experiment but utilized surface water induced aerosols. Using a CH$_4$/N$_2$/H$_2$ (40:30:30) atmosphere over liquid water, they used an ultrasonic wave focused on the water's surface to initiate an aerosol population with an initial droplet size of ~3 μm. The energy source was a standard spark discharge unit, akin to the initial Miller–Urey experiment. The initial results are satisfyingly in line with the advantages predicted by the bubble–aerosol–droplet model. Relative to the standard gas/liquid bulk environments, the introduction of these water-air microenvironments creates:

1. New classes of biochemically important organics (carboxylic and hydroxyl acids along with heterocyclic-like adenine).
2. Chemical yields several *orders of magnitude* greater than comparable experiments lacking the bubble-aerosol-droplets.

Among the highlights of this experiment are the following:

1. The creation of adenine with a yield 100 times that of previous work. Additionally, Ruiz Bermejo et al. (2005) emphasize that it is free adenine in a soluble form as opposed to previous Miller-type experiments where the adenine was in the form of an unknown precursor. They interpret that the high yield of adenine, using aerosols, as being due to the local enrichment of earlier stage compounds in the liquid-gas interface. Furthermore, the authors state that "this represents a reduction of importance of bulk-solution interfering reactions," as is predicted by Lerman's bubble-aerosol-droplet model.
2. Other unexpected heterocyclics are created. For example, Ruiz Bermejo and colleagues clearly demonstrate that hydantoins [C$_3$H$_4$N$_2$O$_2$] are a real possibility on the early Earth offering, as they suggest a potentially critical role as precursor for prebiotic peptides.* This could be prebiotically important as they are created through bubble-aerosol processes in a "more efficient pathway than the classical base catalyzed hydration of nitriles (under similar conditions of pH and temperature to those of our experiment)."
3. Polyhydroxylated acids are created solely in the bubblesol version of the experiments, including several sugar-related compounds (such as tartaric and glyceric acid) that have not been seen in comparable prebiotic synthesis experiments.

---

* Hydantoins can also be precursors to amino acids, but which would racemize easily. That might still be useful in a prebiotic setting... who knows? (Steve Benner, private communication, 2006).

4. A range of potentially critical carboxylic acids are also uniquely created in these experiments (as opposed to the standard bulk reservoir approach). Again, quoting Ruiz Bermejo et al:

> The succinic and malic acids are members of the Krebs cycle.... These carboxylic acids could take part in a primordial variant of the Krebs cycle. This demonstrates that aerosol chemistry also could have contributed to the production of the raw materials for primordial metabolism.

To conclude, Ruiz Bermejo and colleagues state:

> As a general conclusion of the results obtained, we can say that the analysis of organic material obtained in presence of aerosol shows greater amounts and greater diversity of molecules than the material obtained using the same conditions without aerosol. The experimental evidence obtained in this work support the hypothesis made by Lerman, that the aerosol droplet behaves as a microscopic chemical reactor that offers the possibility of the concentration, stabilization and transformation of molecules synthesized by means of the energy supplied by the spark discharge. Therefore, we suggest that aerosols could play a significant role in the origin of molecular diversity, evolution, and the origin of life.

Subsequent experimental work by Ruiz Bermejo and colleagues at the Centro de Astrobiologia (Ruiz Bermejo et al., 2007a, 2007b) nicely confirms and expands upon this early work. The number and diversity of positive results in these early experiments is highly encouraging; leading us to expect an increasing number of experiments further exploring the bubble–aerosol–droplet hypothesis. Indeed, Sections 18.12 and 18.13 discuss more speculative inferences from these results that may have additional significant impact on biogenesis.

18.8 CLIMBING THE LADDER OF CHEMICAL EVOLUTION: AN OUTLINE OF THE APPROACH

We will now deal with the above-stated functional requirements in more detail. Starting with the question of basic existence, we will show that the system had a high probability and one that is semi-closed. We then discuss the resulting microenvironments with respect to the molecular scale issues of concentration, chemical selectivity, energetics, and heterogeneous chemistry. Lastly, we explore, at the system level, the potential role of these microenvironments for enabling the transition from planetary organic chemistry to the existence of biochemistry (i.e., biogenesis).

We will reason by qualitative analogy with contemporary Earth processes. Because these processes take place at the molecular level, and are independent of planetary-scale phenomena or specific details of local chemistry, they are expected to be primary ones for any terrestrial-like planet with water, organics, and heavy metals. Additionally, many of the effects are driven by surface tension and will therefore to a first order be independent of the strength of the planet's gravitational field.

We will address in some detail the extent of these processes on the contemporary Earth, fully expecting them to be just as primary on other water and carbon-rich terrestrial-like planets. Many of our results will be directly applicable to the Archaean and the origin of life on the Earth. Additionally, by applying this work to an early wet Mars, we introduce the idea that the bubble–aerosol–droplet cycle could have been the necessary supporting infrastructure for the origin of life during the Martian Noachian. Obviously, the specific numerical values of an early Martian phenomenology will be different. Work to quantify better these processes for Mars, Titan, Europa, and Enceladus is in progress.

18.9 THE SEMI-CLOSED NATURE OF THE BUBBLE–AEROSOL–DROPLET CYCLE

Because of its surface energy, a lone bubble in a contemporary terrestrial ocean will tend to dissolve on its own. With rise times for such bubbles being 0.13 cm/s, a creation depth of 1 m is the nominal limit from which 100 µm bubbles not stabilized by adsorbed surfactants can be expected to reach the surface. Naturally, smaller bubbles would not be expected to make it at all.

As noted above a bubble becomes a micro-version of the air–sea interface, stabilizing itself through the adsorption of a vast variety of dissolved materials. Blanchard (1975) suggests that a steady state concentration due to this adsorption sets in after about 20–40 s of rising (or traveling) through the water. Most tellingly, for our purposes, in laboratory experiments with artificially made bubbles (<100 μm in sea water) Johnson and Cooke (1980) found that:

> Bubble dissolution *always* resulted in the formation of a particle. To provide a control we tried to produce water in which particles could not be produced but did not succeed regardless of the treatment. . . . As the bubble dissolved, striations appeared on the surface, probably a manifestation of film collapse. These lines are well known in monolayer studies and are the result of folding of the surface film.

Along the way to this dissolution, a separate set of experiments (Johnson and Cooke, 1981) followed the time evolution of this dissolution, and found a stabilized set of microbubbles with lifetimes of about 24 h. Not only are the resulting particles highly enriched in organic molecules, but one has the emergence of a membrane-like (organic-rich) film.

Both the stabilized microbubbles and the resulting organic particles can act either as nucleation sites for new bubble formation, be adsorbed by other bubbles, or aggregate with other such objects. Through several generations of bubbling, much of the particulate matter in a local solution will therefore become adsorbed and aggregated. This is a very efficient process, whereby the majority of dissolved organics are adsorbed and concentrated from the surrounding water.

On the contemporary Earth, these organic rich particles are a principal form of food for marine life from a variety of zooplankton in the upper ocean (Baylor and Sutcliffe, 1963; Wangersky, 1974; Sieburth, 1983; Mitchell et al., 1985) to the largest of whales. But in the prebiotic context, without competition from extant life-forms, bubble-created organic-metal rich particulates (and still separate molecular concentrations) would be free to be serially collected by successive generations of bubbles. While some would sink to the sediment, the majority would be taken to the water's surface. The salient point for cycle continuity is that these same materials will be preferentially formed into the surface of bubbles.

Thus, upon the bubble's bursting the organics and accompanying surfactants will be further concentrated and released into the atmosphere as aerosols and film-cap drops. By conservation of momentum, when bubbles burst on the surface, ejecting jet drops and film cap drops into the atmosphere, concentrated matter is also ejected into the ocean. Much of this downward-oriented matter becomes the next generation of bubbles. As previously asserted, once organic matter, metal ions, and even mineral surfaces enter into the bubble phase of the overall cycle, they tend to be continuously recycled and worked upon.

But what of the atmospheric components of the overall bubblesol cycle? Further cycle continuity follows from the fact that the impact of the precipitation object (raindrop or snowflake) on the water surface preferentially forms bubbles, returning us to the bubble phase of the bubble–aerosol–droplet cycle (Blanchard and Woodcock, 1957). Snowflakes seem to be particularly good bubble makers, producing bubbles <100 μm (Blanchard, private communication, 1986; Blanchard, 1989; Blanchard and Woodcock, 1957).

There is another, equally critical, cycle-continuity process to consider. Raindrops larger than 100 μm typically produce bubbles with diameters less than 200 μm, whereas (water) droplets of less than 100 μm tend not to directly produce bubbles. Since most aerosols are <100 μm, dry deposition of aerosols may not make bubbles efficiently. On the other hand, organic materials from dry deposition will still be preferentially trapped in the sea-surface microlayer, and immediately reenter the bubblesol cycle by adsorption onto those bubbles created by wind-induced whitecaps and breaking waves. Such waves will be readily available, for wave breaking represents the major loss of (wind) momentum flux in momentum transfer (to oceanic waves) at the air–sea interface (Melville and Rapp, 1985). On the contemporary Earth the most important bubble makers are whitecaps forming at wind speeds exceeding 3 m/s, and with a time-averaged whitecap bubble coverage of roughly 1% of the oceans' surface at any time.

In summary, precipitation-induced bubbles are an important link in merging the oceanic and atmospheric subcycles of the entire bubble–aerosol–droplet supercycle, while wind-induced bubbles will tend to collect the rest of the material congregating at a sea surface. Hence once organic matter, inorganic ions, and terrestrial and meteoritic particles enter anywhere into the bubblesol cycle, they tend to be continuously recycled, becoming available for the self-organizing chemical–physical processes discussed in the remainder of this paper.

18.10   ENERGETICS: POTENTIAL MECHANISMS OF BUBBLE-INDUCED CHEMISTRIES

In past modeling of prebiotic chemistry, much attention has been paid to reactions in homogeneous aqueous solutions (Chang et al., 1983). In addition, many studies have concentrated on the catalytic processes on the surfaces of solids, especially clays (Rao et al., 1980). However, in the prebiotic context, very little is known of the reactions that occur at the gas–liquid interface, especially those at the surface of bubbles and droplets. As discussed in Section 18.7, a fluid interface determined by phase boundaries can act as a region where the chemistry may take quite different pathways than in the bulk homogeneous environment. Not only are organics and metals selectively concentrated, but there is the potential for relatively anhydrous and short-lived high temperature nonequilibrium regimes. Taken together, these geophysical/geochemical microenvironments allow for new classes

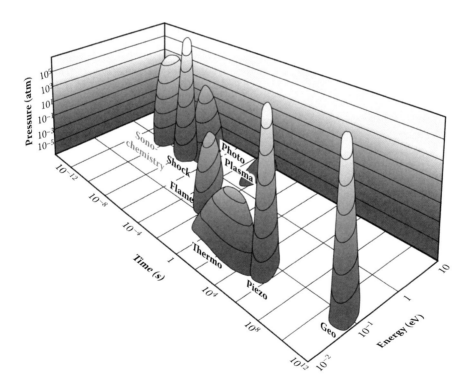

FIGURE 18.5   (See color insert following page 302.) The unique phase-space of sonochemical energetics. On an early terrestrial-like planet there will be great ranges of acoustic cavitation energies, ranging from ocean waves to near-surface geophysically active regions (submarine volcanoes) to meteoritic impacts on an ocean or sea, and of course accompanying all will be the ubiquitous bubble. (From Suslick, K. S., *Sci. Am.* 260:80–86, 1989. With permission.)

of reactions to occur under what are likely prebiotic conditions. There are four different mechanisms by which such chemical changes may be specifically facilitated inside bubbles:

1. Concentration of organic molecules at the interface due to dipolar hydrophobic–hydrophilic interaction with the two phases. This leaves the molecules oriented in a nonrandom fashion, reducing the entropy of reaction activation between two adjacent molecules. Such specific interphase-induced orientations may play a role in the selection between competing chemical reactions.
2. The release of a water molecule from a condensation process into the aqueous phase at the interface may involve a higher gain in enthalpy and free energy, compared with a reaction in a homogeneous aqueous medium, because the reactants were less hydrated to begin with.
3. The surface compression or shrinkage that occurs on condensation of two molecules on the inner surface of a bubble involves a gain in free energy, similar to carrying out a reaction under hydrostatic pressure. This might lower the free energy of activation of condensation while increasing that of the reverse hydrolysis process. This effect might, therefore, drive the chemical system in the desirable direction from the perspective of chemical evolution since polymerization is in essence a dehydrating condensation process.
4. The rapid contraction of bubbles is associated with adiabatic compression of the internal gas and, therefore, with local heating of the inner surface. This heat will accelerate any chemical reactions at the surface. Early experiments by Fitzgerald et al. (1956) suggested regimes of nonequilibrium chemistry with shock temperatures up to 1000s K or pressures in the kilobar range, while Anbar (1968) was the first to suggest its applicability to prebiotic chemistry.

Only in the past few years, however, have these high nonequilibrium temperatures been confirmed (Brenner et al., 2002). In fact, acoustic cavitation creates regimes of nonequilibrium energetics quite unlike any other and deserves serious further investigation as to its likely prebiotic role. Suslick et al. (1999) and others have shown that these sonochemical hot spots, existing in otherwise cold regimes, can have temperatures of 5000 K, pressures of 1000 atm, and heating and cooling rates of 10 billion °C/s (Figure 18.5). As Suslick (1994) says, "For a rough comparison, these are, respectively, the temperature of the sun, the pressure at the bottom of the ocean, the lifetime of a lightning strike, and a million times faster cooling than a red-hot iron rod plunged into water. Thus, cavitation serves as a means of concentrating the diffuse energy of sound into a chemically useful form." Indeed, there are indications that the majority of energy released in energetic bubble burstings goes into chemical reactions (Lohse, 2002; Didenko and Suslick, 2002). Because these are rapidly quenched nonequilibrium processes, conventional high-temperature destruction of organics and complex chemistries need not occur. Indeed, these processes may well expedite the condensation reactions (and other endothermic reactions) of interest in chemical evolution that are facilitated or accelerated by any of the other three mechanisms.

Each of these four processes involves the transduction of mechanical into chemical free energy. The first three processes convert surface free energy, generated from mechanical energy (agitation or turbulence), into chemical free energy. The fourth is a thermal mechanism, not involving surface energy, which may accelerate reactions catalyzed by the air–water interface, solid surfaces, or colloidal particles. The latter mechanism may be important because such catalysts also aggregate at the surface of bubbles. Also important with respect to organic reactions on catalytic mineral surfaces are the surface orientating entropy-reducing effects of the first mechanism. We have, therefore, substantial reasons to expect significant changes in chemical behavior in systems with bubbles.

18.11   THE HETEROGENEOUS CHEMISTRY OF AEROSOLS

A detailed overview of the energetics and heterogeneous chemistry available in this phase of the bubblesol cycle follows. As in the case of bubbles, there are substantial reasons to explore the

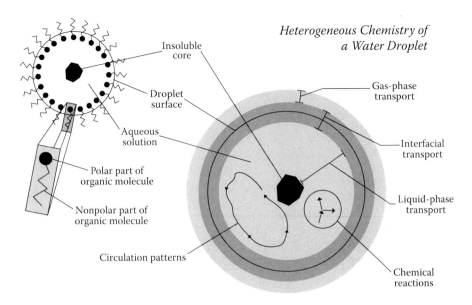

FIGURE 18.6 Heterogeneous regimes of an atmospheric aerosol: organic skin coupled with principle transport regimes. (Transport components are from Turco R. P., O. B. Toon, R. C. Whitten, R. G. Keesee, and P. Hamill. 1982. In *Heterogeneous Atmospheric Chemistry*, ed. D. Schryer. *Geophys. Monogr. Ser.* 26:231–240. With permission; chemical components are from Graedel, T. E., and C. J. Weschler. 1981. *Rev. Geophys. Space Sci.* 19:505–539. With permission.)

special chemical environments associated with these micro-Miller bottles (Figure 18.6). A different chemistry than that of bubble phenomena may be associated with aerosols, making them of additional interest in further explorations of the phase space of combinatorial chemical possibilities.

The mechanisms by which specific chemical changes will occur on the outer surface of droplets and liquid aerosol particles are complex. The only trivial catalytic mechanism is the concentration of solutes as the water evaporates. The shrinkage of a droplet involves an increase in surface free energy that results in the cooling of the droplet. The surface orientation effects described above for the inner surface of bubbles will also exist here, leading to enhanced condensation of appropriately oriented species.

There is also the possibility that the dissipation of the high surface energy of a small droplet on impact with bulk water will produce some other form of free energy than heat. Since these droplets are usually charged, the released charges may induce additional types of redox reactions that may not occur in bubbles. Moreover, in the primordial world there would have been a significant flux of ultraviolet radiation to initiate photochemical processes. The interior of the droplet will typically contain a solid nucleus, able to promote heterogeneous reactions (Table 18.1; Turco et al., 1982), and which may in turn be photochemically induced.

These possibilities are further compounded by the many subcycles of phase transitions that these bodies undergo: on the contemporary Earth the condensation nucleus of an aerosol typically undergoes ~10 hydration–dehydration cycles over a 7-day period (T. Graedel, private communication, 1986).

Neglecting, for the moment, the anoxic environment of the primitive Earth's atmosphere and the still unknown atmospheric chemistry of an early Mars, extrapolation from contemporary aerosols and their chemistry within the aqueous solutions of these atmospheric bodies offers some unique and unusual conditions for metallo-organic chemistry and catalysis:

1. The pH of raindrops is initialized through its equilibrium with atmospheric CO_2 and subsequently modified through the scavenging of aerosols and gases. Sea-salt aerosols ejected from bubbles are slightly alkaline (pH of 7.8–8.3), which is principally due to NaCl.

TABLE 18.1
Classes of Heterogeneous Chemistry of Aerosols and Droplets

Condensation of gases on aerosols

Dissolution of gases in aerosols

Chemical reaction in aerosol

Reactions of gases on solid surface

Vapor nucleation

Dry deposition of gases

Dry deposition of aerosols

Scavenging of gases by cloud drops

Precipitation removal of soluble vapors

Collection of aerosols by hydrometeors

Nucleation of aerosols into cloud drops

Aqueous chemistry in cloud and rain drops

Natural emission and absorption of gases and particles from land and oceans

2. The metal ions accumulated from bubble bursting or atmospheric scavenging, coupled with their relative inability to diffuse out through any organic skins, will lead to the possibility of soaps, foams, and emulsions.

3. Polyvalent metallic ions can increase the surface tension of such films by up 50%, thus making surface-spreadable organic substances, which would not do so on a pure H$_2$O surface.

4. Homogeneous catalysis (including those involving redox reaction) is likely to occur due to transition metals.

5. Heterogeneous catalysis with the solid core is likely to be important. Metal ions as well as clay and mineral particles are all concentrated by bubble processes, and meteoritic particles will be in the core due to atmospheric scavenging. This is substantiated by recent work on aerosols (Murphy et al., 1998) indicating the incorporation of a large range of meteoritic transition metals coupled to high organic content (10–50%). In fact, of the 46 different elements accompanying the organic molecules, the majority of metals were found to be meteoritic as opposed to anthropogenic.

6. The successive hydration–dehydration cycles lead to the likelihood of condensation reactions, accompanied by the possibility of their encapsulation within phospholipid bilayers (as described in the next section).

18.12 ON AMPHIPHILIC BILAYERS AND THINGS TO GO IN THEM

The production of topologically closed amphiphilic bilayers, surrounding a solute rich in other organics and metal ions in nonequilibrium proportions, may result from the rehydration of previously dehydrated organic films (Deamer and Pashley, 1989). Taking the simplest single hydrocarbon chain amphiphiles as a sample case, this requires relatively stringent chemical conditions:

1. An admixture of amphiphiles $> C_{10}$.
2. A pH of 8.5.
3. Hydration time scales sustained for hours to days.
4. Constant high concentrations of amphiphilic compounds, hence requiring a continual source of fresh amphiphilic material (since the decay constant for anything but the simplest molecules is days to weeks).

This is remarkably close to the aerosol-droplet environments described above:

1. Aerosols derived from oceanic bubbles, and with atmospheric scavenging, would be rich in whatever fatty acids, phospholipids, proteins, and other organic compounds were available. Evidence taken from Jaenicke (2005) supports the extremely high percentage of complex organics of all types that will be adsorbed onto (contemporary) aerosols.
2. Such atmospheric bodies would likely have a pH in the range of 7.8–8.3, thus promoting a variety of metallic-amphiphilic soaps.
3. Contemporary aerosols have a lifetime of about 1 week, and in the process go through perhaps 10 cycles of hydration/dehydration.

Recall that in Section 18.9, we saw the formation of a similarly organic rich membrane/film during the bubble phase of the cycle. Therefore, at every stage of the bubble–aerosol–droplet cycle there is the opportunity of not just making a ladder of increasingly complex organic molecules but of "wrapping them up" as well. Following the lead of Deamer and Pashley (1989), subsequent work supports the notion that amphiphilic compounds of meteoritic origin and/or of terrestrial "modification" (in a prebiotic sense) are capable of creating encapsulated microenvironments in which a variety of protometabolic processes and pathways can occur. This is principally due to the work of research groups led by Dave Deamer (Monnard et al., 2002; Dworkin et al., 2001; Apel et al., 2002), Jack Szostak (Szostak et al., 2001; Hanczyc et al., 2003; Hanczyc and Szostak, 2004; Chen and Szostak, 2004; Chen et al., 2004), and Luigi Luisi (Luisi et al., 1993, 1994; Bachmann et al., 1992).

18.13   FROM CHEMICAL EVOLUTION TO BIOGENESIS

Everything above leads directly to the following hypothesis. In spite of considerable effort, no synthesis of RNA has yet been found "in the wild." Hence, it would seem natural to suggest that RNA synthesis, based on the precursors previously created, might have occurred inside the microenvironments and protocells created by the bubble-aerosol-droplet supercycle. This hypothesis is supported by the fact that these microenvironments naturally produce conditions close to the optimum for RNA activity: a pH of 8–9, high salinity, divalent ions such as Mg, and monovalent such as Na and K (Laura Landweber, private communication, 2000). RNA synthesis involving hydration–dehydration–induced condensation reactions (Zaug and Cech, 1985) could then be supported by these dynamic microenvironments.

Of a more speculative nature, these microenvironments may also support a novel idea on the origins of pre-RNA recently put forward by Simon Nicholas Platts (2004). Platts suggests the existence of an early PAH world,* in which the functional structure of an RNA progenitor was templated by a congruent structure of stacked polyaromatic hydrocarbons (PAHs). These PAHs are themselves hypothesized to have been organized by an interfacial phase separation between the discotic (columnar) aromatic cores of the PAHs and the surrounding water (Figure 18.7). One of the most important consequences of their use as molecular scaffolding is the creation (due to π–π interactions) of a spacing distance of 0.34 nm rise per base pair, the same as RNAs.

The underlying concepts of PAH world have not yet been tested experimentally. But they fit in nicely with the ideas we have elaborated here on the critical role of interfaces as the locales for self-organization (Section 18.1), in particular, the role of the water interface in functionally supporting and geometrically organizing prebiotic chemical evolution. With respect to the prebiotic Earth, recall from Figure 18.4b that PAHs are massively concentrated in just these same air–water interfacial microenvironments.

---

* Dubbed "PAH world" by Robert Hazen, an excellent historical overview of the development of this model can be found in his new book, *Genesis: The Scientific Quest for Life's Origins* (Hazen, 2005).

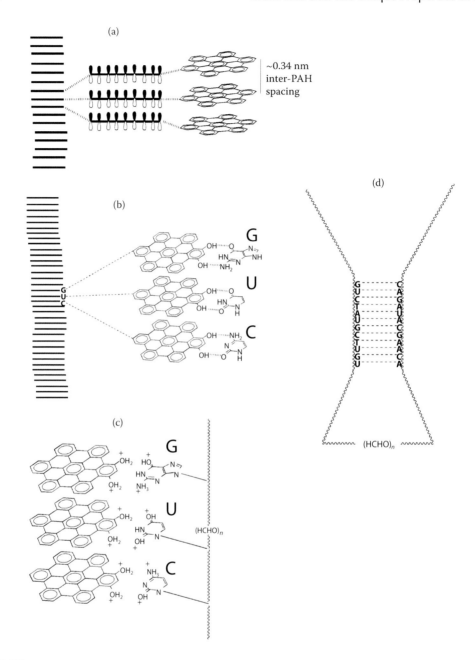

FIGURE 18.7 The PAH world hypothesis of Nick Platts. (a) Nick Platts' PAH world hypothesis utilizes the organizing ability of a water interface to facilitate the self-organization of polycyclic aromatic molecules (PAH) into stacks. (b) Once in a stacked and discotic array, the PAHs attract small flat molecules (notably the bases of DNA and RNA) to the edges. Three prebiotic nucleobases are shown hydrogen bonded to hydroxy functions in the edge structures of derivatized neighboring PAH molecules. (c) A molecular backbone forms linking the bases into a long chain. (d) The RNA-like chain of bases separates from the PAHs and folds into a molecule that carries information. (e) Complex assemblages of these chains have the potential to catalyze reactions. (From Platts, N. 2006. Contributions to chemical questions in origins of life. PhD thesis, Department of Chemistry and Chemical Biology, Rensselaer Polytechnic Institute, Troy, NY. With permission.)

Another possibility that these microenvironments might help explain are those hints of a hydrophobic/hydrophilic codon–anticodon correlation (Lacey and Mullins, 1983). In Lerman (1992), we both address these questions in a broader context and describe a class of early experiments, designed by Anastasia Kanavaroti, to test the possibilities of nucleotide polymerization. Given such a situation, it should be easier to make (and make use of) nucleic acids in the more protected confines of a metabolizing, ATP-utilizing, bilayered protocell. For example, the polymerization of ATP cannot occur in water due to electrostatic hindrances, but can do so on an anhydrous surface.

As we have seen, the bubble–aerosol–droplet cycle provides a sufficiently broad range of mechanisms to support in a real-world context, not just the conventional hypothesis of the creation of a RNA world but also the more recent work by the Deamer, Szostak, and Luisi groups (op cit.). Even more generally, the underlying infrastructure we propose might even support the two-origin theory of life put forward by Dyson (1999) and Shapiro (1999): metabolism first, replication second. Indeed, Ruiz Bermejo et al. (2005) based on their initial experiments in support of the bubblesol hypothesis, speculate as to the implications for an early metabolism scenario:

> Metabolic type reactions could have had a central role in the processes that gave rise to the origin of life. Currently, the issue about if 'metabolic life' could have truly existed and preceded life is controversial. Our experiment with aerosols support the [possibility of the] 'metabolic life' hypothesis and demonstrates that the materials for the development of a proto-metabolic system can be synthesized simultaneously with the structural and information system materials. These materials could establish a cycle of reactions as a precursor to intermediary metabolism. (Ruiz Bermejo et al., 2005)

Crossing the great divide from chemical evolution to biological, we can further speculate about the potential relationship between bubbles, bacteria, and Lynn Margulis' theory of the endosymbiotic origins of eukaryotes (Margulis, 1981). Lerman and Teng (2004a) note that the extraordinary efficiency by which bacteria are collected and concentrated by bubbles (Blanchard and Syzdek 1970a, 1970b) invites the question: could bubblesol phenomena have helped catalyze the endosymbiotic relationship between prokaryotes that resulted in the development of eukaryotes?

In particular, did the hydrophobic/hydrophilic properties of bubbles concentrate the otherwise disparate bacteria leading to unusual populations at the sea surface, in bubble-created marine snow, and in the sea-salt aerosol?

Analogous to the contemporary world, an Archaean version of marine snow would have provided a superior microenvironment for collections of bacteria to survive, and disparate species of bacteria collected onto a jet drop would undergo a variety of environmental assaults ranging from enhanced UV radiation, to repetitive hydration–dehydration and freezing cycles. Early cell membranes could hardly have remained completely intact during all of this, quite possibly leading to a mixing of bacterial components. Mutations as well would more likely have been induced, to the occasional advantage of "interesting" combinations of cellular components.

18.14 DOES THE ELEMENTAL CHEMISTRY OF LIFE MIRROR THE SEA-SURFACE MICROLAYER?

Due to bubble mass transfer from below, and deposition from the atmosphere (both dry and wet), the concentration of selected organics, metals, ions, and particulate matter can be 10^4 (or more) greater than the concentration of these materials in the bulk water. Upon inertial bubble formation, it is this surface layer that becomes the inner surface of the bubble. This leads to an intriguing smoking "bubble" of a phenomenological link, correlating sea-surface microlayer concentrations with the chemical composition of organisms across the breadth of life's kingdoms.

Banin and Navrot (1975) point out that a plot of elemental enrichment factors (the ratio of concentration of an element in an organism to its concentration in the Earth's crust) versus ionic potentials for four major groups of organisms (bacteria, fungi, plants, and land animals) are fascinatingly congruent to that of seawater (Figure 18.8). What is most intriguing is the fact that for ionic potentials greater than 10 (sulfur, carbon, nitrogen, and other of the primary biochemical elements), organisms have enrichments of 10–10,000 over the nominal ocean value. This is also the primary range of enrichments due to bubble processes at the sea surface. Hence, there is the strong hint that the enhanced elemental enrichments of living systems are correlated with the sea surface microlayer enrichments resulting from these bubble processes.

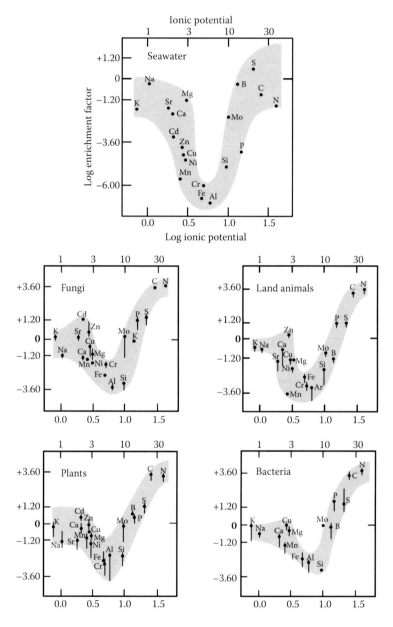

FIGURE 18.8 Elemental enrichment factors (ratio of concentration of an element in an organism to its concentration in the Earth's crust) versus ionic potentials for the four major groups of organisms (bacteria, fungi, plants, and land animals). (From Banin, A., and J. Navrot. 1975. Origin of life: clues from relations between chemical compositions of living organisms and natural environments. *Science* 189(4202):550–551. With permission.)

18.15 DO MARTIAN BLUEBERRIES HAVE PITS? CONSEQUENCES AND ARTIFACTS OF "ORGANIC" WEATHER CYCLES ON AN EARLY MARS

We now turn to the more general question of chemical evolution on another terrestrial-like body, and ask whether this cycle could have supported prebiotic chemical self-organization during the Noachian and Hesperian periods on Mars. A warmer early Mars, one with intermittent water but lacking tectonics, offers more limited geophysical/chemical opportunities to support chemical evolution of the type we believe occurred on the Archaean Earth. But for the same chemical physical reasons as on the early Earth, the bubble–aerosol–droplet cycle would likely have existed on an earlier warmer Mars with at least intermittent freestanding bodies of water (Lerman, 2002a, 2004a, 2004b, 2005a, 2006; Lerman and Teng, 2004b).

That the bubble–aerosol–droplet cycle is the predominant one on the contemporary Earth for the concentration and transport of organics, dependent for its initiation only on the Rayleigh–Taylor instability, is a confidence-inspiring measure of its robustness. The fundamental mechanisms of the bubblesol supercycle require only a liquid water–gas interface, and are therefore independent in basic form of the assumptions of specific atmospheric, oceanic, or geological chemistry. Hence, the model is applicable to any planetary body that has, or had, a liquid water–gas interface. And there is increasing reason to believe that the free running water and the hydrology cycles underlying the bubble–aerosol–droplet cycle may have existed on an earlier warmer and wetter Mars (Figure 18.9). This was becoming evident before the era of the Mars Exploration Rovers (Kargel, 2004a, 2004b; Second Conference on Early Mars, 2004). More recently the extraordinary efforts of the MERs Opportunity and Spirit, and their human teams, have provided dramatic evidence for water rich environments in Mars' geological past (Squyres et al., 2004, 2005, 2008, 2009).

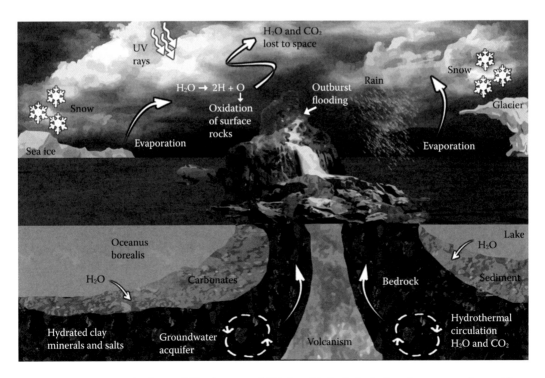

FIGURE 18.9   (See color insert following page 302.) Possible hydrological cycles on an early wet Mars. Compare to the bubble–aerosol–droplet supercycle outlined in Figures 18.1 and 18.3. (Drawing by Jessica Rhodes after Kargel and Strom, 1996.)

Because bubble processes take place near the surface of water, it matters little if the early Martian oceans were kilometers deep or merely meters. Of course the vapor pressure of water in the Noachian atmosphere would have played a large role in the sizes and timescales of early Martian atmospheric phenomena. And a difference in gravity and atmospheric pressure can make a substantial difference on dynamic processes in an early Martian atmosphere. Not only will the terminal velocity of falling precipitation objects be different, but recent work by Xu et al. (2005) shows that for splashing droplets the surrounding air pressure can change the instability of the edge effects (i.e., the existence of a surrounding crown). But this is more important for droplets falling on a hard surface, and seems less likely to qualitatively affect the substance and nature of droplet-induced bubble formation.

More generally, however, on the tectonically simple early Mars (and one having liquid water only intermittently on its surface), such a complex hydrology cycle may have been the only initiator and supporter of the rapid cycles of concentration, hydration, and dehydration necessary for organic polymerization in the "bulk" quantities necessary for chemical evolution to occur (Lerman, 2002c, 2004b, 2005b; Lerman et al., 2004).

In proposing the functional cycle presented above, we are consistent with the game of minimalist prebiotics, making the simplest strong-principle–based assumptions on the early planets. As an example, one can assume the existence of bubbles on the early Earth and Mars with greater confidence than (say) the widespread availability of a particular montmorillonite clay. And because the existence of these bubbles, cavities, and droplets are such common phenomena in nature, their relevance to chemical evolution is potentially critical, whether on the Archaean Earth, the Noachian Mars, or any other terrestrial-like planetary body.

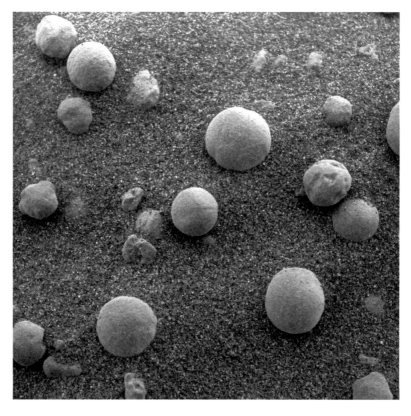

FIGURE 18.10 Martian blueberries (NASA. With permission).

TABLE 18.2
Lifetimes of Atmospheric Particles of the Bubble–Aerosol–Droplet Cycle

Particles	Atmospheric Lifetime
Aitken particles	1–4 h
Aerosols	4–7 days
Fog droplets	3 h
Cloud droplets	7 h
Raindrops	3–15 min
Snowflakes	15–60 min

Source: Gill, S., T. E. Graedel, and C. J. Weschler, *Rev. Geophys. Space Sci.* 21, 903–920, 1983.

One of the more experimentally relevant consequences is the possibility that the Martian "blueberries" (Figure 18.10) discovered by *Spirit* and *Opportunity* (Squyres et al., 2004) are nucleated around organic matter or otherwise mediated by organic rich fluids. These concentrated organics may in turn be the result of Martian analogs to the above-described "organic weather" cycles that follow the bubble-aerosol-droplet supercycle.

From their initial discovery, Martian "blueberries" were linked to terrestrial concretions as their most likely analog; with the strong implication that they were similarly a result of Martian sedimentary processes. If Martian "blueberries" are concretion-like objects, then their ubiquity suggests highly efficient formation processes. On Earth, the most efficient of such processes (for ooids to concretions larger than the size of the Martian blueberries) involve organic nucleation sites or organic coatings of mineral cores accompanied by intermittently agitated water. On Earth many of these organic nucleation sites are of biogenic origin. But organic is all that is actually "necessary." Lerman suggested that this mechanism was also likely for Mars, due to organic complexes likely to have been made by the Noachian equivalent of the bubble–aerosol–droplet cycle (Lerman, 2004a, 2005a, 2006; Lerman et al., 2004; see also Moomaw, 2004).

The potentially crucial role of an abiotic organic component, as opposed to a biological, is highlighted when looking at the terrestrial (Utah) hematite concretions considered closest to the Martian blueberries (Chan et al., 2004, 2005, 2007). A close examination shows that (although a biological nucleation component would not be unexpected) a hydrocarbon-rich fluid component was the likely mediator for their concretions' formation due to groundwater flow through a permeable host rock coupled to a chemical reaction front.

As on the early Earth, basic organic molecules are expected to have existed on the surface of an early Mars due to deposition by meteorites, comets, and asteroids. But to play such a concretion-forming role requires larger-scale organic "clumps" for nucleation purposes or a high fluid density of longer chain amphiphiles for surface mediation of sedimentation. Either possibility requires a chemical evolution of these primordial organics along with the concentration and aggregation of the resulting higher-order organic molecules. As this paper has clearly shown, all of these evolutionary and aggregating processes are first-order consequences of first-order chemical physics.

The time scales for these processes are fast, being quite compatible with even short timescale intermittent bodies of surface water on an early Mars. Bubble processes occur in seconds to hours; and as shown in Table 18.2 terrestrial timescales for the atmospheric portions of the cycle range from minutes to days (data from Gill et al., 1983). If applicable to an early Mars, this would allow even ultrashort periods of turbulent surface water in a warmer wetter Noachian to support many different cycles of concentration, stabilization, and reactions.

It is thus difficult to imagine an earlier, warmer, and wetter Mars where these processes of organic self-organization did not occur. It is also difficult to conceive of likely alternatives to the bubble–aerosol–droplet cycle as a primary organizer of organic compounds on an earlier Mars.

Another natural consequence of the terrestrial bubble–aerosol–droplet supercycle, and its ability to aggregate organics and metals, are objects that when dried appear fascinatingly akin to the so-named "nanobacteria" of ALH 84001 (Lerman 2002a, 2004a, 2005a). These congruent properties include the basic morphology (spheres and sausages), gross chemistry (suites of organics along with metals), and size distributions (nanometers to microns). Whether of biological or bubblesol origin, these striking similarities are due to the universality of the chemical physics involved in the interactions of charge-polarized organic amphiphiles at an air–water interface, coupled to the ubiquity of such scalable macroscopic physical phenomena as the Rayleigh–Taylor instability. The basic morphology and size distributions come from surface tension effects that dominate structure at the micron level while the gross chemistry comes from the concentration effects of the air–water interface for organics, metal ions, and mineral dust. Hence, when designing and interpreting Mars missions looking for evidence of life in Martian paleosols, or when interpreting Martian meteoritic matter akin to the microscopic structures found in ALH84001, the possibility of such abiotic artifacts of past Martian hydrology cycles must be taken into account (Lerman, 2005a).

Thus, it seems likely that the Martian blueberries (and perhaps even the "nanobacteria" of ALH 84001) offer not just evidence of sedimentary processes in a water-rich environment, but hint at the existence of complex past organic weather cycles capable of creating higher-order organics and their aggregates. These same processes of molecular self-organization, which seem to provide the functional requirements for a planetary-scale chemical evolution, are conceivably capable of supporting the autonomous creation of Martian life. Indeed, if the panspermia ideas of Sleep and Zahnle (1998) are correct, that terrestrial life was seeded from an earlier Martian origin, this is all the more relevant. Further work on Mars and the bubble–aerosol–droplet model for prebiotic chemical evolution is under preparation (Lerman, 2006).

18.16 ROBUSTNESS AND DIVERSITY OF EXPLORATIONS THROUGH CHEMICAL PHASE SPACE

This is a new methodological approach, with a strong phenomenological basis in contemporary geophysics/chemistry underlining its potential importance in the reconstruction of chemical evolution. The hypothesis of the fundamental role played by the bubble–aerosol–droplet cycle offers the potential to overcome a number of current stumbling blocks in the field of prebiotic chemistry. In particular, it addresses the problems of selectivity, concentration, and stabilization of organic products in prebiotic chemistry. The bubble–aerosol–droplet cycle affords the possibility of non-equilibrium heterogeneous chemical processes different from conventional solution chemistry, as well as the possibility of coupling a supply of mechanical free energy to chemical processes and reactant concentration mechanisms. It also allows for the possibility of relatively nonaqueous chemical environments within an aqueous medium. Potential condensation reactions are one of the more intriguing consequences. Cycling of hydration–dehydration conditions is easily produced, and with a high surface area to volume ratio for the reactant substrates. A wide range of initial conditions of chemical phase space are sampled, and just as importantly, once organics enter into the bubblesol cycle, they tend to remain within it. All of the above processes are spatially localized and temporally coincident.

Lerman (1986, 1992, 1994a, 1994b, 1996) suggested applying, as global chemical reactor, this contemporary geophysical/geochemical cycle to prebiotic problems. In so doing, it unites and provides a geophysical/chemical basis for a host of other prebiotic studies (Chang, 1993); providing a real-world dynamic framework for the majority of specific chemical model environments developed and suggested by others.

From a historical perspective, Raven and Johnson (2001, p. 66) offer insight on the general concept of bubbles and prebiotic processes. More specifically, Wangersky (1965), in an elegant and far-reaching paper on seawater chemistry, commented in passing that the scavenging ability of bubbles for organics was superior to the similar role played by clays. And Anbar (1968), among others, suggested that bubbles undergoing sonolysis could fix nitrogen from the atmosphere. Polymerization is a fundamental concern, and the soluble salt substrate model of Chang and Lahav (1982), using hydration–dehydration cycles to efficiently drive polymerization, is naturally supported in the bubble–aerosol supercycle. Regarding the atmospheric component of the supercycle presented here, Woese (1979) first suggested that atmospheric droplets might be useful in prebiotic synthesis; and following the work of Lerman, this was expanded upon by Oberbeck et al. (1991). More recently, Vaida and her collaborators have begun to explore the potential role of the aerosol phase (Dobson et al., 2000; Donaldson et al., 2001, 2002; Tervahattu et al., 2005; Vaida and Tuck, this volume).

The breadth and potential completeness of this overall approach overwhelms Darwin's "warm little pond" idea, which has resulted in an overly simplistic idea of chemical evolution occurring somehow, somewhere, sometime in an evaporating tide pool. In contrast, the bubble–aerosol–droplet cycle makes optimal use of all the above, and

- Occurs exclusively at the spatial-temporal scales in which heterogeneous nonequilibrium chemistry is most efficiently supported.
- Is coupled to a rapid and robust exploration of chemical phase space.
- Occurs in a coupled geophysical/chemical supercycle that "had-to-have-been" in widespread existence.

Whether on Earth, Mars, Enceladus, Titan, or Europa, in the search for life and its origins in our solar system the global-scale geochemical reactor that is the bubble–aerosol–droplet cycle must be taken into critical consideration. More abstractly, it all stems from the air–water interface and the symmetry-breaking polar nature of the water molecule. Phenomenologically, it reflects the fundamental role of symmetry-breaking interfaces in the support of self-organizing processes as claimed by the author. For this reason, we further suggest that if water is unavailable elsewhere in the universe as the foundation of chemical evolutionary processes leading to an independent origin of life, then other polar solvents must necessarily assume the role.

ACKNOWLEDGMENTS

To owe so much to so many, and have so little space to list and thank them all! First and foremost, the John Templeton Foundation is thanked for their support of the writing of this document; with Simon Conway Morris being the epitome of a critical yet supportive editor. An unofficial editor of note was the polymathic Steve Benner, whose offer to join him at the newly founded Westheimer Institute for Science and Technology/Foundation for Applied Molecular Evolution has been intellectually most felicitous. The wonderful experimental work of M. Ruiz Bermejo, C. Menor Salvan, S. Osuna Esteban, and S. Veintemillas Verdaguer of Spain's Centro de Astrobiología opens a new and fertile frontier in experimental astrobiology, and they have my deepest appreciation both for the work itself and their making it available to me prior to publication.

NASA's Jim Lawless and Sherwood Chang, at the time of my first germinating ideas of what became the bubble–aerosol–droplet model, encouraged and challenged a beginning student to the never-ending fascination of the origin of life as a legitimate question of scientific exploration. This early student work (Lerman, 1986) eventually morphed into an MS thesis at Stanford (Lerman, 1992); and subsequent opportunities to present and discuss this work at Gordon Conferences, ISSOL, NASA Astrobiology Institute, and Lunar and Planetary Institute (LPI)-sponsored meetings (Lunar and Planetary Science Conference and Early Mars) have tightened the logic immeasurably. The Fannie and John Hertz Foundation is thanked for its remarkably generous support of my early work on self-organizing systems, while Peter Gogarten's offer of the keynote address at the 2003 Gordon Conference on the Origin of Life forced a *self*-organization of this work for the public. "Danke!" Peter for the alliterative opportunities.

The gracious hospitality of Duncan Blanchard, Ferren MacIntyre, and Tom Graedel at their respective institutions introduced me, first hand, to some of the more amazing properties of bubbles and aerosols; while Anastassia Kanavarotti's (UCSC) infectious enthusiasm for experimentally testing these ideas was much appreciated and the sections on physical chemical reaction mechanisms owe much to discussions with Michael Anbar of SUNY-Buffalo.

My warmest thanks for the constancy of their support go to Jerry van Andel (Cambridge), Norm Sleep (Stanford), James Symons (Lawrence Berkeley Laboratory), Paul Bratterman (Univ. of North Texas), and Jennifer Blank (SETI). What organic geochemistry I have learned nucleated around the teachings of Keith Kvendvolden (USGS); and the early and effusive enthusiasms of Bill Schopf for the ideas presented here certainly helped keep my own enthusiasms strong. Michael New (NASA), Danielle Wyrick (Southwest Research Institute), John Lindsey (LPI), and George Lerman (University of Washington) have each provided serious catalysis of insight. All are thanked!

Three wonderful artist/engineers are thanked for their continuing contributions to this project: Prof. Andrew Davidhazy of RIT spent days chasing after the ultimate of bubble-bursting photographs, while Jessica Rhodes did the spectacular representation of an early Martian hydrosphere. Jacqueline Teng created the majority of graphs and figures and as always my deepest appreciation goes to her for her dedicated efforts to improve the clarity of my text, figures, and logic. David Lerman and friends (the Bach-Koon clan) provided all manner and flavors of support; while Sara Kuwabara offered considerable help with bibliographic research and copy editors Linda Turnowski and Wayne Chambliss offered improvements in language and style. To all I am grateful.

Finally, this paper is dedicated to Eliot Kalmbach (1985–2009) … junior research assistant, Princeton undergraduate ('09) in geology and physics, geology expedition junkie, member of the Explorer's Club, and friend. Prost!

REFERENCES

Adamson, A. 1997. *The Physical Chemistry of Surfaces*, 6th ed. New York, NY: Wiley-Interscience.

Anbar, M. 1968. Cavitation during impact of liquid water on water: Geochemical implications, *Science* 161: 1343–1344.

Apel, C., D. W. Deamer, and M. Mautner. 2002. Self-assembled vesicles of monocarboxylic acids and alcohols: Conditions for stability and for the encapsulation of biopolymers *Biochim. Biophys. Acta Biomembr.* 1559:1–9.

Bachmann, P. A., P. L. Luisi, and J. Lang. 1992. Autocatalytic self-replicating micelles as models for prebiotic structures. *Nature* 357:57–59.

Banin, A., and J. Navrot. 1975. Origin of life: clues from relations between chemical compositions of living organisms and natural environments. *Science* 189(4202):550–551.

Baylor, E. R., and W. H. Sutcliffe. 1963. Dissolved organic matter in seawater as a source of particulate food. *Limnol. Oceanogr.* 8:369–371.

Benner, S. A., A. Ricardo, and M. A. Carrigan. 2004. Is there a common chemical model for life in the universe? *Curr. Opin. Chem. Biol.* 8:672–689.

Blanchard, D. C. 1975. Bubble scavenging and the water to air transfer of organic material in the sea. In *Applied Chemistry at Protein Interfaces*, ed. R. Baier. *Adv. Chem.* 145:360–387.

Blanchard, D. C. 1983. The production, distribution, and bacterial enrichment of the sea-salt aerosol. In *Air / Sea Exchange of Gases and Particles* (NATO ASI Series. Series C, Mathematical and Physical Sciences, No. 108), ed. P. Liss and W. Slinn. Dordrecht, The Netherlands: Kluwer Academic Publishers, pp. 407–454.

Blanchard, D. C. 1989. The ejection of drops from the sea and their enrichment with bacteria and other materials: a review. *Estuaries* 12:127–137.

Blanchard, D. C., and L. Syzdek. 1970a. Mechanism for the water-to-air transfer and concentration of bacteria. *Science* 170:626–628.

Blanchard, D. C., and L. Syzdek. 1970b. Concentration of bacteria from bursting bubbles. *J. Geophys. Res.* 77:5087–5099.

Blanchard, D. C., and A. H. Woodcock. 1957. Bubble formation and modification in the sea and its meteorological significance. *Tellus* 9:145–148.

Brenner, M., S. Hilgenfeldt, and D. Lohse. 2002. Single-bubble sonoluminescence. *Rev. Mod. Phys.* 74: 425–452.

Chan, M. A., B. Beitler, W. T. Parry, J. Ormö, and G. Komatsu. 2004. A possible terrestrial analogue for haematite concretions on Mars. *Nature* 429:731–734.

Chan, M. A., B. Beitler Bowen, W. T. Parry, J. Ormö, and G. Komatsu. 2005. Red rock and red planet diagenesis: Comparisons of Earth and Mars concretions. *GSA Today* 15:4–10.

Chan, M. A., J. Ormö, A. J. Park, M. Stich, V. Souza-Egipsy, and G. Komatsu. 2007, Models of iron oxide concretion formation: Field, laboratory, and numerical comparisons. *Geofluids* 7:1–13.

Chang, S. 1993. Prebiotic synthesis in planetary environments. In *The Chemistry of Life's Origins*, ed. J. M. Greenberg, C. X. Mendoza-Gómez, and V. Pirronello. Dordrecht: Kluwer Academic Publishers, pp. 259–299.

Chang, S., and N. Lahav. 1982. The possible role of soluble salts in chemical evolution. *J. Mol. Evol.* 19: 36–46.

Chang, S., D. DeMarais, R. Mack, S. L. Miller, and G. Strarhearn. 1983. Prebiotic organic synthesis and the origin of life. In *Earth's Earliest Biosphere: Its Origin and Early Evolution*, ed. J. Schopf. Princeton, NJ: Princeton University Press, pp. 53–92.

Chen, I. A., and J. W. Szostak. 2004. A kinetic study of the growth of fatty acid vesicles. *Biophys. J.* 87(2):988–998.

Chen, I. A., R. W. Roberts, and J. W. Szostak. 2004. The emergence of competition between model protocells. *Science* 305:1474–1476.

Cincinelli, A., L. Lepri, L. Checchini, and M. Perugini. 1999. The role of the surface microlayer in contaminant concentration and pollutant transport via marine aerosols. In *The Role of Sea Surface Microlayer Processes in the Biogeochemistry of the Mediterranean Sea*, ed. F. Briand CIESM Workshop Series, vol. 9, pp. 23–25.

Cronin, J. R., and S. Chang. 1993. Organic matter in meteorites: Molecular and isotopic analyses of the Murchison meteorite. In *The Chemistry of Life's Origins*, ed. J. M. Greenberg, C. X. Mendoza-Gómez, and V. Pirronello. Dordrecht: Kluwer Academic Publishers, pp. 209–258.

Deamer, D., and R. M. Pashley. 1989. Amphiphilic components of the Murchison carbonaceous chondrite: surface properties and membrane formation. *Orig. Life Evol. Biosph.* 19:21–38.

Didenko, Y., and K. S. Suslick. 2002. The energy efficiency of formation of photons, radicals and ions during single-bubble cavitation. *Nature* 418:394–397.

Dobson, C. M., G. B. Ellison, A. F. Tuck, and V. Vaida. 2000. Atmospheric aerosols as prebiotic chemical reactors. *Proc. Natl. Acad. Sci.* 97:11864–11868.

Donaldson, D. J., A. F. Tuck, and V. Vaida. 2001. Spontaneous fission of atmospheric aerosol particles. *Phys. Chem. Chem. Phys.* 3:5270–5273.

Donaldson, D. J., A. F. Tuck, and V. Vaida. 2002. The asymmetry of organic aerosol fission and prebiotic chemistry. *Orig. Life Evol. Biosph.* 32:237–243.

Duce, R., and E. Hoffman. 1976. Chemical fractionation at the air/sea interface. *Annu. Rev. Earth Planet. Sci.* 4:187–228.

Dworkin, J. P., D. W. Deamer, S. A., Sandford, and L. J. Allamandola. 2001. Self-assembling amphiphilic molecules: Synthesis in simulated interstellar/precometary ices. *Proc. Natl. Acad. Sci. USA* 98:815–819.

Dyson, F. 1999. *Origins of Life*. Cambridge: Cambridge University Press.

Fitzgerald, M. E., V. Griffing, and J. Sullivan. 1956. Chemical effects of ultrasonics—"Hot spot chemistry." *J. Chem. Phys.* 25:926–933.

Gill, S., T. E. Graedel, and C. J. Weschler. 1983. Organic films on atmospheric aerosol particles, fog droplets, cloud droplets, raindrops, and snowflakes. *Rev. Geophys. Space Sci.* 21:903–920.

Graedel, T. E., and C. J. Weschler. 1981. Chemistry within aqueous atmospheric aerosols and raindrops. *Rev. Geophys. Space Sci.* 19:505–539.

Hanczyc, M. M., S. M. Fujikawa, and J. W. Szostak. 2003. Experimental models of primitive cellular compartments: Encapsulation, growth, and division. *Science* 302:618–622.

Hanczyc, M. M., and J. W. Szostak. 2004. Replicating vesicles as models of primitive cell growth and division. *Curr. Opin. Chem. Biol.* 8(6):660–664.

Hazen, R. 2005. *Genesis: The Scientific Quest for Life's Origins*. Washington, D.C.: Joseph Henry Press, pp. 223–232.

Jaenicke, R. 2005. Abundance of cellular material and proteins in the atmosphere. *Science* 308:73–74.

Johnson, B. D., and R. C. Cooke. 1980. Organic particle and aggregate formation resulting from the dissolution of bubbles in seawater. *Limnol. Oceanogr.* 25:653–661.

Johnson, B. D., and R. C. Cooke. 1981. Generation of stabilized microbubbles in seawater. *Science* 213: 209–211.

Kargel, J. S. 2004a. *Mars—A Warmer, Wetter Planet*. Chichester: Springer Praxis Books.

Kargel, J. S. 2004b. Proof for water, hints of life? *Science* 306:1689–1691.

Kargel, J. S., and R. G Strom. 1996. Global climatic change on Mars. *Scientific American* 275:80–88.

Kraus, E. B., and J. A. Businger. 1994. *Atmosphere-Ocean Interaction*. New York, NY: Oxford University Press.

Lacey, J., and D. Mullins. 1983. Experimental studies related to the origin of the genetic code and the process of protein synthesis—A review. *Orig. Life* 13:3–42.

Lemlich, R., Ed. 1972. *Adsorptive Bubble Separation Techniques*. New York, NY: Academic Press.

Lerman, L. 1986. The potential role of bubbles and droplets in primordial and planetary chemistry: Exploration of the liquid-gas interface as a reaction zone for condensation processes. *Orig. Life* 16:201–202.

Lerman, L. 1992. Exploration of the liquid-gas interface as a reaction zone for condensation processes: The potential role of the air-water interface in prebiotic chemistry. MS thesis, Stanford University.

Lerman, L. 1994a. The bubble-aerosol-droplet cycle as natural reactor for prebiotic chemistry (I). *Orig. Life* 24:111–112.

Lerman, L. 1994b. The bubble-aerosol-droplet cycle as natural reactor for prebiotic chemistry (II). *Orig. Life* 24:138–139.

Lerman, L. 1996. The bubble-aerosol-droplet cycle: A prebiotic geochemical reactor (that must have been). *Orig. Life* 26:369–370.

Lerman, L. 2002a. Consequences and artifacts: Terrestrial findings and Martian analogues of an air-water interface. *33rd Lunar and Planetary Science Conference*, March 11–15, 2002, NASA Johnson Space Center (http://www.lpi.usra.edu/meetings/lpsc2002/pdf/2062.pdf).

Lerman, L. 2002b. How the chemical physics of the microscale drives prebiotic evolution. *10th ISSOL Meeting and the 13th International Conference on the Origin of Life*, Oaxaca, Mexico, June 30 through July 5, 2002.

Lerman, L. 2002c. Consequences and artifacts of an intermittent air-water interface on Mars. Presented at *10th ISSOL Meeting and the 13th International Conference on the Origin of Life*, Oaxaca, Mexico, June 30 through July 5, 2002.

Lerman, L. 2002d. On the symmetry of nuclear identity between relativistic primary and secondary nuclei. PhD thesis, Philipps University, Marburg, Germany (http://archiv.ub.uni-marburg.de/diss/z2002/0105/).

Lerman, L. 2003. The primordial bubble: Symmetry-breaking and the origin of structure. *Keynote Address of the 2003 Gordon Conference on the Origin of Life*, Bates College, July 13–18, 2003.

Lerman, L. 2004a. Do Martian blueberries have pits? Artifacts of Martian water past. *2nd Conference on Early Mars*, Lunar Planetary Institute, Jackson Hole, WY, October 11–15, 2004 (http://www.lpi.usra.edu/meetings/earlymars2004/pdf/8063.pdf).

Lerman, L. 2004b. Could Martian strawberries be? Existence of a planetary-scale infrastructure for chemical evolution on an early Mars. *2nd Conference on Early Mars*, Lunar Planetary Institute, Jackson Hole, WY, October 11–15, 2004 (http://www.lpi.usra.edu/meetings/earlymars2004/pdf/8066.pdf).

Lerman, L. 2005a. Do Martian blueberries have pits? Artifacts of an early wet Mars. *Lunar and Planetary Science Conference XXXVI*, March 14–18, 2005, NASA Johnson Space Center (http://www.lpi.usra.edu/meetings/lpsc2005/pdf/2210.pdf).

Lerman, L. 2005b. Could Martian strawberries be? Prebiotic chemical evolution on an early wet Mars. *Lunar and Planetary Science Conference XXXVI*, March 14–18, 2005, NASA Johnson Space Center (http://www.lpi.usra.edu/meetings/lpsc2005/pdf/2317.pdf).

Lerman, L. 2005c. Variational approaches to water and the origin of life. *Invited Lecture to the Workshop on Counterfactual Chemistry and Fine-Tuning in Biochemistry*, Templeton Foundation, Varenna, Italy, April 29–30, 2005.

Lerman, L. 2006. Prebiotic chemical evolution on an early Mars: Consequences and artifacts of "organic" weather cycles in the Noachian. Available at: http://www.lpi.usra.edu/meetings/lpsc2006/pdf/1566.pdf.

Lerman, L., and J. Teng. 2004a. In the beginning: A functional first principles approach to chemical evolution. In *Origins: Genesis, Evolution, and Biodiversity of Life*, ed. J. Seckbach. Dordrecht, The Netherlands: Kluwer Academic Publishers, pp. 35–58.

Lerman, L., and J. Teng. 2004b. Chemical evolution on an early Mars: Existence of a planetary-scale infrastructure. *3rd NASA Astrobiology Science Conference*, NASA-Ames, March 28–April 1, 2004.

Lerman, L., Teng, J., and S. Kuwabara. 2004. Chemical evolution on an early Mars: Implications for mission planning. *3rd NASA Astrobiology Science Conference*, NASA-Ames, March 28–April 1, 2004.

Lohse, D. 2002. Sonoluminescence: Inside a micro-reactor. *Nature* 418:381–383.

Luisi, P. L., P. Vonmont-Bachmann, and M. Fresta. 1993. Self-reproduction of micelles and liposomes and the transition to life. *J. Liposome Res.* 3:631–638.

Luisi, P. L., P. Walde, and Th. Oberholze. 1994. Enzymatic RNA synthesis in self-reproducing vesicles: An approach to the construction of a minimal synthetic cell. *Ber. Bunsenges.Phys.Chem.* 98:1160–1165.

MacIntyre, F. 1970. Geochemical fractionation during mass transfer from sea to air by breaking bubbles. *Tellus* 22:451–462.

MacIntyre, F. 1972. Flow patterns in breaking bubbles. *J. Geophysi. Res.* 77:5211–5228.

MacIntyre, F. 1974a. Chemical fractionation and sea-surface microlayer processes. *The Sea (Mar. Chem.)* 5:245–299.

MacIntyre, F. 1974b Non-lipid related possibilities for chemical fractionation in bubble film caps. *J. Rech. Atmosph.* 7:515–527.

MacIntyre, F. 1974c. The top millimeter of the ocean. *Sci. Am.* 230:62–77.

MacIntyre, F., and J. W. Winchester. 1969. Phosphate ion enrichment in drops from breaking bubbles. *J. Phys. Chem.* 73:2163–2169.

Margulis, L. 1981. *Symbiosis in Cell Evolution: Life & Its Environment on the Early Earth.* W. San Francisco, CA: H. Freeman & Company.

Melville, W., and R. Rapp. 1985. Momentum flux in breaking waves, *Nature* 317:514–516.

Mitchell, J. G., A. Okubo, and J. A. Fuhrman. 1985. Microzones surrounding phytoplankton form the basis for a stratified marine microbial ecosystem. *Nature* 316:58–59.

Monnard, P. A., C. L. Apel, A. Kanavarioti, and D. W. Deamer. 2002. Influence of ionic solutes on self-assembly and polymerization processes related to early forms of life: Implications for a prebiotic aqueous medium. *Astrobiology* 2:213–219.

Moomaw, B. 2004. The geology of Mars mid-'04. *SpaceDaily.* June 8, 2004 http://www.spacedaily.com/news/mars-mers-04zzzzx.html.

Murphy, D. M., D. S. Thomson, and M. J. Mahoney. 1998. In situ measurement of organic, meteoritic material, and other elements in aerosols at 5 to 19 kilometers. *Science* 282:1664–1669.

Noether, E. 1918. Invariante variations probleme, *Nachr. d. König. Gesellsch. Wiss. Göttingen, Math-phys. Klasse,* 235–257; English translation by Travel, M. A. 1971. Invariant variation problems. *Transp. Theory Stat. Phys.* 1(3):183–207.

Oberbeck, V. et al. 1991. Prebiotic chemistry in clouds. *J. Mol. Evol.* 32:296–303.

O'Dowd, C. D., and G. de Leeuw. 2007. Marine aerosol production: A review of the current knowledge. *Phil. Trans. R. Soc. A* 365:1753–1774.

Platts, N. 2006. Contributions to chemical questions in origins of life. PhD thesis, Department of Chemistry and Chemical Biology, Rensselaer Polytechnic Institute, Troy, NY.

Rao, M., D. G. Odom, and J. Oro. 1980. Clays in prebiotic chemistry. *J. Mol. Evol.* 15:317–331.

Raven, P., and G. Johnson. 2001, *Biology,* 5th ed. New York, NY: McGraw-Hill.

Ruiz Bermejo, M., C. Menor Salvan, S. Osuna Esteban, and S. Veintemillas Verdaguer. 2005. *Synthesizing the building blocks for the origin of life by using aerosols for reactors.* Preprint from *Centro de Astrobiologia (CSIC-INTA).*

Ruiz-Bermejo, M., C. Menor-Salván, S. Osuna-Esteban, and S. Veintemillas-Verdaguer. 2007a. Prebiotic microreactors: A synthesis of purines and dihydroxy compounds in aqueous aerosol. *Origins Life Evol. B.* 37:123–142.

Ruiz-Bermejo, M., C. Menor-Salván, S. Osuna-Esteban, and S. Veintemillas-Verdaguer. 2007b. The effects of ferrous and other ions on the abiotic formation of biomolecules using aqueous aerosols and spark discharges. *Origins Life Evol. B.* 37:507–521.

Second Conference on Early Mars. 2004. *Geologic, Hydrologic, and Climatic Evolution and the Implications for Life,* Lunar Planetary Institute, Jackson Hole, WY, October 11–15, 2004; http://www.lpi.usra.edu/meetings/earlymars2004/.

Shapiro, R. 1999. *Planetary Dreams: The Quest to Discover Life beyond Earth.* New York, NY: Wiley.

Sharp, D. 1984. An overview of Rayleigh-Taylor instability. *Physica D* 12D:3–18.

Sieburth, J. 1983. Microbiological and organic-chemical processes in the surface and mixed layers. In *Air/Sea Exchange of Gases and Particles* (NATO ASI Series. Series C, Mathematical and Physical Sciences, No. 108), ed. P. Liss and W. Slinn. Dordrecht, The Netherlands: Kluwer Academic Publishers, pp. 121–172.

Sleep, N. H., and K. Zahnle. 1998. Refugia from asteroid impacts on early Mars and the early Earth. *J. Geophys. Res.* 103:28529–28544.

Smolin, L. 1997. *Life of the Cosmos.* New York, NY: Oxford University Press.

Squyres, S. W., and A. H. Knoll. 2005. Sedimentary rocks at Meridiani Planum: Origin, diagenesis, and impli-
   cations for life on Mars. *Earth Planet. Sci. Lett.* 240:1–10.
Squyres, S.W., J. P. Grotzinger, R. E. Arvidson, J. F. Bell, III, W. Calvin, P. R. Christensen, B. C. Clark, J. A.
   Crisp, W. H. Farrand, K. E. Herkenhoff, J. R. Johnson, G. Klingelhöfer, A. H. Knoll, S. M. McLennan,
   H. Y. McSween, Jr., R. V. Morris, J. W. Rice, Jr., R. Rieder, and L. A. Soderblom. 2004. In situ evidence
   for an ancient aqueous environment at Meridiani Planum, Mars. *Science* 306:1709–1714.
Squyres, S., R. E. Arvidson, S. Ruff, R. Gellert, R. V. Morris, D. W. Ming, L. Crumpler, J. D. Farmer, D. J.
   Des Marais, A. Yen, S. M. McLennan, W. Calvin, J. F. Bell, III, B. C. Clark, A. Wang, T. J. McCoy,
   M. E. Schmidt, and P. A. de Souza, Jr. 2008. Detection of silica-rich deposits on Mars. *Science* 320:
   1063–1067.
Squyres, S. W., A. H. Knoll, R. E. Arvidson, J. W. Ashley, J. F. Bell, III, W. M. Calvin, P. R. Christensen,
   B. C. Clark, B. A. Cohen, P. A. de Souza, Jr., L. Edgar, W. H. Farrand, I. Fleischer, R. Gellert, M. P.
   Golombek, J. Grant, J. Grotzinger, A. Hayes, K. E. Herkenhoff, J. R. Johnson, B. Jolliff, G. Klingelhöfer,
   A. Knudson, R. Li, T. J. McCoy, S. M. McLennan, D. W. Ming, D. W. Mittlefehldt, R. V. Morris, J. W.
   Rice, Jr., C. Schröder, R. J. Sullivan, A. Yen, and R. A. Yingst. 2009. Exploration of Victoria Crater by the
   rover Opportunity. *Science* 324:1058–1061.
Suslick, K. S. 1989. The chemical effects of ultrasound. *Sci. Am.* 260:80–86.
Suslick, K. S. 1994. The chemistry of ultrasound. In *Encyclopedia Britannica Yearbook of Science and the
   Future*. Chicago, IL: Encyclopedia Britannica, pp. 138–155.
Suslick, K. S., Y. Didenko, M. M. Fang, T. Hyeon, K. J. Kolbeck, W. B. McNamara III, N. M. Mdleleni,
   and M. Wong. 1999. Acoustic cavitation and its chemical consequences. *Philos. Trans. R. Soc. A* 357:
   335–353.
Szostak J. W., D. P. Bartel, and P. L. Luisi. 2001. Synthesizing life. *Nature* 409:387–390.
Tervahattu, H., A. F. Tuck, and V. Vaida. 2005. Chemistry in prebiotic aerosols: A mechanism for the ori-
   gin of life. In *Origins: Genesis, Evolution, and Biodiversity of Life*, ed. J. Seckbach. Dordrecht, The
   Netherlands: Kluwer Academic Publishers.
Turco, R. P., O. B. Toon, R. C. Whitten, R. G. Keesee, and P. Hamill. 1982. Importance of heterogeneous pro-
   cesses to tropospheric chemistry: Studies with a one-dimensional model. In *Heterogeneous Atmospheric
   Chemistry*, ed. D. Schryer. Washington D.C.: American Geophysical Union, 231–240.
Vaida, V., and A. F. Tuck. 2010. Water: The tough-love parent of life. In *Water and Life: The Unique Properties
   of H$_2$O*, ed. R. M. Lynden-Bell, S. C. Morris, J. D. Barrow, J. L. Finney, and C. Harper. Boca Raton, FL:
   CRC Press, pp. 231–244 (this volume).
Wangersky, P. 1965. The organic chemistry of sea water. *Am. Sci.* 53:358–374.
Wangersky, P. J. 1974. Particulate organic carbon: Sampling variability. *Limnol. Oceanogr.* 19:980–984.
Woese, C. R. 1979. A proposal concerning the origin of life on the planet Earth. *J. Mol. Evol.* 13:95–101.
Xu, L., W. W. Zhang, and S. R. Nagel. 2005. Drop splashing on a dry smooth surface, Presented at the American
   Physical Society Meeting, Los Angeles, CA, March 21–25, 2005; http://arxiv.org/abs/physics/0501149.
Zaug, A., and T. Cech. 1985. Oligomerization of intervening sequence RNA molecules in the absence of pro-
   teins. *Science* 229:1060–1064.

19 Liquids, Biopolymers, and Evolvability
Case Studies in Counterfactual Water-Life

Wilson C. K. Poon

CONTENTS

19.1 INTRODUCTION: A TAXONOMY OF WATER-LIFE

> Alice laughed. 'There's no use trying,' she said, 'one can't believe impossible things.' 'I daresay you haven't had much practice,' said the Queen. 'When I was your age, I always did it for half-an-hour a day. Why, sometimes I've believed as many as six impossible things before breakfast.'
>
> —*Through the Looking Glass*, **Lewis Carroll**

The title of the symposium at which a version of this chapter was first presented was "Water of Life: Counterfactual Chemistry and Fine-Tuning in Biochemistry." The word "counterfactual" in the subtitle was held by the organizers to be key: we were asked to imagine all kinds of alternative worlds in order to figure out whether it made sense to speak of water being "fine-tuned" for life.

Such counterfactual exercises are well established in cosmology, where our understanding now permits us to imagine universes with different fundamental laws or physical constants [1]. For example, in an alternative universe in which the strength of the strong nuclear force is just 2% higher than our value, the diproton would be strongly bound, and hydrogen would be an explosive nuclear fuel. In this universe, little of the primordial hydrogen would survive the hot primeval phase, leaving us with a helium cosmos. What Paul Davies calls our universe's "most significant macroscopic structures," slow-burning hydrogen stars (like the sun), would not exist; neither would organic materials or, for that matter, water.

Success in such counterfactual cosmology makes it tempting to extend the exercise to other sciences. In this book, we consider the counterfactual chemistry of water in relation to life. In the title of this chapter, I have called this an exercise in counterfactual water-life. The composite noun water-life is cumbersome, but it serves as a mnemonic to remind us of the taxonomy of the issues that we may want to consider:

Real water, Real life	Counterfactual water, Real life
Real water, Counterfactual life	Counterfactual water, Counterfactual life

"Real water, Real life" is conceptually the most straightforward: we inquire how the various properties of the molecule H_2O in its liquid state, such as having a negative thermal expansivity near its freezing point, are well matched to the requirements of life as we know it on this Earth. This exercise has a long history, producing a catalog of fascinating results that have been ably reviewed in Chapter 8.3 of Barrow and Tipler's book [2]. But even here there are hidden dimensions not often included in the usual discussions. Both water and life on the present Earth are found under a strikingly wide range of environments: from subzero to well above 100°C and from atmospheric pressure (1 bar) to at least 1000 atm (1 kbar), not to mention a large range of salinity (up to saturated brine). The conditions on the primitive Earth under which life first arose are still a matter of debate, but phylogenetic analysis of prokaryotes now living under extreme conditions (extremophiles) makes at least a plausible case for life having arisen under conditions of high temperature and probably high pressure [3]. Since the literature on the fine-tuned properties of water to date [2] overwhelmingly focuses on its behavior at or near standard temperature and pressure (STP; 1 atm, 298 K = 25°C), considering how non-STP water supports life, both now and in the geological past, is not without interest in a volume on water of life. I do this in the third section of this chapter. Rather than reviewing the considerable amount of data available on this subject, I will instead suggest that the tunability of the properties of (real) water may have contributed to the evolution of evolvability. Considering non-STP water also leads me to question whether it is not more appropriate to speak of life being fine-tuned to water, rather than water being fine-tuned to life.

Consider now "Real water, Counterfactual life." The issue under this heading is that of rerunning the tape of life: If evolution were to start all over again from identical prebiotic conditions, and if life did originate, would the result be anything like life as we know it? Or would the result be something quite different (counterfactual life supported by water)? This issue has already been well aired in a well-known debate between Stephen Jay Gould [4] and Simon Conway Morris [5]. The latter famously suggests that so many constraints exist along the way that, in fact, biological evolution, if it gets started at all, will probably finish up with something very much like what we have now. For our purposes here, the important thing is that part of the identical prebiotic conditions referred to above includes liquid water. The question then is: what sort of constraints does water place on the course of evolution? In other words, could "wet life" have turned out in a radically different way? I make some very brief remarks about this issue in the second section.

Next, we have "Counterfactual water, Real life": Could we have life as we know it without water? This question can be approached from many different angles. For instance, we can inquire into the possibility of carbon-based life evolving not in an aqueous environment, but in (say) liquid ammonia. On the other hand, we may wish to investigate whether carbon-based life, having evolved in an aqueous milieu, could have survived a subsequent drought (of water) with or without an alternative solvent. These questions are fresh in our minds in the wake of recent advances in Martian exploration. But I would not be discussing them further.

Alternatively, under the heading of "Counterfactual water, Real life," one may wish to engage in the exercise of thinking about slightly different H_2O—and how life as we know it might survive in such counterfactual water. On the most basic level, we ask how the properties of the molecule H_2O predicted by the Schrödinger equation change when the fine structure constant $\alpha = e^2/\hbar c$ and the ratio of the masses of the electron and proton $\beta = m_e/m_p$ take on values other than 1/137 and 1/1837, respectively. On the other hand, it is possible to engage in a somewhat more phenomenological exercise and calculate, using classical molecular dynamics, the properties of various water-like

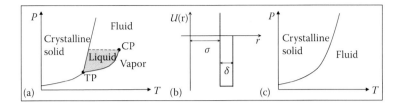

FIGURE 19.1 (a) The pressure–temperature (*P*–*T*) phase diagram of a simple substance. The liquid state (shaded) has a fragile existence between the triple point (TP) and the critical point (CP). Note that this schematic is drawn for the "generic simple substance," and is wrong for water in one important qualitative aspect. One of the unique properties of water is that the slope of the phase boundary separating the solid and liquid states is negative, because it shrinks when it freezes. (b) A model intermolecular potential, *U*(*r*), consisting of a "hard-core" repulsive part of range σ and an attractive "square well" of range δ. A liquid state only exists in a collection of molecules interacting via this potential when δ/σ is large enough. (c) When the range of the intermolecular attraction becomes too short, the vapor–liquid critical point becomes metastable with respect to the fluid–solid transition, and disappears from the equilibrium phase diagram.

molecules as a function of bond angle, dipole moment, etc. Other authors in this volume do indeed take these approaches (see the chapters by Allen and Schaefer and Lynden Bell, respectively). Suffice it to say here that there is the requirement of consistency: not all values of (α,β) would give rise to a universe with atoms in the first place; and counterfactual quantum chemistry within the permitted range of (α,β) would presumably not yield all possible values of the bond angle, etc. In other words, the task of imagining "other water" may be much more difficult than at first sight, because H_2O as we know it may be deeply constrained by the constants of nature.

The heading of "Counterfactual water, Real life" may prompt us to adopt yet a third line of inquiry. The phase diagram of water contains a vapor–liquid critical point, and a region of temperature and density where water vapor coexists with liquid water. In this respect, water is no different from all other simple atomic and small-molecular substances (Figure 19.1a). Indeed, we are so used to the existence of the liquid state that, as far as I am aware, the literature on fine-tuning and fitness for life has not yet addressed the following question: "Is it possible to have a universe in which not only the liquid known as 'water,' but all liquids, do not exist?" And, if so, could life as we know it have evolved and thrived in such a liquidless universe? I address this question in the next section. It turns out that the existence of liquids (of all kinds, not just water) may be a deep emergent property predicated on the constants of nature.

Last in my fourfold taxonomy is "Counterfactual water, Counterfactual life." Under this most ambitious heading, we need to consider whether wholly unfamiliar forms of life (noncarbon-based, etc.) could evolve and be sustained in fluids not made of the molecule H_2O as we know it. I have little to say about this most speculative of questions, except that in the second section below, I do briefly consider one possible constraint on the evolution of life built out of noncarbon-based polymers in nonaqueous solvents.

19.2 THE CURIOUS EXISTENCE OF LIQUIDS

'Is there any other point to which you would wish to draw my attention?'
'To the curious incident of the dog in the night-time.'
'The dog did nothing in the night time.'
'That was the curious incident,' remarked Sherlock Holmes.

—*Silver Blaze*, **Arthur Conan Doyle**

Sherlock Holmes, the fictional Victorian detective, is a master of grasping the significance of the commonplace. The existence of the liquid state for all known chemical elements and for all small-molecular compounds is commonplace. It is so commonplace that its physical basis has seldom been questioned. In particular, in discussions of how the properties of water are fine-tuned for life, the very liquidness of water has always been taken for granted, and never included in the list of those properties "up for grabs" in the counterfactual imagining exercise.

But liquids have a fragile existence. Unlike the vapor and solid states, the liquid state occupies a *finite* region in the phase diagram (Figure 19.1a). At first sight at least, it seems that relatively small changes to "the way the world is" may be sufficient to wipe out this corner of the phase diagram altogether. But investigating this issue would not be easy, because of the three common states of matter, the liquid state is the most difficult to grasp theoretically. That this is so can be seen from this eloquent testimony from a top-ranking theoretical physicist and one-time director of Conseil Européen pour la Recherche Nucléaire, Victor Weiskopf [6]:

> The existence and general properties of solids and gases are relatively easy to understand once it is real-ized that atoms or molecules have certain typical properties and interactions that follow from quantum mechanics. Liquids are harder to understand. Assume that a group of intelligent theoretical physicists had lived in closed buildings from birth such that they never had occasion to see any natural structures. Let us forget that it may be impossible to prevent them to see their own bodies and their inputs and outputs. What would they be able to predict from a fundamental knowledge of quantum mechanics? They probably would predict the existence of atoms, of molecules, of solid crystals, both metals and insulators, of gases, but most likely not the existence of liquids.

It seems that we definitely should *not* take liquids for granted! But before turning to consider the question of what it takes for liquids to exist, we should first be clear about what is meant by the word liquid. Referring to the phase diagram of a simple substance (Figure 19.1a), we see that the term "liquid" only has meaning below the critical point. Above the critical point, only a single kind of noncrystalline state exists, a *fluid*. Below the critical point, however, *two* noncrystalline states may exist, depending on the precise temperature and pressure—liquid and vapor. Both of these states are "fluid," in that they flow under infinitesimal applied shear. We could have called these "fluid 1" and "fluid 2" rather than "vapor" and "liquid." The key aspect of life below the critical point is that, where these two fluids are found together, they are separated by *vapor–liquid interfaces*.

Now we can state the two questions of interest to us in this section more precisely. First, why do liquids exist quite generally in our universe? Second, do we need vapor–liquid interfaces for the evolution of the kind of life that we know about? Below, I will first suggest plausible reasons for answering "yes" to the second question. I will then move on to consider the conditions necessary for the existence of liquids.

19.2.1   Life without Liquids

That liquids and life go together (like the fact that the dog did nothing during the night in Conan Doyle's story) seems so commonplace that few, if any, have ever questioned it. But do we really need liquids for life? Let me discuss this question in two stages. First, and briefly, I think we do need reasonably dense fluids: dense, so that the rate of molecular encounter can be sufficiently high, and fluid, so that diffusion is sufficiently rapid.* But to satisfy these requirements, a dense fluid will do, and we do not need the existence of two different fluid states, vapor and liquid. Physically, what we *do* get extra if the phase diagram has a critical point are vapor–liquid interfaces. But what do these do for life?

To answer this question, consider planetary atmospheres as shown in Figure 19.2. In a universe without liquid–vapor coexistence, a solid planet will simply interface with its gaseous atmosphere (like air above land on Earth). Depending on the size of the planet, the layer of atmosphere immedi-ately next to its solid surface may be a quite dense fluid, but this density will just decrease smoothly

---

* The simplest life we know, the prokaryotes, are diffusion limited [20]. Multicellularity emerged partly to beat this limita-tion. Diffusion in solids is many orders of magnitude slower than in fluids at comparable densities.

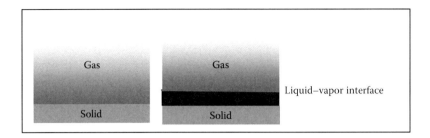

FIGURE 19.2 A planet in a universe without (left) and with (right) liquid–vapor coexistence. The latter has an extra, favorable interface at which prebiotic chemistry could take place.

as a function of height. In contrast, in a universe with liquid–vapor coexistence, it is possible for a solid planet to interface with its gaseous atmosphere via a layer of liquid (like air above the sea on Earth).

It is generically clear that interfaces of some kind are important for life based on chemical reactions, which itself depends on intermolecular encounter. As Delbrück noted long ago [7], the rate of encounter between two molecules undergoing a random walk on a surface is increased significantly compared to two wanderers in three dimensions.* But do we need vapor–liquid interfaces, or would fluid–solid interfaces do? After all, clay surfaces have been proposed to have played an important role in prebiotic chemistry [8]. But vapor–liquid interfaces generically offer significant advantages over fluid–solid interfaces. Species sequestering at a vapor–liquid interface in general remain mobile within the interface; this is not the case for adsorption at a fluid–solid interface. In the latter case, cycles of adsorption, reaction, desorption, diffusion, or convection in bulk, and readsorption are necessary. In fact, there are scenarios for the origin of life on Earth in which reactions on labile interfaces play a key role (e.g., [9]).

Interfaces are also necessary for individuality, or, to put it differently, for confinement. It is almost impossible to imagine life originating as anything other than individual mesoscopic (colloid-size) entities—anything bigger would have such a small surface to volume ratio that an internal circulation system is needed to transport nutrients and waste.† In other words, life must have come from *droplets*, and droplets mean interfaces. A liquid drop within a solid matrix satisfies this requirement; but a protocell of this kind will be severely limited by the very slow rate of diffusion in the exterior, solid matrix. A universe that permits liquid droplets in equilibrium with its own vapor phase with or without contact with a solid phase as well permits the emergence of individuals exchanging material with its surroundings while carrying out a network of chemical reactions that it can call its own.‡

These kinds of consideration (see also Vaida in this volume) suggest that vapor–liquid interfaces are necessary for life as we know it, and possibly even for all kinds of "counterfactual life."

19.2.2 How to Make a Liquid

If it is indeed important to have vapor–liquid interfaces for life, then the next question is whether vapor–liquid interfaces inevitably exist for all possible universes where atoms and molecules exist.

It has been known since van der Waals' 1873 thesis that intermolecular *attraction* is necessary for the existence of the liquid–vapor transition (boiling or condensation, depending on the direction

---

* Their advantage is even greater if we collapse things onto one dimension; hence the importance of both membranes and rigid polymers such as actin and microtubules in cells.
† Note that some authors have speculated about the end point of life as very large, diffused entities, e.g., Freeman Dyson's "sentient black clouds" [10].
‡ Dobson et al. [11] have pointed out the coincidence in size between typical aerosol droplets above the sea surface and the typical bacterium, and suggested a role for these droplets in the genesis of life. This once again points to the importance of vapor–liquid interfaces for confinement and therefore individuality.

of the change) [12]. Research more than a century later has shown that this is only a necessary, but not sufficient, condition. This research initially came from a surprising quarter—colloid science (reviewed in Ref. [13]), but the results are known to be general. It is now known that not "any old attraction" would do, but it has to be attraction of *long enough range* (Figure 19.1b). Decreasing δ/σ from 1 (which models the situation for many real intermolecular potentials), the length of the liquid–vapor coexistence curve, TP–CP, shrinks, and then disappears altogether, leaving only a single fluid phase in the phase diagram, shown in Figure 19.1c.

The precise point where this occurs depends on the actual shape of the intermolecular attraction [14]. For the idealized square well shown in Figure 19.1b, where a precise range can be defined, the critical point disappears at $\delta/\sigma \approx 0.14$. For our purposes, the more interesting result is that for an attractive potential of the "Yukawa" form:

$$U(r) = -\varepsilon \frac{\exp[\kappa\sigma(1 - r/\sigma)]}{r/\sigma}, \quad r \geq \sigma, \tag{19.1}$$

where κ controls the range, σ is the molecular diameter, and $\varepsilon = U(r = \sigma)$. Simulations show that the critical point disappears for $\kappa\sigma > 7$ [15,16]. It can easily be verified by direct plotting that the Yukawa potential for $\kappa\sigma = 7$ corresponds very closely to a power-law attraction

$$U(r) = -\varepsilon \left(\frac{\sigma}{r}\right)^n \tag{19.2}$$

with $n = 9$. In other words, for power-law attractions, $n < 9$ is required for liquids to exist.

The significance for us of this last result is this. Intermolecular interactions in our world are entirely controlled by electrostatics. The fundamental force law here is Coulomb's law, which states that the force between point charges Q_1 and Q_2 in a medium with dielectric constant ϵ and separated by distance r is given by

$$F = k \frac{Q_1 Q_2}{r^m}, \tag{19.3}$$

with the constant $k = 1/(4\pi\epsilon_0\epsilon)$ (where ϵ_0 is the permittivity of free space) and the power $m = 2$. Quantum mechanics then shows us that any two molecules will attract each other via the dispersion or van der Waals force because of their fluctuating electrical dipoles. Coulomb's law dictates that such an intermolecular attraction has $n = 6$. Longer range attraction is possible, e.g., $n = 2$ can be obtained between molecules with (unlike) net charges. When molecules become sufficiently close to each other, specific mechanisms such as hydrogen bonding may also come into play, modifying the shape of the r^{-6} potential.* Nevertheless, the r^{-6} intermolecular van der Waals attraction can never be turned off. Since $6 < 9$, my prediction is that a vapor–liquid critical point generically exists for small molecules of the kind that we think of as solvents for life (real or counterfactual), although the details (the precise critical temperature, and the shape of the entire phase boundary) will depend on details (molecular shape, hydrogen bonding, etc.).

According to current knowledge, the only known exception to this prediction is the class of framework molecules assembled from carbon atoms, known as fullerenes. In these molecules (refer to Figure 19.1b), σ is controlled by the framework architecture, but δ is controlled by the van der Waals attraction between two carbon atoms. This gives rise to a sufficiently small δ/σ that, according to a recent set of simulations, the liquid phase was stable over a mere 70 K temperature range in C$_{60}$, whereas there is no vapor–liquid critical point in C$_{96}$ [17].

---

* These specific mechanisms are also almost invariably anisotropic, i.e., dependent on relative molecular orientations.

Thus, it appears that the existence of simple molecular liquids, including the liquid we call water, is an emergent property based on fundamental physics, including the fact that the force between charges falls off as the inverse second power of distance. Furthermore, we should recall that the inverse square law itself is a consequence of space being three dimensional [18], and that atoms and molecules are unstable in universes with more than three spatial dimensions [19].

Thus, it may be impossible to imagine a counterfactual universe in which there are atoms and molecules but no liquid state, however hard we practice!

19.3 LONG MOLECULES AND THE WATERY TALE OF LIFE

'You promised to tell me your history, you know,' said Alice …
'Mine is a long and sad tale!' said the Mouse, turning to Alice, and sighing.
'It is a long tail, certainly,' said Alice, looking down with wonder at the Mouse's tail; 'but why do you call it sad?'

—*Alice in Wonderland*, **Lewis Carroll**

Kauffman [21] and others have made the case that the sort of complex autocatalytic networks of chemical reactions needed for any kind of life currently conceivable are most naturally implemented with *polymers*—long, chainlike molecules built out of repeatedly bonding single molecular units, or monomers, together. Our biopolymers are carbon-based. The tetravalency of carbon and its other chemical properties, such as the ability to form covalent bonds of different order with oxygen and nitrogen (single, double, and in the case of nitrogen, triple), make it beautifully suited to fulfilling this role. In this brief section, I would like to raise two related questions. First, can counterfactual, noncarbon-based polymers be imagined that can participate in similar autocatalytic networks (life)? Second, what constraints do the properties of (real) water impose on these putative counterfactual biopolymers?

The existence of polymer families based on silicon and phosphorus (see the chapters on the respective elements in Ref. [22] for a concise summary and references) suggests that these may well be worthwhile questions to explore under the heading of "Real water, Counterfactual life." At first sight, the chemistry of silicon-based and (especially) phosphorus-based polymers is significantly less versatile than that of carbon-based ones, but this may simply reflect the large amount of research effort that has been lavished on the latter. Water does impose constraints. Thus, for example, the small-molecule silanes that may form the putative biochemical precursors for silicon-based polymers are unstable in aqueous solution [22]. I do not have the necessary chemical knowledge to explore these issues further here.

19.4 THE MANY FACES OF WATER

My name is Legion: for we are many.

—**Mark's Gospel, Chapter 5**

Discussions on the unique properties of water for life on Earth overwhelmingly focus on liquid water at or near standard temperature and pressure (STP, 1 atm, 298 K = 25°C) (see [2], Chapter 8.3 and references therein). But water on Earth is found under a much broader range of conditions than this. Confining ourselves to liquid water, conditions range from subzero temperatures in the Arctic Ocean to near criticality (the critical pressure and temperature of water are $P_c = 218.3$ bar, $T_c = 374.1°C$) inside hydrothermal vents; the pressure in the deepest oceanic trenches can be 1 kbar or higher. Apart from these extremes in temperature and pressure, extreme concentrations of certain solutes are also known (e.g., saturated brine). In or near almost all of these watery environments, life has been found [3].

What is little discussed in the biological literature is the fact that the properties of water can change quite dramatically when conditions move away from STP, and solutes are found in increasing

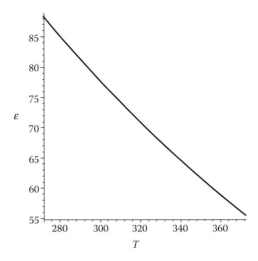

FIGURE 19.3 The dielectric constant of water as a function of temperature at 1 atm pressure. (Data taken from R. C. Weast, D. R. Lide, W. H. Beyer, and M. J. Astle, *CRC Handbook of Chemistry and Physics*, 70th edition, CRC Press, Boca Raton, FL, 1989, p. E-57 [23].)

concentration. Thus, for example, the dielectric constant of water at 1 atm is plotted as a function of temperature in Figure 19.3. The significance of this rather dramatic variation over nearly a factor of 2 can be seen with reference to Coulomb's law, Equation (19.3)—the interaction between two charges scale as 1/ϵ, and many (if not most) biologically significant molecules (both big and small) are charged. To give one indication of the possible biological consequences, we only need to recall work on the condensation of DNA (the transition from open coil to dense globule) in a variety of water + organic liquid mixtures [26]. The authors show that in each case, as more organic liquid is added and the dielectric constant decreases, condensation starts when $\epsilon \lesssim 55$. Recall also that, for genomic DNA (as opposed to short fragments), melting of the double-stranded helix occurs abruptly at $\gtrsim 80°C$ (see, e.g., [25])—reference to Figure 19.3 shows that alternations in the dielectric constant of water need to be taken into account when considering such phenomena.

Thus, it is not clear that the life forms existing within the dramatically different watery environments on Earth would recognize each other's surroundings as being meaningfully "the same liquid." As far as the bacteria living near a hydrothermal vent in a trench at the bottom of the ocean is concerned, the water on the surface of a tropical lake is, if not counterfactual water, at least a species of "quasi-water"! Life clearly has the ability to fine tune itself to the many faces of water.

Is it significant that life on Earth has encountered so many faces of water over its geological history? Perhaps it is. In the literature on evolution, there is now increasing attention paid to the idea of evolvability [26]. The ability to change by evolution is one of the most fascinating and important properties of life as we know it. Evolvability is itself a selectable trait [27], and therefore subject to evolution. But what was the environmental pressure pushing early life to evolve this ability to evolve? I would like to suggest that the very large range of quasi-waters that early life was likely to encounter may well constitute just such a pressure toward evolving evolvability. This is very speculative, but I would like to propose it for discussion.

19.5 CONCLUSION

We are prodding, challenging, seeking contradictions or small, persistent residual errors, proposing alternative explanations, encouraging heresy.

—*The Demon-Haunted World*, **Carl Sagan**

The suggestions I have made in this paper are necessarily very speculative. There are possibly two exceptions to this disclaimer. The first is that in a universe with atoms and molecules, liquids should generically exist, if not for all molecules, then at least for the kind of small molecules that may function as solvent for life. I believe that this result is rather solidly based on recent advances in liquid-state physics and some well-known facts about conditions for the existence of stable atoms and molecules. Thus, if we believe that liquids (or, more precisely, vapor–liquid interfaces) are crucial for life, then this result reveals an emergent aspect of fine tuning that (to my knowledge) has not been suggested before. Second, on the methodological level, the fourfold taxonomy of water-life with which I started may well turn out to be a useful way of organizing future discussions of aqueous fine tuning. It is very simple, but has the virtue of forcing us to be clear about which water and which life we are talking about when discussing whether water is indeed fine-tuned for life.

REFERENCES

[1] P. C. W. Davies, *The Accidental Universe*, Cambridge University Press, Cambridge, 1982.
[2] J. D. Barrow and F. J. Tipler, *The Anthropic Cosmological Principle*, Clarendon Press, Oxford, 1986.
[3] J. Wiegel and M. W. W. Adams, eds., *Thermophiles*, Taylor & Francis, London, 1998.
[4] S. J. Gould, *Wonderful Life*, Penguin, Harmondsworth, 1991.
[5] S. Conway Morris, *The Crucible of Creation*, Oxford University Press, Oxford, 1998; *Life's Solution*, Cambridge University Press, Cambridge, 2003.
[6] V. F. Weisskopf, *Trans. N. Y. Acad. Sci. II* 38 (1977) 202.
[7] G. Adam and M. Delbrück, in *Structural Chemistry and Molecular Biology*, eds. A. Rich and N. Davidson, W. H. Freeman, New York, NY, 1968, pp. 198–215.
[8] A. Cairns-Smith, *Seven Clues to the Origins of Life*, Cambridge University Press, Cambridge, 1985.
[9] D. Segré, D. Ben-Eli, D. W. Deamer, and D. Lancet, *Orig. Life Evol. Biosph.* 31 (2001) 119.
[10] F. Dyson, *Rev. Mod. Phys.* 51 (1979) 447.
[11] C. M. Dobson, G. B. Ellison, A. F. Tuck, and V. Vaida, *Proc. Natl. Acad. Sci. USA* 97 (2000) 11864.
[12] J. D. van der Waals, *Over de continuiteit van den gas-en vloeistoftoestand*, thesis, Leiden, 1873. The translated thesis, with commentary, is available as *On the Continuity of the Gas and Liquid States*, ed. J. S. Rowlinson, North Holland, Amsterdam, 1988. (This book is volume 14 in the series *Studies in Statistical Mechanics*.)
[13] W. C. K. Poon, *J. Phys. Condens. Matter* 14 (2002) R859.
[14] M. G. Noro and D. Frenkel, *J. Chem. Phys.* 113 (2000) 2941.
[15] M. H. J. Hagen and D. Frenkel, *J. Chem. Phys.* 101 (1994) 4093.
[16] M. Dijkstra, *Phys. Rev. E* 66 (2002) art. no. 021402.
[17] B. Chen, J. I. Siepmann, S. Karaborni, and M. L. Klein, *J. Phys. Chem. B* 107 (2003) 12320.
[18] I. Kant, *Thoughts on the True Estimation of Living Forces* (1747). Available in *Kant's Inaugural Dissertation and Early Writings*, trans. J. Handyside, University of Chicago Press, Chicago, IL, 1929.
[19] P. Ehrenfest, *Proc. Amsterdam Acad.* 20 (1917) 200; *Ann. Phys.* 61 (1920) 440.
[20] H. N. Schulz and B. B. Jorgensen, *Annu. Rev. Microbiol.* 55 (2001) 105.
[21] S. A. Kauffman, *The Origins of Order*, Oxford University Press, New York, NY, 1993.
[22] N. N. Greenwood and A. Earnshaw, *Chemistry of the Elements*, 2nd ed. Butterworth Heinemann, Oxford, 1997.
[23] R. C. Weast, D. R. Lide, W. H. Beyer, and M. J. Astle, *CRC Handbook of Chemistry and Physics*, 70th edition, CRC Press, Boca Raton, FL, 1989, p. E-57.
[24] S. M. Mel'nikov et al., *J. Am. Chem. Soc.* 121 (1999) 1130.
[25] V. A. Bloomfield, D. M. Crothers, and I. Tinoco, *Nucleic Acids: Structures, Properties and Functions*, University Science Books, Sausalito, 2000.
[26] An entire double issue of the journal *Biosystems*, edited by C. L. Nehaniv, is devoted to evolvability: *Biosystems* 69 (2–3) (2003) 77–272.
[27] E. J. Earl and M. W. Deem, *Proc. Natl. Acad. Sci. USA* 101 (2004) 11531.

Part V

Water—The Human Dimension

20 Some Early Responses to the Special Properties of Water*

Colin A. Russell

CONTENTS

20.1 INTRODUCTION

Underlying any quest for a counterfactual chemistry, or any inquiry into the peculiar properties of water in relation to the fitness of the environment for life, lurk two usually unspoken questions: "does fitness imply 'design'" and "does 'design' imply a designer?" We would probably not choose to express it publicly in these terms, especially in the United States, where the phrase "intelligent design" has overtones and implications we might wish to avoid. In particular, would we wish to avoid the arguments of so-called "creation science" in America and elsewhere that treats the biblical book of Genesis (for example) as a purveyor of scientific truth. Nevertheless, any form of argument from scientific data to matters of divinity falls into what eighteenth century men of science called natural theology, and the quest "from nature up to nature's God" is far from new.

However, a natural theology based on water may seem novel, and is certainly unusual in several respects. First, it is an argument that turns out to be essentially chemical, not at all like those from the intricate structures of a living organism or the macrostructure of the cosmos. These topics were to become the standard subjects for natural theologians. Second, it focuses on one single, chemically simple substance, so simplicity rather than complexity might seem to be its theme. However, a third characteristic is in strong contrast, for a main preoccupation will not be a study of an object in isolation but rather in its relationships with the rest of the material world. Fourth, it might become possible to submit the scientific conclusions of this natural theology to rigorous experimental testing. It is difficult to see how one could evaluate the testimony of cosmology in this way, and more difficult still to modify a living organism to see if it can still adapt to its surroundings, Attempts have been made in these directions, but they are never easy.

---

* This chapter was previously published in Russell, C. A. 2007. Hydrotheology: towards a natural theology for water. *Sci. & Christian Belief* 19:161–184.

Finally, it is possible to trace the specific trajectory of ideas relating to the natural theology of water over a longer period, and with rather more prospect of success, than in most other cases. The function of this paper is to attempt precisely that. Such an enterprise could, of course, become merely an exercise in antiquarianism, "with one damn thing after another" (which is how some people still perceive history). However, an attempt will be made to do more than that, and to locate events in a wider context and to inquire whether any patterns of development arise.

20.2   THE BEGINNINGS

As far back into antiquity as we can peer, there are signs of an unusual fascination with water and its role in the world. This was partly because there was so much of it.

The Egyptian and Babylonian civilizations were well aware of the importance of water for life, whether in the annual irrigation by the Nile or in the welcome rain that followed months of drought and hunger. Mediterranean civilizations were so obviously dependent on water for human survival. The Babylonians gave water a critical role in the creation process, all the lands being sea until the god Marduk intervened to create dry land. Centuries later, around 700 B.C., the Milesian philosopher Thales marveled at the distribution of water between land and sky, perceived the world as afloat like a leaf upon the ocean, and concluded that the material world was, in fact, made of water. His was no natural theology, however, for he produced "a coherent picture of a number of observed facts *without letting Marduk in*" [1]. So did some of his successors, who maintained the primacy of other "elements" than water. More than a thousand years later, the prophet Mohammed claimed a revelation from God that "We have made of water everything living" [2]. Even to the religious mind water was at most a constituent of the universe or an instrument in the hand of God. By the twelfth century, the same fascination with the role of water in creation was being expressed by Thierry of Chartres. Writing about the third day of the Genesis creation narrative, he noted:

> Observe what happens when the temperature increases in a room with a table in it that has been covered with a uniformly thin plane of water; individual dry spots soon appear on the surface of the table, while in other places the remaining water draws back and collects. (Thierry of Chartres, *Tractatus de sex dierum operibus* – 1; cited in Ref. [3])

Similar sentiments about the priority of water in nature continue beyond this time, up to 2000 years from Thales. The Brussels chemist Joannes Baptista van Helmont (1579–1644) experimented on the basis that "all things" were made of water, and so did some of his followers [4]. Robert Boyle wondered whether water had "a natural temperature" [5], and Richard Bentley spoke of it as "that vital blood of the Earth which composeth and nourisheth all things" [6]. But in no sense do their writings amount to any kind of natural theology, even at the hands of avowedly Christian writers.

From at least Roman times, there was much recourse to spa and other mineral waters. These were generally believed to have been put on the Earth by God in order to cure human ills. Introductions to almost all the books on these topics were couched in terms of a kind of natural theology. In pre-Christian Europe, such theology was not necessarily Christian, but, as paganism gave way to Christianity, beliefs about the healing powers of mineral waters were reinterpreted in Christian terms (Dr. N. G. Coley, personal communication, 1997).

With the coming of Christianity, the attitude of the early Fathers to water was complex. In addition to sharing the general view of water's importance, they brought a distinctively Christian set of values. On one hand, they frequently thought in allegorical terms and saw water as a picture of the cleansing from sin by Christ. This, of course, was sanctioned by dozens of verses from the New Testament. On the other hand, they often thought sacramentally as well, and regarded water that had been set apart by some ecclesiastical process as mysteriously changed in nature, much as they regarded the bread and wine in the communion service. Thus they used "holy" water to cleanse from sin, to sanctify vessels washed in it, and especially to bring spiritual benefits at baptism. The sprinkling of congregations at Mass seems to date from the ninth century.

In such circumstances the "natural" properties of water were of much less interest. However, there was a further tradition that called upon all things created to praise their Maker, and water was one of those things. It is exemplified by the ancient canticle sung until recently in many Anglican churches, the *Benedicite*.* This song enjoins all creation, especially the aquatic parts, to "praise Him and magnify Him for ever," including "waters that be above the firmament," "showers and dews," "dews and frosts," "ice and snow," "seas and floods," and even "ye wells." The later hymn of Francis of Assisi similarly urges: "thou flowing water pure and clear, make music for thy Lord to hear, Allcluia." Water in this context was not invested with special properties but was just one part of the created order. After the Renaissance and Reformation, however, matters were to take quite a new turn.

20.3   THE COMING OF PHYSICOTHEOLOGY

This phenomenon was a response to the rapid rise of a scientific worldview often referred to as the "Scientific revolution" that was essentially mechanical in character. The whole of creation was seen by many as a vast interlocking mechanism that routinely excluded forces that could be deemed as occult or magical. This raised problems about the role of divine Providence, although these were addressed by many writers and did not, in the end, prove to be destructive of faith. But in the short term the challenge was to use the very complexity of the world as a signpost to its Creator. Science was therefore seen not as an adversary, but as an ally, for faith. At this time, when Isaac Newton was still a young man, physicotheology was born, or as we might prefer, a distinctive brand of natural theology.

The marvels of the mechanical universe were interpreted mathematically, rather than in terms of allegory as had been the medieval tradition. They were seen as a wonderful antidote to the atheism that threatened the church at that time. In that spirit, Robert Boyle endowed his famous Lectures,† whose purpose was to prove Christian religion against its enemies, and thus in effect to justify the ways of God to man. This prospect was made even more attractive by discoveries through the microscope, and, in 1691, John Ray's *Wisdom of God Manifested in the Works of Creation* extended the argument to the world of biology [7].

The first specific attempt to focus strongly on water may perhaps be located in 1734. In that year, a book appeared in Hamburg mysteriously entitled *Hydrotheologie* [8], However, its subtitle tells all:

Hydrotheologie
Or an attempt
through observing the properties, distribution and movement of
Water,
to encourage human beings to love and admire
the benevolence of the powerful Creator

The author of this curiously named book was Johann-Alberto Fabricius (1668–1736), professor at the Hamburg Gymnasium, a copious fact-collector, classical scholar, and author of more than a dozen books on bibliography, theology, and history [9]. Now he was to produce a volume of about 450 pages devoted to a topic rarely discussed in theology but frequently in science—or what passed for science in those days.

---

* The book of *The song of the three* in the Apocrypha.
† These lectures, after an absence of well over a century, have recently been revived in the Church of St-Mary-le-Bow, in the city of London.

FIGURE 20.1   Title page of Fabricius, J. A., *Hydrotheologie*, König and Richter, Hamburg, 1734.

Within the volume were three "books" that dealt with, respectively, the properties, distribution, and movement of water. The last two concerned the vast amounts of this fluid on earth and were mainly about geography. The first was still obsessed by water *en masse*, but nevertheless did touch on some matters of chemical interest. These included the viscosity of water, its "wetness and ability to make other bodies wet," its power as a solvent, and its ability to extinguish fires, soften, cleanse, and erode. He concluded that these properties demonstrate "the wisdom and goodness of the great Creator," both separately and together.

It was the integration of water within one vast mechanical universe that so moved Fabricius. Writing of "all parts of nature," he says:

> There is none which does not give us reasons to wonder at the magnitude of the works of the Lord.... But nothing else might move us more in this way than the combined consideration of all the properties of water... and of its beneficial relation to the other creatures. [10]

FIGURE 20.2   The role of water in creation, according to Fabricius.

Here, then, was an early attempt to use water in the service of natural theology. Some historians have consigned it to a footnote with ill-concealed contempt [11], but it is important not to consider this in isolation. In the early eighteenth century, F. C. Lesser (1692–1754) [12] produced works on the theology of stones [13], insects (e.g., [14]), and even one on shells [15]. Far from being an eccentric crank, Lesser was a German naturalist well respected by his scientific community, and probably the greatest bibliographer of the eighteenth century. All these works went through several editions and/or translations and their influence must have been considerable. Before Fabricius wrote his treatise on water, he had also produced *Pyrotheologie* [16], an analogous volume on fire. Yet a further field was pharmacotheology, where numerous workers, including the famous chemist from Halle, F. Hoffmann (1660–1742), argued from natural theology to justify the therapeutic use of herbs and other naturally occurring products [17]. So Fabricius was far from being the isolated figure of fun as it was once common to regard him.

At this stage, there was little understanding of the unique properties of water: it was capable of many uses, there was a great deal of it, and that was about all that could be said. It was just "there." When William Paley wrote his epochal *Natural Theology* in 1802, water was one still commonly regarded as of the four elements, remarkable for its ability to support the life of fish, animals, and plants, and for its absence of taste, which enables us to continue drinking it without boredom. But, with Fabricius, Paley is chiefly impressed by its movement and "incessant circulation."

FIGURE 20.3   William Paley (1743–1805).

ASTRONOMY AND GENERAL PHYSICS

CONSIDERED WITH REFERENCE TO

NATURAL THEOLOGY

BY THE

REV. WILLIAM WHEWELL M.A.

FELLOW AND TUTOR OF TRINITY COLLEGE

CAMBRIDGE

LONDON

WILLIAM PICKERING

1834

FIGURE 20.4 Title page of Whewell, J. A., *Astronomy and General Physics Considered with Reference to Natural Theology.* 4th ed, Pickering, London, 1834.

20.4 FINE-TUNING THE ARGUMENTS: MACROSCALE

These sentiments were still being expressed as late as 1819 [18], although by now chemists and others had learned a great deal more about water. From about 1800 it was widely agreed that water was a compound of hydrogen and oxygen [19]. Latent and specific heats were recognized by now and those for water had been measured [20]. The remarkable expansion of water as it cooled just above its freezing temperature had been known for more than a century, and in 1806 the temperature where this starts was established as 39.5°F (or 4°C) by the Scottish chemist Thomas Charles Hope [21].

In the early nineteenth century another legacy (from the Earl of Bridgewater) led to the publication of a further series of books on natural theology, eight in all, the Bridgewater Treatises [22], first published between 1833 and 1836 and intended to show "the power, wisdom and goodness of God as manifested in the creation." They offered an unprecedented opportunity for fine-tuning the old arguments in the light of fresh discoveries about water. One of these was delivered by the mathematician and philosopher William Whewell, later to become Master of Trinity College, Cambridge.

His brief was "Astronomy and general physics considered with reference to natural theology." He uses the newly discovered "laws of heat with respect to water" [23], although it is only in those terms that he may be said to have given "the first systematic argument for the fitness of water" [24]. Some of the "offices" of this fluid he pronounced to be:

1. Heat is communicated by "internal circulation" (=convection). This helps to moderate differences between hot and cold parts of the earth.
2. Expansion near freezing temperature (from 40 to 32°F), though causing burst pipes in winter, also ensures that ice floats and can therefore be melted in summer, and additionally helps to pulverize the soil. This property is evidently "selected with a beneficent design."
3. Water readily evaporates into the air, a process essential to plant life.
4. Water as clouds helps to modify the fervor of the sun.
5. Water as clouds gives rain to the earth.
6. Snow and ice are bad conductors of heat and thus protect bulbs and roots in the ground.
7. The high latent heat of water avoids quick changes of state.
8. Water supplies springs.

Of these eight functions, only some reflect the rare if not unique physical properties of water, numbers 2, 6, and 7. The rest could apply to other liquids, but only water is present in such large amounts as to demonstrate them.

FIGURE 20.5   William Whewell (1794–1866).

As with earlier writers, Whewell is profoundly impressed by water circulation. He quotes John Dalton to the effect that, of the 39 inches of water that fall annually on Britain (or perhaps Manchester!), 13 inches come by evaporation of the sea, and the remaining 23 inches from the ground. And so there is a continual circulation. Whewell cites Howard's *Climate of London* [25], writing about the frequent introduction of rain by a southeast wind:

> Vapour brought to us by such a wind must have been generated in countries to the south and east of our island. It is therefore, probably, in the extensive valleys watered by the Meuse, the Moselle and the Rhine, if not from the more distant Elbe, with the Oder and the Weser, that the water rises, in the midst of sunshine, which is soon afterwards to form *our* clouds, and pour down *our* thunder-showers. [26]

FIGURE 20.6   William Prout (1785–1850).

CHEMISTRY, METEOROLOGY,

AND THE FUNCTION OF DIGESTION,

CONSIDERED WITH REFERENCE TO

NATURAL THEOLOGY.

BY

WILLIAM PROUT, M.D., F.R.S.

FELLOW OF THE ROYAL COLLEGE OF PHYSICIANS.

THIRD EDITION.

LONDON:

JOHN CHURCHILL, PRINCES STREET, SOHO.

MDCCCXLV.

FIGURE 20.7   Title page of Prout, W., *Chemistry Meteorology and the Function of Digestion, Considered with Reference to Natural Theology*, 3rd ed, Churchill, London, 1845.

It is impossible to avoid the conclusion that the chief fascination of water for those natural theologians was still in connection with the weather.

  This impression is strengthened by another Bridgewater Treatise, that written by William Prout, whose title could hardly be more specific, dealing with chemistry, meteorology, and digestion [27]. He observes that air is hardly ever saturated with water, so condensation when it rises does not inevitably lead to cloud formation. Thus, "by the benevolent arrangement we enjoy," permanent mist is prevented [28]. He has much to say about snow, commenting on its whiteness, which reflects rays and thus prevents rapid melting [29]; its low conducting properties and its lightness (shielding vegetation below); and the ability of snow water to contain much oxygen and to be thus "particularly favourable to vegetation" [30]. All of which leads him to infer from the "great designs" incontrovertible evidence for "the great Author of nature" [31]. But when he develops his third promised theme, the important role of water in digestion rather strangely receives little attention.

A third Bridgewater Treatise was written by John Kidd, regius professor of medicine at Oxford, his theme being "The adaptation of external nature to the physical condition of man." Unlike his colleagues, Kidd plies an argument that was entirely anthropocentric—as his title suggests. Among the general uses for water he mentions its necessity for life, and stresses the consequences of long-term deprivation. It is also used in the manufacture of garments, utensils, etc., and in the preparation of medicines. He revels in the delights of drinking tea and coffee, and (at great length) of taking a bath! Scientific issues are addressed in a section on "the fluidity of water." Its fluid properties enable

ON THE

ADAPTATION OF EXTERNAL NATURE

TO THE

PHYSICAL CONDITION OF MAN:

PRINCIPALLY

WITH REFERENCE TO THE SUPPLY OF HIS WANTS AND THE
EXERCISE OF HIS INTELLECTUAL FACULTIES.

BY

JOHN KIDD, M.D. F.R.S.

REGIUS PROFESSOR OF MEDICINE IN THE UNIVERSITY OF OXFORD.

LONDON:
H. G. BOHN, YORK STREET, COVENT GARDEN.
1852.

FIGURE 20.8 Title page of Kidd, J., *On the Adaptation of External Nature to the Physical Condition of Man*, Bohn, London, 1852.

it to be used to generate power, both by water wheels and by steam engines. Its high specific and latent heats are referred to (although not in those terms), and again the consequences of its anomalous freezing behavior are discussed. Most interestingly, he observes:

> With reference to the constitution of nature, we may more forcibly be impressed with the conviction of its general harmony and subserviency to our wants, by the supposition of its being different from what it is, than by the direct contemplation of its natural state. [32]

This is near as a Bridgewater author gets to contemplating a counterfactual science.

It would be a great mistake to write off the Bridgewater Treatises as merely "worthy" documents, a response to the last will and testament of the earl, and of no significance thereafter. In fact, they had a large circulation, again with several editions for most of them, and many camp followers whose dependence on them was obvious. Just one can be mentioned, Henry Duncan (1774–1846), a church minister in Scotland, whose many admirers ranged from Andrew Carnegie to Rudyard Kipling. He wrote a highly acclaimed work, the *Sacred Philosophy of the Seasons* [33]. The Treatises of both Whewell and William Buckland are cited, and much is made of the role of water in the climatic cycles. The conclusion to his section on rain is entirely Whewellian:

> We see tremendous and destructive forces controlled and regulated with consummate skill, so as to harmonize with the other powers and conditions of the physical world, while these, again, equally harmonize with the circumstances of the moral world—thus forming one amazing but mysterious whole. [34]

20.5 HYDROTHEOLOGY AFTER DARWIN

It is often said that the coming of Darwinian evolution was the death knell for natural theology. Design was to be replaced by chance. The universe was recognized to be far older than was once thought and there was time for all kinds of evolutionary processes to produce life-forms that could adapt to their environment. Thus, it would seem that the appearance of "design" was a chimera, an illusion, and no self-respecting scientist would subscribe again to natural theology.

That statement is in fact a caricature of the true situation, and for several reasons. First, large numbers of leading scientists would have echoed the words of Lord Kelvin who described Paley's *magnum opus* as "that excellent old book"; many prominent physical scientists were far from convinced by Darwin. Second, many people readily accepted both design and evolution, which was conceived to be God's method for accomplishing His design (this happened to be especially true of Christians with evangelical leanings [35]). Others, like Clerk Maxwell, regarded molecules as outside the sphere of natural selection, for their properties were fixed for all time; a clear distinction therefore existed between Darwin's notions and molecular events or properties that "happened" to support life. Third, the notion of a permanent "conflict" between science and religion has been shown to be largely an artifice devised by a few militant agnostics (notably T. H. Huxley) as a means of social control. If people believed that Darwin had really routed the religious establishment they would be more likely to support British science in its quest for both autonomy and government support [36].

In the light of this, it could be argued that the need for natural theology was greater, not lesser, than before. Various works appeared on this subject, some more influential than others. Much of it was clearly derivative, probably from a variety of sources. Thus, the very popular *Theology in Science* [37], by the ubiquitous Dr. Brewer, first appeared in 1860 but was still thriving 30 years later. He expounds the large-scale phenomena of evaporation and circulation of water and draws the usual conclusions. His book also has a specific section on "Deviations from general laws in the case of water" [38]. Brewer deals with two laws: the expansion on cooling and the phenomenon that arise from a high specific heat ("water can receive or lose heat without showing it"). They offer him "proof of design, beyond the possibility of a doubt" [39]. Still, in the spirit of Fabricius, he has a further section on "The circulation of water (as evidence of divine wisdom and goodness)." His conclusion on the watering of England from "the Meuse, the Rhine and the Moselle" comes straight from the pages of Whewell or of his source, Luke Howard.

Clearly, water was a good topic for these apologists, for evolutionary dogma hardly applied to that (chemical evolution was still far in the future, in the late twentieth century). One major effort was by a chemist, Josiah Cooke (1827–1894), professor of chemistry at Harvard, although his book covered the whole of chemistry, and not just water [40]. It was based on lectures delivered in 1861:

> At the time when the lectures were written, Mr. Darwin's book on the Origin of Species, then recently published, was exciting great attention, and was thought by many to have an injurious bearing on the argument for design. It was, therefore, made the chief aim of these lectures to show that there is abundant evidence of design in the properties of the chemical elements alone, and hence that the great argument of Natural Theology rests upon a basis which no theories of organic development can shake. [41]

One of Cooke's 10 chapters is devoted to water. He begins with its role in meteorology, and after discussing the mechanisms producing rain concludes that "science, by discovering these evidences of skilful adaptation" can only fall back on the answers given to Job [42]:

> Does the rain have a father? Who fathers the drops of dew? [Job 38:28]

> How great is God—beyond our understanding! The number of his years is past finding out. He draws up the drops of water, which distil as rain to the streams. [Job 36: 26–27]

Having written at some length about circulation of water in general, he continues:

> Let us consider some of the qualities of water; but at the same time, let us not forget that the strength of our argument lies not so much in the fact that each property has been skilfully adjusted to some special end, as it does in the harmonious working of all the separate details.. . . In these very facts lies the whole force of the argument from design, and it is only the limitations of our knowledge and faculties which weaken the impression on our minds. [43]

This explicitly emphasizes the world as a *system*, in which water plays an important, if not decisive, part. So he announces "a familiar fact, that water is an essential condition of human life," constituting "the greatest part of all organized beings" [44]. The distribution of this vital fluid is afforded by condensation on the mountains, which are among "the most beneficent means in the Divine Providence by which the earth had been fertilized and rendered a fit abode for man" [45]. Condensation by other means is also mentioned, most notably formation of dew absorbed by leaves, which also have other important functions. "If this can result from chance, under its modern name of natural selection, then chance is but a counterfeit name of God" [46]. Water "is the great cleansing agent of the world" [47]. Spurred on by this confidence (and possibly by his own rhetoric), Cooke waxes eloquent over the aesthetic advantages of water (murmuring brooks, roaring torrents, bubbling springs, etc.), marveling again that "The grand result is an harmonious system" [48].

All this may well be very interesting, but it is not chemistry, nor does it address the question as to *why* water should be so necessary. He begins to repair this omission when he goes on to speak of human uses for steam. Steam heating is possible because steam is "peculiarly fitted for this work." Although it "merely acts the part of a common carrier," steam is capable of "holding so large a quantity of heat" [49]. He does not specify latent or specific heats but does claim that "when water is heated through a given number of degrees it absorbs twice as much heat as any other substance" (with one or two exceptions) [50]. Then, once more, we are launched on to a geographical excursus (particularly about the Gulf Stream) from which we return at last to specific reference to the anomalous density of water just above its freezing temperature. Its familiar effects on lakes, fishes, and vegetation are mentioned, and there follows a surprising caution not to make too much of this one anomaly since other fluids may in the future be found to behave in a similar manner. It is the whole interlocking system of effects that so impresses Cooke.

His final point is truly chemical: that "water is the most universal solvent known" [51]. The consequences of this are sketched for life and for inorganic nature, notably the hydrates, rocks, and other constituents of our Earth. He concludes:

> Of all the materials of our globe, water bears most conspicuously the stamp of the Great Designer, and as in the Book of Nature, it teaches the most impressive lesson of His wisdom and power, so in the Book of Grace it has been made a token of God's eternal covenant with man, and still reflects His never-fading promise from the painted bow. [52]

Water featured prominently in several other essays that appeared in the middle and late nineteenth century. One remarkable one was a lecture delivered in 1877 to the Nottingham Literary and Philosophical Society by James M. Wilson, then a mathematics master at Rugby School. Ordained in 1879, he became headmaster of Clifton College and (from 1890) Archdeacon of Manchester. He supported Darwinism but still saw a role for natural theology.

His lecture was entitled "Water: some properties and peculiarities of it; a chapter in natural theology"; it later appeared as the first chapter of his volume, *Essays and Addresses* [53]. Like all his predecessors, he was deeply impressed by the atmospheric and meteorological effects of water circulation. But unlike them, he tried to focus more strongly on the anomalous properties of water. Thus, after referring to climatic consequences of the expansion of water by heat, he strongly asserts that "in one property of expansion, water is unique, absolutely unique" [54]. This is its expansion on cooling below 4°C. The consequences of ice floating on water for marine life and the structures of rocks and soil are clearly displayed.

Wilson goes on to admire the liquidity of water, noting that its existence in that state depends on many parameters and that "these elements are adjusted to one another in such a manner as to produce the actual result we see" [55]. These, rather than anything specific about water, are the cause for wonder. However, he goes on immediately to record that

> Water is once more unique. Its specific heat, as this property is called, is far greater than that of all other bodies, solid or liquid. [56]

The very high specific heat is used to account for the large-scale transference of heat in the oceans, as in the Gulf Stream.

The high latent heat of fusion of ice prevents too rapid a thaw with vast flooding in consequence. When it comes to converting liquid water to vapor by the expenditure of heat, no other liquid "approaches water in the quantity it requires" [57]. And this "extraordinary latent heat of vapour" means that vast stores of heat are transferable across the globe, to be discharged when most needed.

One remarkable property of water, anomalous but not uniquely so, is the power of absorption of radiant heat, explored by John Tyndall, i.e., its ability to trap heat:

> Aqueous vapour will not let heat radiated from the earth pass through. It absorbs it; it serves as a covering, as a light but warm blanket to the earth at night. . . this singular and beautiful property of water. [58]

This must be one of the earliest descriptions of what we now call the "greenhouse effect" of water vapor. It is, in fact, at least as significant as that of CO₂.

Wilson's final point of anomaly is the solvent ability of water. "Its powers of dissolving salts, and of fertilising the soil, are perhaps the most extraordinary" [59], although he had little time to develop the theme. Sadly, he left unexplored the whole range of the reactions of water. The conclusion of the scientific part of his address related to molecules of water, although at that time he had no way of connecting their existence with the phenomena he had described. The conclusion is a lengthy philosophical musing on natural theology in general. While accepting that biological evolution may have led natural theology to change its emphasis, he is adamant that this does not apply to the physical realm. He acknowledges that conclusions about the mutability of organic forms had to change:

> But this can never be the case with inorganic nature. We cannot conceive of a revolution in Science which would overthrow our belief in the permanence of the chemical and physical properties of inorganic elements. And hence the basis of Natural Theology that I am considering this evening is not transient. [60]

A new development came at the very end of the century with a book by the Rev. Joseph Morris from Durham, offering *A New Natural Theology*, but this time "based upon the doctrine of evolution" [61]. Dealing specifically with Wilson's *Address* on water, he admits the arguments are "ingenious to a degree" but focuses on Wilson's admission that all anomalies probably depend on a single property of water (which we do not yet know). These must result either from design or from an evolutionary growth. Morris argues that his opponent "sees only the evolution of life, and is blind to the evolution of matter" [62]. This was indeed the case, but Morris' ideas on this topic were exceedingly vague, and we should certainly not read into them anything like the modern theories of chemical evolution. Nevertheless, his book offers further proof that, contrary to popular opinion, evolution did not kill natural theology, even though it may have transformed it.

One last essay in this genre, concerned in evaluating the impact of evolution on teleology, and thus natural theology, appeared in America just before World War I. This was a book, *The Fitness of the Environment*, whose subtitle defined it as "an inquiry into the biological significance of the properties of matter" [63]. Its author, Lawrence Henderson, was a professor of biological chemistry at Harvard, and his intention was made clear in his opening words:

> Darwinian fitness is compounded of a mutual relationship between the organism and the environment. Of this, fitness of environment is quite as essential a component as the fitness which arises in the process of organic evolution; and in fundamental characteristics the actual environment is the fittest possible abode for life. [64]

Apart from offering an early example of the use of the word "environment" in a book title, Henderson's volume proposes "deep design laws" but draws back from a natural theology in any traditional sense. It is not theistic, though certainly not atheistic. The third of his eight chapters is devoted to water. He refers to the Bridgewater Treatises and also to another chemist at Harvard, J. P. Cooke. He does not appear to have been aware of Wilson's essay. He deals with the anomalous properties of water in a more systematic manner than any of his predecessors.

First, he emphasizes specific heat, showing how it may be derived and enumerating some of the consequences of the unusually high specific heat for water. The effects include a relatively constant temperature in streams, lakes, and oceans, and also a moderation of the summer and winter temperatures on earth. Marine organisms are able to move about, and animals like man maintain constancy of body temperature without too much difficulty since water is a high proportion of their bodies. Moving on to the high latent heat of water, Henderson observes that the large amount of heat required to melt ice ensures the temperature of the underlying water is fairly constant. He continues:

> In view of the other favourable qualities of water it is perhaps not surprising to find that its latent heat of evaporation is by far the highest known. So great, in truth, is this quantity and so important the process that the latent heat of evaporation is one of the most important regulatory factors at present known to meteorologists. [65]

The threefold inheritance of this property can be summarized:

> First, it operates powerfully to equalise and to moderate the temperature of the earth; secondly, it makes possible very effective regulation of the temperature of the living organism; and thirdly, it favours the meteorological cycle. All of these effects are true maxima, for no other substance can in this respect compare with water. [66]

After brief reference to the unusually high thermal conductivity of water, Henderson turns to the anomalous expansion on cooling below 4°C. Without it, life on Earth would have been "very greatly restricted," although he believes Prout overstated the importance of this anomaly [67].

Henderson then becomes much more chemical than Cooke, reflecting no doubt the advances in chemistry in the previous half-century. In referring to the great solvent powers of water, as exemplified by its action on rocks, etc., he depicts it as "a chemically inert solvent," whereas most of his examples arise from chemical action before any act of solution. This is curious, as the facts were very familiar even in Victorian times. He triumphantly vindicates his position by listing more than 50 individual substances known to be present in urine. The fact that water can promote ionization to a unique degree was interpreted in terms of the ideas of Arrhenius and Loeb. At that time, almost nothing was known of reactions in other polar solvents as liquid ammonia and liquid hydrogen fluoride. Finally, the great surface tension of water leads to its rise in capillary systems and to complex phenomena of adsorption, important for its passage through soil and the processes of chemical physiology.

Henderson concludes that

> Water is fit . . . with a fitness no less marvelous and varied than that fitness of the organism which has been won by the process of adaptation in the course of organic evolution.

He acknowledges that the present situation of science differs from that of Whewell "only in the better definition of the issue." However, he claims for himself and his contemporaries "our modern freedom from metaphysical and theological associations" [68], apparently oblivious to the fact that he was no freer from metaphysical assumptions than those before him, and that upon those assumptions depend the role (if any) for theology. As the old proverb has it, "you see what you want to see," and (of course) fail to see what your metaphysics denies. (A further, more philosophical, discussion by the same author may be found in Ref. [69].)

20.6 MODERN EXAMPLES OF FINE-TUNING ON THE MICROSCALE

All of the arguments so far have been extensively concerned with large-scale physical phenomena particularly those of meteorology and geology. In the 1930s, the emphasis began to change. Once more, the role of water in biological processes was discussed. In 1934, a book appeared with the provocative title *The Great Design* [70]. It contained one chemical chapter, from that master of polemics, H. E. Armstrong [71].

Writing on "the chemical romance of the green leaf," he stressed the importance of water (and commended Moses for placing it first in creation). It is, as he said "a material of intense activity, provided it has the proper companionship." All vital changes take place when the water decomposes into hydrogen and hydroxyl. That is the very beginning of the processes leading to the formation of sugar and other organic molecules. Moreover, water has another function, already recognized by Wilson: "it is the buffer to receive the shock of the impinging light waves and transmit to us our share of solar energy" [72]. His conclusion is not so much theological as mystical, calling for a new chemistry, for the old has "no romance in it, no religion" [73]. He had, of course, for many years been demanding a revolution in chemistry teaching and these arguments are probably part of that crusade.

FIGURE 20.9   H. E. Armstrong (1848–1937).

The biological theme was taken further in 1938 by a medical writer, W. O. Greenwood, in a book called *Biology and Christian Belief* [74]. In his final chapter, "On purpose," he chooses water as one of many examples refuting "the hypothesis of a blind chance." Dealing yet again with the 4°C anomaly and its consequences, he considers the presence of water in each living cell and concludes that "of all the thousands of possible liquids other than water, there is not one that would have the faintest possibility of supporting life." Stressing also the critical importance of carbon to life he argued that these two bodies, water and carbon, present "a clear case of purpose—a thought-out plan" [75]. In the case of carbon, adaptation cannot "touch the fringe of an approach to explanation" for the very construction of complex molecules depends on the carbon atom being there. As for water's behavior between 0°C and 4°C, he remarks:

Why this unparalleled exception in the case of water? If built on a mechanistic plan, a fortuitous method, is there any one single reason why water should not behave in this physical respect exactly as other liquids do? [76]

It is obvious that these two illustrations with biological themes related to *origins* of life, not its subsequent development. Moreover, they do not address the problem as to how these beneficial properties may change at very high pressures. This is a complex issue and includes phenomena of bacteria able to withstand quite remarkable extremes of heat as well as pressure.

At this point it is necessary to point briefly to three developments in recent science that place the earliest arguments in a wider context, though by no means destroy their value.

First, there has been a spectacular growth in cosmology and with it a new understanding of how the elements in our universe have been formed. That includes carbon, hydrogen, and oxygen. Natural theologians like Cooke and Wilson recoiled from the notion of evolution of inert matter, but that was hardly relevant to the fitness of water for life processes, if only on grounds of the different timescales. Interestingly, if a resonance level in the carbon nucleus had differed from its actual value by a very small percentage, carbon would have been a rare element, and organic chemistry and life a mere dream. It is reported that Fred Hoyle found nothing that could shake his atheism as much as this discovery [77].

However, modern cosmology has also shown something of the vastness of the universe, and therefore of the probability of millions of planets on which something like life as we know it could have developed. In an Introduction to a late edition of Henderson's book *The Fitness of the Environment*, the biochemist George Wald commented in 1958 that "we now believe that life, being part of the order of nature, must arise inevitably wherever it can, given enough time" [78]. However, it may be observed that scale has nothing to do with teleology, and that the repeated formation of life elsewhere is no more (and no less) wonderful than a one-off appearance on earth. Yet this is a point that many scientists find difficult to grasp.

Second, discoveries with polar nonaqueous solvents have shown that many of the chemical properties of water are not unique. This began with the pioneer work on liquid ammonia by Franklin and others at the end of the nineteenth century. Some have gone so far as to claim that the prolific availability of water has had a bad effect, having inhibited the development of inorganic chemistry [79]! It has certainly opened our eyes to the general importance of nonaqueous polar solvents, and the possibility—however apparently remote—that one or more of them might serve as an alternative basis for life.

Then, third, there is now a gradual realization that "the remarkable behavior of water is due largely to the structure of the water molecule, its dipole character, its small volume, and properties related thereto" [80]. These, in turn, have been at least partially explained in terms of the hydrogen bond developed by Huggins [81] and others in the 1920s, originally in connection with organic tautomerism. Wald wrote that most of the properties described by Henderson [82] in fact arise from a single cause, the hydrogen bond. We now know that without this hydrogen bond, water would boil at 200 K rather than 373 K, the DNA helix would not exist, and life as we know it would be totally impossible [83]. It must again be said that to know a physical cause of a group of phenomena is only problematic to a philosophy of God-in-the-gaps. One could well write a natural theology of the hydrogen bond itself (e.g., [84]).

These and other developments in contemporary science do not invalidate a form of hydrotheology, although they may greatly modify it. So it cannot be surprising that modern authors have written about water as providing evidence of design. They range from short popular treatments [85] to more detailed scholarly examinations of water in the light of modern science, evolution included. One of these is an account in Barrow and Tipler's *Anthropic Cosmological Principle* [86], and another is in a recent book by Michael Denton* that devotes a whole chapter to the subject, one of the most comprehensive statements available today.

---

* It is of course true that hydrogen bonds are also found elsewhere, as in the ferociously reactive hydrogen fluoride and in liquid ammonia (although that substance is a gas at ordinary temperatures and pressures). Other compounds having an −O-H bond also exhibit hydrogen bonding but to nothing like the extent of water. Nor do they possess most of its other highly unusual properties.

Denton begins his argument by stating that "the properties of water in themselves provide as much evidence as physics and cosmology in support of the proposition that the laws of nature are specifically arranged for carbon-based life" [87]. He dates the first significant consideration of water's fitness to the early nineteenth century, after Lavoisier's experiments on the nature of water (though Lavoisier most certainly did not show "that it was made up of two hydrogen atoms combined with one oxygen atom"!). Denton starts with Whewell in 1832, 30 years after "scientific knowledge had rapidly increased" (he does not say how). He also cites Henderson, but goes on to add further characteristics of water that Henderson did not mention. One was its exceptionally low viscosity that seems to be at the right level for supporting life processes, not least those depending on diffusion. Another was proton conductance, highly important in photosynthesis and oxidative phosphorylation. He piles "coincidence upon coincidence" by listing three sets of properties having beneficial consequences. The lists below are his (although the table is mine). The first column contains properties affecting its action on rocks such as weathering and dissolution; the second those which help to keep water in the liquid state; and the third those which work to maintain low body temperatures in animals.

Properties of Water That Enhance Weathering	Properties of Water That, Together, Tend to Preserve it in The Liquid State	Properties of Water Involved in Temperature Regulation
High surface tension	High thermal capacity	Low viscosity
Expansion of freezing	Conductivity of water	High thermal capacity
Solvation power	Expansion of water on freezing	Heat conductivity
Viscosity of ice[a]	Expansion of water below 4°C	Latent heat of evaporation
Low viscosity of water	Low heat conductivity of ice	
Chemical reactivity of water	High latent heat of freezing	
	Relatively high viscosity of ice[a]	

[a] By "viscosity of ice," Denton appears to mean its ability to "creep" or slowly "flow" over rocks as in glaciers. The "high viscosity" in column 2 is, however, replaced by "low viscosity" in the text. It refers to the sliding of high level ice down to warmer seas where it melts and becomes liquid again. Presumably, the "viscosity" is a function of the friction set up between ice and underlying rocks.

Denton's conclusion is surely beyond doubt:

There is indeed no other candidate fluid which is remotely competitive with water as the medium for carbon-based life. If water did not exist, it would have to be invented. Without the long chain of vital coincidences in the physical and chemical properties of water, carbon-based life could not exist in any form remotely comparable with that which exists on earth. And we, as intelligent carbon-based life forms, would almost certainly not be here to wonder at the life-giving properties of this vital fluid. [88]

If Fabricius had been a chemist, he would have been delighted!

Although the author did not set out to develop an argument from design, he confesses that, at the end, it had become "in effect an essay in natural theology in the spirit and tradition of William Paley's *Natural Theology* or the Bridgewater Treatises" [89]. Yet, in fact, Denton's seems to be a rather isolated voice, although his book has been welcomed by advocates of a modern natural theology as John Polkinghorne. We may briefly explore some of the possible reasons.

Natural theology as a whole has taken some hard knocks since World War II, not least from those theologians who, following Karl Barth, say "Nein!" to the whole enterprise. For them, revealed theology is rather less threatening and far more important than natural theology. The latter can only take us so far, leading us to, at best, a limited theism. Christianity, however, has always depended on the Biblical revelation of God in Christ who is presented as Lord and Savior of the world. Partly for this reason, the initiative for the modern revival of natural theology has come more from scientists than from theologians. The Baconian "book of God's works" must always be read in conjunction with the "book of God's Word."

There have also been reactions to what has been seen to be the crude and naive arguments of some writers in the nineteenth century, who ignored the philosophical caveats of Hume and others. More recently, the resumption of natural theology by the proponents of Intelligent Design has actually weakened the support for the arguments from design in general. This is because, especially in the United States, the Intelligent Design movement is seen as "creationist" (in the American sense) and anti-evolution in its approach. There is no necessary reason for intelligent design *not* to be perceived by those who see Genesis in other than literal terms. It seems that "guilt by association" is too readily conferred and natural theology should be evaluated on its own merits, not by asking whether some of its supporters are approved or not.

A second reason may be identified for the low profile for hydrotheology. When natural theology itself is frowned upon for whatever reasons, it is hardly likely that its application to a single chemical substance would enthuse many supporters. There is the sheer technical difficulty in understanding the detailed arguments relating to water chemistry. Even Denton's lucid account would make difficult reading for a nonscientist, whether theologian or not. One may go further, and stress that this is a rare example of arguments coming not from cosmology, or even biology, but from chemistry. Apart from Josiah Cooke, virtually no writer until now has derived a natural theology from that science. If so, hydrotheology is genuinely breaking new ground.

Third, it seems that other, more pressing agendas constrain the scientific and theological communities to place their emphases elsewhere. These are too many, and too obvious, to be listed here (although one is touched upon in the conclusion).

Fourth, the favorable reception of the Anthropic Principle has been seen as a threat to natural theology in anything like its classical sense. There is an inevitability about the universe that precludes an argument from intelligent design. Denton's conclusion (above) is perfectly acceptable provided it is not theocratized. Moreover, a chemistry that explores alternatives to water in the generation and support of life may well be counterfactual, but it need not be counterfeit. And if such chemistry succeeds, where does it leave the proponents of hydrotheology?

It leaves them precisely where they were. Those who used the natural theology of water to "prove" the existence of God as a logical necessity would be still struggling (despite Aquinas). However, those who *for other reasons* had embraced the Christian faith and found in nature facts that were entirely congruent with that faith, would have nothing to fear from any counterfactual chemistry. If, as they believe, the whole universe is God's creation, then He will be the cause of all phenomena, including those we cannot yet understand. Such a God would not be limited by our circumscribed view of what constitutes a necessary condition of life or anything else.

Meanwhile, it seems the time is right not to wind up hydrotheology but to extend it and indeed to proclaim the wonders it discloses. Today, natural theology may have been transformed, but its arguments still possess great force, and its appeal continues. As Polkinghorne wrote in 1988, "Natural theology is currently undergoing a revival, not so much at the hand of the theologians (whose nerve, with some honorable exceptions, has not yet returned) but at the hand of the scientists" [90]. Since then he has written with even greater confidence [91]. Surely the time has come for chemists to take it seriously, and never more effectively than in the study of that utterly extraordinary liquid, water.

In working toward a natural theology for water, however, we are treating the arguments of natural theology in a rather new way. Traditionally, the great aim was to demonstrate the universality of the laws of nature that pointed to a Lawmaker of incredible power and ingenuity. Even Paley did this, as did Fabricius when he discussed the hydrological and geological cycles. But the later, more chemical, aspects of hydrotheology are at the opposite methodological extreme. One single type of molecule is brought into focus, in perhaps the ultimate case of fine-tuning. The question here is even simpler than that posed by Paley's watch. Without design, how can one single substance have such an amazing array of anomalous properties, be so well suited for biological, meteorological, and geological functions, and at the same time be so abundant on Earth? To a nonbeliever, this may have no more, or no less, an apologetic constraint than (say) Newton's inverse square law of gravity. But to those who already hold a theistic position, it can exert a powerful stimulus to marvel and worship. And unlike other arguments, it confirms the concerns of God for the smallest as well as the largest objects in His universe. But then, we already knew that from Jesus.

20.7   CONCLUSION

If one function of a natural theology of water is to stress design and teleology, there is one other that has not yet been addressed and which must merit even a brief discussion here. Many of the earlier natural theologians had more than one agenda. Although many—probably most—of them genuinely wanted to defend Christianity, some were clearly using their arguments for other purposes. There has been endless debate as to how far "social control" was in the minds of those who wished to show that everything in nature was for the best, so we had better conform to the mores of the establishment that thus was a mirror to nature (to give but one example, Ref. [92] argues for birth control and control of immigration). A study of hydrotheology cannot fail to underline the importance of water in our world, and from there it is but one step to the problems of its uneven availability.

Here is one area of contemporary concern to the communities of science and theology. Here they have a common interest and it is one of extreme urgency. This concerns the environment, affected at all points by science and technology, and yet also a matter for deep theological reflection. The environmental crises of pollution, climate change, extinction of species, and rape of the earth must be faced in our own generation. There are obviously reasons of self-interest in such action. The fate of our great-grandchildren hangs in the balance, and possibly ours, too. Yet, it seems that these reasons are not sufficient for politicians and public opinion to move fast enough. An additional spur to action could be provided by the Christian church, although first it has to recognize the many biblical injunctions to responsible care of a world that belongs not to us but to its Creator. If, as Old and New Testaments proclaim, the universe is God's handiwork and we humans have a stewardship to exercise over it, then we need to exercise our responsibilities at once. Never was this truer than in the case of water.

Although water is freely available over the globe, drinkable water is scarce in many lands, and millions in Africa and elsewhere are literally dying from thirst. And in other places excess of water poses immense threats of flooding and devastation. Human compassion by itself may spur some of us to take action, but it is rarely enough. It is here that natural theology surprisingly comes in. According to Moltmann [93], natural theology has three functions. It can be "educative," prompting basic questions about God; it can be "hermeneutical," helping people to understand their faith; and it can be "eschatological," anticipating a knowledge of God in glory [93]. As we have seen, the natural theology of water has consistently pointed to that substance as a marvelous creation, and eminently suited to many of our needs on Earth. Add to that the commands from revealed theology to love our neighbors as ourselves and we have an additional motive for action. How can we deny to our fellow citizens their "divine inheritance . . . in air and water undefiled" [94]? How can we sit idly by when millions in Bangladesh are likely to be rendered homeless by rising water due to global warming? How can we contemplate this amazing product of creation and do nothing about it when millions more need wells, water purifiers, and a whole new water technology?

If those in the West who profess to follow Christ spring into action and share their privileged knowledge with the dying or threatened multitudes, if they bring the "water of life" in a physical as well as a spiritual sense, the natural theologians will not have written in vain. The additional incentive might just be sufficient to supplement action undertaken for other reasons. If so, humanity will gain at every level as the prospects of imminent disaster recede. And then may come to pass the words of the ancient prophecy:

The earth shall be filled with the knowledge of the glory of the Lord, as the waters cover the sea. [95]

REFERENCES

[1] Farington, B. 1953. *Greek Science.* Harmondsworth: Penguin, p. 37 (original in italics).

[2] *Qur'an*, 21: 30.

[3] Speer, A. 1997. The discovery of nature: the contribution of the Chartions to the twelfth-century attempts to find a *scientia naturalis. Tradition* 52: 134–151.

[4] Multhauf, R. P. 1966. *Origins of Chemistry.* London: Oldbourne.

[5] Boyle, R. 1686. *A Free Enquiry into the Vulgarly Received Notion of Nature.* London; reproduced in E. B. Davis and M. Hunter (eds.), 1996. Cambridge: Cambridge University Press, p. 84.

[6] Bentley, R. 1692. *A confutation of atheism from the origin and frame of the world.* London.

[7] Ray, J. 1691. *The wisdom of God manifested in the works of creation.* London.

[8] Fabricius, J. A. 1734. *Hydrotheologie.* Hamburg: König and Richter.

[9] *Neue Deutsche Biographie*, Berlin, 1960.

[10] *Hydrotheologie*, I, 30.

[11] Dillenberger, J. 1961. *Protestant Thought and Natural Science.* London: Collins, p. 150.

[12] *Neue Deutsche Biographie*, Berlin, 1960.

[13] Lesser, F. C. *Lithotheologie*, Hamburg, 1735.

[14] Lesser, F. C., *Insecto-Theologie*, Frankfurt, 1738.

[15] Lesser, F. C. *Testaceotheologie*, Leipzig, 1756.

[16] Fabricius, J.-A. *Pyrotheologie*, Hamburg, 1732.

[17] Krafft, F. 1996. Pharmaco-theology. *Pharmazie* 51:422–426.

[18] Paley, W. 1819. *Natural Theology*, 18th ed., pp. 310–311.

[19] Partington, J. R. 1962. *A History of Chemistry*, vol. iii. London: Macmillan.

[20] McKie, D., and N. H. de V. Heathcote. 1935. *The Discovery of Specific and Latent Heats.* London: E. Arnold.

[21] Hope, T. C. 1805. Experiment on the contraction of water by heat. *Trans. R. Soc. Edinb.* 5:379.

[22] Topham, J. 1992. Science and popular education in the 1830s: the role of the Bridgewater Treatises. *Br. J. Hist. Sci.* 25:397–430.

[23] Whewell, W. 1834. *Astronomy and General Physics Considered with Reference to Natural Theology*, 4th ed. London: Pickering, pp. 80–95.

[24] Denton, M. J. 1998. *Nature's Destiny.* New York, NY: The Free Press, p. 22.

[25] Howard, L. 1818–1820. *The Climate of London*, vol. ii. London: Hutchinson, pp. 216–217.

[26] Whewell, W., op. cit., p. 95.

[27] Prout, W. 1845. *Chemistry, Meteorology and the Function of Digestion, Considered with Reference to Natural Theology*, 3rd ed. London: Churchill.

[28] Ibid., p. 269.

[29] Ibid., p. 23.

[30] Ibid., p. 289.

[31] Ibid., pp. 315–316.

[32] Kidd, J. 1852. *On the Adaptation of External Nature to the Physical Condition of Man.* London: Bohn, p. 108.

[33] Duncan, H. 1838. *Sacred Philosophy of the Seasons: Illustrating the Perfections of God in the Phenomena of the Year*, 3rd ed. Edinburgh: Oliphant.

[34] Ibid., p. 34.

[35] Livingstone, D. N. 1987. *Darwin's Forgotten Defenders: The Encounter between Evangelical Theology and Evolutionary Thought.* Edinburgh: Scottish Academic Press; Livingstone, D. N., D. G. Hart, and M. A. Noll (eds.), 1999. *Evangelicals and Science in Historical Perspective.* New York, NY: Oxford University Press.

[36] Russell, C. A. 1989. The conflict metaphor and its social origins. *Sci. Christ. Belief* 1:3–26.

[37] Brewer, E. C. ca. 1890. *Theology in Science; or the Testimony of Science to the Wisdom and Goodness of God*, 7th ed. London: Jarrold.

[38] Ibid., pp. 236–240.
[39] Ibid., p. 240.
[40] Cooke, J. P. 1864. *Religion and Chemistry*. New York, NY: Little.
[41] Ibid., Preface to the First Edition, reproduced in rev. ed., 1881. London: Macmillan, pp. vii–viii.
[42] Ibid., pp. 125–126.
[43] Ibid., pp. 127–128.
[44] Ibid., p. 129.
[45] Ibid., p. 130.
[46] Ibid., p. 135.
[47] Ibid., p. 137.
[48] Ibid., p. 138.
[49] Ibid., p. 141.
[50] Ibid., p. 143.
[51] Ibid., p. 155.
[52] Ibid., p. 162.
[53] Wilson, J. M. 1887. *Essays and Addresses: An Attempt to Treat Some Religious Questions in a Scientific Spirit*. London: Macmillan, pp. 1–30.
[54] Ibid., p. 4.
[55] Ibid., p. 6.
[56] Ibid., p. 7.
[57] Ibid., p. 9.
[58] Ibid., p. 11.
[59] Ibid., p. 12.
[60] Ibid., p. 27.
[61] Morris, J. 1905. *A New Natural Theology, Based upon the Doctrine of Evolution*. London: Walter Scott.
[62] Ibid., pp. 28–29.
[63] Henderson, L. J. 1913. *The Fitness of the Environment: An Inquiry into the Biological Significance of the Properties of Matter*. New York, NY: Macmillan.
[64] Ibid., p. v.
[65] Ibid., p. 98.
[66] Ibid., p. 105
[67] Ibid., pp. 107–108.
[68] Ibid., p. 131.
[69] Henderson, L. J. 1917. *The Order of Nature*. London: Harvard University Press.
[70] Mason, F., Ed. 1934. *The Great Design: Order and Progress in Nature*. London: Duckworth.
[71] Ibid., pp. 187–206.
[72] Ibid., pp. 204–205.
[73] Ibid., p. 194.
[74] Greenwood, W. O. 1938. *Biology and Christian Belief*. London: SCM Press, p. 180.
[75] Ibid., p. 181.
[76] Ibid., p. 179.
[77] Gingerich, O. 1994. Dare a scientist believe in design. In J. M. Templeton (ed.), *Evidence of Purpose*. New York, NY: Continuum, p. 24.
[78] Wald, G., in Henderson (op. cit.,), p. xxiii.
[79] Addison, C. C. 1960. *Use of Non-Aqueous Solvents in Inorganic Chemistry*. London: Royal Institute of Chemistry.
[80] Jander, J. 1949. *Die Chemie in Wasserähnlichen Lösungsmitteln*. Berlin: Springer Verlag (cited in Addison, above).
[81] Huggins, M. L. 1971. 50 years of hydrogen bond theory. *Angew. Chem. Int. Ed.* 10:147.
[82] Wald, G., in Henderson (op. cit.,), p. xxi.
[83] Pimentel, G. C., and A. L. McClennan. 1960. *The Hydrogen Bond*. San Francisco, CA: Freeman.
[84] Russell, C. A., 1975. A drop in the ocean. In *Wonders of Creation*, ed. Pearman, R., M. Fergus, and P. Alexander. Berkhamsted: Lion, pp. 77–81.
[85] Barrow, J. D., and F. J. Tipler. 1988. *The Anthropic Cosmological Principle*, new edition. New York, NY: Oxford University Press, pp. 524–541.
[86] Denton, M. J. 1998. *Nature's Destiny: How the Laws of Biology Reveal Purpose in the Universe*. New York, NY: The Free Press, pp. 19–46.
[87] Ibid., p. 19.

[88] Ibid., p. 46.
[89] Ibid., p. xii.
[90] Polkinghorne, J. 1988. *Science and Creation*. London: SPCK, p. 15.
[91] Polkinghorne, J. 2006. Where is natural theology today? *Sci. Christ. Belief* 18:169–179.
[92] Cleobury, F. H. 1967. In *A Return to Natural Theology*. London: Clarke, p. 233.
[93] Moltmann, J. 1985. *God in Creation: An Ecological Doctrine of Creation*. London: SCM Press, p. 58.
[94] Morrison, J. 1878. The surprising sentiment of a Victorian chemical manufacturer. *Trans. Tyne Chem. Soc.* 167.
[95] Habakkuk 2:14.

21 Lawrence Henderson's Natural Teleology

Bruce H. Weber

CONTENTS

> *We shall not cease from exploration*
> *And the end of all our exploring*
> *Will be to arrive where we started*
> *And know the place for the first time.*
>
> –T. S. Eliot

21.1 HENDERSON'S VIEW OF WATER'S SPECIAL ROLE IN LIFE

Harvard biochemist Lawrence Henderson, from the perspective of the fledgling field of physical chemistry, provided in *The Fitness of the Environment* an updated extension of the design arguments about the fine-tuning of the physical environment, especially water, presented in the *Bridgewater Treatises* of Whewell and Prout (Whewell, 1833; Prout, 1834; Henderson, 1913).

Henderson in several places wrote explicitly about this forgotten literature of natural theology. It is curious that in a work of science he should have attempted a recovery of this largely outmoded nineteenth-century Anglo-American tradition. This is particularly so since he was an agnostic (John Greene, personal communication, 2002). Henderson did have decided bibliophilic, philosophical, and sociological interests (Parascandola, 1971, 1972: Fry, 1996). Additionally, he was influenced by the writings of Alfred North Whitehead, C. Lloyd Morgan, Jan Smuts, Josiah Royce, and Vilfredo Pareto. Writing with the echo of the *Bridgewater Treatises*, particularly that of Prout, gave a structure for his argument and an organizing rhetorical trope.

Prout had undertaken the first attempt to consider the implications of chemical facts and phenomena for natural theology. Therefore, Prout's volume was a natural compass point for Henderson. Prout had noted that an understanding of the laws of physics could be viewed theistically, and that, following William Paley, there was strong evidence for design in the functional complexity of biology. However, Prout argued that chemical phenomena presented a particular challenge for Paley's perspective on natural theology. Neither the function of its processes nor a law-like basis for its

327

mechanisms seemed obvious. Prout observed that the molecular constitution of matter exhibited aspects that seemed arbitrary to him, they must have arisen either by chance or by a voluntary act of an intelligent will. "With respect to the first of these alternative, viz. chance: *the endless repetition of similar parts* presented by the molecular constitution of matter seems absolutely to preclude this supposition" (Prout, 1834, p. 88; emphasis in original). Indeed, chance was "too monstrous to be entertained for a moment by any rational being" (Prout, 1834, p. 89), so there must be a Being with intelligence responsible for the order seen in chemistry. He concludes:

> Now if we judge of the molecular constitution of matter by this rule [argument from design] we shall find that there is not only the most extraordinary fitness and adaptation to circumstances displayed in its arrangements, as far as we can understand them, but evidently much further; that is to say, the maker of this system must not only have personal intelligence, but infinitely surpassing our own. (Prout, 1834, p. 89)

So chemistry indeed provided evidence on Prout's reckoning, especially in the properties of the atmosphere and water, for design. Most likely what struck Henderson in passages such as this was the notion of a fitness and adaptation of chemical arrangements and properties for the conditions of Earth, including those needed for life. Of course, after 1859, fitness and adaptation had Darwinian connotations for Henderson. Consequently, he articulated an argument that not only did life conform to the demands of the environment to be fit, but that the environment itself was reciprocally fit for life. Henderson does not cite Alfred Russel Wallace's *Man's Place in the Universe* (Wallace, 1903). However, Wallace had argued for a similar view, even to the point of considering the sensitivity of life to the specific properties of water and the atmosphere. He concluded that life could only exist in the finely tuned environment of the earth. Wallace argued that "very similar, if not identical, conditions must prevail wherever organic life is or can be developed" (Wallace, 1903, p. 313).

In his treatment of the special ways in which water is uniquely fit for life, Henderson examined properties, some of which had been considered by Prout and Whewell, and also Wallace, such as the anomalous expansion of water on freezing, the specific heat of water, the latent heat of fusion, the anomalous melting point, and the dielectric constant. To this discussion, Henderson was able to bring not only greater quantitative data but also a deeper understanding of the laws of physical chemistry. He concluded that no other substance could substitute for water if life were to be possible. "A half a century has passed since Darwin wrote *On the Origin of Species*. And once again, but with a new aspect, the relation between life and the environment presents itself as an unexplained phenomenon" (Henderson, 1913, p. 274). The fitness of the environment thus presented a challenge to science—a challenge that tended to be ignored since it had no obvious explanation.

Resurrecting old arguments and terminology by providing a novel trope for a new scientific notion (e.g., Lovelock's use of Gaia) has its dangers. Readers could have discounted the attempt as retrograde. However, early responses to Henderson's arguments seem mostly to have been positive. A yearlong lecture series sponsored by the Paleontology Club at Yale (during the 1922–1923 academic year) focused on the role of the environment in organic adaptation (Thorpe, 1924). In particular, Lorande Woodruff built the argument of his paper on Henderson's concept of the fitness of the environment:

> Since living involves a constant interplay of matter and energy between protoplasm and the environment, this indicates, as has been recently emphasized by Henderson, that the fitness of the organisms implies a reciprocal fitness of their milieu. (Woodruff, 1924, p. 48)

Woodruff continued by quoting passages from Henderson on the unique properties of water and carbonic acid that Henderson argued are essential for life and represent fit characteristics of the chemical environment. Woodruff concluded, "The properties of matter and the course of cosmic evolution are intimately related to the structure of living beings and to their activities" (Woodruff, 1924, p. 49). During the 1920s and 1930s, two well-known, philosophically inclined biochemists, J. B. S. Haldane and Joseph Needham, commented favorably on Henderson's ideas but otherwise there were few who took Henderson's work seriously (Needham, 1936; Haldane, 1937; see also Denton, 2005; Barrow, 2006). A noteworthy exception was Harold Blum, who explicitly based his approach on Henderson and stated at the outset that he was building upon Henderson's concept of fitness. Blum had first encountered Henderson's concept in 1933. Subsequently, he worked with it. One of his major chapters in *Time's Arrow and Evolution* is headed "Fitness of the Environment" (Blum, 1951). Blum brought Henderson's argument up-to-date generally and extended Henderson's use of thermodynamics to include irreversible phenomena more specifically.

Nearly a century later, we know much more than Henderson about the physical and chemical properties of water and how they can be described by quantum mechanics. Furthermore, we know much more about the ways in which water provides the milieu for life. The aqueous environment places chemical constraints on, and provides opportunities for, the types of compounds and reactions essential for living systems found on earth (Williams and Fraústo da Silva 1996, 1999, 2003, 2006a, 2006b; Weber, 2007b). The physical chemistry of Henderson's day could not explain the reasons why water had the structure and properties that gave rise to the phenomena that so fascinated him, hence the appeal for him of design or teleological fine-tuning arguments. Today, we understand that, in addition to the full set of particle properties and physical laws at the basic level, the Schrödinger equation, the Pauli principle, and the Born–Oppenheimer separation of nuclear and electron motion are sufficient to account for all the unusual properties of water that are deemed so essential for life (Eisenberg and Kauzmann, 1969; Franks, 1971, 2000; Allen and Schaefer, this volume). Physical chemistry is constrained not to allow any free tuning of molecular properties without the whole edifice of the underlying physics being changed in ways that are not at present comprehensible. Thus, the special properties of an individual water molecule, or the supervenient properties of ensembles of water molecules, are not in any sense obviously crafted, but rather flow naturally from the sum of all properties of the constituent elements, and the laws of quantum mechanics and statistical mechanics. That, as Henderson argued, other possible solvents, including liquid ammonia, lack one or more of the particular properties of water necessary for living systems also follows from the underlying chemistry and physics. Thus, the unique role that water plays to make carbon-based life possible in an appropriate environment and history, as here on Earth, does not demonstrate fine-tuning caused by a "designer" in producing the characteristics of water. However, the nature of water could, of course, reflect a deeper-level propensity or teleology in natural laws and processes.

Barrow suspects that use of design teleology based on Prout and other Bridgewater authors was responsible for lessening the impact that Henderson's ideas might otherwise have had (Barrow, 2006). We have to be careful here, however. Design and teleology do not necessarily entail each other, although they are often conflated. The distinction Henderson ultimately sought was between design, as conscious intent and artifice applied from without to nature to achieve some end or external goal, and teleology, as function or purpose that can exist within nature that is the expression of natural laws and natural order. In a section below on uncoupling teleology and design, I examine the shift Henderson made in his argument from design to teleology just a few years after publishing *The Fitness of the Environment*. I also examine the nuanced enthusiasm of Needham for Henderson's use of a "physical teleology" or more generally "natural teleology." And finally, I situate Henderson in a long-standing discourse about the legitimate uses of teleology within science.

21.2 HENDERSON, DESIGN, AND TELEOLOGY

In many respects, current knowledge about the properties of water, the biological roles of water, and the insight that water's special characteristics arise from fundamental physical law, gives more power and precision to Henderson's main contention (see, e.g., contributions to this volume). However, the shift of the causality to deep laws leaves less space for design with regard to specific single-property details. This does not have to be a problem for Henderson's thesis if design and teleology are distinguished and teleology is used instead. This was already realized by William Whewell, who in his *Bridgewater Treatise* rejected the design argument, whose value for natural theology he doubted, but rather used teleology, which he saw in the laws governing nature and in the "arbitrary magnitudes which such laws involve" (Whewell, 1833, p. 18). Indeed, Henderson ultimately did not follow the natural theology of the other *Bridgewater Treatises*, nor the type of natural theology that has currently been promoted by the intelligent design (ID) movement. Rather, he posited that the uniqueness of water for life was a consequence of deep natural laws that ultimately led to life under the right types of circumstances. Henderson argued that "if life has originated by an evolutionary process from dead matter, that is surely the crowning and most wonderful instance of teleology in the whole universe" (Henderson, 1913, p. 310). He concludes with: "... the whole evolutionary process, both cosmic and organic, is one, and the biologist may now rightly regard the universe in its very essence as biocentric" (Henderson, 1913, p. 312). The propensity of matter to organize in specific ways was a fact about nature that had to be faced, even if there were no obvious explanations why this might be so, "Matter and energy have an original property, assuredly not by chance, which organizes the universe in space and time" (Henderson, 1913, p. 308). It is in this sense of natural organization resulting from deep laws that Henderson deploys the term teleology. The shift from design arguments to teleological ones, already observed here, was fully developed in Henderson's *The Order of Nature* (Henderson, 1917). Deep sources of order and organization may more appropriately be considered to reflect an even deeper purpose or *telos* and as such the biocentric nature of the universe would be less an issue of design than of teleology. But before disentangling design and teleology, let us consider the nature of the argument of design and its role in the natural theology that intrigued Henderson.

21.3 ARGUMENTS TO AND FROM DESIGN

The argument *to* and *from* design has a long history. When considering the organized complexity of the cosmos, or just our solar system and planet, it seems natural to appeal to some sort of apparent design in such organized complexity that reflects the action of an intelligent designer. The basic argument goes back to the ancients, as in Cicero's *De Natura Deorum*, and has not changed in its basic structure, even as subsequent defenders have updated the science used to make the argument. As John Barrow and Frank Tipper have pointed out, Cicero even used a clock metaphor (Barrow and Tipler, 1986). Cicero wrote, "An, cum machinatione quadam moveri aliquid vedemus, ut sphaeram ut horas ut alia permulta, non dubitamus quin illa opera sint rationis. . ." (When we see something moved by machinery, like an orrery or clock or many other such things, we do not doubt that these contrivances are the work of reason) (Cicero, 1933, pp. 216–217). Cicero used many metaphors for design, such as a statue, or a building, or a city, all of which could not arise by chance but clearly were products of planning and craft. Such arguments, albeit in inchoate form, were extended by Cicero to encompass the functional complexity of living beings.

The project of natural theology, especially in England, first used the argument to design and, from thence, to a designer. Subsequently, from examining the organized and functional complexity of nature, conclusions about the character and intentions of the deity were deduced. This was a rich tradition, stretching from John Ray through William Paley and the authors of the *Bridgewater Treatises* (Ospovat, 1978, 1981; Moore, 1979; Gillespie, 1990; Fyfe, 1997, 2002; Topham, 1998; Ruse, 2002; Grene and Depew, 2004; Weber and Depew, 2004; Depew, in press; Weber, in press).

Hume raised objections to such a project in his *Dialogues Concerning Natural Religion*, demonstrating the logical problems with design arguments. He argued that there is a disanalogy between natural organization and human artifacts, between human craftsmen and the divine. [Polkinghorne echoes this distinction, pointing out that scripture distinguishes between *asah*, human making, and *bara*, divine creative activity (Polkinghorne, 2004).] Hume goes on to speculate that there may be a natural source of order innate in matter:

> . . . order, arrangement, or the adjustment of final causes is not, in itself, any proof of design but only as far as it has been experienced to proceed from that principle. For aught we can know a priori, matter may contain the sources or spring of order originally, within itself, as well as mind does; and there is no more difficulty in conceiving that the several elements, from an internal unknown cause, may fall into the most exquisite arrangement, that to conceive that their ideas, in the great, universal mind, from an internal, unknown cause, fall into that arrangement. (Hume, 1779 [1994], pp. 736–737)

Hume also revived the Aristotelian notion that the universe might be more like an organism than an artifact and that embryological development might be an alternative analogy to design. Later, Hume argues that a ship might appear to be a perfect example of design, but in fact it could be the product of a "stupid mechanic" who copied others and/or worked by trial and error. Hume then made the analogy explicit, giving it, what we today would regard, a multiverse, and even evolutionary twist:

> Many worlds might have been botched and bungled, throughout an eternity, ere this system was struck out: much labor lost: many fruitless trials made: and a slow, but continued improvement carried on during infinite ages in the act of world-making. (Hume, 1779 [1994], p. 752)

Furthermore, since there is just one world, he argued, and one instance of life, we cannot know whether the conditions obtained are uniquely needed. Hume thus appealed to a type of multiverse argument when he claimed that apparent design could be accounted for by an ensemble of worlds (existing simultaneously or a temporal sequences of worlds) in which, by chance, one happened to have the right mix of properties to sustain life (Hume, 1779 [1994]; Oppy, 1996). Hume ultimately lets his spokesman conclude that, despite the problems with the logic of design arguments, such arguments remain rational as an explanatory choice if the *only* alternative is pure chance.

Despite Hume, the tradition of natural theology continued into nineteenth-century Britain, where it deepened its arguments by focusing on the biological realm (Sober, 2004; Grene and Depew, 2004). William Paley's *Natural Theology* reasserted the argument from design applied to biology. He also presented an argument for design based on the laws of nature (Paley, 1802). In addition to using the famous clock analogy, Paley relied on the cumulative weight of examples from biology, for which natural causes at the time seemed implausible. Paley does not cite Hume specifically, although it is clear that he had read Hume and in part responds to him (Grene and Depew, 2004). Later in the book, Paley includes a counterfactual argument about water: If water had other properties, such as being acidic or oily, or those of alcohol, life would not be possible (Paley, 1802, Chapter 22). Paley then addressed the laws of nature and observed that, "out of an infinite number of possible laws, those which were admissible for the purpose of supporting heavenly motions, lay within certain narrow limits . . . " (Paley, 1802, p. 390). Paley continued by considering counterfactual possibilities for the law of gravitational attraction:

> That the subsisting law of attraction falls within limits which utility requires, when these limits bear so small a proportion to the range of possibilities upon which chance might equally have cast it, is not, with any appearance of reason, to be accounted for, by any other cause than a regulation proceeding from a designing mind . . . . No other law would have answered the purpose intended. (Paley, 1802, p. 395)

Paley seemed to have Hume in mind when he wrote that every argument against design,

> leads to the inference, not only that the present order of nature is insufficient to prove the existence of an intelligent Creator, but also that no imaginable order would be sufficient to prove it; that *no* contrivance, were it ever so mechanical, ever so precise, ever so clear, ever so perfectly like those which we ourselves employ, would support this conclusion. A doctrine, to which, I conceive, no sound mind can assent. (Paley, 1802, p. 415)

Paley went on to argue against laws as secondary causes, asserting that laws by themselves have no power except that given to it by an agent. In effect, Paley gives a "dare" to anyone to provide a naturalistic explanation for the phenomena he has reviewed (see Depew and Weber, 1995).

Paley was widely read in nineteenth-century England. Indeed, Darwin read Paley with approval as an undergraduate (Desmond and Moore, 1991; Browne, 1995). The Earl of Bridgewater's bequest, which made possible the *Bridgewater Treatises*, was motivated by concern that advances in geology in the early nineteenth century, which pointed toward a very old earth riddled with fossils and undergoing episodic if not continual change, was undermining the argument for a divine craftsman. Leading lights in science were recruited to update Paley's design argument. The treatises were an enormous success. They ran into many editions (ultimately, more than 60,000 copies were published), and were widely reviewed in the 1830s. Topham (1998) cites 120 such reviews published in more than 40 periodicals. Although "radical artisans" argued that "science, properly understood, served a materialist and antireligious end" (Topham, 1998, p. 259), the majority popular opinion was that Paley had the better of Hume. Elite opinion was more guarded. Whewell was not a fan of Paley's use of the design argument. He sought rather to bridge the physics–biology gap by conceiving the laws of physics as having purpose, developing a line of reasoning similar to the contemporary anthropic literature (Grene and Depew, 2004). There were also theological objections to Paley. Most notable were Samuel Taylor Coleridge and John Henry Newman, well before Darwin published his theory (Levere, 1981; McGrath, this volume).

Of course, a watershed occurred in 1859 with the publication of Charles Darwin's *On the Origin of Species*. Darwin took up Paley's dare by providing a natural explanation for biological functional complexity and adaptation (Darwin, 1859). In his next book on orchids, Darwin showed how natural selection could explain the exquisite coadaptations of insects and orchids without the need of, and indeed contrary to, the use of design arguments (Darwin, 1862; Ghiselin, 1969). Darwin's deep insights put him forward as a "Newton of a blade of grass" (Depew and Weber, 1995). Now there was an alternative to choosing between design and chance. Darwin provided the argument Hume lacked. Darwin's "long argument" offered a convincing explanation of apparent biological design and of adaptation, and by extension, doing away with the need for a designer. For many Christians who had problems with Paley's approach, as for example his notion of perfect adaptation, Darwin's theory of adaptation through natural selection provided a corrective position. It also made the whole process of the transformation of species subject to general laws of nature (see, e.g., Powell, 1860; Moore, 1979; McGrath, this volume). Already started before 1859, the shift against design arguments accelerated.

An interesting, early dissenter to the "death of design" was Harvard chemist Josiah Cooke. Cooke delivered a lecture series in 1861 on the relation of science and religion soon after the appearance of Darwin's magnum opus in the United States. Cooke was not bothered by the idea of evolution (organic development as Cooke called it), because he felt that the more crucial issue had to do with the physical and chemical nature of the world:

> At the time the lectures were written, Mr. Darwin's book the Origin of Species, then recently published, was exciting great attention, and was thought by many to have an injurious bearing on the argument for design. It was, therefore, made the chief aim of these lectures to show that there is abundant evidence for design in the properties of the chemical elements alone, and hence that the great argument of Natural Theology rests upon a basis which no theories of Organic development can shake. (Cooke, 1865, p. vii)

Cooke proceeded to consider the special properties of oxygen, nitrogen, carbonic acid, and water, and how they were adapted to life, reflecting the design of the Divine Intelligence. Water, in particular, was a showcase for Cooke's design argument:

> Attempt to find a liquid, which, if in sufficient quantity, might supply its place, and you will be still further impressed by this evidence of intelligence and of thought. Of all the materials of our globe, water bears most conspicuously the stamp of the Great Designer . . . . (Cooke, 1865, p. 170)

Cooke's target was not Darwin but the philosophy of materialism:

> Development is the pet word of this philosophy, and it constantly aims to show how the whole scheme
> of nature, with all its adaptations, might have been evolved through the concurrent action of various
> unintelligent causes alone. (Cooke, 1865, p. 257)

Cooke assumed Darwin correct with regard to common descent, but pointed out that design can unfold over time. It was natural selection that caused Cooke's concern: "The difficulty to my mind in Mr. Darwin's particular theory is not its developmental feature, but in the fact that he refers the development to what I can understand only as an unintelligent cause" (Cooke, 1865, p. 263). But even if Darwin and his followers advocated such an unintelligent cause (which Cooke saw as threatening the longevity of Darwin's theory more than that of natural theology), it did not impair the evidence for design, for the evidence from cosmology and chemistry were, to his analysis, impervious to such arguments:

> Before the first organic cell could exist, and before Mr. Darwin's principle of natural selection could
> begin that work of unnumbered ages which was to end in developing a perfect man, nay, even before
> the solid globe itself could be condensed from Laplace's nebula, the chemical elements must have been
> created, and endowed with those properties by which alone the existence of that cell is rendered possible. (Cooke, 1865, p. 265)

Cooke's lectures were thin on specific scientific content. They were delivered to nonscientific audiences at the Brooklyn Institute and at the Lowell Institute. The basic argument foreshadows Henderson. However, it lacks Henderson's nuanced subtlety and chemical detail. Interestingly, Henderson does not mention Cooke in any of his writings and it is not clear if he was aware of his predecessor's book. Cooke felt he had to defend natural theology, but Henderson's program was rather different. And Henderson changed his argument within a few years.

21.4   UNCOUPLING DESIGN AND TELEOLOGY

Design arguments as noted above, including Prout's, went into abeyance as Darwin's theory of natural selection became accepted as a plausible explanation for apparent design in biology. Henderson's revival of the arguments of natural theology, in the context of evolutionary explanations and concepts, showed a possible consilience of these projects by pushing design deeper into nature, but he ultimately abandoned design language for that of teleology. We need now to address why Henderson made such a shift and consider how wedded teleology as a concept is to that of design.

An argument for design appeared in Plato's *Laws*. Aristotle countered with an explanation of the functional complexity observed in animals, which involved purpose but did not invoke an external cause, a "natural teleology." Aristotle's position was that the cause of apparent design was within the organism itself. *Telos* existed within the natural order. God, for Aristotle, was the "unmoved mover," the ultimate cause but not an external causal crafter. It was Cicero's *De Natura Deorum* that presented the Stoic argument from design. It was through Cicero that, until recently, educated Europeans most often encountered the argument. However, as Grene and Depew point out, this Stoic argument was read through a filter of Christian theology. In *De Natura Deorum*,

> the Stoic representative of the argument portrays the world we see as a single living substance integrated and organized by a 'world soul.' This being assumed, there is no pressing need to distinguish between external and internal teleology . . . this organicist view was rendered problematic, however, by pluralism about substances and even more by Christian creationism, which made God transcendent to the world. . . . By this route, teleology came to mean intentional design by an agent who is separate from his product. Final causes were thereby removed from the organism, their primary locus for Aristotle, and ascribed to a maker—with the result that living things were assimilated to craft-objects, and ultimately to machines. (Grene and Depew, 2004, pp. 184–185)

In this way, design and teleology came to be seen as wedded and virtually the same thing. However, what if natural causes can explain away claims to detect apparent design? Are teleological concepts thereby also excised?

Darwin answered Paley's dare and provided a natural explanation of biological apparent design. However, teleology, as adaptation produced by natural selection, did have an implicit survival in Darwinism (Moore, 1979; Brandon, 1981; Lennox, 1994; Depew, in press). When Asa Gray commented that Darwin's great service to science was to wed morphology and teleology, Darwin's famous reply was that Gray was always a man to "hit the nail on the head" (Darwin, 1887, p. 367; Lennox, 1994). Neither was talking about design arguments per se (Depew, in press). Rather, Gray was trying to restore Cuvier's Aristotelian functional morphology. This view had been receding from the forefront of biology around the middle of the nineteenth century. It was supplanted by Geoffroyian structuralism. Gray saw Darwin's explanation of adaptation as restoring notions of function and purpose to biology. It was in this sense he that meant teleology. Although Gray would have liked to assimilate this argument to one of design, he ultimately realized the problems for theology in the design argument; besides his real concern was with Cuvierian final cause, which resided in the internal relationships within organisms. Gray saw Darwin as expanding Cuvierian final cause to include the relationships of organisms and their environments. Eventually, Gray came also to accept that variation could not be directed, even if this raised problems of waste and profligate death, that is, in theological terms the problem of natural evil. In 1876, Gray published the essay "Evolutionary Teleology" in his *Darwiniana: Essays and Reviews Pertaining to Darwinism.* Here, he presented his most systematic thought on teleology. Gray points out that design need not be specific, but rather it can be the result of universal principles. He asserts that Darwin does not abandon purpose and that Darwin's use of teleology per se does not argue for or against a deity. Gray states explicitly that, "Darwinian evolution. . . is neither theistical or atheistical" (Gray, 1876, p. 379). This point is often forgotten in today's heated debates about evolution. Darwin appreciated an ally who saw his theory as appealing neither to chance alone nor to design, but to the functional teleology of adaptation through natural selection, although Darwin himself eschewed using the term "teleology" in his own writing. Earlier, in a letter to Jeffries Wyman in 1860, Darwin wrote, "No other person understands me so thoroughly as Asa Gray. If I ever doubt what I mean myself, I think I shall ask him!" (Burkhardt et al., 1993, p. 405).

In 1883, Lewis Ezra Hicks, professor of geology at Denison University in Ohio, published *A Critique of Design Arguments*, in which he reviewed many of the problems of design for science, philosophy, and theology (Hicks, 1883). He warned of the ambiguity of conflating design seen as created contrivances with intent or purpose resulting from the action of natural law. He saw defenders of design logically and unjustifiably gliding from the first usage to the second in many of their arguments. For this reason alone, he thought design arguments should be banished. Furthermore, many of the arguments presupposed their conclusion in their major premise (apparent design supposes a designer). He agreed with Whewell's argument in *The Philosophy of the Inductive Sciences* that a final cause is not deduced from the phenomena associated with "organized bodies" but rather is assumed. Hicks, however, was wary of the possible confusion of final cause with functionality. Rather, the phenomena indicated to Hicks that *order* is not only the result of natural law but that "natural law . . . is merely one form of order; it is order itself by another name" (Hicks, 1883, p. 22). He terms this type of reasoning as "eutaxiological" from *eutaxy* or established order. Hicks criticized Cooke's use of design and suggested that Cooke's argument instead was one of eutaxiology. For Hicks, theism did not need to be defended from Darwin since evolution "has simply reduced the phenomena of biology under the reign of natural law, where vastly the greater part of the universe was already; thus making the problem of the relation of the cosmos to the creator the same uniform problem for all sorts of phenomena" (Hicks, 1883, p. 114). Hicks agreed with Asa Gray that invocation of special creation created a burden for teleology both in terms of imperfections and sheer magnitude of detail, which Darwin relieved by making teleology the result of natural law, not design. Hicks concluded that, "a belief in evolution leaves theism intact" (Hicks, 1883, p. 331).

And he called for a "new teleology" based on order to replace the "old teleology" of mechanics and design. Cooke accepted Hicks's criticism and shifted his argument away from design to that of order and the action of natural law (Cooke, 1888). Again, it is not clear if Henderson was aware of Hicks's work or either of Cooke's volumes, even though Henderson journeyed the same conceptual path.

Near the end of *The Fitness of the Environment*, Henderson quotes from Du Bois-Reymond that the problem for biologists is that if one does not give overall causality to Epicurean chance, then one inevitably slides into Paley's discarded natural theology. Neither does justice to the internal causes and purposes of organisms. Darwin, however, offers a *via media* for philosophical natural-ists. Henderson is not certain that natural selection alone has enough causal power, citing Hugo de Vries's mutation theory. This was a time when even someone sympathetic to Darwinism could be skeptical about the causal sufficiency of gradual natural selection to produce speciation (Depew and Weber, 1995). Henderson uses Teleology as the heading of the last section of his book. He makes it clear that he has no intention of endorsing either a "vitalist teleology" or a "metaphysical teleology." By putting the "old teleology" to death, he argues, it is possible for science to use a "new teleology" or a "physical teleology" or what we might choose to call a type of natural teleology based on the concept of reciprocal fitness within nature as a guide to future research (see also Fry, 1996, for a discussion on this point).

Henderson returns to this argument in his 1917 book, *The Order of Nature*. The first part of the book is a review of past teleological arguments, from Aristotle through Hume and Kant. Nowhere is mention made of Paley or of the *Bridgewater Treatises*. Henderson is at pains to distance himself from the classical design argument to a designer, and from natural theology in general. He praises Hume for eliminating dogmatic theology and for laying the groundwork for Kant's teleology that views the parts of an organism as both cause and effect. Henderson wrote, "Hume's discussion of teleology is in many respects decisive. It remains for the man of science, on the whole, the best treatment of the subject, for it is clear, specific, and single-minded" (Henderson, 1917, pp. 52–53). Henderson accepted that organisms are not like machines, because they have "formative powers" in addition to mechanism. With Hume and especially Kant, Henderson also accepts that the teleology of organisms is dissimilar to that of artifacts. Henderson is more interested in general laws govern-ing evolution. He considers the only two candidates to be the second law of thermodynamics and the ideas of Herbert Spencer. "Spencer perceived that we can know a complex phenomenon only when we understand both its elements and how these elements cooperate in order to produce it" (Henderson, 1917, p. 121). Despite cautionary statements about what Spencer actually achieved, it is clear that Henderson was looking for something more satisfying than the mechanism of natural selection. He was favorably inclined toward attempts to apply the thermodynamics of J. Willard Gibbs to living cells, a position already articulated by Frederick Gowland Hopkins some years pre-viously (Hopkins, 1913). Yet, Henderson was left with a sense that there was something incomplete. ". . . [T]he laws of nature provide an imperfect yet intelligible account of certain general character-istics of the orderliness in the phenomena of nature and the products of evolution. These principles, however, give no account of the origin of diversity" (Henderson, 1917, pp. 180–181). Nor did they give an explanation for the teleological appearance of nature as exemplified by its propensity to become more organized. (On the issue of propensities in nature, see Popper, 1990, and Ulanowicz, 2004.) Thus, the teleological concept of reciprocal fitness of life and its environment provided, at the minimum, a heuristic, a way forward even in the absence of a complete set of laws of nature. Henderson states:

> The inorganic, such as it is, imposes certain conditions upon the organic. Accordingly, we may say that the special characteristics of the inorganic are the fittest for those general characteristics of the organic which the general characteristics of the inorganic impose upon the organic. This is one side of reciprocal biological fitness. The other side may be similarly stated: through adaptation the specific characteristics of the organic come to fit the special characteristics of a particular environment, to fit, not any planet, but a litter corner of the earth. (Henderson, 1917, pp. 186–187)

The unique properties of the elements of carbon, nitrogen, and oxygen, as well as water and carbonic acid provide the "fittest ensemble of characteristics" for the possibility of life. Henderson continues, "This environment is indeed the *fittest*" (Henderson, 1917, p. 185, emphasis in original). He concludes that, "we are obliged to regard this collation of properties as in some intelligible sense a preparation for the processes of planetary evolution . . . . Therefore the properties of the elements must for the present be regarded as possessing a teleological character" (Henderson, 1917, p. 192). Henderson clarifies once again in what sense he is using the term teleological, "Thus we say that adaptation is teleological, but do not say that it is the result of design or purpose" (Henderson, 1917, p. 204). Henderson argues at the end that we, as Darwin and Hume, cannot admit *external* design or purpose in nature. Yet we cannot escape the pressure of evidence of a form of teleology that is apparent *within* nature. Later in life, Henderson came to regret his use of the design argument in *The Fitness of the Environment*. He, however, continued to assert a kind of secularized teleology, which accepted "biocentric" fitness as axiomatic, as a basic, even if inexplicable, fact about nature (Parascandola, 1971, 1972).

As Henderson feared, not everyone who read his works made the distinction he had come to make between design and teleology. Michael Denton, for example, claims that J. B. S. Haldane had commented that Henderson's *The Fitness of the Environment* provided convincing evidence of design, and goes on to claim it as "a great work of teleology" and "one of the most important books of the twentieth century" (Denton, 2005, pp. 169–170). However, examination of what Haldane actually wrote shows a more cautious approach that suggested to him that the facts assembled by Henderson "are *at present* more readily conformable with design theory than with any other" (Haldane, 1932 [1985], p. 96, emphasis added). Denton not only exaggerates Haldane's approbation of Henderson's design argument, but he also does not address Henderson's change of mind in *The Order of Nature*. Henderson clearly distinguished design from teleology, rejected design arguments, and shifted his argument to a type of natural teleology.

Joseph Needham, who also was influenced by Whitehead, and was aware of Henderson's shift, commented in 1936 on *The Fitness of the Environment*:

> Here, as will be familiar to all, the evolutionary process was shown to not be a matter of chance, but inevitable, granting the general principle of biological organisation and the properties of the chemical elements—a conclusion at least as acceptable to dialectical materialism as to orthodox theology. *Vitalism was thus dissolved in universal teleology . . .* (Needham, 1936, p. 15, emphasis added)

Needham went on to argue that this teleological move gives space for biologists to explore why biological organization has the characteristics it has and how these characteristics arise. During the 1930s, Needham and his associate Conrad Waddington brought together scientists and philosophers in the "biotheoretic gathering" at Cambridge, pursuing the project of developing a theory of biological organization as well as setting the conceptual foundation for the yet-to-be-born science of molecular biology (Abir-Am, 1987). Needham argued that, just as the physical chemist takes the periodic table to be axiomatic but provides an explanation of it in terms of the electronic structure of atoms, so too can biochemists hope to explain organized, functional complexity, even if they have to accept it *pro tempore* as axiomatic. Needham had strong historical and philosophical interests. He played an important role in the conceptual development of biochemistry and theoretical biology in the first half of the twentieth century. He clearly was attuned to the logic of Henderson's teleology. Through Needham, Henderson influenced on a number of students in the biochemistry department at Cambridge University through the 1960s (Abir-Am, 1987; Weatherall and Kamminga, 1992; Prebble and Weber, 2003). Although not cited often, Henderson's teleological thinking contributed to the development of the research approaches of influential scientists such as Peter Mitchell (Peter Mitchell, personal communication, 1982; Prebble and Weber, 2003).

Harold Blum discussed both of Henderson's books in his class at the University of California at Berkeley during the 1930s. His interest in Henderson's ideas became the starting point for Blum's program of extending Henderson's notion of fitness and the role of thermodynamics, as mentioned above. Blum went along with the use of the term teleology. For him, it captured the apparent purpose and epigenetic complexification in nature that he was trying to understand as an action of the "times arrow" property of the second law. He noted that purpose was understood within a specific context. "To Lawrence J. Henderson teleology involved the development of existing order and fitness through an evolutionary process, participated in by natural selection but also involving other strictly physical features" (Blum, 1951, p. 210). Blum also makes a very Hendersonian point that the physical and chemical properties of matter and the laws of thermodynamics were "angled with" the beginnings of life, asserting that "living systems today are inextricably tangled with their origin and evolution, and with the origin and evolution of this earth, and of the universe" (Blum, 1951, p. 209). For Blum, this entanglement and increase of organization, this form of natural teleology needed explanation. Blum believed that Henderson made the first step by calling attention to the problem. Blum thought of himself as pointing toward the next step in suggesting a way to develop a theory of organization by using irreversible thermodynamics.

Teleology was, of course, to become one of the "hot topics" in the philosophy of biology during the latter third of the twentieth century. This discourse arose, not from a consideration of Henderson or Needham, but rather from addressing issues of function and adaptation in evolutionary biology. (For a brief review, see the anthology edited by Allen et al., 1998, and Grene and Depew 2004, pp. 313–321.) Some scholars proposed to excise the notion of purpose from teleology by coining the term teleonomy, defined as apparently purposive behavior resulting from evolutionary adaptations and/or "programmed" in genes. Others considered a "consequence etiological teleology" that emphasized function of a trait or organ, and then sought an evolutionary causal account of its origin. However, such accounts are often not available. Often, biologists have to be satisfied with answers to "How does it work?" and "What is it for?" Grene and Depew conclude that teleonomy simply sidesteps the difficult conceptual problems and that the etiological arguments ultimately lack traction in practice. Unfortunately, this debate on neoteleology was limited to discussions of how problems were framed within the neo-Darwinian evolutionary theory. At present, the time may well be ripe for reexamining the kind of teleology that Henderson and Needham advocated. This view accepts natural organization as axiomatic. How conceptually coherent is such an approach? Can it provide a reliable guide for developing research programs? Or, more strongly, do such notions engender important conceptual challenges for contemporary philosophers to pursue? In attempts to develop a science of complexity, will agendas toward developing a theory of emergence and self-organization engage the issue of the axiomatic appearance of natural teleology?

During the past 15 or so years, a particular version of design teleology has been resurrected using concepts of "irreducible complexity" and "explanatory filters" (Behe, 1996, 2007; Dembski, 1998). This current ID movement is conceptually a revival of the earlier, pre-Henderson natural theology tradition. It attempts to update Paley and the *Bridgewater Treatises* in terms of design in nature as analogous to that of artifacts (Gillespie, 1990). However, ID writers do not deploy those specific arguments of Paley and Whewell that appealed to the deep character of natural law rather than to artifact-like, crafted design. Modern design theorists assume once again that design is inseparable from teleology. In contrast, versions of the anthropic approach appeal to deep order in natural law and are more in the spirit of Henderson's distinction of design from teleology. Furthermore, the robust success of modern evolutionary biology and its accelerating research advances (particularly in evo-devo; see, e.g., Carroll, 2005, as well as Carroll et al., 2008) casts serious doubt about the scientific credibility of ID and similar attempts to revive conventional natural theology (Weber and Depew, 2004; Weber, 2007a, in press). More generally, what are the implications, scientific, philosophical, and theological for deploying a teleology of the sort Henderson and Needham advocated? Will future scientific efforts be fruitful in looking at the deepest levels, even searching for new laws that might explain why the universe seems to be so constituted as to allow life and intelligence to emerge through the action of natural processes and laws? The philosophical analysis of the anthropic principle and its relation to various versions of teleology perhaps remains in relatively early stages. Developments in science that embrace the behavior of complex systems may provide a richer and more fecund context for such analyses.

21.5   EMERGENT COMPLEXITY

During the last quarter of the twentieth century, conceptual and computational developments gave rise to what may be characterized as the sciences of complexity or complex systems dynamics, combining nonlinear dynamics, nonequilibrium thermodynamics, and information theory, inter alia (Prigogine, 1981; Brooks and Wiley, 1986; Wicken, 1987; Pagels, 1988; Swenson, 1989; Waldrup, 1992; Casti, 1994; Depew and Weber, 1995; Weber and Depew, 1996; Ulanowicz, 1997; Bar-Yam, 1997; Dewar, 2003; Taylor, 2003; Kleidon and Lorenz, 2005; Schneider and Sagan, 2005; Weber, 2007b). Exploration of the dynamics of complex systems has reinvigorated notions of, and interest in, the concept of emergence, which seems to be intimately connected with the self-organizational properties observed in some complex systems (Kauffman, 1993, 2000; Ulanowicz, 1997; Corning, 2002; Morowitz, 2002; Silberstein, 2002; Deacon, 2003; Clayton, 2004; Sharma and Annila, 2007; Weber, 2007b). The prospect of a successful pursuit of such a research program focused on explaining the logic of emergent behavior offers some hope for progress toward a theory of organization for which Henderson and Needham recognized the need but lacked the science and conceptual tools to be able to develop. Is it possible that by the centenary of Henderson's books (ca. 2015), we may be able to see the outlines of such a theory?

If realized, the development of a general theory of organization and emergence probably would substantially change how we frame questions about the roles of chance and design with respect to biochemical as well as cosmological phenomena. Just as Darwin's theory provided a mechanism that allowed a type of teleology to arise from the interplay of chance and necessity to produce apparent design, so we might hope that a theory of organization might play a similar role in addressing issues of fine-tuning raised by Henderson.

Shifting the discourse about chance, necessity, design (apparent or real), and teleology in nature from assumptions of a closed universe, in which initial and boundary conditions determine the future, to that of emergent complexity in an open universe, in which genuine novelty can emerge, would constitute a paradigm shift in which there might be a "teleology without teleology" (i.e., teleology without external design or purpose) as Paul Davies suggested (Davies, 1998). Philip Clayton similarly suggests that there can be a protopurpose or "*purposiveness without purpose*" (Clayton, 2005, p. 97, emphasis in original). Recent speculations point to such a possibility, that the universe is radically nonergodic in principle and that genuine, unexpected novelty has and will continue to arise by the processes of nature (Kauffman, 2000, 2004; Conway Morris, 2003; Deacon, 2003; Polkinghorne, 2004; Ulanowicz, 2002, 2004; Weber and Depew, 2004; Clayton, 2004; Barbour, 2005; Weber, 2007b).

What happens to design and to teleology in such a view of the universe informed by complexity theory and emergentist concepts? Paul Davies suggests that such developments are leading us back to a concept of Aristotle's metaphor of the universe as organism, not in the sense that it is alive but that it has "purposes," that is, that there are functional relationships among its parts. (Actually, this notion of a living universe was that of the Stoics rather than Aristotle's.) This also seems to bring us back to the type of teleology that Darwin and Gray espoused of final causality within such relationships (organs within a body, between organisms and environments, etc.), but without the fixed point of Aristotle's final causality (Davies, 2005; for further consideration of organic relationship in ecosystems, see Ulanowicz, 1997). Emergent structures may give the appearance of well-crafted design, but "it is entirely the result of natural process. In effect, it renders a teleology without providing one" (Davies, 1998, p. 151). This "teleology without teleology" is, I believe, what Henderson and Needham had in mind: a teleology of emergent organization and complexity. It is not the teleology of the classical design arguments of Cicero, Paley, or Prout, but rather more like the teleology of Gray or Darwin, with causality within organisms or in their relationship to the environment, or even of Whewell, with his emphasis on the properties of deep laws. In such a teleology, chance is not the enemy but a collaborator in creation.

As we have seen, Henderson did not have a metaphysical or theological agenda. But as Gray pointed out about Darwinism, it in itself does not either support or deny theism. Conway Morris admits that any directionality or teleology observed in evolution is consistent with theism but that it does not provide any proof of it (Conway Morris, 2003). On one hand, having a natural explanation for something does not logically require denying that God exists or plays a role in bringing forth and sustaining the cosmos (Cobbe, 1872). Methodological naturalism requires that science be practiced *as if* it is without a priori metaphysical commitments, just methodological ones, such that the same science can be done and debated by theists, agnostics, and atheists. Philosophers and theologians, however, can and should consider the metaphysical implications of the science and engage in dialogue with scientists.

21.6   END POINT

Henderson drew attention to the biocentric properties of water. He thereby reopened the issues of design and teleology within science. Henderson's speculations evolved to a position in which he carefully distinguished between the two, favoring teleology over design. However, today's ID theorists, as in an earlier time Paley, Prout, and Cooke, do not distinguish design from teleology. Whewell (and Hicks) by contrast rejected design *per se*, but accepted a *telos* or end point—humankind. Henderson recapitulated Whewell's critical position on design. He also accepted teleological arguments. However, he saw the end point as life itself rather than our species.

Henderson's natural teleology reminds us that there are some interesting features of the universe that require explanation, such as the emergence of novel structure, phenomena, and organization generally, and specifically of the universe itself, life, and mind, through natural processes. In this view, there is purpose but no fixed endpoint. Thus, purpose is expressed via an open-ended process of emergent complexity and novelty. Perhaps in time science may develop a robust theory of emergent organization such that we need not take such phenomena as axiomatic on one hand or as merely epiphenomenal on the other. We are only at the beginning of fulfilling this desideratum. At the beginning of the last century, Henderson called our attention to this matter. For that, we can be grateful.

ACKNOWLEDGMENTS

I wish to thank David Depew and Terry Deacon for sharing manuscripts in preparation and for discussions on teleology and emergence. I also wish to thank John Barrow and Colin Russell for drawing my attention to Josiah Cooke, and William Buckley, David Depew, James Moore, and Robert Ulanowicz for helpful suggestions on an earlier draft.

REFERENCES

Abir-Am, P. G. 1987. The biotheoretical gathering, transdisciplinary authority and the incipient legitimation of molecular biology in the 1930s: new perspective on the historical sociology of science. *Hist. Sci.* 25:1–70.

Allen, C., M. Bekoff, and G. Lauder. 1998. *Nature's Purposes: Analysis of Function and Design in Biology.* Cambridge, MA: MIT Press.

Barbour, I. G. 2005. Evolution and process thought. *Theol. Sci.* 3:161–178.

Barrow, J. D. 2006. *Chemistry and Sensitivity, in Fitness of the Cosmos for Life: Biochemistry and Fine-Tuning.* Cambridge: Cambridge University Press.

Barrow, J. D., and F. J. Tipler. 1986. *The Anthropic Cosmological Principle.* Oxford: Oxford University Press.

Bar-Yam, Y. 1997. *Dynamics of Complex Systems*. Reading, MA: Perseus.

Behe, M. J. 1996. *Darwin's Black Box: The Biochemical Challenge to Evolution*. New York, NY: The Free Press.

Behe, M. J. 2007. *The Edge of Evolution: The Search for the Limits of Darwinism*. New York, NY: Free Press.

Blum, H. 1951. *Time's Arrow and Evolution*. Princeton, NJ: Princeton University Press.

Brandon, R. 1981. Biological teleology: questions and explanations. *Stud. Hist. Philos. Sci.* 12:91–105.

Brooks, D. R., and E. O. Wiley. 1986. *Evolution as Entropy: Toward a Unified Theory of Biology*. Chicago, IL: University of Chicago Press.

Browne, J. 1995. *Charles Darwin: Voyaging*. New York, NY: Knopf.

Burkhardt, F., D. M. Porter, J. Browne, and M. Richmond. 1993. *The Correspondence of Charles Darwin, Volume 8, 1860*. Cambridge: Cambridge University Press.

Carroll, S. B. 2005. *Endless Forms Most Beautiful: The New Science of Evo-Devo*. New York, NY: W. W. Norton & Co.

Carroll, S. B., B. Prud'homme, and N. Gompel. 2008. Regulating evolution. *Sci. Am.* 298(5):60–67.

Casti, J. L. 1994. *Complexification: Explaining a Paradoxical World Through the Science of Surprise*. New York, NY: HarperCollins.

Cicero, M. T. 1933. *De Natura Deorum*, H. Rackham (trans). London: William Heinemann.

Clayton, P. 2004. *Mind & Emergence: From Quantum to Consciousness*. Oxford: Oxford University Press.

Cobbe, F. P. 1872. *Darwinism in Morals and Other Essays*. London: Williams and Norgate.

Conway Morris, S. 2003. *Life's Solutions: Inevitable Humans in a Lonely Universe*. Cambridge: Cambridge University Press.

Cooke, J. P. 1865. *Religion and Chemistry: Or Proofs of God's Plan in the Atmosphere and Its Elements*, 2nd ed. New York, NY: Griggs.

Cooke, J. P. 1888. *The Credentials of Science; The Warrant of Faith*. New York, NY: Robert Carter & Bros.

Corning, P. A. 2002. The re-emergence of "emergence": a venerable concept in search of a theory. *Complexity* 7(6):18–30.

Darwin, C. 1859. *On the Origin of Species by Means of Natural Selection, or the Preservation of Favoured Races in the Struggle for Life*. London: John Murray. Available also as a facsimile reprint (1964), with an introduction by Ernst Mayr, Cambridge, MA: Harvard University Press.

Darwin, C. 1862. *On the Various Contrivances by which British and Foreign Orchids are Fertilised by Insects and on the Good Effects of Intercrossing*. London: John Murray.

Darwin, F., Ed. 1887. *Life and Letters of Charles Darwin*, 2 volumes. New York, NY: Appleton.

Davies, P. 1998. Teleology without teleology: purpose through emergent complexity. In *Evolutionary and Molecular Biology: Scientific Perspectives on Divine Action*, ed. R. J. Russell, W. R. Stoeger, and F. J. Ayala. Vatican City: Vatican Observatory Publications, pp. 151–162.

Davies, P. 2004. Emergent complexity, teleology, and the arrow of time. In *Debating Design: From Darwin to DNA*, ed. W. A. Dembski and M. Ruse. Cambridge: Cambridge University Press, pp. 191–209.

Davies, P. C. W. 2005. The universe—what's the point. In *Spiritual Information*, ed. C. L. Harper Jr. Philadelphia, PA: Templeton Foundation Press.

Deacon, T. W. 2003. The hierarchic logic of emergence: untangling the interdependence of evolution and self-organization. In *Evolution and Learning: The Baldwin Effect Reconsidered*, ed. B. H. Weber and D. J. Depew. Cambridge, MA: MIT Press, pp. 273–308.

Dembski, W. A. 1998. *The Design Inference: Eliminating Chance through Small Probabilities*. Cambridge: Cambridge University Press.

Denton, M. J. 2005. Henderson's "fine-tuning argument" time for rediscovery. In *Spiritual Information: 100 Perspectives on Science and Religion*, ed. C. L. Harper Jr. Philadelphia, PA: Templeton Foundation Press.

Depew, D. J. 2008. Consequence etiology and biological teleology in Aristotle and Darwin. *Stud. Hist. Philos. Biol. Biomed. Sci.* 39(4):379–390.

Depew, D. J., and B. H. Weber. 1995. *Darwinism Evolving: Systems Dynamics and the Genealogy of Natural Selection*. Cambridge, MA: MIT Press.

Desmond, A., and J. Moore. 1991. *Darwin*. London: Michael Joseph.

Dewar, R. 2003. Information theory explanation of the fluctuation theorem, maximum entropy production and self-organized criticality in non-equilibrium stationary states. *J. Phys. A: Math. Gen.* 36:631–641.

Eisenberg, D., and K. Kauzmann. 1969. *The Structure and Properties of Water*. New York, NY: Oxford University Press.

Franks, F. 1971. *Water: A Comprehensive Treatment*. New York, NY: Plenum.

Franks, F. 2000. *Water: A Matrix of Life*, 2nd ed. Cambridge: Royal Society of Chemistry.

Fry, I. 1996. On the biological significance and properties of matter: L. J. Henderson's theory of the fitness of the environment. *J. Hist. Biol.* 29:155–196.

Fry, I. 2000. *The Emergence of Life on Earth: A Historical and Scientific Overview.* New Brunswick, NJ: Rutgers University Press.

Fyfe, A. 1997. The reception of William Paley's "natural theology" in the University of Cambridge. *Br. J. Hist. Sci.* 30:321–335.

Fyfe, A. 2002. Publishing and the Classics: Paley's "natural theology" and the nineteenth-century scientific canon. *Stud. Hist. Philos. Sci.* 33:729–751.

Ghiselin, M. T. 1969. *The Triumph of the Darwinian Method.* Chicago, IL: University of Chicago Press.

Gillespie, N. C. 1990. Divine design and the industrial revolution: Williams Paley's abortive reform of natural theology. *Isis* 81:214–229.

Gray, A. 1876. Evolutionary teleology. In *Darwiniana: Essays and Reviews Pertaining to Darwinism.* New York, NY: Appleton, pp. 356–390.

Grene, M., and D. Depew. 2004. *The Philosophy of Biology: An Episodic History.* Cambridge: Cambridge University Press.

Haldane, J. B. S. 1932. God-makers. In *The Inequality of Man and Other Essays,* London: Chatto & Windus. Reprinted in J. B. S. Haldane, 1985. *On Being the Right Size,* ed. J. Maynard Smith. New York, NY: Oxford University Press, pp. 85–100.

Haldane, J. B. S. 1937. Physical science and philosophy. *Nature* 139:1002.

Henderson, L. J. 1913. *The Fitness of the Environment: An Inquiry into the Biological Significance of the Properties of Matter.* New York, NY: Macmillan.

Henderson, L. J. 1917. *The Order of Nature.* Cambridge, MA: Harvard University Press.

Hicks, L. E. 1883. *A Critique of Design Arguments: A Historical Review and Free Examination of the Methods of Reasoning in Natural Theology.* New York, NY: Charles Scribner's & Sons.

Hopkins, F. G. 1913. The dynamic side of biochemistry. Reports of the British Association 1913: 652. Reprinted (1949) in *Hopkins & Biochemistry,* ed. J. Needham. Cambridge: Heffer, pp. 136–159.

Hume, D. 1779 (1994). *Dialogues Concerning Natural Religion.* Reprinted in *The English Philosophers from Bacon to Mill,* ed. E. A. Burtt. New York, NY: Random House Modern Library.

Kauffman, S. A. 1993. *The Origins of Order: Self-Organization and Selection in Evolution.* New York, NY: Oxford University Press.

Kauffman, S. A. 2000. *Investigations.* New York, NY: Oxford University Press.

Kauffman, S. A. 2004. Autonomous agents. In *Science and Ultimate Reality: Quantum Theory, Cosmology and Complexity,* ed. J. D. Barrow, P. C. W. Davies, and C. L. Harper Jr. Philadelphia, PA: Templeton Foundation Press.

Kleidon, A., and R. D. Lorenz. 2005. *Non-equilibrium Thermodynamics and the Production of Entropy: Life, Earth, and Beyond.* Heidelberg: Springer Verlag.

Lennox, J. G. 1994. Darwin was a teleologist. *Biol. Philos.* 8:405–421.

Levere, T. H. 1981. *Poetry Realized in Nature: Samuel Taylor Coleridge and Early Nineteenth-Century Science.* Cambridge: Cambridge University Press.

Moore, J. R. 1979. *The Post-Darwinian Controversies: A Study of the Protestant Struggle to Come to Terms with Darwin in Great Britain and America, 1870–1900.* Cambridge: Cambridge University Press.

Morowitz, H. J. 2002. *The Emergence of Everything: How the World Became Complex.* New York, NY: Oxford University Press.

Needham, J. 1936. *Order and Life.* New Haven, CT: Yale University Press.

Oppy, G. 1996. Hume and the argument for biological design. *Biol. Philos.* 11:519–534.

Ospovat, D. 1978. Perfect adaptation and the teleological explanation: approaches to the problem of the history of life in the mid-nineteenth century. *Stud. Hist. Biol.* 2:33–56.

Ospovat, D. 1981. *The Development of Darwin's Theory: Natural History, Natural Theology, and Natural Selection.* Cambridge: Cambridge University Press.

Pagels, H. R. 1988. *The Dreams of Reason: The Computer and the Rise of the Sciences of Complexity.* New York, NY: Simon and Schuster.

Paley, W. 1802. *Natural Theology, or Evidences of the Existence and Attributes of the Deity Collected from the Appearances of Nature.* London: Fauldner.

Parascandola, J. 1971. Organismic and holistic concepts in the thought of L. J. Henderson. *J. Hist. Biol.* 4: 63–113.

Parascandola, J. 1972. Lawrence Joseph Henderson. In *Dictionary of Scientific Biography VI,* ed. C. C. Gillispie. New York, NY: Scribner's Sons, pp. 260–262.

Polkinghorne, J. 2004. The inbuilt potentiality of creation. In *Debating Design: From Darwin to DNA*, ed. W. A. Dembski and M. Ruse. Cambridge: Cambridge University Press, pp. 246–260.

Popper, K. 1990. *A World of Propensities*. Bristol: Thoemmes.

Powell, B. 1860. On the study of the evidences of Christianity, In *Essays and Reviews*, ed. J. W. Parker. London: Parker & Son, pp. 94–144.

Prebble, J., and B. Weber. 2003. *Wandering in the Gardens of the Mind: Peter Mitchell and the Making of Glynn*. New York, NY: Oxford University Press.

Prigogine, I. 1981. *From Being to Becoming: Time and Complexity in the Physical Sciences*. New York, NY: Freeman.

Prout, W. 1834. *Climate, Meteorology and the Function of Digestion Considered with Reference to Natural Theology*. London: William Pickering.

Ruse, M. 2002. *Darwin and Design*. Cambridge, MA: Harvard University Press.

Schneider, E. D., and D. Sagan. 2005. *Into the Cool: Energy Flow Thermodynamics and Life*. Chicago: University of Chicago Press.

Sharma, V., and A. Annila. 2007. Natural processes—natural selection. *Biophys. Chem.* 127:123–128.

Silberstein, M. 2002. Reduction, emergence and explanation. In *The Blackwell Guide to the Philosophy of Science*, ed. P. Machamer and M. Silberstein. Malden, MA: Blackwell, pp. 80–107.

Sober, E. 2004. The design argument. In *Debating Design: From Darwin to DNA*, ed. W. A. Dembski and M. Ruse. Cambridge: Cambridge University Press, pp. 98–129.

Stewart, I. 2003. The second law of gravitics and the fourth law of thermodynamics. In *From Complexity to Life: On the Emergence of Life and Meaning*. New York, NY: Oxford University Press, pp. 114–150.

Swenson, R. 1989. Emergent attractions and the law of maximum entropy production: foundations to a theory of general evolution. *Syst. Res.* 6:187–197.

Taylor, M. C. 2003. *The Moment of Complexity: Emerging Network Cultures*. Chicago, IL: University of Chicago Press.

Thorpe, M. R. 1924. *Organic Adaptation to Environment*. New Haven, CT: Yale University Press.

Topham, J. R. 1998. Beyond the "common context": the production and reading of the *Bridgewater Treatises*. *Isis* 89:233–262.

Ulanowicz, R. E. 1997. *Ecology the Ascendant Perspective*. New York, NY: Columbia University Press.

Ulanowicz, R. E. 2002. Ecology, a dialog between the quick and the dead. *Emergence* 4(1):34–52.

Ulanowicz, R. E. 2004. Ecosystem dynamics: a natural middle. *Theol. Sci.* 2(2):231–253.

Waldrup, M. M. 1992. *Complexity: The Emerging Science at the Edge of Order and Chaos*. New York, NY: Simon and Schuster.

Wallace, A. R. 1903. *Man's Place in the Universe: A Study of the Results of Scientific Research in Relation to the Unity or Plurality of Worlds*. London: Chapman and Hall.

Weatherall, M., and H. Kamminga. 1992. *Dynamic Science: Biochemistry in Cambridge 1898–1949*. Cambridge: Wellcome Unit for the History of Medicine.

Weber, B. H. 2007a. Fact, phenomenon, and theory in the Darwinian research tradition. *Biol. Theory* 2:168–178.

Weber, B. H. 2007b. Emergence of life. *Zygon* 42:837–856.

Weber, B. H. in press. Design and its discontents. *Synthese*.

Weber, B. H., and D. J. Depew. 1996. Natural selection and self-organization: dynamical models as clues to a new evolutionary synthesis. *Biol. Philos.* 11:33–65.

Weber, B. H., and D. J. Depew. 2004. Darwinism, design and complex systems dynamics. In *Debating Design: From Darwin to DNA*, ed. W. A. Dembski and M. Ruse. Cambridge: Cambridge University Press, pp. 173–190.

Whewell, W. 1833. *Astronomy and General Physics Considered in Reference to Natural Theology*. London: William Pickering.

Wicken, J. S. 1987. *Evolution, Information and Thermodynamics: Extending the Darwinian Program*. New York, NY: Oxford University Press.

Williams, R. J. P., and J. R. R. Fraústo da Silva. 1996. *The Natural Selection of the Elements*. Oxford: Oxford University Press.

Williams, R. J. P., and J. R. R. Fraústo da Silva. 1999. *Bringing Chemistry to Life: From Matter to Man*. Oxford: Oxford University Press.

Williams, R. J. P., and J. R. R. Fraústo da Silva. 2003. Evolution was chemically constrained. *J. Theor. Biol.* 220:323–343.

Williams, R. J. P., and J. R. R. Frausto da Silva. 2006a. Evolution revisited by inorganic chemists. In *Fitness of the Cosmos for Life: Biochemistry and Fine-Tuning*, ed. J. D. Barrow, S. Conway Morris, S. J. Freeland, and C. L. Harper Jr. Cambridge: Cambridge University Press.

Williams, R. J. P., and J. R. R. Frausto da Silva. 2006b. *The Chemistry of Evolution*. Amsterdam: Elsevier.

Woodruff, L. L. 1924. The protozoa and the problem of adaptation. In *Organic Adaptation to Environment*, ed. M. R. Thorpe. New Haven, CT: Yale University Press, pp. 45–66.

22 Water
A Navigable Channel from Science to God?

Alister E. McGrath

CONTENTS

22.1 INTRODUCTION

December 1970 seems a long time ago. I was then a schoolboy at the Methodist College in Belfast, Northern Ireland. I had developed a deep love for chemistry and was entranced by the spectacular intellectual rigor of the discipline. An inherent fascination for the deeper questions that it raised thrilled my young mind. It seemed clear to me then that I wanted to spend the rest of my life pursuing research in chemistry. The best place to study the subject in depth was the University of Oxford. Oxford's reputation in this field back in the late 1960s was outstanding. It seemed as if my dream might soon be realized—I had been called to Oxford for an interview. I was both excited and frightened.

Oxford is a collegiate university. I had chosen to apply to Wadham College. In part, this was because of the very high reputation of one of its chemists: R. J. P. Williams, alongside his gifted colleague C. S. G. Phillips, had published a wonderful text on inorganic chemistry [1]. I found it to be sheer intellectual delight. It was also very hard going. For that reason, I had made a mental note to try to avoid any questions about inorganic chemistry during the interview and instead concentrate on what was for me the relative safety of organic chemistry.

I arrived at Oxford the evening before the interview in the middle of a power outage and had to be shown to my room by torchlight up four flights of creaking wooden stairs. Having not brought candles with me, I was unable to do any serious preparation for the interview. The next morning, I arrived at the appointed time. I had managed to grab a few hours' study of a standard text—Fieser and Fieser on organic reaction mechanisms—hoping to be quizzed on these. I soon found myself confronted by three Oxford dons. The one in charge introduced himself as Jeremy Knowles, one of Wadham College's tutors in chemistry and later to become Dean of Arts and Sciences at Harvard, a post from which he retired recently. He explained that he taught organic chemistry. He looked at me with the earnestness of a gun dog, ready for the kill.

"So what aspects of chemistry interest you?" he asked. Discombobulated and impulsive from nervous energy, and fearful of Knowles' penetrating stare, I blurted out, "inorganic chemistry."

My plan was now totally inverted. Knowles turned to the man sitting beside me. "Well, I'm sure Bob Williams would like to ask you some questions," he said.

The next 20 minutes offered many lessons in humility. The experience seemed like being mentally undressed in public and then left standing naked. I discovered many things, not least that Bob Williams was not at all impressed at having his own ideas quoted back to him. I discovered new depths of ignorance about the Jahn–Teller effect. We ended in a heated discussion on why certain inorganic ions behaved so strangely in aqueous solution. We considered how these properties seemed to have such momentous implications for water-based life. At the end of the interview, I felt utterly defeated intellectually. Yet at the same time, I was exhilarated at having had such an interesting conversation. It just seemed a pity that so much depended on it.

Shortly thereafter, a handwritten envelope arrived at my home in Northern Ireland. Glancing at it, I realized that it bore an Oxford postmark. I tore the envelope open. It was from Jeremy Knowles. He told me that Oxford would be offering me a major scholarship to Oxford in chemistry. My delight knew no bounds.

As I studied at Oxford, I began to wonder what topic to research in due course. I often found myself drawn to the possibility of working on the inorganic chemistry of biological systems. As it happened, I ended up developing new physical methods of exploring biological systems [2–4] under Professor Sir George Radda, who later went on to become chief executive of the Medical Research Council.

But by that point, I had decided that I wanted to explore the relations between science and religion and that I would not pursue scientific research beyond the doctoral level. I realized that I should immerse myself in the study of Christian theology if I were to make a significant contribution within the arena of the science–religion interaction.

Yet Bob Williams' emphasis on the incredibly complex nature of biological chemistry, and the critical role of water, remained with me. It was a topic that I was able to explore with him personally on a number of occasions. What would have happened if its physical properties had been minutely different? What were its implications for catalysts in biological systems? For organic redox reactions? For the folding of proteins? For the distinctive shape of DNA? And so on. Williams himself set these ideas out in a number of works, including the 1996 book *The Natural Selection of the Chemical Elements* and the 1999 volume *Bringing Chemistry to Life*, both cowritten with J. J. R. Fraústo da Silva [5, 6].* To put it crudely, the emergence of life on Earth seemed exceptionally dependent on certain particular, and perhaps peculiar, physical, and chemical properties of water— such as the critical role of hydrogen bonding in the appearance of liquid water and its role in determining the folding of proteins and polynucleotides. Bob was always willing to ask the big questions while being aware of his own personal limits and those of his discipline.

I met Bob again 25 years after I had begun to study chemistry at a college reunion for the chemists who had studied at Wadham over a period of nearly 40 years. It was a remarkable event, partly because it turned out that hardly any of my fellow students had remained in chemistry—most were now working in the financial or service sectors. Bob spoke about his reflections on his life's work. "I'm not a religious man," he told me. "But there's something funny going on here. Why is there anything there at all? And why can we make so much sense of it?"

It is a point that has often been made, especially by those active in research. This point is made more and more by those who realize the limitations of earlier reductionist approaches, in which the properties of biological systems were often held to be predetermined by physical parameters [7–12]. Yet the growing awareness of the phenomenon of emergence has raised some fascinating questions, even if the answers to many of them still seem a considerable way off. Might a new awareness of the phenomenon of emergence open up some important debates ranging far beyond the sciences—

---

* R. J. P. Williams and J. J. R. Fraústo da Silva contributed the chapter "Evolution revisited by inorganic chemists" to the predecessor to this book, *Fitness of the Cosmos for Life* (Cambridge University Press, 2007).

including fundamentally religious questions, which are often regarded as lying beyond the pale of the orthodox scientific method?

This brings me to the question I wish to focus on in this chapter—whether, how, and to what extent we can develop pathways from science to God [13–18]. Or, to put it another way, in what way does the natural world itself lead us to suppose that there may be something God-like beyond it—or within it? What is there about the world that puzzles us—and by doing so, alerts us to the possibility of new ways of looking at things and seeing them in a new light?

22.2 NATURAL THEOLOGY

By one of those pleasing accidents of history, the "Water of Life" symposium on which this book is based was held on almost exactly the 200th anniversary of the death of William Paley (1743–1805), widely regarded as one of the most influential writers on science and religion. Paley's reputation rests chiefly on an image. It was not especially original; yet it proved, in Paley's hands, to have considerable popular appeal. The image is that of the so-called divine watchmaker [19–22].

As I propose to be critical of Paley, it is important that I first acknowledge some of his achievements. Paley lived in an age of educated gentlemen, not professional scientists. His knowledge of both science and theology is derivative, based on what might reasonably be said to be the best publications of his day. His achievement was to rescue, although as events proved only temporarily, the idea of natural theology—the idea that the scientific exploration and interpretation of nature itself could be used as an argument for the existence of God.

The American philosopher of religion William P. Alston offers a definition of natural theology as "the enterprise of providing support for religious beliefs by starting from premises that neither are nor presuppose any religious beliefs" [23]. This is a convenient starting point for our reflections in this essay. Such an approach to natural theology as a means of finding one's way to religious belief through reflection on the natural order began to blossom in England during the seventeenth century. This was partly in response to political and intellectual developments that had created unease about, and occasionally suspicion of, traditional Christian approaches to revelation.

Several factors appear to have shaped this new interest in "natural theology" (often referred to at the time as "physical theology," from the Greek *physis* for "nature") and "natural religion" at this time in England [24]. We may note three:

1. *The rise of biblical criticism.* This called into question the reliability or intelligibility of Scripture and thereby motivated in a compensatory manner a new interest in the revelatory capacities of the natural world.
2. *A growing distrust of ecclesiastical authority.* This distrust led some to explore sources of knowledge that were seen to be independent of ecclesiastical control, such as an appeal to reason or to the natural order.
3. *A new mentality of dislike for organized religion and for Christian doctrines.* This caused many to seek a simpler "religion of nature," in which nature was valued as a source of revelation.

The fundamental assumption underlying the turn toward natural theology is that nature can be "read" in such a way as to disclose the existence and, within limits, the nature of God without the need for any specifically theological or religious traditions, authorities, or dogmatic assumptions. To use a textual metaphor, the "book of nature" can be read on its own, without the need for theistic presuppositions. The "two books" tradition tended to minimize the need for interpretation, generally regarding the natural world as publicly accessible and not requiring any hermeneutical devices other than human reason [25]. Nature is thus to be regarded as a "universal and public manuscript" [26] capable of being interpreted and appreciated on the basis of assumptions that were not specific to the Christian tradition, but that were rather part of the common intellectual and cultural fabric of western civilization.

If one may speak of a "golden age" of natural theology, this may reasonably be argued to occupy a period of a century and a half, beginning in the late seventeenth century and ending in the first half of the nineteenth [27]. It is not difficult to understand why. The rise of the Newtonian worldview gave natural theology a new lease on life, as the celebrated "Boyle lectures" make clear. Shortly before his death in 1691, Robert Boyle—unquestionably one of England's greatest scientists of that age [28–30]—added a codicil to his will, by which he bequeathed a sum of money that was to endow a series of lectures to be devoted to "proving the Christian Religion against notorious Infidels." Boyle's understanding of "Infidels" was comprehensive and not a little mystifying, including "Atheists, Theists, Pagans, Jews and Mahometans." Four trustees (John Evelyn, Thomas Tenison, Sir Henry Ashurst, and Sir John Rotherham) were named to administer the fund and appoint the lecturers. The lectures rapidly became the bulwark of the Church of England's campaign against the growing rise of skepticism within society at large. Boyle himself seemed to see natural theology as the outcome, not the foundation, of his faith. Yet he was not unaware of the apologetic implications of such a natural theology and its relevance to the worsening intellectual and cultural reputation of the Church of England at that time [31].

In the year 2000, the first 21 Boyle lectures were republished in a facsimile edition. The lectures, delivered over the period 1692–1732, are widely regarded as the most significant public demonstration of the "reasonableness" of Christianity in the early modern period, characterized by that era's growing emphasis on rationalism and increasing suspicion of ecclesiastical authority. These sermons occupy 1500 pages, spread over four volumes. They are an absolute delight to a historical theologian like me. They provide a snapshot of a lost and bygone era, when it was still possible to offer a publicly persuasive "confutation of atheism"—the title of the first series of Boyle Lectures, delivered in 1692 by Newton's spokesman Richard Bentley, that inaugurated the golden years of natural theology.

Yet the Boyle lectures signally failed to accomplish the goal for which they had been established. As the eighteenth century unfolded, the "reasonable" Christianity of the Boyle lecturers came to be seen as increasingly simplistic and vulnerable to criticism from all sides—scientific, philosophical, and theological. Their abandonment was inevitable. By about 1750, it was obvious that they had had their day and had come to be seen as a liability rather than as an asset for the apologetic task of the church. Far from persuading their audiences of the intellectual robustness of the Christian faith, they functioned directly to sow the seeds of doubt.

The rise of the Newtonian worldview gave natural theology a new lease on life, as the original Boyle lectures make clear. This led to a new interest in a specific form of natural theology that went far beyond anything known to the first 16 centuries of Christian theological reflections. Yet this renewal proved to be temporary rather than permanent. Initially, the apologetic prospects for this approach seemed extremely positive. However, by the middle of the eighteenth century, it was increasingly recognized as a dead end. Why this transition?

First, natural theology seemed to lead to a generic belief in God—that is, to Deism, rather than to orthodox Christianity. Matthew Tindal's *Christianity as Old as Creation* (1730) [32], for example, argued that Christianity was nothing other than the "republication of the religion of nature." God is understood as the extension of accepted human ideas of justice, rationality, and wisdom. This apologetic approach does not lead *away* from Christianity; nevertheless, it is certainly not well disposed toward Christian specifics [33]. Alarmingly, at least to the defenders of traditional Christian orthodoxy, some of the most influential Boyle lecturers were Arians, committed to a thoroughly rationalist understanding of Christ [34]. The common sense underlying the "natural theology" developed by William Whiston and Samuel Clark extended to their Christology. To the orthodox, a second transition therefore seemed necessary: from the deistic God posited by natural theology to the rather more specific God revealed and proposed by Christianity itself.

Second, natural theology tended to erode the conceptual space traditionally occupied by God. The amalgam of Newtonian natural philosophy and certain forms of Anglican theology proved popular and plausible in postrevolutionary England. Nevertheless, it was an unstable amalgam. It was not long before the "estrangement of celestial mechanics and religion" began to set in [35]. Celestial mechanics seemed to many to suggest that the world was a fully self-sustaining mechanism. It seemed to have no need for direct intervention, divine governance, or sustenance for its day-to-day operation.

This had been recognized at an early stage by one of Newton's interpreters, Samuel Clarke—himself a Boyle lecturer. In his correspondence with Gottfried Leibniz, Clarke expressed concern over the potential implications of the growing emphasis on the regularity of nature. It seemed to him that the idea of the world as a great machine merely eliminated conceptual space for God. The image of God as a "clockmaker" (and the associated natural theology that appealed to the regularity of the world) was thus seen as potentially leading to a purely naturalistic understanding of the universe, in which God had no continuing role to play. By the end of the eighteenth century, it seemed to many that natural theologies based on the Newtonian worldview probably led to atheism, heresy, or agnosticism rather than to Christian faith. The deathblow perhaps came a few decades later, when the poet Percy Bysshe Shelley famously remarked that "the consistent Newtonian is necessarily an atheist."

22.3 PHYSICAL THEOLOGY AND THE DIVINE WATCHMAKER

Yet in the eighteenth century, natural theology underwent a new development. In the enthusiastic hands of William Paley, arguments once deployed in relation to the physical world were now given a new lease on life by being transposed to the biological level. Paley's *Natural Theology or Evidences of the Existence and Attributes of the Deity, Collected from the Appearances of Nature* (1802) [36]—had a profound influence on popular English religious thought in the first half of the nineteenth century. It is known to have been read with great interest by Charles Darwin [22].

The opening paragraphs of Paley's *Natural Theology* set out the analogy for which Paley became famous [37]:

> In crossing a heath, suppose I pitched my foot against a *stone*, and were asked how the stone came to be there. I might possibly answer, that for any thing I knew to the contrary it had lain there for ever; nor would it, perhaps, be very easy to show the absurdity of this answer. But suppose I had found a *watch* upon the ground, and it should be inquired how the watch happened to be in that place. I should hardly think of the answer which I had before given, that for any thing I knew the watch might have always been there. Yet why should this answer not serve for the watch as well as for the stone; why is it not admissible in the second case as in the first?

Paley then offers a detailed description of the watch, noting in particular its container, coiled cylindrical spring, many interlocking wheels, and glass face. Having carried his readers along with this careful analysis, Paley turns to draw his critically important conclusion:

> This mechanism being observed—it requires indeed an examination of the instrument, and perhaps some previous knowledge of the subject, to perceive and understand it; but being once, as we have said, observed and understood, the inference we think is inevitable, that the watch must have had a maker—that there must have existed, at some time and at some place or other, an artificer or artificers who formed it for the purpose which we find it actually to answer, who comprehended its construction and designed its use.

The analogy, like most of Paley's work, was borrowed, and the scholarship decidedly second rate. Paley had ruthlessly plagiarized John Ray's writings in his quest to develop a new natural theology. Although a derivative thinker, Paley was still an excellent communicator. What he communicated so effectively, however, was an outmoded way of thinking. Nature, Paley argued, shows signs of "contrivance"—that is, purposeful design and fabrication. More particularly, nature bears witness to a series of biological structures that, he argues, must be considered to have been constructed with a clear purpose in mind. "Every indication of contrivance, every manifestation of design, which existed in the watch, exists in the works of nature."

Now we need to be clear that this famous image is only one element—although by far the best known—in Paley's cumulative argument for the plausibility of a theist worldview. Although his *Natural Theology* is almost entirely devoted to exploration of the biological world, we also find a brief, and somewhat unconvincing, appeal to astronomy. Alongside this, we find Paley setting out another line of argument—namely, that the existence and shape of the laws of nature points to a lawmaker or lawgiver. Naturally, Paley interprets this in a theistic direction. It is not his preferred mode of argument. Yet, interestingly, it is this line of thought that may prove to be of more lasting significance. Why is there something rather than nothing? Why is nature something profoundly and sublimely ordered in a rational manner described by mathematically complex elegant formulations? Consequently, some commentators have argued that *The Anthropic Cosmological Principle* by John D. Barrow and Frank J. Tipler [38] is to be seen as the twentieth century's alternative to William Paley's classic nineteenth century text [39].

Yet Paley's vision of the specific manner in which God's creative activity was to be conceptualized caused far more difficulties than it solved. Before Darwin's new theory made its appearance, a growing body of informed theological opinion urged the abandoning of Paley's ideas—or at least their significant modification. In 1852, John Henry Newman was invited to give a series of lectures in Dublin on "the idea of a university," which allowed him to explore the relationship between Christianity and the sciences, and especially the "physical theology" of William Paley [40]. Newman was scathing about Paley's approach. He lambasted it as "a false gospel." Far from being an advance on the more modest apologetic approaches adopted by the early church, he argued, it represented a degradation of those views.

The nub of Newman's criticism of Paley's natural theology can be summarized in a single sentence: "[I]t has been taken out of its place, has been put too prominently forward, and thereby has almost been used as an instrument against Christianity." In Newman's view, Paley's "physical theology" was a liability that should be abandoned before it discredited Christianity:

> Physical Theology cannot, from the nature of the case, tell us one word about Christianity proper; it cannot be Christian, in any true sense, at all. . . . Nay, more than this; I do not hesitate to say that, taking men as they are, this so-called science tends, if it occupies the mind, to dispose it against Christianity.

Seven years before Darwin had subverted Paley's approach on scientific grounds through his theory of natural selection, Newman—widely regarded as the most important English theologian of the nineteenth century—had repudiated Paley as an outdated theological liability.

Newman's critique does not arise from an awareness of a new crisis of faith about to be precipitated by the publication of Darwin's work in 1859. His views rest solely on his belief that Paley's approach fails in what it sought to deliver, trapping Christian theology in an apologetic that can only go wrong. It was not the first time Christian apologetics had taken a disastrous wrong turn. An immediate correction was, in Newman's view, long overdue.

Paley's vision for natural theology was simply that of a rational exercise of sense-making. Where was there to be found any sense of transcendence, mystery, glory, or love in such a scientifically derivative notion of God? Paley's image might appeal to the banalities of rationalism. But what about the imagination? Or the emotions? A concept of God that failed to excite the human imagination—for example, by moving us to worship or prayer—is seriously deficient. For Newman, Paley tended to proclaim a cold, distant, mechanical God—a lawgiver rather than a savior—contrary to the heart of the Christian view. For Newman, any authentic vision of the Christian God arrested people in their tracks. God, if truly known, compelled a response of worship, adoration, and existential transformation. Newman believed that Paley left people with little more than a vague sense of intellectual satisfaction, perhaps to be compared with that experienced after the successful completion of a crossword puzzle.

Here we observe a theological objection registered well before Darwin's theory became public, and raised by perhaps England's greatest theologian of the nineteenth century. Actually, most theologians have generally tended to follow Newman in expressing similar concerns. Yet such theological critiques were totally eclipsed in the latter half of the nineteenth century, as Darwin's views on natural selection became increasingly accepted within Western intellectual culture. Their implications for Paley's approach could hardly be more obvious. Paley's argument depended on a static worldview. It simply could not cope with the dynamic worldview underlying Darwin's biological insights.

Today, Paley's most famous critic is, of course, the renowned biologist Richard Dawkins. In his *Blind Watchmaker*, Dawkins relentlessly points out the failings of Paley's viewpoint and the explanatory superiority of Darwin's approach—especially as it has been modified through the neo-Darwinian synthesis [41]. Dawkins argues that Paley's static view of the world is rendered obsolete by science. Dawkins himself is eloquent and generous in his account of Paley's achievement, noting with appreciation his "beautiful and reverent descriptions of the dissected machinery of life." Without in any way belittling the wonder of the mechanical "watches" that so fascinated and impressed Paley, Dawkins argues that Paley's case for God—although made with "passionate sincerity" and "informed by the best biological scholarship of his day"—is "gloriously and utterly wrong." The "only watchmaker in nature is the blind forces of physics." Paley is typical of his age; his ideas are entirely understandable, given his historical location before Darwin. But nobody, Dawkins argues, could share these ideas now. Paley is obsolete. And although Paley retains supporters today within some of the more conservative and scientifically illiterate sections of American Protestantism, that seems to be the consensus of the day.

Yet the question of whether reflection on the natural world points to belief in a god remains immensely important, especially in light of the continuing influence of what I am going to loosely refer to as the culture and ideology of "scientific atheism" [42]. Richard Dawkins is an excellent example of this "scientistic" worldview, one that has many disciples. I myself used to be one of them. When I was growing up in Belfast during the 1960s, I came to the view that God was an infantile illusion, suitable for the sentimentalist elderly, the intellectually feeble, and the fraudulently religious. With hindsight, I admit that this was a rather arrogant view, and one that I now find somewhat embarrassing. "Scientific atheism" is hardly more sophisticated intellectually than young creationism. My excuse for this intellectual haughtiness then is little more than that a lot of other people felt the same way. It was the perceived wisdom of the day that religion was a reactionary phenomenon that was on its way out. A glorious progressive godless dawn was just around the corner.

Northern Ireland was then infamous for its notorious tensions between Protestants and Catholics. A joke captures this well. An Englishman goes to visit some friends in Belfast. Late at night, he finds himself cornered by a street gang. Menacingly, the leader asks: "Are you a Protestant or a Catholic?" Aware that his answer might have significant implications, he replies: "I'm an atheist." After a brief pause came the second question: "Are you a *Protestant* atheist or a *Catholic* atheist?"

22.4   THE SECOND EDUCATION OF A CHEMIST

Now I would say that God is the oxygen of my existence, giving a sense of direction to my life, and is a highly productive explanatory framework by which I can make sense of the world both intellectually and existentially. Part of the reasoning that led me to this conclusion was based on the natural sciences. I had specialized in mathematics and science during high school in preparation for going to Oxford to study chemistry. Although my primary motivation for studying the sciences was the fascinating insights into the wonderful world of nature they allowed, I also found my links with the sciences to be a convenient ally in my critique of religion. Atheism and the natural sciences seemed to me at the time to be coupled together by the most rigorous of intellectual bonds. And there things rested until I arrived at Oxford in October 1971.

Chemistry proved to be intellectually exhilarating. At times, I found myself overwhelmed with an incandescent enthusiasm as more and more of the complexities of the natural world seemed to fall into place. I chose to specialize in quantum theory and found this to be equally rewarding. Although fascinated by the quantum universe, I found myself increasingly drawn to the biological world, intrigued by the complex chemical patterns of natural organisms. Yet, alongside this growing delight in the natural sciences, which exceeded anything I could have hoped for, I found myself rethinking my atheism. It is not easy for anyone to subject their core beliefs to criticism. My reason for doing so was the growing realization that things were not quite as straightforward as I had once thought [43]. A number of factors had converged to bring about what I suppose I could reasonably describe as a crisis of faith.

Atheism, I began to realize, rested on a less-than-satisfactory evidential basis. The arguments that had once seemed bold, decisive, and conclusive increasingly appeared to be circular, tentative, and uncertain. The opportunity to talk to Christians about their faith revealed to me that I understood relatively little about Christianity, which I had come to know chiefly through the biased and often inaccurate descriptions of leading critics such as Bertrand Russell and Karl Marx. Perhaps more important was that I began to realize that my assumption of the automatic and inexorable link between the natural sciences and atheism was naive and uninformed. One of the most important things I had to sort out after my conversion to Christianity was the systematic uncoupling of this bond. I discovered that it was possible to see the natural sciences from a Christian perspective in very interesting and deeply illuminating ways having nothing to do with the typical relatively shallow debates involving the defeat of naive creationism by evolution.

I also set out to try to understand why others did not share this perspective. In 1977, I read Dawkins' *The Selfish Gene*, which had appeared the previous year [44]. It was a fascinating book, brimming with ideas and showing a superb ability to put difficult concepts into words. I devoured it. And I longed to read more of his work. However, as his subsequent works appeared, I became increasingly puzzled by what I considered to be a rather superficial atheism, one not adequately grounded in serious scientific arguments. Atheism seemed to be tacked on with intellectual Velcro, rather than demanded by the scientific evidence Dawkins assembled. Although a brilliant scientific popularizer, I believed him also to be an aggressive propagandist.

There is no doubt that Dawkins' lucid and aggressive atheism—evident especially in his recent book *A Devil's Chaplain*—has done much to shape public perceptions of the credibility of Christian faith [45]. Believing in God, Dawkins argues, is like believing in Santa Claus or the Tooth Fairy. It cannot be sustained when we grow up and learn the realities of the scientific method. And by and large, popular culture seems to have accepted this glib diagnosis. But is there really a substantive scientific logic underlying this critique? While we rightly criticize Paley for what we now tend to see as a misguided natural theology argument, there can be no doubt of his effectiveness in shaping broad cultural perceptions in his own time. Indeed, as a historian I might suggest that the perception that Darwinism undermined faith may have been shaped to no small extent by Paley's success as an apologist, which unwittingly (and, I think, unwisely) linked the public credibility of faith with a very specific understanding of the notion of the relation between God and the biological world of nature, involving the position known as "special creation."

Yet one may ask whether Paley's approach should be dismissed entirely or whether it could be revised with a different structure of argument, such as, for example, the idea of "general creation" through the laws of nature, which underlies Darwin's own idea of evolution. Clearly, we cannot use Paley's argument in its original form. So what may be necessary for this approach to have a new lease on life? And in what ways might it potentially tie in with the theme of this book?

In the remainder of this chapter, I will explore questions that I hope will simply open up a wider discussion.

22.5   THE DILEMMA OF EXISTENCE

What kind of explanation of the world does religion offer us? Paley's broad argument is that it is impossible to offer a compelling explanation of the way the world is without proposing a divine creator. Today, the force of that argument, on the basis of the evidence that he offered and believed to be conclusive is much reduced—some would say to vanishing point. Others hold out the possibility of restatement, while being aware of the difficulties that it must face. Religious apologists have tended to adopt positions that can be broadly grouped under two headings: (1) the "God of the gaps" approach and (2) the "big picture" approach.

The first, which one tends to find in more popular writings, including the movement widely known as "Intelligent Design," argues that science is unable to offer a complete account of the world. There are significant gaps in our understanding. These explanatory deficits, it is argued, can and should be remedied by an appeal to God. Thus God "fills" the gaps where science is presumed to lack explanatory reach. Now I must confess that I am so persuaded of the combined theological and scientific deficiencies of this approach that I would not myself defend it in any form whatsoever. Of course, the inexorable advance of the scientific enterprise is such that gaps tend to get filled, thus eliminating this putative gap-filling God. The core assumptions of this flawed approach inevitably state that God is little more than a fiction of ignorance.

Paley himself experiences considerable difficulties at this point, even though some of his critics pointed out that his approach could be salvaged. For example, James Moore has shown in his massive and definitive account of Christian responses to Darwin that many believed that the obvious deficiencies in Paley's account of biological life—most notably, the notion of "perfect adaptation"—were actually *corrected* by Darwin's notion of natural selection [46]. More important is that a series of writers discarded Paley's interest in specific "adaptations" (to use a Darwinian term unknown to him) and preferred to focus on the fact that evolution appeared to be governed by certain quite definite laws. This proposed modification simply represents the incorporation of the emerging new biology within the general approach developed in the Middle Ages by Thomas Aquinas.

There is a second, quite different alternative that I believe to be much more resilient and interesting. This notion builds on a point that we find in many twentieth century writers, such as Albert Einstein and Ludwig Wittgenstein. The argument here is that the intelligibility of the universe itself requires explanation. It is therefore not the gaps in our understanding of the world that point to God, but rather the very comprehensiveness observed in the success of scientific and other forms of understanding that requires an explanation. In brief, the argument is that *explicability itself requires explanation.*

In my view, this approach is much preferred. It avoids the obvious problem of historical erosion: What apparently cannot be explained today can be explained tomorrow. However, my reasons for preferring this option are not ultimately pragmatic in nature. My interest is rooted in the notion that belief in God is possessed of explanatory vitality. "I believe in Christianity," wrote C. S. Lewis, "as I believe that the Sun has risen, not only because I see it, but because by it I see everything else" [47]. In concluding his essay "Is theology poetry?" with these words, Lewis was highlighting one of the many difficulties associated with extending a scientific worldview into a full-blown philosophy of life—that the philosophical presumption of the scientific enterprise was, in effect, obliged to presuppose its conclusions. The ordering of the world demands to be explained, certainly; it is also a fundamental assumption of the scientific method itself. For Lewis, the Christian faith offered illumination of the world that permitted it to be seen in a certain way—and by being seen in this way, to open up ways of exploring and examining it that resonated with reality.

As has often been observed, there seems to be something about human nature that prompts us to ask questions about the world. And there seems to be something about the world that allows answers to those questions to be given. This seemingly trivial observation is actually of considerable importance. It lay behind R. J. P. Williams' comments, noted earlier, that proved very formative to my own thinking. The former theoretical physicist John Polkinghorne is one of many to make this important point, and he does so with great clarity [48]:

> We are so familiar with the fact that we can understand the world that most of the time we take it for granted. It is what makes science possible. Yet it could have been otherwise. The universe might have been a disorderly chaos rather than an orderly cosmos. Or it might have had a rationality which was inaccessible to us. . . . There is a congruence between our minds and the universe, between the rationality experienced within and the rationality observed without.

There is a deep-seated congruence between the rationality present in our minds and the rationality—the *orderedness*—that are found to be present in the world when we apply the methods and aims of science to nature. But why? Why is it that the abstract structures of pure mathematics—which are meant to be a free creation of the human mind—provide such important clues to understanding the world?

Yet our interest extends beyond the intelligibility of the world and its possible theological implications. The more we understand of the structures of the world, the more we find ourselves raising legitimate "biocentric" or "anthropocentric" questions about why life exists at all. Certain aspects of the chemistry of water cause us to reflect on questions concerning the emergence of life on Earth. For example, life on Earth is water-based. But what if the physical chemistry of water were very slightly different? What if it did not expand on freezing? Or if the van der Waals forces were weaker or stronger than those observed? Would life then simply exist under a different adaptedness? Or would life not be possible at all?

Many of my colleagues are familiar with Lawrence J. Henderson, professor of biological chemistry at Harvard University, who raised the question of whether the nature of the physical world could be described as "biocentric." Henderson's "biocentricity" hypothesis appeared in his book, *The Fitness of the Environment: An Inquiry into the Biological Significance of the Properties of Matter* [49]. For Henderson, the whole process of evolution appeared to possess a "biocentric" character. "The whole evolutionary process," he wrote, "both cosmic and organic, is one, and the biologist may now rightly regard the universe in its very essence as biocentric." Although Henderson used the term "teleology" to designate this aspect of the environment, he understood this term to refer to the inherent harmony of nature, and he assiduously avoided any suggestion of "purpose" within the natural process. Henderson's ideas were explored in depth at a Templeton-sponsored workshop at Harvard in 2003.*

Such themes may be scientifically fruitful for advancing our understanding of the deep logic of life—how life arose and evolved, and perhaps most important, how it can "work" at all. But, in terms of theology, do such reflections *prove* anything? I think not. They are suggestive—but hardly conclusive. Thomas Aquinas, probably the greatest scholastic theologian, was quite clear in the thirteenth century that one could not prove the existence of God with certainty by argument (*demonstratio*). What one could do was to identify "reasons" (*rationes*) that suggested this or that provided an explanatory framework for these observations. In modern philosophy of science, this approach has been reframed through Gilbert Harman's notion of "inference to the best explanation" [51]. Harman's approach naturally raised the question of how one might determine which is indeed the "best" explanation—a matter that remains to be resolved, many believe, simply because it cannot be resolved [52]. So rather than wasting time on the essentially futile question of whether these counterfactuals *prove* anything (in the strict sense of that word), we might explore whether they alter the weight of balance in competing explanations of things. Do they tip the balance in favor of the existence of a god? And, if so, which one?

So what of the theme of this book? Might it add further to such suggestions of "fine-tuning" within the biological, chemical, or biochemical domains of relevance to this debate? Or might it

---

* See http://www.templeton.org/archive/biochem-finetuning/. The book based on that symposium, and the predecessor to the current volume, is dedicated to Lawrence J. Henderson [50].

force further refinement and qualification of the controversial, yet immensely interesting, concept of "fine-tuning?" My intention here is simply to raise some questions on the possible implications of the chemistry of water, while marking an important and interesting landmark in the history of the interaction of science and religion in the person of William Paley. The two topics are not unrelated. I have sought to make Paley out neither to be a fool nor some kind of hero. As I read his works, he appears to me to be a rather pious, socially conservative person that enjoyed reading books about science and thinking about the deeper implications of what he read. He sought to make connections across disciplines and, above all, to link the quest for knowledge with the quest for wisdom. And at least in that regard, he has much to say to us today.

At the very least, the conference on which this book is based, and the volume itself, have brought home to us how very remarkable a thing water really is. Yet perhaps we have still to answer the question of whether this observation represents a gateway to the transcendent or is simply a clarification of the natural. Water may not be a theological Panama Canal—a direct and secure link between God and science that can bear heavy intellectual traffic—but it certainly makes us think, ask questions, and reflect on why things are the way they are. And that, I think, is why so many are interested in exploring the relationship between science and religion. The linking of these largely distinct worlds gives us permission to ask important, difficult, and perplexing questions that may prove to be important for science, questions that perhaps otherwise we might never dare ask.

REFERENCES

[1] Williams, R. J. P., and C. S. J. Phillips. 1969. *Inorganic Chemistry*. Oxford: Oxford University Press.
[2] de Kreef, B., A. E. McGrath, C. G. Morgan, et al. 1977. In *Nobel Foundation Symposium: Biological Membranes and Their Models*, ed. S. Abrahamsson and I. Pascher. New York, NY: Plenum Press, pp. 389–407.
[3] McGrath, A. E., C. G. Morgan, and G. K. Radda. 1976. Photobleaching. A novel fluorescence method for diffusion studies in lipid system. *Biochim. Biophys. Acta* 426:173–185.
[4] McGrath, A. E., C. G. Morgan, and G. K. Radda. 1976. Positron lifetimes in phospholipids dispersions. *Biochim. Biophys. Acta* 466:367–372.
[5] Williams, R. J. P., and J. J. R. Fráusto da Silva. 1997. *The Natural Selection of the Chemical Elements*. Oxford: Oxford University Press.
[6] Williams, R. J. P., and J. J. R. Fráusto da Silva. 1999. *Bringing Chemistry to Life*. Oxford: Oxford University Press.
[7] Luisi, P. P. 2002. Emergence in chemistry: chemistry as the embodiment of emergence. *Found. Chem.* 4:183–200.
[8] Klee, R. 1984. Micro-determinism and concepts of emergence. *Philos. Sci.* 51:44–63.
[9] Bedau, M. A. 1997. Weak emergence. In *Philosophical Perspectives: Mind, Causation, and World*, ed. J. Tomberlin, vol. 2. Oxford: Blackwell, pp. 375–399.
[10] Clayton, P. 2004. *Mind and Emergence: From Quantum to Consciousness*. Oxford: Oxford University Press.
[11] Holland, J. H. 2000. *Emergence: From Chaos to Order*. Oxford: Oxford University Press.
[12] Morowitz, H. J. 2002. *The Emergence of Everything: How the World Became Complex*. Oxford: Oxford University Press.
[13] Kretzmann, N. 1997. *The Metaphysics of Theism: Aquinas's Natural Theology in Summa Contra Gentiles I*. Oxford: Clarendon Press.
[14] Kretzmann, N. 1999. *The Metaphysics of Creation: Aquinas's Natural Theology in Summa Contra Gentiles II*. Oxford: Clarendon Press.
[15] Barr, J. 1993. *Biblical Faith and Natural Theology*. Oxford: Clarendon Press.
[16] Brooke, J. H. 1989. Science and the fortunes of natural theology: some historical perspectives. *Zygon* 24:3–22.
[17] Fisch, H. 1953. The scientist as priest: a note on Robert Boyle's natural theology. *Isis* 44:252–265.
[18] Wilkinson, D. A. 1990. The revival of natural theology in contemporary cosmology. *Sci. Christ. Belief* 2:95–115.
[19] Gillespie, N. C. 1990. Divine design and the industrial revolution: William Paley's abortive reform of natural theology. *Isis* 81:214–229.

[20] LeMahieu, D. L. 1976. *The Mind of William Paley: A Philosopher and His Age*. Lincoln, NE: University of Nebraska Press.

[21] Clarke, M. L. 1974. *Paley: Evidences for the Man*. London: SPCK.

[22] Fyfe, A. 1997. The reception of William Paley's "natural theology" in the University of Cambridge. *Br. J. Hist. Sci.* 30:321–335.

[23] Alston, W. P. 1991. *Perceiving God: The Epistemology of Religious Experience*. Ithaca, NY: Cornell University Press.

[24] Westfall, R. S. 1992. The scientific revolution of the seventeenth century: a new world view. In *The Concept of Nature*, ed. J. Torrance. Oxford: Oxford University Press, pp. 63–93.

[25] Howell, J. K. 2002. *God's Two Books: Copernican Cosmology and Biblical Interpretation in Early Modern Science*. Notre Dame, IN: University of Notre Dame Press.

[26] Browne, T. 1962. *Religio Medici*. Boston: Ticknor & Fields, p. 32.

[27] McGrath, A. E. 2006. Towards the restatement and renewal of a natural theology: a dialogue with the classic English tradition. In *The Order of Things: Explorations in Scientific Theology*, ed. A. E. McGrath. Oxford: Blackwell Publishing, pp. 63–96.

[28] Wojcik, J. W. 1997. *Robert Boyle and the Limits of Reason*. Cambridge: Cambridge University Press.

[29] MacIntosh, J. J. 1992. Robert Boyle's epistemology: the interaction between scientific and religious knowledge. *Int. Stud. Philos. Sci.* 6:91–121.

[30] Jacob, J. R. 1972. The ideological origins of Robert Boyle's natural philosophy. *J. Eur. Stud.* 2:1–21.

[31] Dahm, J. J. 1970. Science and apologetics in the early boyle lectures. *Church Hist.* 39:172–186.

[32] Tindal, M. 1730. Christianity as Old as Creation. London.

[33] Sullivan, R. E. 1982. *John Toland and the Deist Controversy: A Study in Adaptations*. Cambridge, MA: Harvard University Press.

[34] Wiles, M. 1996. *Archetypal Heresy: Arianism through the Centuries*. Oxford: Clarendon Press.

[35] Odom, H. H. 1966. The estrangement of celestial mechanics and religion. *J. Hist. Ideas* 27:533–558.

[36] Paley, W. 1802. *Natural Theology or Evidences of the Existence and Attributes of the Deity, Collected from the Appearances of Nature*. London: Charles Knight, pp. 25–26.

[37] Paley, W. 1849. *Works*. London: Wm. Orr & Co, pp. 25–28.

[38] Barrow, J., and F. J. Tipler. *The Anthropic Cosmological Principle*. Oxford: Oxford University Press, 1986.

[39] Craig, W. L. 1988. Barrow and Tipler on the anthropic principle vs. divine design. *Br. J. Philos. Sci.* 38:389–395.

[40] Newman, J. H. 1907. *The Idea of a University*. London: Longmans, Green, & Co, p. 454.

[41] Dawkins, R. 1986. *The Blind Watchmaker: Why the Evidence of Evolution Reveals a Universe Without Design*. New York, NY: W. W. Norton.

[42] McGrath, A. E. 2004. *Dawkin's God: Genes, Memes and the Meaning of Life*. Oxford: Blackwell Publishing.

[43] McGrath, A. E. 2004. *The Twilight of Atheism: The Rise and Fall of Disbelief in the Modern World*. New York, NY: Doubleday.

[44] Dawkins, R. 1989 OR 1976. *The Selfish Gene*. Oxford: Oxford University Press.

[45] Dawkins, R. 2003. *A Devil's Chaplain: Selected Writings*. London: Weidenfield & Nicholson.

[46] Moore, J. R. 1979. *The Post-Darwinian Controversies: A Study of the Protestant Struggle to Come to Terms with Darwin in Great Britain and America, 1870–1900*. Cambridge: Cambridge University Press.

[47] Lewis, C. S. 2000. *C. S. Lewis: Essay Collection*. London: Collins, pp. 1–21.

[48] Polkinghorne, J. 1988. *Science and Creation: The Search for Understanding*. London: SPCK, p. 20.

[49] Fry, I. 1996. *J. Hist. Biol.* 29:155–196. L. J. Henderson. *The Fitness of the Environment: An Inquiry into the Biological Significance of the Properties of Matter*. New York, NY: MacMillan, 1913; repr. Boston, MA: Beacon Press, 1958; Gloucester: Peter Smith, 1970.

[50] Barrow, J. D., S. Conway Morris, S. J. Freeland, and C. L. Harper Jr., Eds. 2007. *Fitness of the Cosmos for Life: Biochemistry and Fine-Tuning*. Cambridge: Cambridge University Press.

[51] Harman, G. 1965. The inference to the best explanation. *Philos. Rev.* 74:88–95.

[52] Lipton, P. 2004. *Inference to the Best Explanation*. London: Routledge.

Index

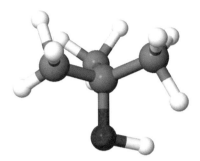

FIGURE 11.2 A tertiary butanol molecule oriented with its polar alcohol tail group pointing downward and its nonpolar head group pointing upward. (With acknowledgment to Dr. Daniel Bowron, ISIS Facility, Rutherford Appleton Laboratory, UK.)

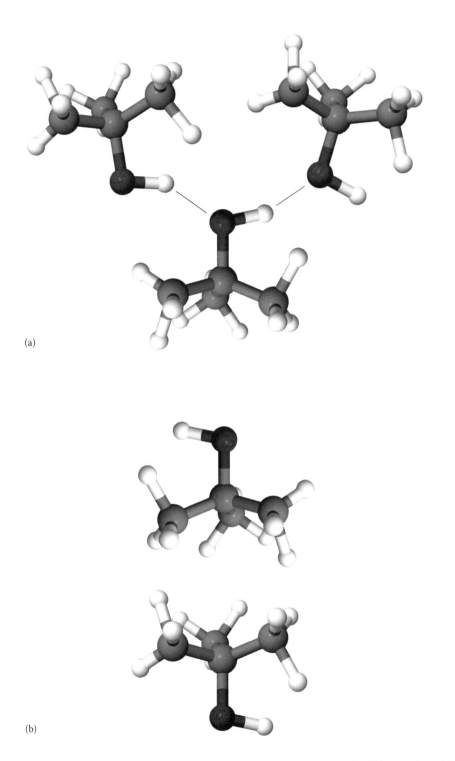

(a)

(b)

FIGURE 11.3 Hypothetical arrangement of (a) three *t*-butanol molecules in the liquid interacting with each other through hydrogen bonding (indicated by the two linking lines) of their alcohol tails and (b) two *t*-butanol molecules in the liquid interacting with each other through nonpolar head group interactions. (With acknowledgment to Dr. Daniel Bowron, ISIS Facility, Rutherford Appleton Laboratory, UK.)

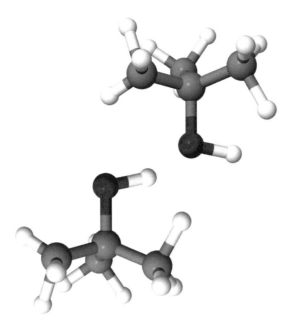

FIGURE 11.4 Schematic of the actual average intermolecular hydrogen bonding situation found in liquid *t*-butanol at room temperature. (With acknowledgment to Dr. Daniel Bowron, ISIS Facility, Rutherford Appleton Laboratory, UK.)

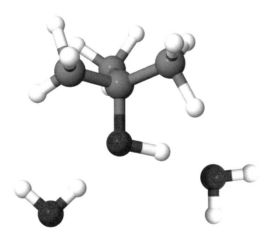

FIGURE 11.5 Schematic picture of the actual average hydrogen bonding situation observed in all but the most concentrated aqueous solutions of *t*-butanol in which there is no hydrogen bonded interaction between solute molecules. The water molecules take up the hydrogen bonding capability of the alcohol's hydroxyl group. (With acknowledgment to Dr. Daniel Bowron, ISIS Facility, Rutherford Appleton Laboratory, UK.)

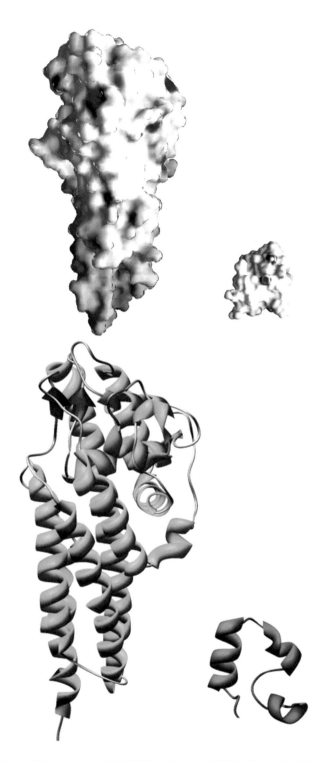

FIGURE 12.1 (a) Space filling model of BBSP341 (left) and VHP36 (right). (b) Chimera ribbon diagram of BBSP341 (left) and VIIP36 (right) showing the α-helices and β-sheets (Pettersen et al., 2004). The arrows show the β-sheets. The structures are from the Protein Data Bank files 1L8W for BBSP341 and 1VII for VHP36.

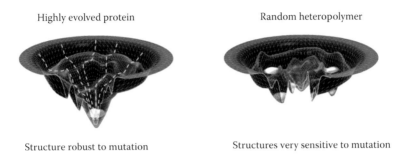

Highly evolved protein

Random heteropolymer

Structure robust to mutation

Structures very sensitive to mutation

FIGURE 14.1 The energy landscape of an evolved protein is funneled (left). Random sequences have rugged surface with structurally disparate minima (right).

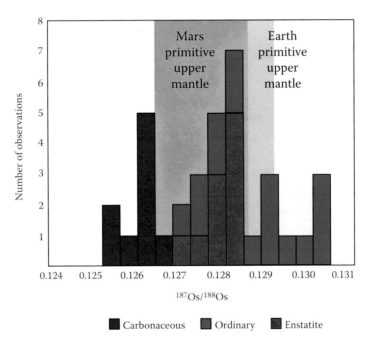

FIGURE 15.4 $_{187}Os/_{188}Os$ ratios in carbonaceous, ordinary, and enstatite chondrites, and in the Earth's primitive upper mantle (PUM), are distinct and are diagnostic of the nature of the Earth's "late veneer." In particular, Earth's PUM is different from water-bearing carbonaceous chondrites. The $_{187}Os/_{188}Os$ ratios in Mars's PUM have been recently revised down (Muralidharan et al., 2008) and, although still uncertain, now overlap with those for Earth's PUM (gray area of the figure). Earlier estimates of the Martian mantle were probably compromised by incomplete dissolution of sample. (A. Brandon and R. Walker, personal communications, March 2008. Modified from Drake, M. J., and Righter, K., *Nature*, 416, 39–44, 2002.)

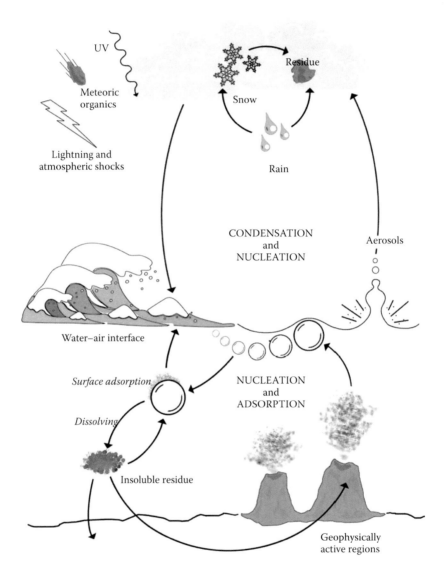

FIGURE 18.1 The bubble–aerosol–droplet cycle. There is every reason to believe that this cycle also existed on the early Earth and possibly Mars.

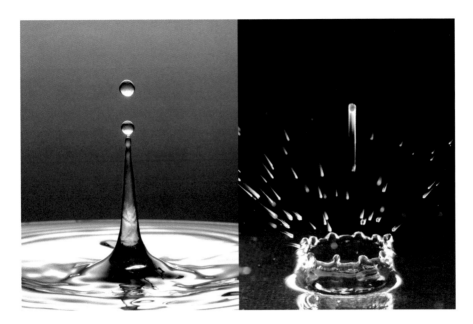

FIGURE 18.2 Formation of "jet drops" and film cap drops from a bursting bubble. Left: For small bubbles (<0.5 mm) surface tension–driven "jet drops" are formed. The material making up successive "jet drops" from a single bursting bubble comes from successively microtomed layers of its air–water interface (MacIntyre, 1974c). Right: For larger bubbles a very large number of "film cap drops" are produced by a gravitationally-induced instability mediated through the bubble cap's draining. (Photographs by Prof. Andrew Davidhazy, 2009. With permission.)

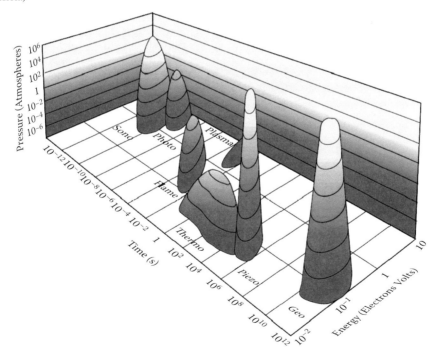

FIGURE 18.5 The unique phase-space of sonochemical energetics. On an early terrestrial-like planet there will be great ranges of acoustic cavitation energies, ranging from ocean waves to near-surface geophysically active regions (submarine volcanoes) to meteoritic impacts on an ocean or sea, and of course accompanying all will be the ubiquitous bubble. (From Suslick, K. S., *Sci. Am.* 260:80–86, 1989. With permission.)

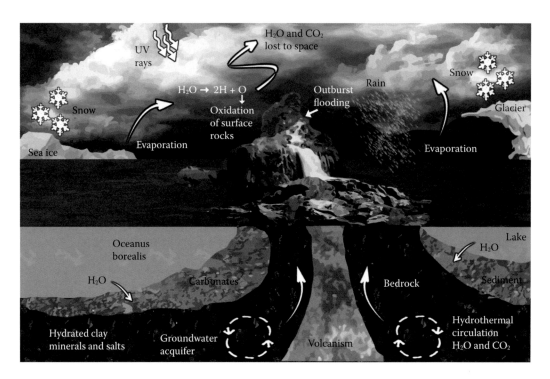

FIGURE 18.9 Possible hydrological cycles on an early wet Mars. Compare to the bubble–aerosol–droplet supercycle outlined in Figures 18.1 and 18.3. (Drawing by Jessica Rhodes after Kargel and Strom, 1996.)

*For Product Safety Concerns and Information please contact
our EU representative GPSR@taylorandfrancis.com Taylor & Francis
Verlag GmbH, Kaufingerstraße 24, 80331 München, Germany*

T - #0072 - 160425 - C8 - 254/178/22 [24] - CB - 9781439803561 - Gloss Lamination